AF368924

Agroecological Approaches for Soil Health and Water Management

Agroecological Approaches for Soil Health and Water Management

Editors

Rajan Ghimire
Bharat Sharma Acharya

MDPI • Basel • Beijing • Wuhan • Barcelona • Belgrade • Manchester • Tokyo • Cluj • Tianjin

Editors

Rajan Ghimire
New Mexico State University
Agricultural Science Center
at Clovis
Clovis, NM 88101
United States

Bharat Sharma Acharya
Texas A&M AgriLife Research
and Extension Center
Corpus Christi, TX 78406
United States

Editorial Office
MDPI
St. Alban-Anlage 66
4052 Basel, Switzerland

This is a reprint of articles from the Special Issue published online in the open access journal *Sustainability* (ISSN 2071-1050) (available at: www.mdpi.com/journal/sustainability/special_issues/Agroecological_Soil_Water).

For citation purposes, cite each article independently as indicated on the article page online and as indicated below:

LastName, A.A.; LastName, B.B.; LastName, C.C. Article Title. *Journal Name* **Year**, *Volume Number*, Page Range.

ISBN 978-3-0365-4566-0 (Hbk)
ISBN 978-3-0365-4565-3 (PDF)

Contents

About the Editors

Rajan Ghimire

Dr. Rajan Ghimire is an Associate Professor of Cropping Systems at New Mexico State University. Dr. Ghimire has a multidisciplinary, externally funded, nationally and internationally recognized research program on soil health and soil organic carbon and nitrogen cycling in agroecosystems. He authored/co-authored 205 scientific publications, delivered >20 invited talks on symposia, and served on the scientific and organizing committee of several national and international meetings. Dr. Ghimire also served in various regional, national, and international professional organizations and advisory councils. Dr. Ghimire's contributions have been highly recognized by his peers. His paper in the Journal of Integrative Agriculture received the Top Article Award in 2020. He is also a recipient of the Emerging Scientist Award from the Western Society of Crop Science in 2020 and the Outstanding Associate Editor Award from Agronomy Journal in 2021. He holds a BS and MS in Agriculture from Tribhuvan University, Nepal, a Ph.D. in Soil Science from the University of Wyoming, and Postdoctoral Training from Oregon State University.

Bharat Sharma Acharya

Dr. Bharat Sharma Acharya is a hydrologist and soil scientist with interests in vadose zone hydrology, water quality, irrigation systems, and unmanned aerial vehicles. His research focuses on interactions between plant, soil and water, with particular emphasis on woody plant encroachment in grassland, cover cropping, and dedicated bioenergy crops. He hods a BS in Agriculture from Tribhuvan University, Nepal, a MS in Agro-Environmental Management from Aarhus University, Denmark and a Ph.D. in Natural Resource Ecology and Management from the Oklahoma State University, USA. He has authored/co-authored over two dozen scientific publications and delivered >50 talks on several national and international conferences. Currently, he is an Associate Editor for the Soil Science Society of America Journal and Guest Editor for the Water Journal. He is also a recipient of the 2022 Education and Public Service Award from Universities Council on Water Resources (UCOWR) and 2021 Outstanding Associate Editor Award from the Agronomy Journal.

Preface to "Agroecological Approaches for Soil Health and Water Management"

Soils provide the foundation for food production, soil water and nutrient cycling, and soil biological activities. With land use and land cover changes over the last century, soil fertility depletion, greenhouse gas emissions, irrigational water scarcity, and water pollution have threatened agricultural productivity and sustainability. An improved understanding of biochemical pathways of soil organic matter and nutrient cycling, and microbial community involved in regulating soil health and soil processes associated with water flow and retention in soil profile helps design better agricultural systems and ultimately support plant growth and productivity. This book, Agroecological Approaches in Soil and Water Management, presents a collection of original research and review papers studying physical, chemical, and biological processes in soils and discusses multiple ecosystem services, including carbon sequestration, nutrients and water cycling, greenhouse gas emissions, and agro-environmental sustainability. The 15 chapters in this book cover various topics related to soil organic matter and nutrient cycling, soil water dynamics, and related hydrological processes across multiple soils, climate, and management. Several chapters highlight the impacts of land use, landscape position, and land-cover change on soil health and plant productivity. It also has chapters on greenhouse gas emissions as affected by agricultural management, roles of soil amendments like biochar and micronutrients. Novel water management strategies, including the use of coalbed methane co-produced water, biodegradable hydrogels, and livestock-integrated cropping to improve soil health are also discussed. The book further incorporates modeling studies on yield and greenhouse gas emissions and presents a review of sustainable agricultural and water management practices.

Soil microbial communities are sensitive to adopting sustainable management practices. Chapter 1 (Coller et al., 2021) investigates the response of soil living communities to change in farming practices from integrated pest management to organic orchard management in northeastern Italy. This study highlights agricultural strategy that could impact the edaphic community within the first few years of land management transition. While fungi responded quickly to the changes, bacteria and microarthropods had lesser impacts from management and higher from abiotic factors. Landscape positions also affect soil properties, and crop production. Chapter 2 (Sun et al., 2011) presents the effects of landscape positions and types on soil properties and the chlorophyll content of citrus. Soil nutrient content was higher in the footslope and terraces and lower in the upper- and mid-slope positions. However, citrus chlorophyll content was higher in the middle and upper landscape position compared to the footslope. Chapter 3 (Ojha et al., 2021) studied the effect of different landscape positions (ridge, midslope, and valley) on total soil carbon (TC), total nitrogen (TN), and carbon-nitrogen (CN) ratio, and soil pH in a mountainous district in central Nepal. The isotopic signature of the natural abundance $\delta 13C$ and $\delta 15N$ were used to identify the source of C and N. Valley soil had higher TN, CN, and soil pH values than the ridge and midslope soils. Further, the valleys had more positive $\delta 15N$ signatures than ridge and midslope, indicating higher inorganic and organic N fertilizer inputs compared to other landscape positions. Therefore, Chapters 2 and 3 suggest the possibility of improving soil quality and agricultural sustainability through targeted land management measures. In line with this study, Khokhar et al. (2021) in Chapter 4 reported the effect of land slopes and maize and cowpea strip-intercropping on productivity and soil erosion in the Shivalik foothills in northwest India. Results indicated significantly higher maize and cowpea yield on a 1% and 2% slope than on steeper slopes. Runoff, soil, and nutrient losses were lower on 1%

and 2% slopes than on 3% slopes. The study suggested the adoption of a strip-intercropping system with a 4.8 m maize strips and 1.2 m cowpea strips resulted in 24% more yield over sole maize and cowpea, increased net return, and reduced runoff and soil loss by 10.9%, and 8.3%, respectively, than sole maize crop.

The impact of changes in land management on soil and environmental quality depend on land-use history. Ren et al. (2020), in Chapter 5, reported an increase in soil N_2O emissions following forestland conversion to cropland in a subtropical Southwest China. The conversion leads to an increase in the annual cumulative N_2O flux by 76–491%. N_2O emissions from croplands with tillage and fertilization were 94% and 235% higher than those from croplands with tillage and no fertilization, in the short-term and long-term, respectively. The relative contribution of fertilizer and tillage was different. Fertilization increased N_2O emissions by 63% and 84% in the short-term and long-term, while tillage contributed to 37% and 16% increased emissions in the short-term and long-term. Chief factors affecting N_2O emissions were soil NO_3, NH_4^+ availability, and water-filled pore spaces. Tillage disturbs soil structure and increases soil aeration to facilitate soil organic N mineralization. Further, mineral N fertilization will lead to an increase in soil NO_3 and NH_4^+ availability. Chapter 6 (Mahmud et al., 2021) reviews different pathways of N losses such as ammonia volatilization (NH_3), nitrous oxide (N_2O) emissions, and nitrate leaching (NO_3), and suggests potential mitigation strategies such as fertilizer placement, best management practices, livestock management, use of carbon-rich sources, growth-promoting microbial consortia, organic farming, crop diversification, genetic improvements, and site-specific nutrient management for improving agricultural sustainability. The role of livestock on soil health is discussed in Chapter 7 (Dahal et al., 2021). This chapter highlights complex interrelationships between different soil health indicators in pasture lands with inorganic or broiler litter fertilization history. Results discerned a strong positive relationship of active carbon (POXC) with N and potentially mineralizable N, indicating the ability of active carbon fraction to influence nitrogen cycling dynamics. Information on POXC appears highly valuable in determining optimum nitrogen fertilizer recommendations for sustainably managing grazing systems and improving soil health properties.

Biochar has been increasingly considered an ecological tool for soil health and water management. Ayaz et al. (2021) reviewed the implications of using biochar as a soil amendment on soil health and crop productivity in Chapter 8 and suggested improvements in soil physical, chemical, hydrological, and microbial properties with biochar application, the magnitude of impact varying with biochar type, climate, and soil management. Chapter 9 (Yu et al., 2021) studies the effect of corn straw biochar application on soil and water losses during the spring thawing period in northeast China. Biochar application at rates 6 and 12 kg m^{-2} increased soil saturated water content by 24.17 and 42.91% and field capacity by 32.44 and 51.30%, respectively. An increase in biochar application rate was translated into decreased runoff and soil erosion. Biochar application can reduce soil and water losses in sloping farmlands.

Another ecological approach to soil and water management is discussed in Chapter 10 (Čechmánková et al., 2021). This chapter evaluated the impact of a novel, biodegradable hydrogel on soil chemical and hydrological properties. The hydrogel had a positive effect on soil water-holding capacity and the availability of nutrients under controlled settings. For example, 3% whey-based hydrogel application increased the available level of phosphorus and potassium by up to 50 and 84%, respectively. While the gel appears to be promising for soil amendment, field experiments and studies on plant health and growth attributes may prove beneficial. These innovations are more important in water-limited environments. Chapter 11 (Poudyal and Zheljazkov, 2021) studies another innovative approach for water management using coalbed methane co-produced water (CBMW). This study evaluated various blending ratios of CBMW with fresh water on the yield and quality of alfalfa and oat forages. While irrigating forages with different levels of blending increased soil pH and sodium adsorption ratio, there was no significant effect on the nutritive value of both forages. Long-term studies are needed to fully understand the soil and agronomic effects of CBMW and biodegradable hydrogel.

Soil organic matter, major nutrients, and key physical and chemical properties are often emphasized while discussing soil health in agroecosystems. However, Chapter 12 (Thapa et al., 2021) reviews the effects of micronutrients on soil health and soybean production in the Midwest USA. The studies reported inconsistent soil health and yield response; several factors like climate, soil pH, cultivar, irrigation, soil organic matter, and application type (foliar vs. band) influenced plant and soil response to micronutrients. More long-term field studies are necessary to understand better the management of micronutrients for soil health and productivity.

Biogeochemical and process-based models provide a broader perspective on sustainability and resilience. Chapter 13 (Jiang et al., 2021) discusses soil organic carbon and rice yield in Kunshan, China, under variable water and carbon management using a modified Denitrification Decomposition (DNDC) model. The authors used four future climate scenarios (RCP 2.6, RCP 4.5, RCP 6.0, and RCP 8.5) to understand soil organic carbon under climate change scenarios. Climate scenarios significantly affected rice yield but not soil carbon. For example, rice yield decreased by 18.41%, 38.59%, 65.11%, and 65.62% for all RCP 2.6, RCP 4.5, RCP 6.0, and RCP 8.5 scenarios, respectively the 2090s. There was a minimal effect of irrigation and conventional fertilizer application on soil carbon irrespective of RCP scenarios. However, controlled irrigation with straw returning appears to increase both SOC and rice yield in long-term simulation. Chapter 14 (Hang et al., 2021) uses the carbon footprint method to calculate greenhouse gas emissions in a circular farm's planting and breeding system modules. Further, a multi-objective linear programming model is used in determining the optimal structure and scale of growing crops and raising farm animals in circular agriculture in relation to economic and environmental impacts and farm waste utilization. Results showed that greenhouse gas emissions occurred primarily from manure management in the livestock industry. After optimization, agriculture income increased by 64%, and greenhouse gas emissions increased by only 12.3%. Indeed, carbon reduction measures must rely on measures for optimizing the management of manure and adjusting feed structures within the circular agricultural framework.

Finally, the benefits and trade-offs of different sustainable practices, namely conservation agriculture, crop diversification, organic farming, and agroforestry, adopted in Europe and North Africa, are discussed in Chapter 15 (Choden and Ghaley, 2021). While adoption of such practices could increase crop yield, soil and water conservation, and sustainable food production to ensure food security, it largely depends on local conditions and their effectiveness for farmers. As such, investment in crop-system modeling is deemed necessary.

The chapters are primarily compiled to help and motivate students, researchers, scientists, land managers, and policymakers in environmental science, soil science, agronomy, hydrology, and water resources. The book will be useful for anyone interested in soil health and water management across the globe. The editors acknowledge and thank the authors, reviewers, and editorial assistants for their help, support, contribution, and enthusiasm. Any error in facts, data, and interpretation remains the sole responsibility of the authors.

Rajan Ghimire and Bharat Sharma Acharya

Editors

 *sustainability*

Article

Soil Communities: Who Responds and How Quickly to a Change in Agricultural System?

Emanuela Coller [1], Claudia Maria Oliveira Longa [2,*], Raffaella Morelli [3], Sara Zanoni [3], Marco Cristiano Cersosimo Ippolito [3], Massimo Pindo [4], Cristina Cappelletti [3], Francesca Ciutti [3], Cristina Menta [1], Roberto Zanzotti [3] and Claudio Ioriatti [3]

[1] Department of Chemistry, Life Sciences and Environmental Sustainability, University of Parma, Parco Area delle Scienze 11/A, 43124 Parma, Italy; coller.emanuela@gmail.com (E.C.); cristina.menta@unipr.it (C.M.)

[2] Department of Sustainable Agroecosystems and Bioresources, Research and Innovation Centre, Edmund Mach Foundation, Via E. Mach 1, 38098 San Michele all'Adige, Italy

[3] Technology Transfer Centre, Edmund Mach Foundation, Via E. Mach 1, 38098 San Michele all'Adige, Italy; raffaella.morelli@fmach.it (R.M.); sara.zanoni@fmach.it (S.Z.); marco.cersosimoippolito@fmach.it (M.C.C.I.); cristina.cappelletti@fmach.it (C.C.); francesca.ciutti@fmach.it (F.C.); roberto.zanzotti@fmach.it (R.Z.); claudio.ioriatti@fmach.it (C.I.)

[4] Computational Biology, Research and Innovation Centre, Edmund Mach Foundation, Via E. Mach 1, 38098 San Michele all'Adige, Italy; massimo.pindo@fmach.it

* Correspondence: claudia.longa@fmach.it; Tel.: +39-0461-615508

Citation: Coller, E.; Oliveira Longa, C.M.; Morelli, R.; Zanoni, S.; Cersosimo Ippolito, M.C.; Pindo, M.; Cappelletti, C.; Ciutti, F.; Menta, C.; Zanzotti, R.; et al. Soil Communities: Who Responds and How Quickly to a Change in Agricultural System? *Sustainability* 2022, 14, 383. https://doi.org/10.3390/su14010383

Academic Editors: Rajan Ghimire and Bharat Sharma Acharya

Received: 28 October 2021
Accepted: 27 December 2021
Published: 30 December 2021

Publisher's Note: MDPI stays neutral with regard to jurisdictional claims in published maps and institutional affiliations.

Abstract: The use of conservation and sustainable practices could restore the abundance and richness of soil organisms in agroecosystems. Fitting in this context, this study aimed to highlight whether and how different soil living communities reacted to the conversion from an integrated to an organic orchard. The metataxonomic approach for fungi and bacteria and the determination of biological forms of diatoms and microarthropods were applied. Soil analyses were carried out in order to evaluate the effect of soil chemical features on four major soil living communities. Our results showed that the different taxa reacted with different speeds to the management changes. Fungi responded quickly to the changes, suggesting that modification in agricultural practices had a greater impact on fungal communities. Bacteria and microarthropods were more affected by abiotic parameters and less by the management. The diatom composition seemed to be affected by seasonality but the highest H' (Shannon index) value was measured in the organic system. Fungi, but also diatoms, seemed to be promising for monitoring changes in the soil since they were sensitive to both the soil features and the anthropic impact. Our study showed that soil biodiversity could be affected by the conversion to sustainable management practices from the early years of an orchard onwards. Therefore, better ecological orchard management may strengthen soil sustainability and resilience in historically agricultural regions.

Keywords: soil biodiversity; bacteria; fungi; microarthropods; diatoms; metataxonomic assays

1. Introduction

Soil hosts the most representative fraction of the agroecosystem biodiversity. Its communities are among the more diversified and comprise a wide range of living organisms. They are involved in a large number of ecological processes and play key roles for human populations and agriculture [1]. The long history of intensive agriculture and the consequent recurring use of pesticides, herbicides and mineral fertilizers have compromised soil quality and biological diversity [2]. The use of chemical products against plant pathogens, pests and weeds was shown to affect the chemical and biological fertility of soils in several cases, including several potential adverse effects versus non-target organisms [3]. In addition, the long-term and over-application of pesticides cause severe effects on soil ecology and can drastically modify the structure of soil microbial communities, which may further impact their functions in soil [4]. Furthermore, the frequent application of mineral fertilization aggravates the decline of soil organic matter, can significantly reduce soil pH, accelerate soil acidification and affects the nitrogen cycle and the diversity and composition of the microbial community [5,6]. In contrast, sustainable and less invasive practices could reduce the negative impact on agroecosystems, restoring the environment and increasing the abundance and richness of groups of microorganisms in soil [7]. In this context, the characterization of soil communities and their interactions are important topics for defining soil health and quality. Edaphic microorganisms are considered good indicators of soil quality because they are very dynamic and respond quickly to changes in soil management, even before most physico-chemical properties, which take longer to change [8]. Within the soil microbiota, fungi and bacteria are among the most dominant components and are affected by seasonality [9], geographical position [10], soil chemical parameters and agronomic managements [11,12]. They are able to overcome environmental modifications by adjusting their metabolic activity, biomass and community structure [13]. As is already known, long-term organic soil management affects not only primary decomposers, namely, bacteria and fungi, but also other representative trophic levels [14]. Among these, soil microarthropods are often used as indicators of soil health owing to their ability to respond to variations in environmental conditions, soil properties and changes in land management [15–18]. Soil arthropods are highly sensitive to the effects of agricultural management practices and there is evidence that organically managed fields contain a greater abundance and diversity of arthropods than conventionally managed ones [19]. In this regard, some authors have recently proposed the use of terrestrial algae communities for the ecological monitoring and assessment of soil quality [20–23]. A previous study reported differences in the community structure of terrestrial diatoms in response to different farming systems [24]. Diatoms have high biological diversity, are sensitive to anthropic disturbances and environmental parameters and respond to agricultural practices [20,24,25].

Knowledge about the impact of agricultural management systems on the soil biotic components is largely available for arable farming, whereas long-term monocultures, including apple orchards, are less studied. The present study aimed to evaluate (1) whether sustainable agricultural management improves soil biological health and quality, and (2) how fast the edaphic community (fungi, bacteria, diatoms and microarthropods) responds to the modifications induced by agronomic practices. For these purposes, a two-year trial was carried out on an apple orchard in northeastern Italy during the conversion from integrated pest management (IPM) to organic management.

2. Materials and Methods

2.1. Site Description and Agronomic Managements

The trial was carried out in an apple orchard located in the Trentino region (northeastern Italy) (46°02′47″ N, 11°28′19″ E; 400 m a.s.l.) during two years of experimentation (2018–2019). The field is located on a dejection conoid and detrital, gravitational and colluvial deposits. The parent material is represented by gravels mixed with sands and blocks of limestone-dolomitic nature. The apple orchard was planted with Gala cultivar (*Malus* × *domestica* Borkh), clone Buckeye® trained on an M9 rootstock and it was planted according to a tree spacing of 3.60 m × 1.00 m (2778 trees/ha) in 2015. The field was managed with integrated pest management (IPM, which is a form of agriculture aimed at minimizing the use of inputs from outside the farm by implementing a variety of production enterprises) until 2017. The management was characterized by spring mineral fertilization (NPK 14-7-17, 45 kg N/ha), chemical pest and weed control of the rows and a subsequent autumnal mechanical weeding. In April 2018, it was divided into two plots, one of which, formed by seven rows in one block of 2500 m², was converted to organic management according to Reg. EU 2018/848. The other one consisted of nine rows in one block of 3200 m² and was maintained with integrated management. The different practices were fertilization and weeding of the row (Table S1). In the integrated management plot, mineral fertilizer (NPK 14-7-17) was supplied once a year in spring at a dose of 45 kg N/ha.

Matured cattle manure was applied in spring 2018 on the organic management plot in an amount that can release 90 kg N/ha in an available form over two years and bring about 3.2 t/ha of organic carbon in the soil. In the rows of the integrated plot, chemical weed control was applied during the season, followed by mechanical mowing in autumn. In contrast, only mechanical mowing throughout the season was applied on the rows of the organic plot. There were no differences in Inter-row management between the Integrated and Organic systems. In this area, natural vegetation (permanent grass) was allowed to grow, which was regularly mowed yearly during the spring–summer season.

2.2. Soil Sampling

Soil samples were collected along the row (Row) and between the rows (Inter-row) for both management systems at three different times: T0, before the conversion in 2018 and immediately before fertilization (19 April 2018), T1 after six months (17 October 2018) and T2 after 18 months from the conversion (3 October 2019). Ten replicates were collected for each management, position and time. In total, 120 soil samples were analyzed for chemical parameters, fungi, bacteria and microarthropods. Thirty Inter-row samples were collected for diatoms: five soil samples for both managements and for each collection time. The abbreviations for the different plots are as follows: Integrated Inter-row (Int-Ir), Integrated Row (Int-R), Organic Inter-row (Org-Ir) and Organic Row (Org-R).

2.3. Soil Physicochemical Properties

For the physicochemical assays, soil was collected from the 0–20 cm layer, where each soil sample consisted of three cores, which were homogenized to obtain a representative sample. Physicochemical analyses were carried out on the air-dried fine earth fraction (<2 mm). Soil texture was determined as a percentage of sand (2 mm–50 μm), silt (50 μm–2 μm) and clay (<2 μm) via wet sieving and a hydrometric assay of soil dispersant solution (sodium hexametaphosfate). The pH was measured using a potentiometer in a soil–water suspension (w/v 1:2.5). Total carbonates were determined on powdered samples using the volumetric method, measuring CO_2 evolved after the addition of HCl (ISO 10693:1995). The soil organic matter (SOM) was calculated using the organic carbon (conversion factor: 1.724), which was obtained using the difference between the total carbon, measured using Dumas combustion of powdered soil and TCD detection, and total carbonates (ISO 10694:1995). The total nitrogen was measured simultaneously with the total carbon (ISO 13878:1998). The available fraction of elements was extracted in a DTPA solution and Cu, Zn, Pb, Fe and Mn were detected using ICP-OES. The cation exchangeable capacity (CEC) was evaluated using ICP-OES determination of Mg adsorbed on the soil exchangeable surface after monosaturation with Ba and exchange with Mg added as $MgSO_4$. The exchangeable fraction of Mg, K and Ca was extracted in ammonium acetate (pH 7.00) and detected using ICP-OES. The assay of assimilable P was carried out using the Olsen method, providing solubilization of P in a $NaCO_3$ solution and determination using spectrophotometry with the ascorbic acid method. Soluble B was extracted using a hot treatment with a diluted solution of CaCl (w/v 1:2) and determined using ICP-OES.

2.4. Soil Living Communities

For the metataxonomic analysis, soil samples were freeze-dried and sieved with a 0.2 mm mesh size and stored at $-80\ ^\circ$C until the DNA extraction. Total DNA was extracted from 0.25 g of each composite soil sample using the PowerSoil DNA isolation kit (MO BIO Laboratories Inc., Carlsbad, CA, USA) according to the manufacturer's instructions. Total genomic DNA was amplified using primers that are specific to either the bacterial and archaeal 16S rRNA gene or the fungal ITS1 region. The specific bacterial primer set 515F (5'-GTGYCAGCMGCCGCGGTAA-3') and the 806R (5'-GGACTACNVGGGTWTCTAAT-3') were used [26] with degenerate bases suggested by Apprill et al. [27] and Parada et al. [28]. Although no approach based on PCR amplification is free from bias, this primer pair was shown to guarantee good coverage of known bacterial and archaeal taxa [29]. For the identification of fungi, the internal transcribed spacer 1 (ITS1) was amplified using the primer ITS1F (5′-CTTGGTCATTTAGAGGAA GTAA-3′) [30] and ITS2 (5′-GCTGCGTTCTTCATCGAT GC-3′) [31]. All the primers included the specific overhang Illumina adapters for the amplicon library construction. DNA purification, indexing, quantification, library preparation for the Illumina MiSeq sequencing (PE300), preprocessing data and subsequent taxonomic classifications of the OTUs were carried out as described previously by Coller et al. [10]. Filtered sequences were clustered into operational taxonomic units (OTUs) at 97% identity using the de novo greedy algorithm available in MICCA. OTUs were taxonomically classified using the Ribosomal Database Project (RDP) Classifier v2.11 [32]. Raw overlapping ITS paired-end reads were merged and merged sequences with an overlap length smaller than 100 bp and with more than 32 mismatches were discarded. After the primer trimming, merged reads shorter than 150 bp and with an expected error rate higher than 0.5% were removed. Filtered sequences were clustered at 97% identity using the de novo greedy algorithm and OTUs were taxonomically classified as described by Coller et al. [10]. To compensate for different sequencing depths, samples were rarefied to an even depth of 24,180 reads for 16S and 18,645 for ITS sequences (Supplementary material, Figure S1). Samples with less than the minimum number of reads were discarded.

For microarthropod investigation, $10 \times 10 \times 10$ cm^3 soil cubes were collected. The extraction of microarthropods and determination of biological forms was conducted as reported by Parisi et al. [33]. The microarthropod community was assessed depending on the taxa composition, abundance (number of individuals/m^2) and an index of soil biological quality, namely, QBS-ar (acronym of Soil Biological Quality) [33]. This index is an expeditious tool used to evaluate soil biological quality in agricultural ecosystems, woods, degraded lands and recovery areas [33,34]. It is a metric index based on the concept that the number of microarthropod groups morphologically well-adapted to soil is higher in high-quality soils than in low-quality soils. This index was developed to combine two key aspects regarding soil microarthropods: (i) their presence in the soil, i.e., biodiversity, and (ii) their disappearance in degraded conditions, i.e., sensitivity.

For the diatom analysis, soil cores were collected using a plastic ring (diameter 5.7 cm, depth 4 cm). Diatoms were extracted by rinsing the soil surface with sparkling water to detach the diatoms [23] for a total volume of 50 mL and preserved in 70% ethanol according to the standard method proposed by Barragàn et al. [21]. Samples were then treated via oxidation with hydrogen peroxide 35% to obtain clean frustules and mounted with Naphrax® for slide preparation (EN 13946, European Committee for Standardization 2003). The analysis was conducted at the genus level by counting 200 valves using Olympus BX51 light microscope (LM) at 1000× magnification and DIC microscopy due to the presence of small species that required scanning electron microscopy SEM to be identified.

2.5. Statistical Analysis

The statistical analyses were performed using R software version 3.6.0 [35] and the *vegan* R package. The Wilcoxon non-parametric statistical test with the Benjamini–Yekutieli FDR correction (p-value ≤ 0.05) was run in order to identify which parameters significantly differed between the different conditions. For the different communities, richness estimators (Species number (S) and S.Chao1) and evenness indices (Shannon (H) and Pielou's (J)) were determined using *specnumber*, *estimateR* and *diversity* functions respectively. In order to obtain the overall variance in life-form compositions, the similarities in the OTU composition of bacteria and fungi, the taxa of microarthropods and the genera of diatoms across samples were visualized using nonmetric multidimensional scaling (NMDS) ordinations based on Bray–Curtis dissimilarity with 999 permutations (*metaMDS* function). An analysis of similarities (ANOSIM) with 999 permutations based on the Bray–Curtis dissimilarity was conducted to identify differences in organism composition at the three times of sampling in different conditions (*vegdist* function) (value 0 = identical samples, value 1 = completely disjointed samples). The correlation analysis was used to determine the biotic–biotic interactions and biotic–abiotic interactions. In detail, the Mantel tests were performed to evaluate the Spearman rank correlations between each two-distance matrices or between one single factor and a matrix (Euclidean dissimilarities distance for chemical variables and Bray–Curtis distance for the life-form community composition).

3. Results and Discussion

3.1. Soil Physico-Chemical Properties

Some physicochemical properties exhibited statistically significant differences between the organic and integrated plots already at T0 (Supplementary material, Table S1) in the Inter-row and Row positions. According to the USDA (U.S. Department of Agriculture) texture classification, the soil from the Integrated plot was silty-loam (SiLo), whereas the Organic soil was sandy-loam (SaLo). The textural composition, particularly clay and silt, may affect microbial composition [36] and thereby influence the restoration of the microbial community [33]. Some differences were observed in the soil chemical characteristics between management systems and position over time. The available Mn was significantly different in the two management systems (Int = 6.1 ± 0.5 mg/kg d.w., Org = 7.7 ± 0.4 mg/kg d.w., *p*-value 0.0410) in the Inter-row at T2. In the Row, but not in the Inter-row, the assimilable P (Int = 184 ± 14 mg/kg d.w., Org = 97 ± 7 mg/kg d.w.) differed statistically (*p*-value = 0.0005) at T0 and T2. The available Cu concentration was similar between management systems and positions at T0 (Inter-row *p*-value = 0.8499, Row *p*-value = 0.5204). Due to the more frequent treatment with Cu-based products on the canopy like fungicides in the organic plot, the Cu available concentration increased and it was significantly different (*p*-value = 0.0376) in Org-R (42 ± 4 mg/kg d.w.) compared with the Integrated plot (32 ± 2 mg/kg d.w.) at T2 (Supplementary material, Table S2). No differences were observed for the other chemical elements between management systems, positions and times.

3.2. Effects of Agricultural Practices on Fungal Soil Communities

In all investigated soils, 267 fungi genera were found, where the more representative genera were *Mortierella* (6.3%), *Tetracladium* (2.8%), *Guehomyces* (2.1%), *Coprinellus* (1.7%), *Cylindrocarpon* (1.6%), *Nectria* (1.6%), *Ilyonectria* (1.5%), *Myrothecium* (1.2%) and *Cladorrhinum* (1.1%) (Figure 1a), which are the most characteristic groups found in exploited agricultural soils. Indeed, *Mortierella*, *Tetracladium*, *Cylindrocarpon* and *Nectria* are among the main agents that cause replanting diseases, root rot and cancer in apple orchards [12]. At time T0, the Bray–Curtis distance between Org and Int for fungal communities was higher than the management distance for bacterial communities (Inter-row: fungi = 0.41, bacteria = 0.31; Row: fungi = 0.46, bacteria = 0.31) (Figure 2). These data suggested that fungi were more susceptible to the soil structure (chemical parameters) than bacteria, which seemed more stable in the soil than the fungi. Different texture compositions of soil might explain much of the variation seen in the structure and diversity of the fungi under these two types of management. Indeed, some soil properties, such as soil texture, pH and element concentrations chemical properties, including Cu and P, were implicated as important factors in shaping microbial communities [37,38]. Moreover, Whalen et al. [11] showed that the availability of some elements promotes variability in fungal community composition (e.g., Mn). In this study, we detected differences between the Organic and Integrated plots for Mn and Cu. At T0, neither of these elements differed between the two management methods, but the Cu concentration increased during the experimentation, in particular in Row, and at T2, we detected significant differences between Org and Int. The same trend was observed for Mn (Supplementary material, Table S1). The increase in these two elements and the consequent selection action on fungi community [11,35] could explain the Bray–Curtis distance increment from T0 to T1 and T2 (Figure 2). The analysis of the biodiversity indices (Table 1) did not show any differences between Int and Org in the Inter-row position. In the Row, all indices decreased in Int-R, but only S.Chao1 showed a significant reduction over time (Table 1). In contrast, in Org-R, the indices S, H' and J increased significantly over time (Table 1). The opposite trends observed between Org-R and Int-R could have been due to the two different soil managements and practices. Int-R was characterized by mineral fertilization and chemical weeding with glyphosate and other herbicides. Soil features, such as CEC, clay content and SOM, affect the glyphosate degradation rate and its persistence in soil, while its effects on soil microbes are doubtful and conflicting [39]. However, tillage and other mechanical practices are the primary drivers of microbial composition rather than glyphosate application [39]. In this study, the Organic plot was managed by mechanical mowing and the application of organic amendments to the Row. The use of organic amendments enhances the microbial diversity in the soil and, particularly, affects the soil fungal communities [12]. In our research, unlike what was observed in Integrated management, the fungal community of the Organic-Row plot appeared to move close to the fungal community of the Inter-row positions in T2 (Figure 3), suggesting an initial approach to a less impacted agroecosystem.

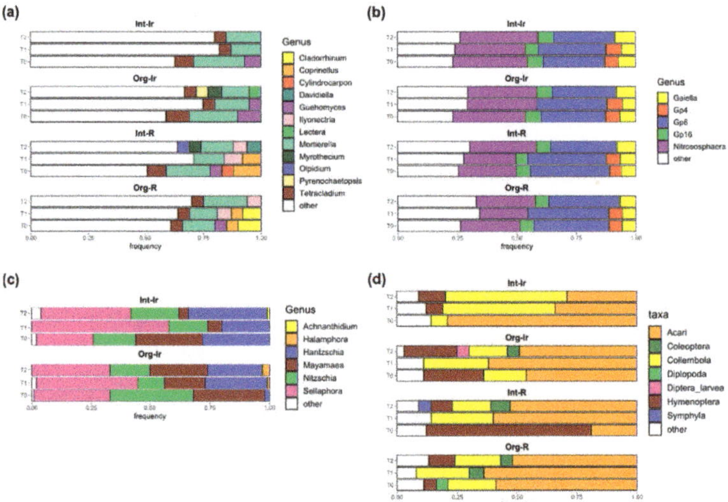

Figure 1. Frequency plots. Representative community distributions under different management systems (Organic and Integrated) on the Row and Inter-row, from T0 to T2. (**a**) Fungi genera: 3,958,526 sequences (ITS region) were rarefied to 18,645 reads per sample and clustered into 8606 OTUs. A total of 1552 sequences provided taxonomic information on 267 genera (18%). (**b**) Bacteria genera: 5,405,403 sequences (V4 region of 16S) were generated. After rarefying, 24,180 reads out of 43,394 OTUs were obtained. A total of 8992 OTUs were associated with 408 genera (20.7%). (**c**) Diatoms genera (frequency: at least 5% of the total and 1% of the total for diatoms) and (**d**) microarthropod taxa.

3.3. Soil Properties Mainly Affected Bacterial Communities

Overall, 408 bacteria genera were found within the soil samples. *Nitrososphaera* (10.3%) and *Gp16* (1.3%) were the most characteristic genera (Figure 1b). Our more representative genera among the dominant groups in the agricultural soils were classified as oligotrophic bacteria [40]. The P/A ratio (P = copiotrophic phyla of Proteobacteria, A = oligotrophic phyla of Acidobacteria) is indicative of the trophic level of the investigated soil [34] and its high values ($\geq$0.5) suggests a sufficient availability of nutrients in soil [33]. In this work, the P/A ratio was similar between Integrated and Organic plots in either position at the beginning of the trial (T0) (Int-IR = 0.69, Org-IR = 0.69, Int-R = 0.77, Org-R = 0.77). At T2, this ratio was higher in the Organic than in the Integrated plot (Int-IR = 0.65, Org-IR = 0.72, Int-R = 0.78, Org-R = 0.82), indicating an improvement of soil nutrient availability in organic management. The dissimilarity of bacteria was higher at T1 than at T0 and T2 in the Inter-row and was higher at T1 and T2 than at T0 in the Row position (Figure 2). Bacterial populations respond quickly to nutrient addition [41], but this response decreases over time [42]. In this study, the samplings were collected distant from cultivar operations (fertilization and weeding); this can account for the lower values of bacterial Bray–Curtis distance relative to those of the fungal community. The bacterial biodiversity indicators decreased from T0 (spring) to T2 (autumn) in both positions and this reduction was statistically significant in the Row position, showing the seasonal impact on the bacterial community [43] (Table 1). In the Inter-row position, the seasonal effect on the bacterial composition (Table 1) was reduced by the permanent grass cover, which might have limited temperature and moisture variation in soil [44]. Additionally, bacterial biodiversity indices and total carbonate content, which were higher in the Inter-row than in the Row positions, were positively correlated (Supplementary material, Table S2); this result underlines the positive effect of carbonates on bacterial biodiversity [45]. The bacterial community composition showed a separation between the Inter-row and Row positions for both management systems, except for the Organic system at T2 (Figure 3).

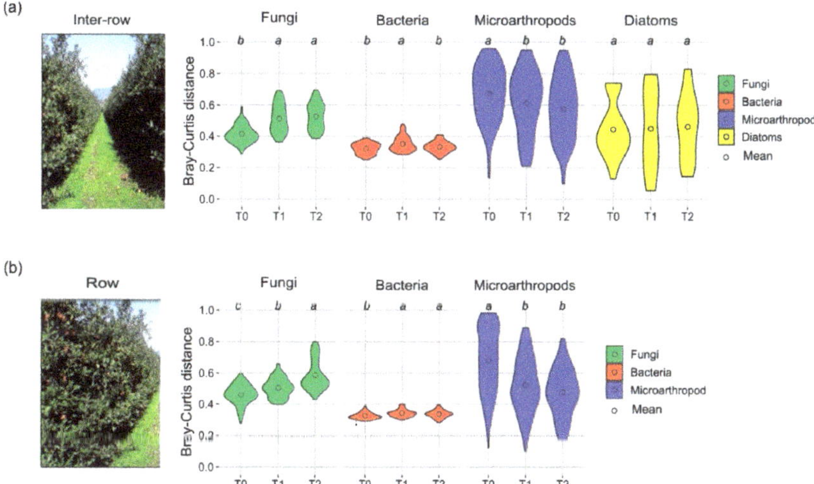

Figure 2. Violin plot of the Bray–Curtis distance between Integrated and Organic managements for soil community: (**a**) Inter-row position and (**b**) Row position. The Bray–Curtis dissimilarity was used to quantify the difference/distance between treatments in the same position. This measure took on values from 0 (identical samples) to 1 (completely disjointed samples).

8

Table 1. Table of biodiversity indices (mean ± standard error). Letters in parentheses represent the results of the non-parametric Wilcoxon test with a BY correction between times in relation to treatment and position. Different letters indicate significant differences between times for the same position and management ($p \leq 0.005$).

Organism	Index	Int-Ir-T0	Int-Ir-T1	Int-Ir-T2	Org-Ir-T0	Org-Ir-T1	Org-Ir-T2	Int-R-T0	Int-R-T1	Int-R-T2	Org-R-T0	Org-R-T1	Org-R-T2
Fungi	S	104 ± 16	1046 ± 37	1035 ± 20	1003 ± 19	976 ± 37	998 ± 30	967 ± 21	994 ± 26	896 ± 45	902 ± 30 (b)	968 ± 27 (ab)	977 ± 47 (a)
	S.Chao1	1703 ± 40	1755 ± 82	1768 ± 47	1680 ± 25	1612 ± 66	1689 ± 60	1722 ± 40 (a)	1696 ± 40 (ab)	1518 ± 77 (b)	1572 ± 52	1631 ± 43	1662 ± 73
	H	4.7 ± 0.1	4.7 ± 0.1	4.6 ± 0.1	4.6 ± 0.1	4.3 ± 0.2	4.6 ± 0.1	4.5 ± 0.1	4.6 ± 0.1	4.3 ± 0.2	4.3 ± 0.1 (b)	4.5 ± 0.1 (ab)	4.6 ± 0.2 (a)
	J	0.67 ± 0.01	0.68 ± 0.01	0.67 ± 0.01	0.66 ± 0.01	0.63 ± 0.03	0.67 ± 0.01	0.66 ± 0.01	0.66 ± 0.01	0.64 ± 0.02	0.64 ± 0.01 (b)	0.66 ± 0.01 (ab)	0.67 ± 0.02 (a)
Bacteria	S	5452 ± 127	5508 ± 177	5243 ± 119	5379 ± 179	5499 ± 102	5477 ± 108	5916 ± 92 (a)	5802 ± 162 (ab)	5463 ± 149 (b)	5991 ± 74 (a)	5863 ± 105 (ab)	5686 ± 103 (b)
	S.Chao1	10,945 ± 284	10,439 ± 44	10,504 ± 343	10,824 ± 47	10,651 ± 33	10,906 ± 30	11,784 ± 212 (a)	11,385 ± 403 (ab)	10,562 ± 34 (b)	11,933 ± 180	11,244 ± 26	11,204 ± 253
	H	6.99 ± 0.06 (ab)	7.14 ± 0.06 (a)	6.92 ± 0.05 (b)	7.02 ± 0.07	7.16 ± 0.06	7.10 ± 0.06	7.33 ± 0.03 (a)	7.38 ± 0.04 (a)	7.16 ± 0.06 (b)	7.33 ± 0.05 (ab)	7.42 ± 0.04 (a)	7.28 ± 0.04 (b)
	J	0.81 ± 0.01 (ab)	0.83 ± 0.01 (a)	0.81 ± 0 (b)	0.82 ± 0.01	0.83 ± 0.01	0.82 ± 0.01	0.84 ± 0.00 (b)	0.85 ± 0.00 (a)	0.83 ± 0.01 (b)	0.84 ± 0.00 (b)	0.86 ± 0.00 (a)	0.84 ± 0.00 (b)
Microartrhopods	Abundance	8146 ± 4454 (ab)	3261 ± 1183 (b)	7643 ± 1665 (a)	4401 ± 740	6070 ± 2928	4974 ± 1288	9776 ± 4563	3503 ± 1041	3611 ± 415	9751 ± 3804	6643 ± 1719	7305 ± 1197
	QBS	85 ± 7 (a)	65 ± 11 (b)	82 ± 7 (ab)	79 ± 11	83 ± 10	65 ± 10	114 ± 10 (a)	76 ± 10 (b)	118 ± 10 (a)	127 ± 9 (ab)	116 ± 10 (b)	148 ± 9 (a)
	S	6.0 ± 0.6 (ab)	5.4 ± 0.8 (b)	7.0 ± 0.4 (a)	7.0 ± 0.6	6.0 ± 0.7	6.0 ± 0.8	8.0 ± 0.7 (a)	6.0 ± 0.6 (b)	8.0 ± 0.5 (a)	8.7 ± 0.7 (ab)	7.5 ± 0.5 (b)	9.9 ± 0.5 (a)
	S.Chao1	9 ± 1	6 ± 1	8 ± 1	8 ± 1	7 ± 1	7 ± 1	11 ± 2	7 ± 1	10 ± 1	10 ± 1	11 ± 2	11 ± 1
	H	1.2 ± 0.1	1.1 ± 0.1	1.2 ± 0.1	1.1 ± 0.1	1.0 ± 0.1	1.1 ± 0.1	1.3 ± 0.1 (ab)	0.9 ± 0.2 (b)	1.4 ± 0.1 (a)	1.4 ± 0.1	1.1 ± 0.1	1.4 ± 0.1
	J	0.70 ± 0.08	0.67 ± 0.06	0.63 ± 0.04	0.60 ± 0.05	0.62 ± 0.05	0.59 ± 0.08	0.62 ± 0.07	0.53 ± 0.06	0.65 ± 0.03	0.66 ± 0.06	0.57 ± 0.05	0.63 ± 0.04
Diatoms	S	7 ± 2	6 ± 1	8 ± 2	5 ± 1	6 ± 1	6 ± 1						
	S.Chao1	7 ± 2	6 ± 1	8 ± 2	5 ± 1	7 ± 1	7 ± 1						
	H	1.20 ± 0.07 (a)	0.84 ± 0.05 (b)	1.12 ± 0.07 (a)	1.16 ± 0.03 (b)	1.1 ± 0.1 (ab)	1.31 ± 0.03 (a)						
	J	0.65 ± 0.05	0.50 ± 0.04	0.58 ± 0.06	0.73 ± 0.02	0.64 ± 0.07	0.71 ± 0.03						

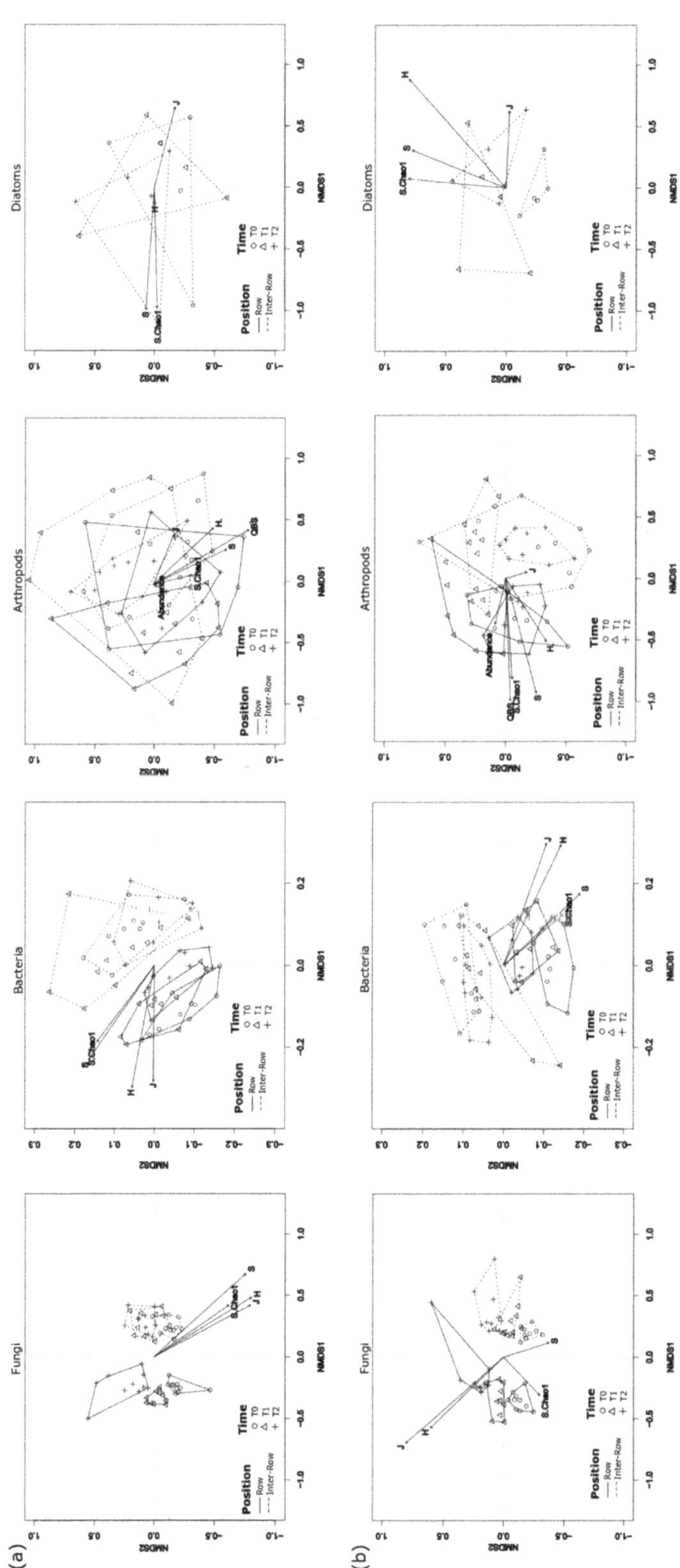

Figure 3. Results of the NMDS analyses based on the rare OTUs of the fungi and bacteria, taxa of arthropods and genera of diatoms. (**a**) NMDS analysis for the Integrated management. The stress values for fungi, bacteria, arthropods and diatoms were 0.1520679, 0.2089021, 0.2565353 and 0.1171184, respectively. (**b**) NMDS analysis for the Organic management. Fungi stress value = 0.1466064, bacteria stress value = 0.1993824, arthropod stress value = 0.2528133 and diatom stress value = 0.1440569. The Bray–Curtis dissimilarity index was used to determine the dissimilarities between the composition communities of different kingdoms. A stress value <0.05 provides an excellent representation in reduced dimensions, <0.1 is great, <0.2 is good and stress <0.3 provides a poor representation. Continuous and hatched areas were Row and Inter-row positions respectively. The shapes indicate the different times: circle—T0, triangle—T1 and cross—T2.

3.4. Diatoms as a Promising Community for Indicating Soil Quality

The investigation of the soil diatom community revealed a low number of taxa (28 genera), as observed by Foets et al. [17], and a high variability within replicates. *Sellaphora* (38.0%), *Hantzschia* (21.7%), *Nitzschia* (19.5%) and *Mayamaea* (18.0%) were the most abundant genera (Figure 1c). When the treatments were considered, *Mayamaea* had a significantly higher presence in the Organic management plot (p-value = 0.0201) than in the Integrated one, whereas the significant variations over time also affected the genera *Hantzschia* (increased frequency) and *Nitzschia* (decreased frequency) (Figure 1c), probably due to the different sensitivities of these three genera to farming practices [20]. Differences in the diatom community between the two management systems were already high at time T0 but increased over time, though not significantly (Figure 2). The biological indices increased over time in both management systems, but only for H' was the difference statistically significant. It increased from T0 to T2 in Org, and at T2, it was also statistically different between management systems (Table 1). The higher value of this index in the Organic system as against the Integrated one agreed with the study of Heger et al. [24]. The change in the agricultural system affected the diatom community composition; indeed, it was different at T0 relative to T1 and T2 only in the Organic system (Figure 3). These results could be explained by the change in agricultural systems, but may also hide seasonal variations, as observed for agricultural soil [17]. The soil diatom variability significantly correlated with organic matter and total N in both management systems and with CSC, total C and exchangeable Mg in the Integrated system (Supplementary material, Table S2), confirming the effects of nutrients, ionic strength and other chemical elements on soil diatom distribution [25]. Terrestrial diatom species are not so well known, and some applications based on them use diatom indices developed for freshwater ecosystems [21]. More studies are needed to produce a "soil diatom index" that is able to evaluate soil quality and anthropic disturbance.

3.5. Effect of Seasonality on Microarthropods

A total of 11789 microarthropod specimens belonging to 16 taxa were collected. Acari were the most dominant group (50.1%), followed by Collembola (21.1%) and adults of Hymenoptera (15.0%) (Figure 1d). The total microarthropod abundance decreased over time in both management systems along the Row, and there was a significant reduction from T0 and T1 in the Int-Ir plot (Table 1). For the microarthropod community, the Bray–Curtis distance between the two treatments decreased from T0 to T1 and T2 in both positions (Figure 2). Moreover, at all sampling times, the variability of arthropods in the distances was much higher than fungi, bacteria and diatoms. In the short-term, seasonality and differences in soil properties had a more significant role in determining arthropod community compared to agronomic management [46]. Temperature and moisture fluctuations induce vertical migration through the soil profile, causing a variable distribution of soil microarthropods in soil [47]. The richness and the evenness indices showed some significant differences between the times under different conditions and the QBS-ar index highlighted differences between T1 and the other two times (Table 1). At T1, the biodiversity indices (H' and J) showed the lowest values. The climatic conditions at T1, such as the absence of precipitation in a month and the lowering of the temperature, could have contributed to this decrease (Supplementary material, Figure S2). These results agree with other studies that showed the strong effect of seasonality on soil arthropods [48]. In this study, no differences were found between the Integrated and Organic plots in the biodiversity indices of the microarthropod community (Figure 3). We hypothesized that two years after the conversion is not enough time to detect a significant change in the soil arthropod community.

4. Conclusions

This study considered three sampling time points and four different soil communities (fungi, bacteria, arthropods and diatoms) to evaluate the differences during the conversion from integrated to organic practices in an apple orchard. The fungal communities reacted quickly to the change in the management system and the values of the biodiversity indices in the organic plot increased compared to the values in the Integrated one. When the change in the agricultural system was first implemented, bacterial and microarthropod communities were influenced more by the seasonality, texture and chemical properties of soil, rather than by distinct farming practices. Diatoms were found to be promising for monitoring changes in the soil since they were sensitive to both the nature of the soil and anthropic disturbance at the genus level. This study highlighted that different components of the soil living community responded in different ways to the change in management system and this result is particularly important in soil monitoring plans. The monitoring of the situation is still ongoing to better characterize the change in the soil microbiota during an agroecosystem conversion.

Supplementary Materials: The following are available online at https://www.mdpi.com/article/10.3390/su14010383/s1, Figure S1: Rarefaction curves of alpha diversity of the (A) number of observed bacterial OTUs, (B) number of observed fungal OTUs. Figure S2: Plots of the temperature and rain from T0 to the 30 days before sampling: (A) daily rainfall, (B) accumulated amount of rain and (C) gradient of soil temperature (geom_ridgeline_gradient {ggridges} of the R program). Meteorological data from the sampling areas were recorded daily in four stations located in proximity to the apple orchard. Table S1: Concentrations of the sand, silt, clay, heavy metals and other chemical parameters according to the management types, positions and times. Different letters in parentheses indicate significant differences between times for the same position and management (Wilcoxon test with BY correction, p-value ≤ 0.05). Table S2: Spearman rank correlations (R values) of single chemical variables with matrices of bacterial, fungal and microarthropod community composition and soil processes based on Mantel tests. Significance levels: *** $p < 0.001$, ** $p < 0.01$ and * $p \leq 0.05$.

Author Contributions: Conceptualization, C.I., C.M.O.L. and R.Z.; formal analysis, E.C.; funding acquisition, C.I.; investigation, R.M., S.Z., F.C., C.C., M.C.C.I. and C.M.O.L.; data curation, M.P. and E.C.; writing-original draft preparation, E.C., C.M.O.L., C.M. and R.M. All authors have read and agreed to the published version of the manuscript.

Funding: This research (project MELOBIO 2018/2020) was sponsored by the Autonomous Province of Trento (PAT) to support research and divulgation in sustainable agriculture, new biotechnologies and organic agriculture.

Institutional Review Board Statement: Not applicable.

Informed Consent Statement: Not applicable.

Data Availability Statement: Data are available upon request.

Acknowledgments: We thank the Azienda Agricola Fondazione E. Mach and the Fondazione de Bellat for providing access to the apple orchard. We are grateful to Enzo Mescalchin for his support and Silvia Gugole for helping with fieldwork.

Conflicts of Interest: The authors declare no conflict of interest. The funders had no role in the design of the study; in the collection, analyses, or interpretation of data; in the writing of the manuscript; or in the decision to publish the results.

References

1. Lavelle, P.; Decaëns, T.; Aubert, M.; Barot, S.; Blouin, M.; Bureau, F.; Margerie, P.; Mora, P.; Rossi, J.P. Soil invertebrates and ecosystem services. *Eur. J. Soil Biol.* **2006**, *42*, S3–S15. [CrossRef]
2. Zimmerer, K.S. Biological diversity in agriculture and global change. *Annu. Rev. Environ. Resour.* **2010**, *35*, 137–166. [CrossRef]
3. Vischetti, C.; Casucci, C.; De Bernardi, A.; Monaci, E.; Tiano, L.; Marcheggiani, F.; Ciani, M.; Comitini, F.; Marini, E.; Taskin, E.; et al. Sub-Lethal Effects of Pesticides on the DNA of Soil Organisms as Early Ecotoxicological Biomarkers. *Front. Microbiol.* **2020**, *11*, 1892. [CrossRef]
4. Yang, T.; Lupwayi, N.; Marc, S.A.; Siddique, K.H.M.; Bainard, L.D. Anthropogenic drivers of soil microbial communities and impacts on soil biological functions in agroecosystems. *Glob. Ecol. Conserv.* **2021**, *27*, e01521. [CrossRef]
5. Ouyang, Y.; Norton, J.M. Short-term nitrogen fertilization affects microbial community composition and nitrogen mineralization functions in an agricultural soil. *Appl. Environ. Microbiol.* **2020**, *86*, e02278-19. [CrossRef]

6. Qaswar, M.; Dongchu, L.; Jing, H.; Tianfu, H.; Ahmed, W.; Abbas, M.; Lu, Z.; Jiangxue, D.; Khan, Z.H.; Ullah, S.; et al. Interaction of liming and long-term fertilization increased crop yield and phosphorus use efficiency (PUE) through mediating exchangeable cations in acidic soil under wheat–maize cropping system. *Sci. Rep.* **2020**, *10*, 19828. [CrossRef]

7. Tully, K.L.; McAskill, C. Promoting soil health in organically managed systems: A review. *Org. Agric.* **2020**, *10*, 339–358. [CrossRef]

8. Garciá-Orenes, F.; Morugań-Coronado, A.; Zornoza, R.; Scow, K. Changes in soil microbial community structure influenced by agricultural management practices in a Mediterranean agro-ecosystem. *PLoS ONE* **2013**, *8*, e80522. [CrossRef]

9. Luo, X.; Wang, M.K.; Hu, G.; Weng, B. Seasonal change in microbial diversity and its relationship with soil chemical properties in an orchard. *PLoS ONE* **2019**, *14*, e0215556. [CrossRef]

10. Coller, E.; Cestaro, A.; Zanzotti, R.; Bertoldi, D.; Pindo, M.; Larger, S.; Albanese, D.; Mescalchin, E.; Donati, C. Microbiome of vineyard soils is shaped by geography and management. *Microbiome* **2019**, *7*, 140. [CrossRef]

11. Whalen, E.D.; Smith, R.G.; Grandy, A.S.; Frey, S.D. Manganese limitation as a mechanism for reduced decomposition in soils under atmospheric nitrogen deposition. *Soil Biol. Biochem.* **2018**, *127*, 252–263. [CrossRef]

12. Liang, B.; Ma, C.; Fan, L.; Wang, Y.; Yuan, Y. Compost amendment alters soil fungal community structure of a replanted apple orchard. *Arch. Agron. Soil Sci.* **2020**, *67*, 739–752. [CrossRef]

13. Schloter, M.; Dilly, O.; Munch, J.C. Indicators for evaluating soil quality. *Agric. Ecosyst. Environ.* **2003**, *98*, 255–262. [CrossRef]

14. Harkes, P.; Suleiman, A.K.A.; van den Elsen, S.J.J.; de Haan, J.J.; Holterman, M.; Kuramae, E.E.; Helder, J. Conventional and organic soil management as divergent drivers of resident and active fractions of major soil food web constituents. *Sci. Rep.* **2019**, *9*, 13521. [CrossRef] [PubMed]

15. Menta, C.; Conti, F.D.; Fondón, C.L.; Staffilani, F.; Remelli, S. Soil arthropod responses in agroecosystem: Implications of different management and cropping systems. *Agronomy* **2020**, *7*, 982. [CrossRef]

16. Mantoni, C.; Pellegrini, M.; Dapporto, L.; Del Gallo, M.; Pace, L.; Silveri, D.; Fattorini, S. Comparison of Soil Biology Quality in Organically and Conventionally Managed Agro-Ecosystems Using Microarthropods. *Agriculture* **2021**, *11*, 1022. [CrossRef]

17. Joimel, S.; Schwartz, C.; Bonfanti, J.; Hedde, M.; Krogh, P.H.; Pérès, G.; Pernin, C.; Rakoto, A.; Salmon, S.; Santorufo, L.; et al. Functional and Taxonomic Diversity of Collembola as Complementary Tools to Assess Land Use Effects on Soils Biodiversity. *Front. Ecol. Evol.* **2021**, *9*, 630919. [CrossRef]

18. Simoni, S.; Caruso, G.; Vignozzi, N.; Gucci, R.; Valboa, G.; Pellegrini, S.; Palai, G.; Goggioli, D.; Gagnarli, E. Effect of long-term soil management practices on tree growth, yield and soil biodiversity in a high-density olive agro-ecosystem. *Agronomy* **2021**, *11*, 1036. [CrossRef]

19. Meyer, S.; Kundel, D.; Birkhofer, K.; Fliessbach, A.; Scheu, S. Soil microarthropods respond differently to simulated drought in organic and conventional farming systems. *Ecol. Evol.* **2021**, *11*, 10369–10380. [CrossRef]

20. Foets, J.; Wetzel, C.E.; Teuling, A.J.; Pfister, L. Temporal and spatial variability of terrestrial diatoms at the catchment scale: Controls on productivity and comparison with other soil algae. *PeerJ* **2020**, *8*, e8296. [CrossRef]

21. Barragán, C.; Wetzel, C.E.; Ector, L. A standard method for the routine sampling of terrestrial diatom communities for soil quality assessment. *J. Appl. Phycol.* **2018**, *30*, 1095–1113. [CrossRef]

22. Bérard, A.; Rimet, F.; Capowiez, Y.; Leboulanger, C. Procedures for Determining the Pesticide Sensitivity of Indigenous Soil Algae: A Possible Bioindicator of Soil Contamination? *Arch. Environ. Contam. Toxicol.* **2004**, *46*, 24–31. [CrossRef] [PubMed]

23. Antonelli, M.; Wetzel, C.E.; Ector, L.; Teuling, A.J.; Pfister, L. On the potential for terrestrial diatom communities and diatom indices to identify anthropic disturbance in soils. *Ecol. Indic.* **2017**, *75*, 73–81. [CrossRef]

24. Heger, T.J.; Straub, F.; Mitchell, E.A.D. Impact of farming practices on soil diatoms and testate amoebae: A pilot study in the DOK-trial at Therwil, Switzerland. *Eur. J. Soil Biol.* **2012**, *49*, 31–36. [CrossRef]

25. Zhang, Y.; Ouyang, S.; Nie, L.; Chen, X. Soil diatom communities and their relation to environmental factors in three types of soil from four cities in central-west China. *Eur. J. Soil Biol.* **2020**, *98*, 103175. [CrossRef]

26. Caporaso, J.G.; Lauber, C.L.; Walters, W.A.; Berg-Lyons, D.; Lozupone, C.A.; Turnbaugh, P.J.; Fierer, N.; Knight, R. Global patterns of 16S rRNA diversity at a depth of millions of sequences per sample. *Proc. Natl. Acad. Sci. USA* **2011**, *108*, 4516–4522. [CrossRef] [PubMed]

27. Apprill, A.; Mcnally, S.; Parsons, R.; Weber, L. Minor revision to V4 region SSU rRNA 806R gene primer greatly increases detection of SAR11 bacterioplankton. *Aquat. Microb. Ecol.* **2015**, *75*, 129–137. [CrossRef]

28. Parada, A.E.; Needham, D.M.; Fuhrman, J.A. Every base matters: Assessing small subunit rRNA primers for marine microbiomes with mock communities, time series and global field samples. *Environ. Microbiol.* **2016**, *18*, 1403–1414. [CrossRef]

29. Walters, W.; Hyde, E.R.; Berg-lyons, D.; Ackermann, G.; Humphrey, G.; Parada, A.; Gilbert, J.A.; Jansson, J.K. Transcribed Spacer Marker Gene Primers for Microbial Community Surveys. *mSystems* **2016**, *1*, e0009-15. [CrossRef]

30. Gardes, M.; Bruns, T.D. ITS primers with enhanced specificity for basidiomycetes—Application to the identification of mycorrhizae and rusts. *Mol. Ecol.* **1993**, *2*, 113–118. [CrossRef]

31. White, T.J.; Bruns, T.; Lee, S.; Taylor, J. Amplification and Direct Sequencing of Fungal Ribosomal RNA Genes for Phylogenetics. In *PCR Protocols: A Guide to Methods Applications*; Innis, M.A., Gelfand, D.H., Snisky, J.J., White, T.J., Eds.; Academic Press: New Yourk, NY, USA, 1990; pp. 315–322.

32. Wang, Q.; Garrity, G.M.; Tiedje, J.M.; Cole, J.R. Naïve Bayesian classifier for rapid assignment of rRNA sequences into the new bacterial taxonomy. *Appl. Environ. Microbiol.* **2007**, *73*, 5261–5267. [CrossRef] [PubMed]

33. Parisi, V.; Menta, C.; Gardi, C.; Jacomini, C.; Mozzanica, E. Microarthropod communities as a tool to assess soil quality and biodiversity: A new approach in Italy. *Agric. Ecosyst. Environ.* **2005**, *105*, 323–333. [CrossRef]
34. Menta, C.; Conti, F.D.; Pinto, S.; Bodini, A. Soil Biological Quality index (QBS-ar): 15 years of application at global scale. *Ecol. Indic.* **2018**, *85*, 773–780. [CrossRef]
35. R Core Development Team. *R: A Language and Environment for Statistical Computing*; R Foundation for Statistical Computing: Vienna, Austria, 2019.
36. Seaton, F.M.; George, P.B.L.; Lebron, I.; Jones, D.L.; Creer, S.; Robinson, D.A. Soil textural heterogeneity impacts bacterial but not fungal diversity. *Soil Biol. Biochem.* **2020**, *144*, 107766. [CrossRef]
37. Cloutier, M.L.; Murrell, E.; Barbercheck, M.; Kaye, J.; Finney, D.; García-González, I.; Bruns, M.A. Fungal community shifts in soils with varied cover crop treatments and edaphic properties. *Sci. Rep.* **2020**, *10*, 6198. [CrossRef] [PubMed]
38. Bach, E.M.; Baer, S.G.; Meyer, C.K.; Six, J. Soil texture affects soil microbial and structural recovery during grassland restoration. *Soil Biol. Biochem.* **2010**, *42*, 2182–2191. [CrossRef]
39. Kepler, R.M.; Epp Schmidt, D.J.; Yarwood, S.A.; Cavigelli, M.A.; Reddy, K.N.; Duke, S.O.; Bradley, C.A.; Williams, M.M.; Maula, J.E. Soil microbial communities in diverse agroecosystems exposed to the herbicide glyphosate. *Appl. Environ. Microbiol.* **2020**, *86*, e01744-19. [CrossRef] [PubMed]
40. Yang, J.; Duan, Y.; Zhang, R.; Liu, C.; Wang, Y.; Li, M.; Ding, Y.; Awasthi, M.K.; Li, H. Connecting soil dissolved organic matter to soil bacterial community structure in a long-term grass-mulching apple orchard. *Ind. Crops Prod.* **2020**, *149*, 112344. [CrossRef]
41. Liang, B.; Ma, C.; Fan, L.; Wang, Y.; Yuan, Y. Soil amendment alters soil physicochemical properties and bacterial community structure of a replanted apple orchard. *Microbiol. Res.* **2018**, *216*, 1–11. [CrossRef]
42. Girvan, M.S.; Bullimore, J.; Ball, A.S.; Pretty, J.N.; Osborn, A.M. Responses of active bacterial and fungal communities in soils under winter wheat to different fertilizer and pesticide regimens. *Appl. Environ. Microbiol.* **2004**, *70*, 2692–2701. [CrossRef]
43. Smit, E.; Leeflang, P.; Gommans, S.; Van Den Broek, J.; Van Mil, S.; Wernars, K. Diversity and Seasonal Fluctuations of the Dominant Members of the Bacterial Soil Community in a Wheat Field as Determined by Cultivation and Molecular Methods. *Appl. Environ. Microbiol.* **2001**, *67*, 2284–2291. [CrossRef]
44. Whitelaw-Weckert, M.A.; Rahman, L.; Hutton, R.J.; Coombes, N. Permanent swards increase soil microbial counts in two Australian vineyards. *Appl. Soil Ecol.* **2007**, *36*, 224–232. [CrossRef]
45. Guo, A.; Ding, L.; Tang, Z.; Zhao, Z.; Duan, G. Microbial response to $CaCO_3$ application in an acid soil in southern China. *J. Environ. Sci.* **2019**, *79*, 321–329. [CrossRef] [PubMed]
46. Gkisakis, V.D.; Kollaros, D.; Bàrberi, P.; Livieratos, I.C.; Kabourakis, E.M. Soil Arthropod Diversity in Organic, Integrated, and Conventional Olive Orchards and Different Agroecological Zones in Crete, Greece. *Agroecol. Sustain. Food Syst.* **2015**, *39*, 276–294. [CrossRef]
47. Sheikh, A.A.; Rehman, N.; Bhat, T.A.; Sofi, M.A.; Bhat, M.A.; Un Nabi, S.; Aijaz, C.; Sheikh, A.; Lone, G.; Sofi, A. Vertical distribution of soil arthropods in apple ecosystem of Kashmir. *J. Entomol. Zool. Stud. JEZS* **2017**, *843*, 843–846.
48. Mantoni, C.; Di Musciano, M.; Fattorini, S. Use of microarthropods to evaluate the impact of fire on soil biological quality. *J. Environ. Manag.* **2020**, *266*, 110624. [CrossRef] [PubMed]

sustainability

Article

Effects of Landscape Positions and Landscape Types on Soil Properties and Chlorophyll Content of Citrus in a Sloping Orchard in the Three Gorges Reservoir Area, China

Siyue Sun [1], Guolin Zhang [2], Tieguang He [3,*], Shufang Song [4] and Xingbiao Chu [1,*]

[1] College of the Arts, Guangxi University, Nanning 530004, China; 2024393014@st.gxu.edu.cn
[2] College of Architecture and Landscape, Peking University, Beijing 100080, China; gl.zhang@pku.edu.cn
[3] Agricultural Resources and Environment Research Institute, Guangxi Academy of Agricultural Sciences, Nanning 530007, China
[4] School of Information and Statistics, Guangxi University of Finance and Economics, Nanning 530003, China; ssffsong@gxufe.edu.cn
* Correspondence: htg118@gxaas.net (T.H.); 20130112@gxu.edu.cn (X.C.)

Abstract: In recent years, soil degradation and decreasing orchard productivity in the sloping orchards of the Three Gorges Reservoir Area of China have received considerable attention both inside and outside the country. More studies pay attention to the effects of topography on soil property changes, but less research is conducted from the landscape. Therefore, understanding the effects of landscape positions and landscape types on soil properties and chlorophyll content of citrus in a sloping orchard is of great significance in this area. Our results showed that landscape positions and types had a significant effect on the soil properties and chlorophyll content of citrus. The lowest soil nutrient content was detected in the upper slope position and sloping land, while the highest exists at the footslope and terraces. The chlorophyll content of citrus in the middle and upper landscape position was significantly higher than the footslope. The redundancy analysis showed that the first two ordination axes together accounted for 81.32% of the total variation, which could be explained by the changes of soil total nitrogen, total phosphorus, total potassium, available nitrogen, available potassium, organic matter, pH, and chlorophyll content of the citrus. Overall, this study indicates the significant influence of landscape positions and types on soil properties and chlorophyll content of citrus. Further, this study provides a reference for the determination of targeted land management measures and orchard landscape design so that the soil quality and orchard yield can be improved, and finally, the sustainable development of agriculture and ecology can be realized.

Keywords: agriculture landscape; chlorophyll content of citrus; landscape position; soil properties; terraces

Citation: Sun, S.; Zhang, G.; He, T.; Song, S.; Chu, X. Effects of Landscape Positions and Landscape Types on Soil Properties and Chlorophyll Content of Citrus in a Sloping Orchard in the Three Gorges Reservoir Area, China. *Sustainability* 2021, 13, 4288. https://doi.org/10.3390/su13084288

Academic Editors: Bharat Sharma Acharya and Rajan Ghimire

Received: 11 March 2021
Accepted: 6 April 2021
Published: 12 April 2021

Publisher's Note: MDPI stays neutral with regard to jurisdictional claims in published maps and institutional affiliations.

1. Introduction

The Three Gorges Reservoir Area of China is one of the most suitable ecological areas for the growth of citrus. The citrus industry has achieved a dominant position in the development of mountainous agriculture and rural economy in this area [1]. Currently, the mountainous agriculture in this area is transforming from conventional farming systems to suburban modern agriculture and leisure and sightseeing agriculture, and the ecosystem service function contained in it has begun to attract attention [2]. Citruses are planted on the sloping land along the Yangtze River and its tributaries because of cultivated land tension in this area. For a long time, orchard managers have been ignoring soil conservation practices, and indiscriminately using large amounts of fertilizers without considering soil differences, which has led to serious problems such as chemical fertilizer pollution, soil degradation, and orchard production reduction, and brought great harm to agricultural production and ecological environment in the area [3–5].

Soil properties, the most important factor determining soil quality, not only affect crop output, but also have a significant impact on the cultivated land use and soil environmental protection [6–8]. Several previous studies have confirmed the effects of terrain, land use, hedgerows, and other environmental factors and management measures on soil quality in the Three Gorges Reservoir Region [9]. Teng Mingjun et al. [10] found that topographical factors are the main factors that cause the spatial heterogeneity of soil organic carbon in the reservoir area. Shen Zhenyao et al. [11] found that load intensities of nitrogen, phosphorus, and other non-point source pollutants were significantly different in soils with different land-use patterns. Xu Feng et al. [12] showed that slope ecological engineering with contour hedgerows could effectively control slope erosion and nutrient loss.

However, in the Three Gorges Reservoir Region, a fragile ecoregion and a developing leisure and sightseeing agricultural area, there are a few studies on the change of soil properties caused by the landscape. Each soil property has a respective spatial distribution in the landscape. The landscape position affects the process of soil formation, so it is considered to be one of the key factors affecting the changes of soil properties [13,14]. Simultaneously, landscape types also cause changes in soil spatial distribution. Arnaz's [15] study found that when sloping land transformed into terraced land, the slope's length and angle decreased significantly, resulting in a decrease in soil erosion and sediment yield. Although soil properties are influenced by many factors, such as climate, parent material, and biological factors, the influence of landscape types and landscape positions cannot be ignored on the regional scale [16]. Therefore, in this area, soil-landscape analysis is crucial to understand the spatial variation law of soil properties for determining targeted land management interventions to improve soil quality, form a charming farmland landscape, and achieve sustainable agricultural development [17].

Additionally, more researchers believe that it is best to combine soil analysis with leaf analysis to comprehensively diagnose the soil quality and citrus nutritional status to guide rational fertilization, improve citrus quality, and increase citrus yield. Mohesh et al. [18,19] Showed that the chlorophyll content of plant leaves determines the photosynthetic capacity and nutritional status of leaves, which can be used as an indicator of plant health. Haboudane et al. [20–22] point out that chlorophyll content in crops plays a key in precision agriculture because it is related to nitrogen concentration in the leaf of a crop. It reflects how the crop responds to nitrogen application, as well as being an important indicator of photosynthetic activity, which determines crop yield. Therefore, it is necessary to assess the comprehensive impact of environmental factors on soil and citrus trees in combination with the changes of the chlorophyll content of citrus (CCC).

The purposes of this study are: (1) to evaluate the effects of landscape position and landscape type on soil properties, including soil total nitrogen, total phosphorus, total potassium, available potassium, available nitrogen, available phosphorus, soil organic matter, and pH; and (2) to evaluate the effects of landscape position and landscape type on CCC. This information provides a reference for knowing how the local ecosystem works and assessing the impact of future landscape changes. It is not only helpful for the determination of targeted orchard land management measures and the formation of a good agricultural landscape but also related to the ecological environment safety and sustainable development of the agricultural economy in the middle and upper reaches of the Yangtze River.

2. Materials and Methods

2.1. Study Area

The study area is in the citrus orchard in Guo Jiagou, Fengjie County, Chongqing, China (31°06′ N, 109°27′ E, Figure 1). Fengjie County, with an altitude ranging from 86 to 2123 m above sea level, is a mountainous landform in the eastern Sichuan Basin and the mountainous area accounts for 88.3% of the total (Figure 1D). The Yangtze River runs through the middle of Fengjie County, stretching 41.5 km, with Mei Xi River, Da Xi River, Shi Sun River, Cao Tang River, Zhu Yi River, and other rivers. Fengjie County is a typical subtropical monsoon climate with four distinct seasons, abundant rainfall, and long sunshine hours. Due to the influence of topography and landform, the vertical change of climate is more obvious and forms a typical three-dimensional climate. The frost-free period is about 287 days; average annual temperature is 16.3°, average precipitation is about 1150 mm, and average sunshine duration is 1639 h.

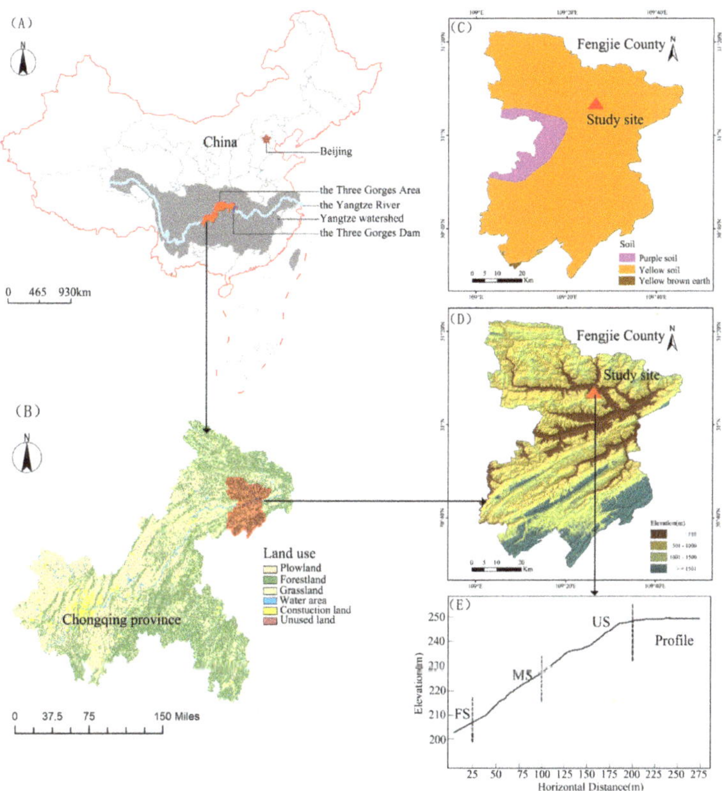

Figure 1. (**A**) The geographical location of the Yangtze watershed in China. (**B**) The land use map of Chongqing province. (**C**) The soil distribution map of study area. (**D**) The geographical structure of study area. (**E**) The geomorphic profile of the study site.

Fengjie County is located in the core area of the citrus industrial belt in the middle and upper reaches of the Yangtze River. As of March 2021, the citrus planting area of Fengjie County has reached 246.7 km^2, the output has reached 370,000 tons, and the comprehensive output value has exceeded 3.5 billion yuan, accounting for about 20% of the total agricultural output value. This has led to the income increase of 0.3 million people in 24 towns and 70,000 households, and the employed population accounts for 28.4% of the total population of Fengjie County. Citrus trees in the study area were planted in 1980; the variety is Feng Yuan 72–1, the row spacing of citrus plants is 4 m × 4 m, and the height of the canopy is 3–4 m. In 1990, part of the sloping land was changed to contour terraces, and the slope of the surface was 5 degrees, showing a trend of high inside and low outside. At present, in the study area the most agricultural landscape is sloping landscape, with a few being terraced landscape.

The soil distribution in Fengjie County is shown in Figure 1C and Table 1. The soil in the study area is mainly yellow soil, and the profile configuration is A–B–C type. The thickness of the soil layer is generally more than 60 cm, and the color is yellow or brownish yellow. The content of silt is 36.9–44.38%, clay is 30%, and the texture is loamy clay. This kind of soil has a deep soil layer, heavy texture, total porosity of about 59%, and strong water and fertility conservation.

Table 1. Information about the geological structure, soils, and land use of the analyzed catchment.

Item	Classification	Area (km²)	Proportion (%)
Soil	Purple soil	297.3	7.3
	Yellow soil	3796.7	92.7
	Yellow brown earth	5.1	0.1
Geography	≤500 m	636	14.1
	501–1000 m	1820.1	40.3
	1001–1500 m	1520.2	33.7
	≥1501 m	537.3	11.9
Land use	Plowland	37,504	45.5
	Forestland	33,575	40.7
	Grassland	7608	9.2
	Water area	333	0.4
	Construction land	2399	2.9
	Unused land	12	0.01

The data of soil and topography is from Fengjie County. The data of land use is from Chongqing.

2.2. Experimental Design

This study was conducted in August 2020 in the citrus orchard operated by Fengjie County Agriculture Development Ecology Co., Ltd., Chongqing, China. Two adjacent slopes with different landscape types were chosen to explore the effects of landscape positions and landscape types on soil properties and CCC in the study area. One of the slopes is a sloping landscape, the other is a contour terrace in the upper slope position, and the middle slope and footslope positions are sloping landscapes. Refer to the research methods of Brubaker et al. to divide slope landscape position [23]. The research is divided into three parts (Figure 1E): the upper landscape position (US), the middle landscape position (MS), and the foot of slope (FS). Landscape types are divided into sloping landscape and terraced landscape. A total of representative 24 plots (4 m × 4 m) were selected based on landscape positions and landscape types: 8 in the US, 8 in the MS, and 8 in the FS. The landscape position, landscape type, and geographical location of each sampling plot were recorded, and slope gradient, slope aspect, and elevation were measured. The basic information of sample plots in different landscape positions is shown in Table 2.

Table 2. The basic information of sample plots in different landscape positions.

Landscape Position	Geographic Position	Elevation (m)	Slope (°)	Slope Aspect (°)	Number of Plots	Dominant Landscape Position
FS	31°6′55″ N 109°27′24″ E	210	19–23	302–355	8	slopes
MS	31°6′52″ N 109°27′25″ E	225	30–36	78–355	8	slopes
US	31°6′49″ N 109°27′26″ E	240	5–21	347–352	8	terraces, slopes

FS = the footslope position, MS = the middle slope position, US = the upper slope position.

2.3. Soil Sampling and Chlorophyll Content of Citrus

In August 2020, soil samples were collected for 24 plots by the diagonal sampling method. After clearing the top litter, three individual soil samples (0–15 cm, one from the center of the field, two from diagonal corners) were taken from a plot and mixed to get 1 sample. The samples were air-dried and sent to the Guangxi Academy of Agricultural

Sciences to determine the contents of soil total nitrogen, total phosphorus, total potassium, available nitrogen, available phosphorus, available potassium, organic matter, and pH.

The CCC, expressed as a chlorophyll content index (CCI) was measured using the CCM-200 plus Chlorophyll Content Meter (OPTI–SCIENCES, Hudson, NH, USA) between 8:00 and 10:00 a.m. on 28 and 29 August 2020 (sunny cloudless). The measurement method was to select well-developed and fully developed leaves from the upper, middle, and lower locations of the south of a citrus tree. After measuring each leaf four times, the mean value of chlorophyll content in the upper, middle, and lower tree locations was taken.

2.4. Data Analysis
2.4.1. Statistics Analysis

Statistical analyses were conducted using Excel 2016 (Microsoft, Redmond, WA, USA) and SPSS 24.0 (IBM, Armonk, NY, USA). A descriptive statistic was performed to describe the soil properties and CCC. Then, Pearson correlation analysis was used to show correlations between soil properties. One-way ANOVA was used to examine the effects of landscape position and landscape type on soil properties and CCC, and the least significant difference method (LSD) was used to compare the mean values.

2.4.2. Redundancy Analysis Method

To find the most important environmental factors that affect the soil properties and CCC in the experimental area, redundancy analysis (RDA) was carried for a constrained ranking analysis based on the experimental data. A Monte Carlo permutation test was used to find the relative importance of each environmental factors in explaining changes in soil properties. Redundancy analysis is a direct gradient analysis technique. Through community ranking, the community sample plots (species) investigated in an area were arranged according to their similarity to analyze the relationships between various species and the environment and effectively evaluate the impact of environmental variables on species [24]. In this research, RDA was applied to find the relationship between species variables (soil properties and CCC) and environmental variables (landscape positions and landscape types, slope surface, gradient, and aspect).

It is worth noting that the gradient length should be measured by detrended correspondence analysis (DCA) on the sample before constraint analysis. Since the first gradient length was 0.6 < 3.0, RDA is the most appropriate method [25]. Before RDA analysis, two data matrices (species data and environmental data) were built, and the environmental data were encoded. In this study, landscape types were divided into two types: 1 represents the sloping landscape and 2 the terraced landscape. Landscape position can be divided into FS, MS, and US, represented by 1, 2, and 3, respectively. The experimental slope surface was divided into the terraced field and sloping field, represented by 1 and 2, respectively. The actual measured values were used to represent the slope and aspect.

3. Results and Discussion
3.1. The Description of Soil Properties and CCC

Table 3 shows the statistical variables of soil properties and CCC under different landscape positions and landscape types. The content ranges of soil properties and CCC were as follows: CCC: 66.03–163.40, total nitrogen: 0.55–2.70 g/kg, total phosphorus: 0.19–1.75 g/kg, total potassium: 11.76–35.11 g/kg, available nitrogen: 30.60–185.50 mg/kg, available phosphorus: 0.30–50.70 mg/kg, available potassium: 50.50–273.40 mg/kg, soil organic matter: 9.30–51.00 g/kg, and soil pH: 5.73–7.54. The average content of soil properties followed a decreasing order: available potassium > available nitrogen > available phosphorus > soil organic matter > total potassium > total nitrogen > total phosphorus. The coefficient of variation (CV) of soil properties and CCC ranged from 27.60% to 83.82% (Table 3), showing medium variation (10% $\leq$ CV $\leq$ 100%), indicating that the soil properties and CCC varied greatly in different landscape positions and types.

Table 3. Descriptive statistics for soil properties and the chlorophyll content of citrus (CCC).

Item	CCC	TN (g/kg)	TP (g/kg)	TK (g/kg)	AN (mg/kg)	AP (mg/kg)	AK (mg/kg)	SOM (g/kg)	pH
Range	97.37	2.15	1.56	23.35	154.90	50.40	222.90	41.70	1.81
Minimum	66.03	0.55	0.19	11.76	30.60	0.30	50.50	9.30	5.73
Maximum	163.40	2.70	1.75	35.11	185.50	50.70	273.40	51.00	7.54
Mean	119.59	1.30	0.77	16.45	76.39	16.83	157.25	21.40	6.81
SD	33.72	0.60	0.51	4.54	38.67	16.40	63.66	11.67	0.58
CV (%)	28.20	46.15	66.23	27.60	50.62	61.40	40.48	54.53	83.82

SD = standard deviation, CV = coefficient of variation, TN = total nitrogen, TP = total phosphorus, TK = total potassium, AN = available nitrogen, AP = available phosphorus, AK = available potassium, SOM = soil organic matter, CCC = chlorophyll content of citrus.

3.2. Soil Properties Correlations

A significant correlation was found between many soil properties measurements (Table 4). The content of total nitrogen (TN) was significantly positively correlated with total phosphorus (TP), available nitrogen (AN), available phosphorus (AP), available potassium (AK), and soil organic matter (SOM). SOM was significantly positively correlated with AN, TP, AP ($p < 0.01$), and AK ($p < 0.05$). The increasing SOM can promote the activity of soil enzymes, and promote the decomposition of plant and animal residues and humus, thus releasing nitrogen and potassium [26]. Moreover, SOM has a significant promoting effect on soil nutrient retention and nutrient supply capacity, which also suggests that the future fertilization should be based on organic fertilizer, supplemented by fertilizer, to maintain anthropogenic mellowing of soil, to meet the nutritional requirements of stable yield, high yield, and good quality of citrus. Soil pH was negatively correlated with TN, AN, AK, and SOM, and it was an important index reflecting soil parent material properties and weathering and leaching conditions [27].

Table 4. Pearson correlation coefficients between soil properties.

	TN	TP	TK	AN	AP	AK	SOM
TP	0.687 **						
TK	−0.058	0.096					
AN	0.930 **	0.584 **	−0.075				
AP	0.744 **	0.876 **	0.089	0.737 **			
AK	0.607 **	0.458 *	−0.094	0.660 **	0.653 **		
SOM	0.954 **	0.699 **	−0.120	0.876 **	0.744 **	0.475 *	
pH	−0.537 **	−0.297	0.089	−0.582 **	0.539 *	−0.611 **	−0.420 *

* = $0.01 < p < 0.05$, ** = $0.001 < p < 0.01$.

3.3. The Effect of Landscape Positions on Soil Properties and CCC

Most soil properties and CCC were significantly different in landscape positions (Table 5). The results of multiple comparisons showed that the CCC among the three landscape positions showed the order of MS > US > FS, and the CCC at the FS was significantly different from the US and MS ($p \leq 0.001$, Figure 2). Vladimir et al. [28] showed that under the same photosynthetic photon flux density, the fluorescence intensity excited by blue light is 2.5 to 3 times that of red light. Blue light significantly promoted the formation and accumulation of chlorophyll [29,30]. In the mountainous environment, the landscape position directly affects the illumination conditions, and the solar radiation at the FS is lower than the MS and US. Therefore, the CCC at the US and MS is extremely significantly higher than that the FS. At the same time, sunlight may also affect the formation of local microclimates, and differences in microclimates can affect the distribution of plant communities, which in turn affects the growth and development of citrus trees [31]. Additionally, a study by Qiang Fu et al. [32,33] showed that in the soil at a depth of 0–20 cm, the higher the landscape position, the greater the soil moisture content, and chlorophyll content is positively related to soil moisture. Therefore, the significant differences of CCC in landscape positions may be caused by many reasons.

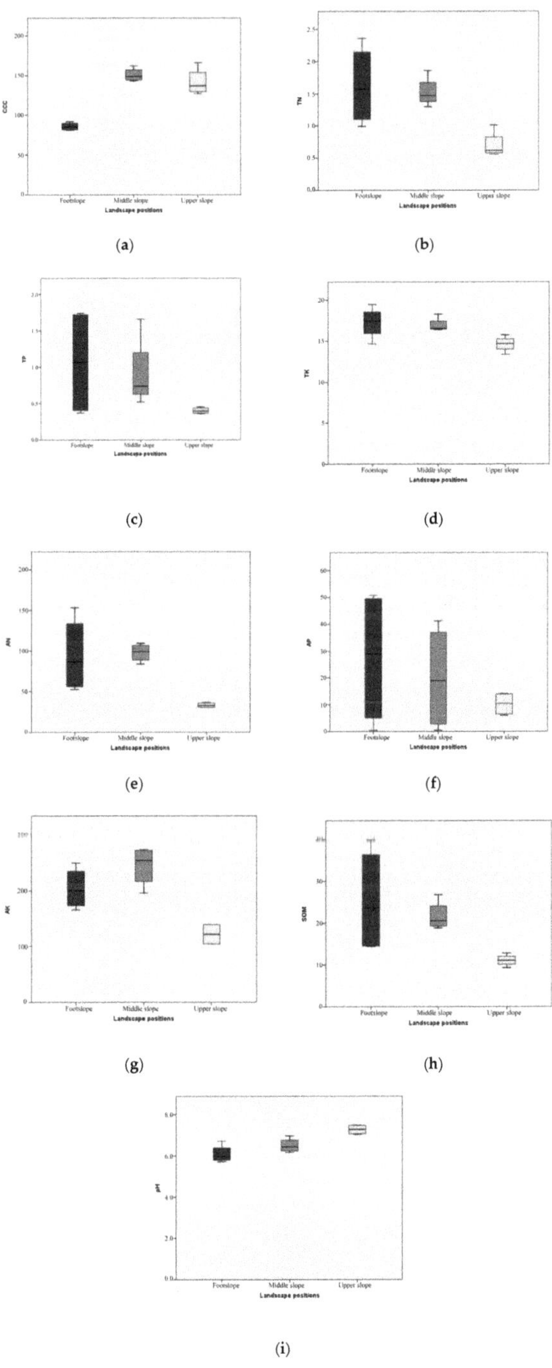

Figure 2. Variations of soil properties and CCC in different landscape positions. (**a**) Changes of citrus chlorophyll content; (**b**) Changes of total nitrogen content; (**c**) Changes of total phosphorus content; (**d**) Changes of total potassium content; (**e**) Changes of available nitrogen content; (**f**) Changes of available phosphorus content; (**g**) Changes of available potassium content; (**h**) Changes of soil organic matter content; (**i**) Changes of pH.

Table 5. Mean comparisons of soil properties by landscape positions of the study site.

LS	CCC	Soil Properties							PH
		TN (g/kg)	TP (g/kg)	TK (g/kg)	AN (mg/kg)	AP (mg/kg)	AK (mg/kg)	SOM (g/kg)	
FS	85.80	1.62	1.06	17.25	94.93	27.25	204.30	25.40	6.10
MS	150.88	1.53	0.91	16.96	98.08	19.85	244.40	21.73	6.51
US	141.96	0.70	0.40	14.64	33.25	10.18	121.85	11.05	7.27
F	37.57 ***	6.15 *	1.73	4.36 *	6.64 *	0.80	14.90 **	3.67	11.53 **

$* = 0.01 < p < 0.05$, $** = 0.001 < p < 0.01$, $*** = p < 0.001$. Abbreviations as in Table 3.

The soil pH value is extremely significantly correlated with the landscape position ($p < 0.01$), manifested as a decrease along the downslope. Due to the citrus orchard being located on the slope of the Three Gorges Reservoir area, the topography has an impact on soil nutrient loss. The leaching or loss of base ions, especially calcium ions, makes the soil acidified [34]. On the contrary, the TN and TK accumulate along the downslope. The multiple comparison results showed that the contents of total nitrogen and total potassium were the lowest in the US position and tended to be the highest at FS position. The content in the US is significantly lower than the FS ($p < 0.05$). This result is consistent with the research conclusion of Hu Chenxia et al. and may be due to the nutrient deposition in the US and the production and residue of plants [35]. The contents of AN and AK were the highest content in the MS, and significantly higher than those in the US ($p < 0.05$, $p < 0.01$, respectively). It could be due to better temperature and moisture in the MS than in the US, or better light conditions than at the FS [36]. Although the difference of TP, AP, and SOM among landscape positions was not significant ($p > 0.05$), they generally tended to accumulate along the downslope, which was consistent with the fact that TN, AN, TK, and AK had lower values on the US and higher values at the FS. Zhang Jianhui et al. [37] showed that the main erosion process in a medium-long slope (40–110 m) was water erosion, and soil nutrients were carried from the upper slope to the foot slope by surface runoff, leading to a decrease in soil quality in the US position.

In general, most soil properties and CCC in orchards have significant correlations among landscape positions. Specifically, most soil properties show a higher value in the MS and FS and lower in the US [38]. The lowest levels of other soil properties, such as soil organic carbon, were also usually found in the US position [39]. The soil properties of the US were worse than those of the MS and FS position. However, unsustainable land use in the upper landscape position has an impact on the lower slope area. The soil nutrient loss is the main reason for the decline of soil quality and non-point source pollution. Therefore, the balance and improvement of soil quality is the critical factor to achieve sustainable agricultural development [40–42]. Xue Zhijing et al. [43] suggested that the high vegetation cover can reduce runoff and soil loss, and then maintain soil nutrients better. Additionally, the effects of tillage erosion cannot be ignored. One solution is to reduce tillage on uphill sites. The other solution is more appropriate to this area to improve the soil fertility by intercropping and applying green manure or organic manure in the US. Applying biological organic fertilizer instead of chemical fertilizer can reduce the emission of nitrogen dioxide and thus improve the acidification and eutrophication of surface water [44]. Therefore, the effects of landscape positions on soil properties and CCC should be further studied.

3.4. The Effect of Landscape Types on Soil Properties and CCC

Some soil properties of different landscape types in the same landscape position (US) in the orchard were significantly different (Table 6). The mean value of TN and TK ranged from 0.70–1.08 g/kg and 14.64–17.68 g/kg, respectively. Multiple comparison results showed that the TN and TK contents in terraced fields were significantly higher than in sloping fields ($p < 0.05$). The contents of AN, AP, and other nutrients that could be directly absorbed by crops in terraced fields were significantly higher than those in sloping ($p \leq 0.001$), and the contents of AN and AP in terraced fields were 77.08% and

217.58% higher than those in sloping, respectively. Although there was no significant difference in TP, AK, SOM, and pH among different landscape types ($p < 0.05$), further observation showed that their contents in terraced fields were still higher than those in sloping, indicating that the difference of landscape types did lead to the changes of soil properties.

The research conclusion is consistent with Fu Bojie et al. [45]. The soil properties of sloping land being lower than terraces may be the combined action of water erosion and tillage erosion. Xu Chang et al. [46,47] showed that rainfall and surface runoff were the impetus of soil nutrient loss. Compared with the sloping field, the slope length and angle of the terraced orchard were significantly reduced, which resulted in the reduction of soil erosion caused by the topography and the better preservation of soil nutrients. At the same time, it reduces water body pollution caused by soil particles, nitrogen, phosphorus, and other elements in farmland under the scouring effect of rainwater and runoff. Furthermore, effects of contour tillage on soil movement (translocation and erosion) were examined by Zhang Jianhui et al. in the steep hillslopes of the Sichuan basin using a physical tracer method. The results showed that tillage significantly affects soil migration on sloping land. The tillage erosion rate under contour tillage was 77% lower than that under downslope tillage [48]. In the future farming of sloping, especially at the top and upper slopes with high soil property variability, it is necessary to consider the cultivation method of contour terraced fields, which is conducive to soil conservation and has a charming farmland landscape [49]. Bo Sun et al. [50,51] indicated that excessive application of nitrogen fertilizer and pesticides is considered to cause water pollution in the Yangtze River basin. The solution is to reduce the loss of soil nutrients and maintain a good farmland ecosystem. Therefore, compared with slope land, terrace farming with good ecological, landscape, and economic benefits is a more suitable farming method for farmland in the Yangtze River Basin. At that time, the charming terraced orchard and the beautiful Yangtze River will become a brilliant and unique scene, which will make the leisure and sightseeing agriculture in this region develop better. Moreover, Shimbahri et al. [52] conducted a study on the effect of terraces on soil water content in an arid area of Ethiopia. The results showed that terraces do have good performance in soil and water conservation. Thus, we should also pay attention to the impact of landscape types on soil water content in future research, which is very important for arid and semi-arid regions in the world.

Table 6. Mean comparisons of soil properties by landscape types.

LT	CCC	Soil Properties							PH
		TN (g/kg)	TP (g/kg)	TK (g/kg)	AN (mg/kg)	AP (mg/kg)	AK (mg/kg)	SOM (g/kg)	
Slope	141.96	0.70	0.40	14.64	33.25	10.18	121.85	11.05	7.27
Terrace	139.83	1.08	0.99	17.68	58.88	32.33	127.05	19.28	7.49
F value	0.05	6.43 *	10.41	7.17 *	30.69 ***	45.22 ***	0.10	3.97	3.02

* = $0.01 < p < 0.05$, *** = $p < 0.001$.

3.5. Redundancy Analysis (RDA)

Through redundancy analysis, we obtained the relationship among soil properties and CCC and environmental variables. The significance of the constraint ordination was tested by the Monte Carlo permutation test (499 permutations were performed). The results showed that the tests on the first and second constraint axes were obvious ($p = 0.028$ and 0.002, respectively). The first and second constraint axes together explain 81.32% of the relationship between the species variables and environmental variables. Therefore, we chose the first two constraint axes with high and significant eigenvalues to draw a biplot for observation and then tried to explain it.

The RDA ordination diagram is explained below: environmental variables (explanatory variables) in red arrows indicate species variables (response variables) with blue. The angle between the arrow of the environmental variables and response variables reflects the correlation (but not the meaning of the angle between the response variable): when the angle is acute, the correlation is positive; correlation is negative when the angle is greater than 90 degrees. The length of the line between the red arrow (environmental variables) and the origin is directly proportional to the degree of correlation between an environmental factor and the distribution of community and species. The angle between the arrow of the environmental variable and the constraint axis represents correlation. The smaller the angle is, the greater the correlation will be. If it is orthogonal, it will be irrelevant. The blue arrow (species variable) points from the origin to the corresponding coordinate of the species score. The direction the arrow points to indicates the direction in which the abundance of the species has increased. The correlation between species and environmental variables was displayed by a perpendicular projection of the species arrow-tips onto the line overlaying the environmental arrow. The longer the projection, the higher the correlation.

The results of the RDA ordination diagram (Figure 3) show that Axis1 is positively correlated with slope gradient and aspect. The first constraint axis is mainly interpreted as slope gradient and aspect because of the length of the arrow. The second ordination axis (Axis2) has a great negative correlation with landscape position. Therefore, the second ordination axis is mainly interpreted as landscape position. The CCC is highly correlated with landscape position, followed by slope gradient, and negatively with aspect. The highest negative correlation with landscape positions is TN, followed by TK, AN, SOM, and TP. Although most soil properties are significantly affected by landscape positions, the influence of slope and aspect cannot be ignored.

In the RDA ordination diagram, slope gradient and aspect are the main determinants of Axis 1 (Figure 3). TN, TP, AN, AK, and SOM are positively correlated with slope gradient and negatively with aspect, which is in keeping with the research conclusions of Holden et al. [53]. Although the heterogeneity of soil properties is influenced by many factors, such as climate, soil parent material, and biology, many soil properties changes can be attributed to topography [54]. The movement and accumulation of soil solutions are significantly affected by slope gradient and aspect, leading to spatial differences in soil properties [55].

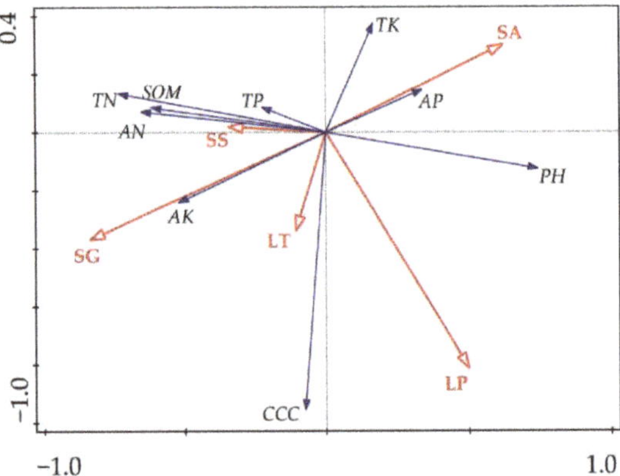

Figure 3. Ordination biplot of redundancy analysis (RDA) displaying the effects of the selected environmental variables on soil properties and CCC. LT = landscape type, LP = landscape position, SG = slope gradient, SA = slope aspect, SS = slope surface, CCC = chlorophyll content of citrus.

The correlation between CCC and slope gradient and aspect may be related to solar radiation. The slope aspect, a topographic factor that changes regional microclimate, determines the solar radiation amount received by the slope surface [56]. Smith's [57] research shows that many environmental changes are related to solar radiation. Sun Ying et al.

showed that the chlorophyll content in plants also had a relationship with light intensity [58,59]. Therefore, in this study, with the change of slope gradient and aspect, the amount of solar radiation probably changes accordingly, which could lead to the CCC changes.

4. Conclusions

In conclusion, our study clearly shows significant changes in soil properties and CCC among different landscape positions and landscape types in the Three Gorges Reservoir Area. Most soil properties showed the highest content in the footslope and terraced landscape and the lowest in the upper slope and sloping. The CCC in the footslope was significantly less than in the MS and US location. In addition to the strong effect of landscape, the well-known principle that spatial heterogeneity of soil properties is affected by topographic factors such as slope gradient and aspect was also confirmed in this research. These results indicated that the changes of soil properties and CCC in this area were mainly affected by landscape position, landscape type, and topography. For Fengjie County to develop suburban modern agriculture and sightseeing agriculture, determining the targeted land management measures of orchards to change the farmland landscape and orchard planting layout is more in line with the requirements of regional development [60]. It can not only improve the soil quality and yield of orchards and reduce the unnecessary nutrient waste and non-point source pollution caused by orchards, but also provide ideas for the landscape design of orchards to realize the sustainable development of agriculture and ecology in this area.

It is necessary to carry out similar and larger scale research in other catchments such as the Yellow River basin or citrus planting areas such as Southern Jiangxi, China to determine the complex influence law between soil–landscape–crop under different soil parent materials, climate, geographical conditions, etc., and formulate the applicable regional eco-agricultural transformation scheme. In addition, this study has the same reference value for the development of sustainable agriculture in other tropical and subtropical countries such as India and Nigeria. However, our research method has some limitations, such as the research on the spatial distribution of soil properties being weak. Therefore, we summarized several lessons for future researchers to conduct further study. 1. More soil physical properties, such as soil bulk density, aeration, permeability, and adhesion, can be included in soil analysis. 2. More response variables can be included in leaf analysis, such as chlorophyll a, chlorophyll b, carotenoids, leaf area, specific leaf index, etc.

Author Contributions: Conceptualization and methodology, G.Z. and S.S. (Siyue Sun); software, validation, investigation and formal analysis, S.S. (Siyue Sun) and S.S. (Shufang Song); resources and data curation, T.H.; writing—original draft preparation, S.S. (Siyue Sun); writing—review and editing, and supervision, G.Z. and X.C.; visualization, S.S. (Siyue Sun); project administration, X.C.; funding acquisition, T.H., G.Z., and S.S. (Shufang Song). All authors have read and agreed to the published version of the manuscript.

Funding: This research was funded by the National Natural Science Foundation of China, grant number: 51808215, Advantage Research Team Project of Guangxi Academy of Agricultural Sciences, grant number: 2021YT037, and Basic Ability Improvement Project for Young and Middle-aged Teachers in Guangxi Universities and Research and Development Fund for Young Teachers in Guangxi University of Finance and Economics of China, grant numbers: 2018KY0505, 2018QNB10.

Acknowledgments: We sincerely thank Fengjie County Agricultural Development Ecology Co., Ltd. for providing us with experimental base and some help. We are also grateful to the journal editors and anonymous reviewers for their constructive comments.

Conflicts of Interest: The authors declare no conflict of interest.

References

1. Wen, Z.F. Thoughts on the Construction of Chongqing Citrus Industry Development Technology System. *Southwest Hortic.* **2002**, *30*, 15–17+20.

2. Liu, Y.S.; Zhang, Z.W.; Wang, J.Y. Regional differentiation and comprehensive regionalization scheme of modern agriculture in China. *Acta Geogr. Sin.* **2018**, *73*, 203–218.

3. Zhang, X.; Zhang, J.; Li, L.; Zhang, Y.; Yang, G. Monitoring Citrus Soil Moisture and Nutrients Using an IoT Based System. *Sensors* **2017**, *17*, 447. [CrossRef] [PubMed]

4. Abdollahnejad, A.; Panagiotidis, D.; Shataee Joybari, S.; Surový, P. Prediction of Dominant Forest Tree Species Using QuickBird and Environmental Data. *Forests* **2017**, *8*, 42. [CrossRef]

5. Wan, S.-Z.; Gu, H.-J.; Yang, Q.-P.; Hu, X.-F.; Fang, X.-M.; Singh, A.N.; Chen, F.-S. Long-term fertilization increases soil nutrient accumulations but decreases biological activity in navel orange orchards of subtropical China. *J. Soils Sediments* **2017**, *17*, 2346–2356. [CrossRef]

6. Jiang, P.; Thelen, K.D. Effect of soil and topographic properties on crop yield in a north-central corn-soybean cropping system. *Agron. J.* **2004**, *96*, 252–258. [CrossRef]

7. Kravchenko, A.N.; Bullock, D.G. Correlation of corn and soybean grain yield with topography and soil properties. *Agron. J.* **2000**, *92*, 75–83. [CrossRef]

8. Jiang, Y.; Sun, K.; Guo, X.; Ye, Y.; Rao, L.; Li, W. Prediction of spatial distribution of soil properties based on environmental factors and neighbor information. *Res. Environ. Sci.* **2017**, *30*, 1059–1068.

9. XiaoHua, X.; Xin-Fa, X.; Sheng, L.; ShaSha, F.; Gaowei, W. Soil Erosion Environmental Analysis of the Three Gorges Reservoir AreaBased on the "3S" Technology. *Procedia Environ. Sci.* **2011**, *10*, 2218–2225. [CrossRef]

10. Teng, M.; Zeng, L.; Xiao, W.; Huang, Z.; Zhou, Z.; Yan, Z.; Wang, P. Spatial variability of soil organic carbon in Three Gorges Reservoir area, China. *Sci. Total Environ.* **2017**, *599*, 1308–1316. [CrossRef] [PubMed]

11. Shen, Z.; Chen, L.; Hong, Q.; Qiu, J.; Xie, H.; Liu, R. Assessment of nitrogen and phosphorus loads and causal factors from different land use and soil types in the Three Gorges Reservoir Area. *Sci. Total Environ.* **2013**, *454*, 383–392. [CrossRef]

12. Xu, F.; Cai, Q.G.; Wu, S.A.; Zhang, G.Y.; Ding, S.W.; Cai, C.F. Research on the Control of Soil Nutrient Loss by Slope Land Ecological Engineering in the Three Gorges Reservoir Area—Take contour hedges as an example. *Geogr. Res.* **2000**, *19*, 303–310.

13. Hook, P.B.; Burke, I.C. Biogeochemistry in A Shortgrass Landscape: Control by Topography, Soil Texture, and Microclimate. *Ecology* **2000**, *81*, 2686–2703. [CrossRef]

14. Ovalles, F.A.; Collins, M.E. Soil-landscape Relationships and Soil Variability in North Central Florida. *Soil Sci. Soc. Am. J.* **1986**, *50*, 401–408. [CrossRef]

15. Arnaez, J.; Lasanta, T.; Errea, M.P.; Ortigosa, L. Land abandonment, landscape evolution, and soil erosion in a Spanish Mediterranean mountain region: The case of Camero Viejo. *Land Degrad. Dev.* **2011**, *22*, 537–550. [CrossRef]

16. Wang, J.; Fu, B.; Qiu, Y.; Chen, L. Soil nutrients in relation to land use and landscape position in the semi-arid small catchment on the loess plateau in China. *J. Arid Environ.* **2001**, *48*, 537–550. [CrossRef]

17. Park, S.J.; Vlek, P.L.G. Environmental correlation of three-dimensional soil spatial variability: A comparison of three adaptive techniques. *Geoderma* **2002**, *109*, 117–140. [CrossRef]

18. Gogoi, M.; Basumatary, M. Estimation of the chlorophyll concentration in seven Citrus species of Kokrajhar district, BTAD, Assam, India. *Trop. Plant Res.* **2018**, *5*, 83–87. [CrossRef]

19. Shaahan, M.M.; El-Sayed, A.A.; Abou El-Nour, E.A.A. Predicting nitrogen, magnesium and iron nutritional status in some perennial crops using a portable chlorophyll meter. *Sci. Hortic.* **1999**, *82*, 339–348. [CrossRef]

20. Haboudane, D.; Miller, J.R.; Tremblay, N.; Zarco-Tejada, P.J.; Dextraze, L. Integrated narrow-band vegetation indices for prediction of crop chlorophyll content for application to precision agriculture. *Remote Sens. Environ.* **2002**, *81*, 416–426. [CrossRef]

21. Shi, D.-Y.; Liu, Z.-X.; Jin, W.-W. Biosynthesis, catabolism and related signal regulations of plant chlorophyll. *Hereditas* **2009**, *31*, 698–704. [CrossRef]

22. Esposti, M.D.D.; de Siqueira, D.L.; Pereira, P.R.G.; Venegas, V.H.A.; Salomão, L.C.C.; Filho, J.A.M. Assessment of Nitrogenized Nutrition of Citrus Rootstocks Using Chlorophyll Concentrations in the Leaf. *J. Plant Nutr.* **2003**, *26*, 1287–1299. [CrossRef]

23. Brubaker, S.C.; Jones, A.J.; Lewis, D.T.; Frank, K. Soil Properties Associated with Landscape Position. *Soil Sci. Soc. Am. J.* **1993**, *57*, 235–239. [CrossRef]

24. Tong, L.L. Application of RDA analysis in Canoco4.5 in water ecological evaluation. *Technol. Wind* **2018**, *2*, 47.

25. Zhai, J.; Song, Y.; Entemake, W.; Xu, H.; Wu, Y.; Qu, Q.; Xue, S. Change in Soil Particle Size Distribution and Erodibility with Latitude and Vegetation Restoration Chronosequence on the Loess Plateau, China. *Int. J. Environ. Res. Public Health* **2020**, *17*, 822. [CrossRef]

26. Liu, Y.; He, N.; Wen, X.; Yu, G.; Gao, Y.; Jia, Y. Patterns and regulating mechanisms of soil nitrogen mineralization and temperature sensitivity in Chinese terrestrial ecosystems. *Agric. Ecosyst. Environ.* **2016**, *215*, 40–46. [CrossRef]

27. Weng, H.L.; Ci, E.; Li, S.; Lian, M.S.; Chen, L. Pedogenetic Process and Taxonomy of Yellow Soil in Chongqing, China. *Acta Pedol. Sin.* **2020**, *57*, 579–589.

28. Lysenko, V.; Kosolapov, A.; Usova, E.; Tatosyan, M.; Varduny, T.; Dmitriev, P.; Rajput, V.; Krasnov, V.; Kunitsina, A. Chlorophyll fluorescence kinetics and oxygen evolution in Chlorella vulgaris cells: Blue vs. red light. *J. Plant Physiol.* **2021**, *258–259*, 153392. [CrossRef]

29. Chen, H.; Li, Q.-P.; Zeng, Y.-L.; Deng, F.; Ren, W.-J. Effect of different shading materials on grain yield and quality of rice. *Sci. Rep.* **2019**, *9*, 9992. [CrossRef]

30. Richter, G.; Wessel, K. Red light inhibits blue light-induced chloroplast development in cultured plant cells at the mRNA level. *Plant Mol. Biol.* **1985**, *5*, 175–182. [CrossRef]
31. Sigua, G.C.; Coleman, S.W. Spatial distribution of soil carbon in pastures with cow-calf operation: Effects of slope aspect and slope position. *J. Soils Sediments* **2010**, *10*, 240–247. [CrossRef]
32. Guo, X.; Fu, Q.; Hang, Y.; Lu, H.; Gao, F.; Si, J. Spatial Variability of Soil Moisture in Relation to Land Use Types and Topographic Features on Hillslopes in the Black Soil (Mollisols) Area of Northeast China. *Sustainability* **2020**, *12*, 3552. [CrossRef]
33. Jia, D.; Qian, W.; Hui, Y.; Shouren, Z. Effects of topographic variations and soil characteristics on plant functional traits in a subtropical evergreen broad-leaved forest. *Biodivers. Sci.* **2011**, *19*, 158–167. [CrossRef]
34. Tang, X.F.; Wang, Y.Q.; Wang, Y.J.; Guo, P.; Hu, B.; Sun, S.Q. Influence of hydrological processes on the runoff variation of base cations in Jinyun Mountain. *Acta Ecol. Sin.* **2014**, *34*, 7047–7056.
35. Hu, C.; Wright, A.L.; Lian, G. Estimating the spatial distribution of soil properties using environmental variables at a catchment scale in the loess hilly area, China. *Int. J. Environ. Res. Public Health.* **2019**, *16*, 491. [CrossRef]
36. Ni, S.J.; Zhang, J.H. Variation of chemical properties as affected by soil erosion on hillslopes and terraces. *Eur. J. Soil Sci.* **2007**, *58*, 1285–1292. [CrossRef]
37. Su, Z.A.; Zhang, J.H.; Nie, X.J. Effect of Soil Erosion on Soil Properties and Crop Yields on Slopes in the Sichuan Basin, China. *Pedosphere* **2010**, *20*, 736–746. [CrossRef]
38. Clemens, G.; Fiedler, S.; Nguyen, D.C.; Nyuyen, V.D.; Schuler, U.; Stahr, K. Soil fertility affected by land use history, relief position, and parent material under a tropical climate in NW-Vietnam. *Catena* **2010**, *81*, 87–96. [CrossRef]
39. Hao, Y.; Lal, R.; Owens, L.B.; Izaurralde, R.C.; Post, W.M.; Hothem, D.L. Effect of cropland management and slope position on soil organic carbon pool at the North Appalachian Experimental Watersheds. *Soil Tillage Res.* **2002**, *68*, 133–142. [CrossRef]
40. Ao, C.; Yang, P.; Ren, S.; Xing, W. Mathematical model of ammonium nitrogen transport with overland flow on a slope after polyacrylamide application. *Sci. Rep.* **2018**, *8*, 6380. [CrossRef]
41. Doni, F.; Isahak, A.; Che Mohd Zain, C.R.; Wan Yusoff, W.M. Physiological and growth response of rice plants (Oryza sativa L.) to Trichoderma spp. inoculants. *AMB Express* **2014**, *4*, 45. [CrossRef]
42. Naseem, F.; Zhi, Y.; Farrukh, M.A.; Hussain, F.; Yin, Z. Mesoporous ZnAl2Si10O24 nanofertilizers enable high yield of Oryza sativa L. *Sci. Rep.* **2020**, *10*, 10841. [CrossRef]
43. Xue, Z.; Cheng, M.; An, S. Soil nitrogen distributions for different land uses and landscape positions in a small watershed on Loess Plateau, China. *Ecol. Eng.* **2013**, *60*, 204–213. [CrossRef]
44. Lin, W.; Ding, J.; Xu, C.; Zheng, Q.; Zhuang, S.; Mao, L.; Li, Q.; Liu, X.; Li, Y. Evaluation of N2O sources after fertilizers application in vegetable soil by dual isotopocule plots approach. *Ecol. Eng.* **2020**, *188*, 109818.
45. Wang, J.; Fu, B.-j.; Qiu, Y.; Chen, L.-d. The effects of land use and its patterns on soil properties in a small catchment of the Loess Plateau. *J. Environ. Sci.* **2003**, *15*, 263–266.
46. Xu, C.; Xie, D.T.; Gao, M.; Tao, C.; Yu, L. Study on the Nitrogen and Phosphorus Loss Characteristics from Sloping Uplands in Small Watershed of Three Gorges Reservoir Region. *J. Soil Water Conservation.* **2011**, *25*, 1–5+10.
47. Shigaki, F.; Sharpley, A.; Prochnow, L.I. Rainfall intensity and phosphorus source effects on phosphorus transport in surface runoff from soil trays. *Sci. Total Environ.* **2007**, *373*, 334–343. [CrossRef] [PubMed]
48. Zhang, J. Assessment of tillage translocation and tillage erosion by hoeing on the steep land in hilly areas of Sichuan, China. *Soil Tillage Res.* **2004**, *75*, 99–107. [CrossRef]
49. Papiernik, S.K.; Lindstrom, M.J.; Schumacher, T.E.; Schumacher, J.A.; Malo, D.D.; Lobb, D.A. Characterization of soil profiles in a landscape affected by long-term tillage. *Soil Tillage Res.* **2007**, *93*, 335–345. [CrossRef]
50. Sun, B.; Zhang, L.; Yang, L.; Zhang, F.; Norse, D.; Zhu, Z. Agricultural Non-Point Source Pollution in China: Causes Mitigation Measures. *AMBIO* **2012**, *41*, 370–379. [CrossRef]
51. Ding, X.; Liu, L. Long-Term Effects of Anthropogenic Factors on Nonpoint Source Pollution in the Upper Reaches of the Yangtze River. *Sustainability* **2019**, *11*, 2246. [CrossRef]
52. Mesfin, G.S.; Almeida, O.; Yazew, E.; Bresci, E.; Castelli, G. Spatial Variability of Soil Moisture in Newly Implemented Agricultural Bench Terraces in the Ethiopian Plateau. *Water* **2019**, *11*, 2134. [CrossRef]
53. Moges, A.; Holden, N.M. Soil fertility in relation to slope position and agricultural land use: A case study of Umbulo catchment in southern Ethiopia. *Environ. Manag.* **2008**, *42*, 753–763. [CrossRef] [PubMed]
54. Gessler, P.; Chadwick, O.; Chamran, F.; Althouse, L.; Holmes, K. Modeling soil–landscape and ecosystem properties using terrain attributes. *Soil Sci. Soc. Am. J.* **2000**, *64*, 2046–2056. [CrossRef]
55. Tsui, C.-C.; Chen, Z.-S.; Hsieh, C.-F. Relationships between soil properties and slope position in a lowland rain forest of southern Taiwan. *Geoderma* **2004**, *123*, 131–142. [CrossRef]
56. Sigua, G.C.; Coleman, S.W.; Albano, J.; Williams, M. Spatial distribution of soil phosphorus and herbage mass in beef cattle pastures: Effects of slope aspect and slope position. *Nutr. Cycl. Agroecosyst.* **2011**, *89*, 59–70. [CrossRef]
57. Chin, G. Of Sunlight, Water, and Trees. *Science* **2005**, *310*, 19. [CrossRef]
58. Sun, Y.; Shi, J.A.; Shao, X.P.; Li, F.; Li, Q.; Gao, X.; Zhong, Q.; Zhao, W.Z. Effects of light intensity on photosynthetic pigment content and flowering of Jacaranda mimosifolia D. Don in different habitats. *Chin. J. Appl. Environ. Biol.* **2015**, *21*, 1150–1156.

59. Yan, X.; Wang, D. Effects of shading on leaf characteristics and photosynthetic characteristics of Kuding tea tree. *Acta Ecol. Sin.* **2014**, *34*, 3538–3547. [CrossRef]
60. Wei, J.B.; Xiao, D.N.; Zeng, H.; Fu, Y.K. Spatial variability of soil properties in relation to land use and topography in a typical small watershed of the black soil region, northeastern China. *Environ. Geol.* **2008**, *53*, 1663–1672. [CrossRef]

sustainability

Article

Carbon and Nitrogen Sourcing in High Elevation Landscapes of Mustang in Central Nepal

Roshan Babu Ojha [1,†] , Sujata Manandhar [2,†], Avishesh Neupane [3] , Dinesh Panday [3,*] and Achyut Tiwari [4,†]

1   National Soil Science Research Center, Nepal Agricultural Research Council, Khumaltar, Lalitpur 44700, Nepal; roshanbachhan@gmail.com
2   Centre of Research for Environment, Energy and Water, Kathmandu 25563, Nepal; sujatamanandhar@gmail.com
3   Department of Biosystems Engineering and Soil Science, University of Tennessee-Knoxville, Knoxville, TN 37996, USA; aneupane@utk.edu
4   Central Department of Botany, Tribhuvan University, Kirtipur, Kathmandu 44600, Nepal; achyutone@gmail.com
*   Correspondence: dpanday@utk.edu or agriculturenepal@gmail.com
†   These authors contributed equally to this work.

Abstract: Mustang valley in the central Himalaya of Nepal is a unique landscape formed by massive soil mass during a glacial period, which is attributed to a mix of vegetations and long agricultural history. Soil nutrients and their sourcing is highly important to understand the vegetation assemblage and land productivity in this arid zone. Twenty soil samples (from 0 to 20 cm depth) were collected from three landscape positions in Mustang district: valley, ridge, and midslope. We explored nutrient sourcing using natural abundance carbon (δ^{13}C) and nitrogen isotope (δ^{15}N) employing isotope ratio mass spectrophotometry. The results showed that the total soil carbon (TC) and total nitrogen (TN) ranged from 0.3 to 10.5% and 0.3 to 0.7%, respectively. Similarly, the CN ratio ranged from 0.75 to 15.6, whereas soil pH ranged from 6.5 to 7.5. Valley soil showed higher values of TN, CN, and soil pH than the ridge and midslope soils. The valleys had more positive δ^{15}N signatures than ridge and midslope, which indicates higher inorganic and organic N fertilizer inputs in the valley bottom than in the midslope and ridge. This suggests that a higher nutrient content in the valley bottom likely results from agro-inputs management and the transport of nutrients from the ridge and midslope. Soil pH and CN ratio were a non-limiting factor of nutrient availability in the study regions. These findings are crucial in understanding the nutrient dynamics and management in relation to vegetation and agricultural farming in this unique topography of the Trans-Himalayan zone of Mustang in central Nepal.

Keywords: carbon; isotopic signature; Mustang; natural abundance; nitrogen; nutrient sourcing

Citation: Ojha, R.B.; Manandhar, S.; Neupane, A.; Panday, D.; Tiwari, A. Carbon and Nitrogen Sourcing in High Elevation Landscapes of Mustang in Central Nepal. *Sustainability* 2021, 13, 6171. https://doi.org/10.3390/su13116171

Academic Editors: Bharat Sharma Acharya and Rajan Ghimire

Received: 11 April 2021
Accepted: 27 May 2021
Published: 30 May 2021

Publisher's Note: MDPI stays neutral with regard to jurisdictional claims in published maps and institutional affiliations.

1. Introduction

Nepal exhibits unique topographic features with a great variation in climate and biodiversity observed in every five kilometers across the longitude [1]. The origin of the Nepal Himalaya started from the Miocene period (50 million years ago), throughout which a constant weathering of soil parent materials occurred [2,3]. Mustang geology is believed to have originated around the Plio-Pleistocenous age and is well known as a Thakkhola formation [4]. The evolution of the Thakkhola formation aligns with major Himalayan uplift events that are set on unique geomorphic and climate patterns of Mustang compared to other parts of Nepal [4].

Mustang is located in the high mountains, where weathering is mainly constrained by climate. Overall, the climate of Mustang is characterized by low temperatures and dry seasons with high wind speed. Specifically, the northern part of Mustang represents the rain shadow area of Nepal [5], locally referred to as the Trans-Himalayan zone. The southern part is relatively more humid than the northern part and is covered with forest area which is only 3.3% (12,324 ha) of total landmass [6]. Low temperature and scant precipitation decelerate the weathering process [7] and result in fragile, weak aggregates, and shallow soil mass (i.e., skeletal soil) in Mustang [8]. High wind speed, however, accelerates the physical weathering of rocks and minerals. Therefore, Mustang exhibits unique topographic and climatic features, and its soil behaves differently compared to other parts of Nepal.

The altitude of Mustang district ranges from 2010 to 8167 m above sea level (masl). It is covered by 57.7% barren land, 30.3% grassland, 5.6% forest and bushes, 2.7% sand and cliffs, 2.1% water bodies, and 1.6% cultivated areas [8]. The geomorphology of Mustang is composed of high peaks, ridges, midslopes and valley bottoms attributing to different landscape positions [5]. The following three landscape positions are found in the hillslope. The ridge is the peak of the hill of the sloped land, midslope is in the middle part between the ridge and the valley bottom, and the basal part of the hill is valley bottom which is generally flat land located near the river channels. These landscape positions are characterized by their own specific micro-climate, micro-relief, aspect, and soil type. Generally, the valley bottoms are relatively warmer, moist, fertile, and have a lower slope than the midslope and ridge.

Most of the cultivated area in Mustang is occupied with apple (*Malus domestica*) orchards, one of the major income sources of Mustang residents. It covers around 72% of the district's total fruit production and is mainly dominant in the lower part of Mustang [9]. Besides apples, crops such as maize (*Zea mays* L.), wheat (*Triticum aestivum* L.), buckwheat (*Fagopyrum esculentum*), barley (*Hordeum vulgare*), naked barley (*Hordeum vulgare* ssp. Vulgare), pea (*Pisum sativum*), mustard (*Brassica* sp.), potato (*Solanum tuberosum*), vegetables, and other temperate fruits (apricot and walnut) are grown in Mustang. Generally, orchards are planted in the ridge and midslope areas and crops/vegetables are grown in the valley bottoms. Most parts of the central and southern Mustang and a few villages of the northern Mustang harvest two crops each year.

Carbon (C) and nitrogen (N) are two fundamental nutrients that serve as key soil fertility indicators [10]. The availability of nutrients for crops is governed by soil pH [11] and the Carbon–Nitrogen (CN) ratio [12]. The stable isotopes (natural abundance) of carbon (δ^{13}C) and nitrogen (δ^{15}N) and the CN ratio have been increasingly used to identify organic matter origin, mixing, and transformations in soil and their cycling in the atmosphere [13–16]. A lower CN ratio indicates a higher mineralization of organic matter and vice versa [17]. The signature of δ^{13}C differs with different vegetation assemblage; higher in C_4 vegetation (−9 to −17‰) and lower in C_3 vegetation (−23 to −30‰) [18]. Measuring the natural abundance of δ^{13}C provides information about the sourcing and migratory nature of soil organic carbon (SOC) in soil, while the natural abundance of δ^{15}N tells us about the biological tracking of the nutrients [16,19]. The isotopic signature of δ^{13}C and δ^{15}N and the CN ratio can be used to track the processes and mechanisms related to organic matter origin, formation, and turnover [17,20].

The availability of plant nutrients in higher elevation soil is mostly linked to the vegetation types and their photosynthetic pathways. As the elevation gradient changes, a shift in the isotopic signature of C and N occurs, and the change in vegetation types governs these dynamics [21]. For instance, the natural abundance of C (δ^{13}C) increases in the foliage of plants along with the increase in elevation [21] but there is a large variation in soil's δ^{13}C values [22,23]. Plant leaves or litters are one of the major contributors to soil organic matter formation through microbial decomposition [24,25]. However, the mixing and fractionation of stable isotopes in the soil during the decomposition process results in a larger variation of the isotopic composition [26]. The elevation and the rate of litterfall in natural ecosystems strongly influence the SOC sourcing and mixing [21]. Similarly, the natural abundance of nitrogen (δ^{15}N) isotope is enriched in an intensively managed environment. The relationship between the δ^{13}C and δ^{15}N with soil nutrients provides important information about the sourcing of nutrients in the higher elevation soils and such information is very limited in the mountainous country of Nepal.

Plant nutrients in the cultivated soils of Mustang are typically low [27], which is attributed to a low mineralization rate, low mobility, and low exchange potential. With an increase in elevation, the availability of plant nutrients are limited due to reduced mineralization rates [21,26]. Nutrient availability in the higher elevation is mostly constrained by the climate, vegetation type, and input management [21,23]. In Mustang, animal husbandry is closely linked with agriculture [28]. There is a significant transfer of biomass from the forest and rangeland to the cropland as fodder and roughages for livestock, ultimately ending in cropland as manure [29]. In addition, as livestock graze, there is also a reciprocal exchange of nutrients from crop residues back to the rangeland and forest through their excrement. This favours the nutrient source mixing in between cropping land and nearby native vegetation coupled with erosion-led nutrient transport [30]. The soils in the agricultural land of the Mustang region are poorly investigated and the existing plant nutrients status is not well known. Furthermore, soil test-based plant nutrient management is barely practised at the local level. The identification of the source, status, dynamism, and retention of those nutrients in the upper elevation soils is critically important to know the managerial aspects of the nutrients in a sustainable manner. We aimed to explore the TC, TN, CN ratio, and soil pH along the transect of different landscape positions (ridge, midslope, and valley) where cropping or orchard plantation is common in Mustang with surrounding natural vegetation. We further analysed the isotopic signature of the natural abundance δ^{13}C and δ^{15}N to identify the source of C and N where the nutrients source mixing is common in higher elevations of Mustang, Nepal.

2. Materials and Methods

2.1. Location and Climate

Mustang is one of the mountainous districts in central Nepal. It is located in the rain shadow of the world's 7th and 10th highest mountains (Dhaulagiri and Annapurna standing 8168 and 8137 masl, respectively) and receives on an average <400 mm annual rain with relatively higher rainfall in the southern part of the district. It presents a diversity of climates ranging from tundra, arid types in the higher elevations above 4500 masl, to alpine and cold temperate in 3000 to 4500 masl and 2000 to 3000 masl, respectively [31]. Furthermore, it is a deeply incised valley of the Kali Gandaki river with an arid valley bottom and characteristic diurnal wind system. It is divided into upper Mustang (above 3800 m) and lower Mustang (below 3800 m), the two divisions differing from each other with respect to the prevailing climatic conditions.

2.2. Soil Sampling Point Determination

We selected 20 sampling points along the Kaligandaki corridor from the southern (Tukuche) to northern part (Korala) of the Mustang district, considering a vertical transect to capture the best possible landscape positions (Figure 1 and Table 1). Out of 20 points, we collected four samples from the midslope, five samples from the ridge, and eleven samples from the valley in October 2011. Difficulties in accessing the varied topographic or landscape positions resulted in uneven sampling points. The details of sampling points are given in Table 1.

Table 1. Detailed information on sampling points in the Mustang district of Nepal.

Sampling Points	Latitude	Longitude	Elevation, masl	Location	Micro-Relief	Site Characteristics	Nearby Dominant Vegetation
1	28.71225	83.64908	2628	Tukuche	Midslope	Orchard	Juniper, Pine
2	28.83692	83.78242	2837	Kagbeni	Valley	Cropped land	Juniper
3	28.80392	83.77322	2852	Between Kagbeni and Lupra	Valley	Cropped land	Juniper
4	28.80389	83.77322	2852	Between Kagbeni and Lupra	Valley	Cropped land	Juniper
5	28.92494	83.82758	2963	Tsungsang	Valley	Cropped land	Juniper shrub, grasses
6	28.80244	83.79028	2997	Lupra	Midslope	Orchard	Juniper shrub, grasses
7	28.80211	83.78958	3017	Lupra	Midslope	Apple orchard	Juniper shrub, grasses
8	28.88406	83.80836	3092	Thangbe	Valley	Cropped land	Juniper shrub
9	28.96161	83.80847	3447	East of Samar	Valley	Cropped land	Pine, Juniper
10	28.81758	83.84944	3524	Jharkot	Ridge	Orchard	-
11	28.94964	83.80181	3560	South of Samar	Midslope	Orchard	Pine, Juniper
12	29.06139	83.87169	3579	Ghami	Valley	Cropped land	Planted Populas
13	28.96169	83.80142	3606	Samar	Ridge	Orchard	Pine, Juniper
14	28.99114	83.83819	3778	Syanboche	Valley	Cropped land	-
15	29.18361	83.95714	3823	Lomanthang	Valley	Cropped land	Planted Populas
16	29.18272	83.95711	3825	Lomanthang	Valley	Cropped land	Planted Populas
17	29.25469	83.96025	4027	South of Chonup, North of Lomanthang	Valley	Cropped land	Grasses
18	29.30347	83.96836	4612	North of Chonup	Ridge	Orchard	Juniper
19	29.30347	83.96836	4612	North of Chonup	Ridge	Orchard	Juniper and grasses
20	29.30347	83.96836	4612	North of Chonup	Ridge	Orchard	Grasses

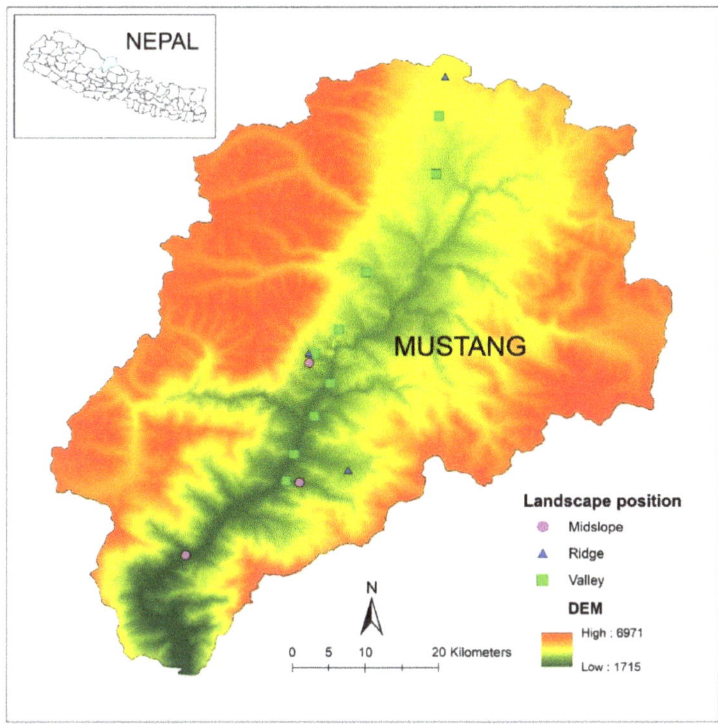

Figure 1. Distribution of soil sampling points showing in the digital elevation model (DEM) map across the transects in the study area, the Mustang district of Nepal.

2.3. Soil Sampling and Laboratory Analysis

We collected soil samples with the help of auger from 0 to 20 cm depth from selected points. A composite sample was taken, which was then spread in a sample box. Roots, undecomposed plant debris, and gravel were removed in the field. Upon returning to the lab, the soil was air dried, ground in mortar and pestle to break aggregates, and sieved to a 2 mm mesh size, which was then subjected to lab analysis.

The total soil carbon (TC) and total nitrogen (TN) were analysed using the dry combustion method for which soil was ground to 0.5 mm in size. The soil pH was determined in a 1:5 soil to water ratio with a digital soil pH meter. The isotopic signature of carbon (δ^{13}C) and nitrogen (^{15}N) were obtained from isotopic ratio mass spectrometry (Thermo Fisher Scientific, Analyzer: FLAS 2000-Conflo IV-Delta V Advantage) for which soil was ground to 0.1 mm in size. The soil sample was replicated twice to determine each of the carbon and nitrogen signatures and the average value of the replicates was reported. Soil standards and reference samples were placed after every 12 samples. The standard error of soil standards/reference sample was <0.23‰.

2.4. Data Analysis

Initially, data were entered in MS Excel and then imported to R studio for descriptive analysis. Standard least square analysis of variance (ANOVA) models were used to investigate the effects of landscape position on soil C and N. The data were tested for ANOVA assumptions prior to the analysis, and they met these assumptions. We employed log and square root transformations to fit data into a normal distribution curve before subjecting them to ANOVA. Tukey means separation tests were used for post-hoc comparisons of the soil parameters amongst landscape positions. A correlation between the variables was calculated at 5% level of significance. Graphs of variables (TC, TN, CN ratio, pH, $\delta^{13}C$, and $\delta^{15}N$) were prepared in R-studio [32] using ggplot2 package [33]. The relationship between different variables were performed in ggpairs function which is an extension of the ggplot2 package [33]. The statistical significance was determined as $p < 0.05$, unless otherwise noted. Analyses were conducted using 15.0.0 JMP SAS software. To elucidate the relationship between the geographical coordinates and the soil properties, a multiple linear regression model was built where soil properties were kept as a dependent factor and geographical coordinates as independent factors.

3. Results

The TC was significantly higher in the valley (6.2%) than in the ridge (2.8%) while the midslope had the intermediate values (4.7%) (Figure 2). The maximum and minimum TC in the valley was 10.5 and 4.0%, the midslope was 5.8 and 2.4%, and the ridge was 6.6 and 0.3%, respectively. Although there was no significant difference in the TN contents between these slope locations, the average TN content showed a decreasing trend from the valley (0.8%) to the midslope (0.6%) and ridge (0.5%). The maximum and minimum TN in the valley was 0.7 and 0.3%, the midslope was 0.9 and 0.4%, and the ridge was 0.7 and 0.3%, respectively.

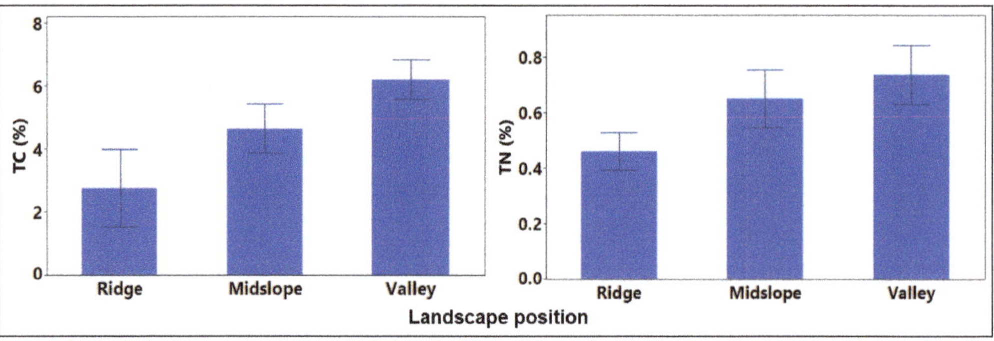

Figure 2. Total soil carbon (TC) and total nitrogen (TN) variation (mean ± SE) at different landscape positions in Mustang district of Nepal. Significant differences between the landscape positions are shown with differing letters.

The $\delta^{15}N$ values in soil were significantly different across the landscape positions (Figure 3). On average, the valley soil contained the positive $\delta^{15}N$ values, whereas the midslope and ridge soils contained negative $\delta^{15}N$ values. However, the range of $\delta^{15}N$ signature showed both positive and negative values at all landscape positions. The maximum and minimum natural abundance of $\delta^{15}N$ in the valley was +7.0 and −15.2‰, the midslope was +7.3 and −16.1‰, and the ridge was +5.4 and −14.9‰, respectively. The soil $\delta^{13}C$ value did not differ significantly between the landscape positions, although the average value showed a slightly increasing trend from the valley to the ridge (Figure 3). The maximum and minimum natural abundance of $\delta^{13}C$ in the valley was −6.6 and −23.6‰, the midslope was −7.6 and −19.4‰, and the ridge was +1.1 and −22.0‰, respectively.

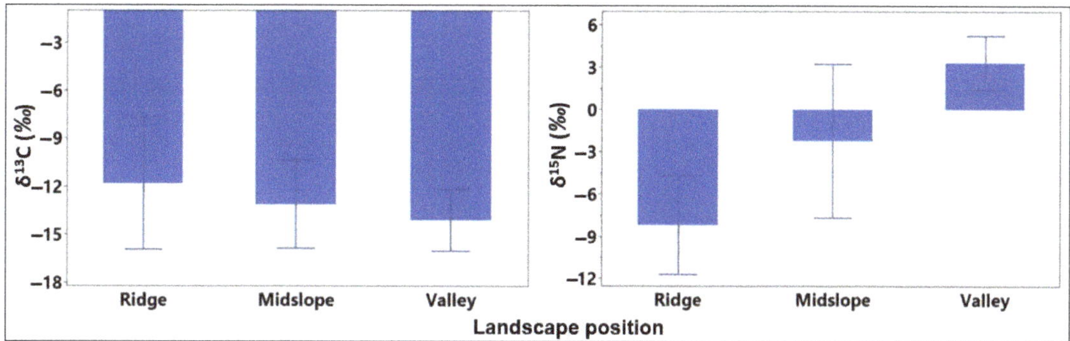

Figure 3. δ^{13}C and δ^{15}N variation (mean $\pm$ SE) in soil at different landscape positions in Mustang district of Nepal. Significant differences between the landscape positions are shown with differing letters.

The CN ratio did not differ significantly between the landscape positions, although there was an indication of a higher CN ratio in the valley than in the midslope and the ridge (Figure 4). A significantly higher soil pH was observed in the valley than in the ridge, while the midslope had intermediate values (Figure 4). The maximum and minimum CN ratio in the valley was 12.0 and 6.0, the midslope was 10.0 and 6.0, and the ridge was 16.0 and 1.0, respectively. Similarly, in the valley and midslope, we found the same soil pH range (7.0 to 7.5) and in the ridge soil the pH range was 7.0 to 6.5.

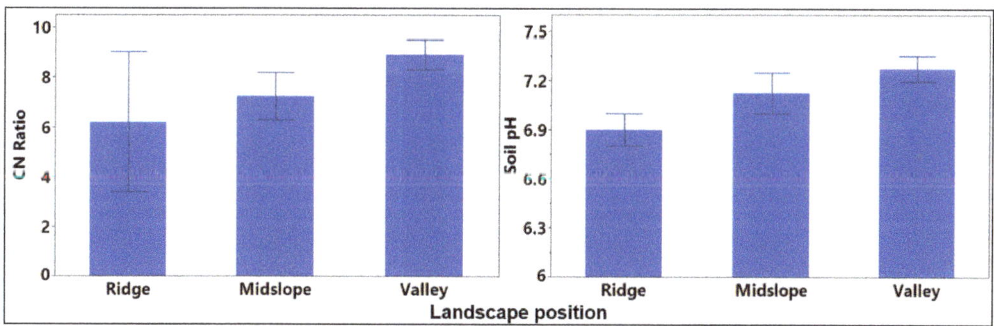

Figure 4. The CN ratio and soil pH variation (mean $\pm$ SE) at different landscape positions in Mustang district of Nepal. Significant differences between the landscape positions are shown with differing letters.

Both a positive and a negative correlation was observed between the soil parameters (Figure 5). A significant positive correlation was found between the TC and TN ($r = 0.7$, $p < 0.001$); the TC and CN ratio ($r = 0.6$, $p < 0.001$); the TC and δ^{15}N ($r = 0.7$, $p < 0.001$); the TN and δ^{15}N ($r = 0.5$, $p = 0.019$); and a significant negative correlation was found between the TN and the δ^{13}C ($r = -0.6$, $p = 0.004$). A detailed correlation matrix including all variables is provided in Appendix A Figure A1. We did not find any significant correlation between geographical co-ordinates (latitude, longitude, and elevation) and soil parameters (Appendix A Figure A1). Furthermore, there was no significant effect of latitude, longitude, and elevation on soil properties as suggested by multiple linear regression models (results not shown), which is in compliance with correlation results.

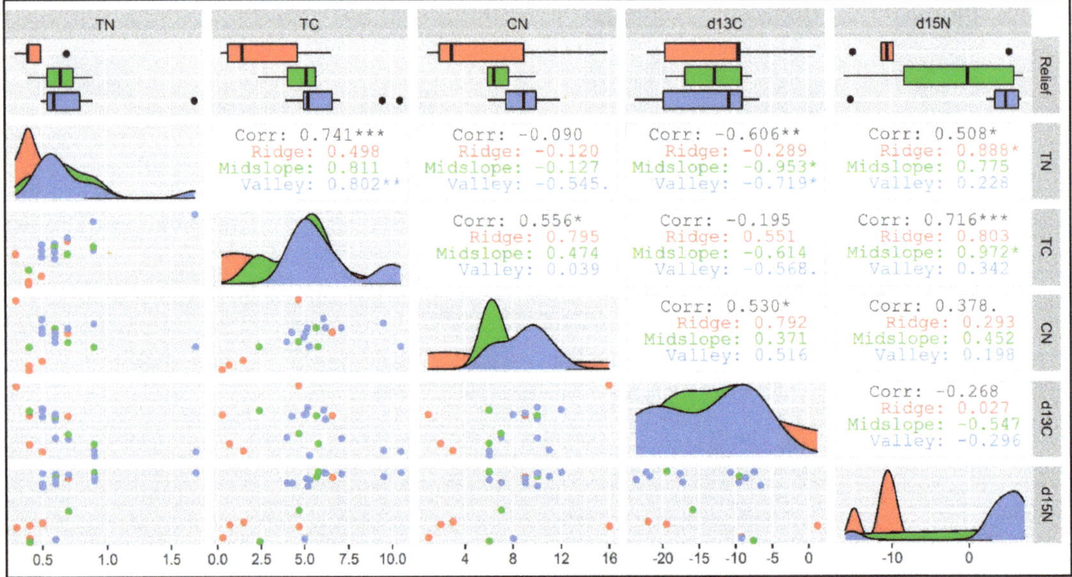

Figure 5. A correlation matrix between the selected variables, scatter points on the lower panel, and correlation values and their 5% significance level on the upper panel. *** *p*-value < 0.001; ** *p*-value < 0.01, * *p*-value < 0.05.

4. Discussion

Our current study showed that the valley bottom had a higher TC and TN concentration compared to the midslope and ridge. There is a progressive increment in the TC and TN concentration from the ridge to the valley bottom. This implies that the valley bottom is more fertile than the ridge and the midslope. The transport of the nutrients from the upper slope (ridge and midslope) towards the lower slope (valley bottom) due to soil erosion may have resulted higher concentration of the TC and TN [34] in the valley bottom. The deposition of sediment in the valley bottom and nutrient transport from the upper slope and forest litter [30] coupled with heavy textured soil [27] resulted in higher TC and TN. Lü et al. [35] reviewed the nutrient transport process associated with rainfall-runoff events and reported their results in terms of factors, forms, carriers, and sources of nutrient transport. Lü et al. [35] concluded that during the erosion process water is a carrier of soil nutrients in the soluble form. The dissolved organic C and available form of N resulting from litter decomposition from the forest is the primary source of nutrient transport, along with eroding water and sediments from the upper slope to the lower slope [36,37]. The lower slope, where the deposition of sediments occurs, are generally high in soil carbon and nitrogen compared to the eroding upper slope [38–40]. Thus, in the current study area, this erosion-led nutrient transport is dominant where the nutrients from the upper slope of the ridge and the midslope are deposited in the valley bottom.

The low CN ratio in the soil that we observed in the current study might be due to presence of inorganic carbon (natural carbonates) or inorganic nitrogen (input management) or both. The presence of inorganic forms of C and N alters the CN ratio [41]. The ratio of total organic C to N is the indicator of the decomposition rate of organic matter with an inverse relationship [42]. The CN ratio at any landscape position does not affect the nutrient availability [43] in the Mustang soil. Similarly, the average soil pH of the study area was 6.9 ± 0.5 units, which is the most favourable range for the nutrient availability. Many of the nutrients essential for plants are available in the pH range of 6.5 to 7.5 [44]. There, the CN ratio and soil pH are found to be non-limiting factors for nutrient availability in the study area.

The range of $\delta^{13}C$ value from -23.6 to $+1.1‰$ in the current study indicates the presence of both inorganic and organic C in soil. The C content of few soil samples, particularly from the ridge, was dominated by inorganic C (i.e., carbonate mineral). Plant photosynthesis discriminates against the heavier C isotope, and the degree of discrimination mainly varies with the type of photosynthetic pathways (C_3, C_4, CAM). The $\delta^{13}C$ values of C_3 and C_4 plants generally lie between -20 to $-40‰$, and -9 to $-17‰$, respectively, while $\delta^{13}C$ values in CAM typically lie in between -10 to $-20‰$ [45]. Garzione et al. [46] reported that C_3 plants, dominantly trees, shrubs, and cool-growing-season grasses in the region produce soil respiration with $\delta^{13}C$ values of about -22 to $-32‰$, whereas C_4 plants, dominantly warm-growing-season grasses, produce soil respiration with $\delta^{13}C$ values between -10 and $-15‰$. The average observed range of soil carbonate formed in equilibrium with C_3-respired CO_2 is $\delta^{13}C = -13$ to $-9‰$, whereas soil carbonates formed in the presence of C_4 plants have $\delta^{13}C = +1$ to $+3‰$ [47].

Galy et al. [48] reported the presence of $\delta^{13}C$ value of -0.4 to $+1.9‰$ in the carbonates of the bedload sediment around the Lomangthang and Kagbeni regions of the Mustang district. The presence of carbonates in the Tethyan sedimentary series, which also includes the Mustang district, comprises of Paleozoic–Mesozoic carbonates and clastic sediments that have $\delta^{13}C$ values ranging from -2.5 to $0‰$ [49]. The $\delta^{13}C$ value in Paleosol carbonates of the Thakkhola formation is between $-5.6‰$ and $+3.5‰$ with a mixed C_3 and C_4 plant species, but predominantly C_4 species [46]. An arid environment like Mustang is likely to have higher $\delta^{13}C$ values resulting from a low respiration rate or the dominance of C_4 plants. Garzione et al. [46] collected different species of grass in between 3000 and 4000 masl and found $\delta^{13}C$ values from -12.3 to $-12.8‰$ in these grass species. They concluded that the higher value of $\delta^{13}C$ in the valley floor of the Thakkhola formation deposition is from paleosol carbonates with the presence of both C_3 and C_4 vegetation. However, the presence of only C_3 vegetation in the lower elevations (Tetang formation), yielded $\delta^{13}C$ values from -21.9 to $-26.5‰$. Szpak et al. [50] demonstrated that foliar ^{13}C values increased with a site's altitude, which is in agreement with our data trend of greater soil ^{13}C enrichment in the ridge compared to the valley. Therefore, the sourcing of soil C can be attributed to the mixture of soil organic matter and paleosol carbonates with the increasing influence of organic matter (mostly from C3 vegetation) as we move from the ridge towards the valley floor of the Mustang district.

The natural abundance of $\delta^{15}N$ in soil represents an integrated signal of the ecosystem's N processes that help constrain N budgets, identify sources, and their fates. The range of $\delta^{15}N$ values in our study sites is between -16.1 to $+7.0‰$. Since Mustang is located in an arid climate, the $\delta^{15}N$ enrichment in soil and plants is expected. The loss of ^{14}N and enrichment of $\delta^{15}N$ values in soil and vegetation samples in dry regions have been reported previously [51,52]. Zhou et al. [53] reported increasing $\delta^{15}N$ values with decreasing rainfall in the Qinghai–Tibetan Plateau, a region similar to our site. The authors suggested that the precipitation and temperatures that influence the CN content and ratio are also the primary factors determining the patterns of soil $\delta^{15}N$ on a regional scale [53]. In a landscape scale similar to our site, variability in $\delta^{15}N$ was positively related to moisture availability, soil fertility, and vegetation cover [54]. Szpak et al. [50] demonstrated that foliar $\delta^{15}N$ values decreased with an increase in altitude, similar to our trend observed in soil $\delta^{15}N$ values. The higher value of $\delta^{15}N$ in valleys in our sites could also be associated with the use of organic fertilizer ($\delta^{15}N$ value of around 2 to 30‰) and synthetic N fertilizer ($\delta^{15}N$ value of around -4 to $4‰$) [55]. Regmi et al. [56] reported the use of synthetic N (23 to 280 kg urea ha^{-1}) and farmyard manure (2.1 to 5.3 t ha^{-1}) in apple orchards of the Mustang district. Hence, the enrichment of $\delta^{15}N$ in the Mustang district and the variations we observed between various sites might come from the agro-inputs or enrichment due to micro and macro scale topographic and climatic variations in these sites.

Furthermore, the higher positive correlation of the TC with the TN, $\delta^{15}N$, and CN ratio; $\delta^{13}C$ with CN ratio; TN with $\delta^{15}N$; and the higher negative correlation of the TN with $\delta^{13}C$ (Figure 5) indicates the interdependence of sourcing between the TC and the TN. This indicates that the source mixing between soil organic matter is C_4 and C_3 vegetation. The nutrient source mixing is further augmented by the transfer of forest litter as manure to farmland by the farmers of the Mustang valley [34]. The use of fresh animal manure in the valley bottom might result in higher $\delta^{15}N$ [57]. There is a mixing of isotopically distinctive carbon and nitrogen in the study area as a weak and negative correlation between $\delta^{13}C$ and $\delta^{15}N$ (Figure 5) [58]. Hence, the isotopic techniques are useful in organic matter source identification, their mixing, and stability in the soil.

5. Conclusions

Our study discloses the sourcing and availability of C and N in higher elevation soils of the Mustang District, Nepal using stable isotope techniques, CN ratio, and soil pH. The findings suggest that the landscape positions strongly influence the nutrient sourcing and mixing in higher elevation soils. As our data is limited to cover the broader geographic range, we do not find a relation of C and N with longitude, latitude, and elevation. C and N sourcing are specific to different landscape positions in Mustang. The ridge and midslopes are dominant with the litter decomposition, either of the forest trees species or of fruit orchards. Valley slopes are mostly dominant with the fresh organic and inorganic substrates through agro-input management by farmers, along with the source mixing of C and N transfer associated with erosion-led nutrient transport from the ridge and midslopes. Hence, we emphasize that the cultivated croplands in the valley or orchards in the midslope and the ridge should be managed according to the nutrient sourcing in the region for sustainable land management. It is important to consider landscape position, broad geographic co-ordinates, and micro-relief to study C and N sourcing in future studies. Further research is necessary to study the micro-climate, decomposition rate constant, and the microbial diversity to understand the cycling of C and N in the higher landscapes.

Author Contributions: Conceptualization—S.M.; funding acquisition—D.P.; methodology—S.M., R.B.O., and A.T.; investigation—S.M.; data curation—R.B.O. and A.N.; formal analysis—A.N. and R.B.O.; visualization—R.B.O., A.N., and D.P.; software—A.N. and R.B.O.; validation—R.B.O., A.N., D.P., S.M., and A.T.; writing—original draft preparation—R.B.O.; writing—review and editing—A.N., A.T., R.B.O., D.P., and S.M. All authors have read and agreed to the published version of the manuscript.

Funding: The MEXT and the Global COE Program at the University of Yamanashi (UoY), Japan supported/funded this study.

Institutional Review Board Statement: Not applicable.

Informed Consent Statement: Not applicable.

Data Availability Statement: The data presented in this study are available on request from the corresponding author.

Acknowledgments: The authors would like to sincerely thank Futaba Kazama for her support and acknowledge the MEXT and the Global COE Program at the University of Yamanashi (UoY), Japan for funding this study. We are thankful to Takashi Nakamura from Interdisciplinary Centre for River Basin Environment, UoY, Japan for his support in the lab analysis. We also extend our thanks to Devraj Chalise for the study area map. We would like to thank three anonymous reviewers and an academic editor for their valuable comments and suggestions, which helped us in improving this paper.

Conflicts of Interest: The authors declare no conflict of interest.

Sample Availability: Not applicable.

Appendix A

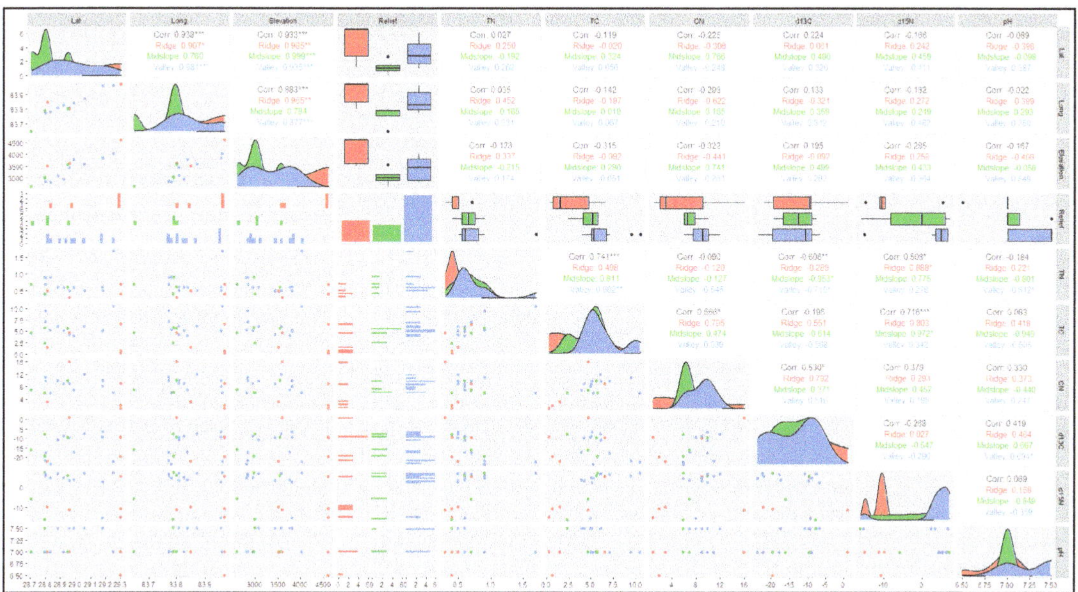

Figure A1. A correlation matrix showing the correlation between all variables used in the study, scatter points on the lower panel and correlation value and their 5% significance level on the upper panel. *** *p*-value < 0.001; ** *p*-value < 0.01, * *p*-value < 0.05.

References

1. Joshi, B.K.; Acharya, A.K.; Gauchan, D.; Chaudhary, P. *The State of Nepal's Biodiversity for Food and Agriculture*; Ministry of Agriculture and Livestock Development: Kathmandu, Nepal, 2017; Volume 1, p. 78.
2. Dhital, M.R. *Geology of the Nepal Himalaya: Regional Perspective of the Classic Collided Orogen*; Springer International Publishing: Berlin, Germany, 2015.
3. Valdiya, K.S. Evolution of the Himalaya. *Tectonophysics* **1984**, *105*, 229–248. [CrossRef]
4. Fort, M.; Freytet, P.; Colchen, M. Structural and sedimentological evolution of the Thakkhola Mustang graben (Nepal Himalayas). *Z. Geomorphol.* **1982**, *42*, 75–98.
5. Fort, M. Natural hazards versus climate change and their potential impacts in the dry, northern Himalayas: Focus on the upper Kali Gandaki (Mustang District, Nepal). *Environ. Earth Sci.* **2015**, *73*, 801–814. [CrossRef]
6. Government of Nepal. *Mustang District Profile*; District Statistical Office, Government of Nepal: Mustang District, Nepal, 2010; p. 87.
7. Norton, K.P.; Molnar, P.; Schlunegger, F. The role of climate-driven chemical weathering on soil production. *Geomorphology* **2014**, *204*, 510–517. [CrossRef]
8. KC, K.; Poudel, K.; Paudel, N.; Pokharel, R.; Koirala, S. *Resource Mapping Report*; District Development Committee: Mustang, Nepal, 2014.
9. Gauchan, D.; Yokoyama, S. *Farming Systems Research in Nepal: Current Status and Future Agenda; National Research Institute of Agricultural Economics*; Ministry of Agriculture, Forestry, and Fisheries: Tokyo, Japan, 1999.
10. Lal, R. Soil carbon sequestration impacts on global climate change and food security. *Science* **2004**, *304*, 1623–1627. [CrossRef]
11. Alam, S.M.; Naqvi, S.S.M.; Ansari, R. Impact of soil pH on nutrient uptake by crop plants. *Handb. Plant Crop Stress* **1999**, *2*, 51–60.
12. Truong, T.H.H.; Marschner, P. Plant growth and nutrient uptake in soil amended with mixes of organic materials differing in C/N ratio and decomposition stage. *J. Soil Sci. Plant Nutr.* **2019**, *19*, 512–523. [CrossRef]
13. Peterson, B.J.; Fry, B. Stable isotopes in ecosystem studies. *Annu. Rev. Ecol. Syst.* **1987**, *18*, 293–320. [CrossRef]
14. Gill, R.; Burke, I.C.; Milchunas, D.G.; Lauenroth, W.K. Relationship between root biomass and soil organic matter pools in the short grass steppe of eastern Colorado. *Ecosystems* **1999**, *2*, 226–236.
15. Poage, M.A.; Feng, X.H. A theoretical analysis of steady state delta ^{13}C profiles of soil organic matter. *Glob. Biogeochem. Cycles* **2004**, *18*, GB2016. [CrossRef]

16. Cai, Y.; Guo, L.; Wang, X.; Aiken, G. Abundance, stable isotopic composition, and export fluxes of DOC, POC, and DIC from the Lower Mississippi River during 2006–2008. *J. Geophys. Res. Biogeosci.* **2015**, *120*, 2273–2288. [CrossRef]

17. Xia, S.; Song, Z.; Wang, Y.; Wang, W.; Fu, X.; Singh, B.P.; Kuzyakov, Y.; Wang, H. Soil organic matter turnover depending on land use change: Coupling C/N ratios, δ^{13}C, and lignin biomarkers. *Land Degrad. Dev.* **2021**, *32*, 1591–1605. [CrossRef]

18. Boutton, T.W. Stable isotope ratios of natural materials: II. Atmospheric, terrestrial, marine, and freshwater environments. In *Carbon Isotope Techniques*; Coleman, D.C., Fry, B., Eds.; Academic Press: New York, NY, USA, 1991; pp. 173–185.

19. Ramaswamy, V.; Gaye, B.; Shirodkar, P.V.; Rao, P.S.; Chivas, A.R.; Wheeler, D.; Thwin, S. Distribution and sources of organic carbon, nitrogen and their isotopic signatures in sediments from the Ayeyarwady (Irrawaddy) continental shelf, northern Andaman Sea. *Mar. Chem.* **2008**, *111*, 137–150. [CrossRef]

20. Middelburg, J.J.; Herman, P.M.J. Organic matter processing in tidal estuaries. *Mar. Chem.* **2007**, *106*, 127–147. [CrossRef]

21. Gerschlauer, F.; Saiz, G.; Schellenberger Costa, D.; Kleyer, M.; Dannenmann, M.; Kiese, R. Stable carbon and nitrogen isotopic composition of leaves, litter, and soils of various ecosystems along an elevational and land-use gradient at Mount Kilimanjaro, Tanzania. *Biogeosciences* **2019**, *16*, 409–424. [CrossRef]

22. Bird, M.I.; Pousai, P. Variations of δ^{13}C in the surface soil organic carbon pool. *Glob. Biogeochem. Cycles* **1997**, *11*, 313–322. [CrossRef]

23. Ortiz, C.; Vázquez, E.; Rubio, A.; Benito, M.; Schindlbacher, A.; Jandl, R.; Butterbach-Bahl, K.; Díaz-Pinés, E. Soil organic matter dynamics after afforestation of mountain grasslands in both a Mediterranean and a temperate climate. *Biogeochemistry* **2016**, *131*, 267–280. [CrossRef]

24. Cotrufo, M.F.; Ranalli, M.G.; Haddix, M.L.; Six, J.; Lugato, E. Soil carbon storage informed by particulate and mineral-associated organic matter. *Nat. Geosci.* **2019**, *12*, 989–994. [CrossRef]

25. Lavallee, J.M.; Soong, J.L.; Cotrufo, M.F. Conceptualizing soil organic matter into particulate and mineral-associated forms to address global change in the 21st century. *Glob. Chang. Biol.* **2019**, *26*, 261–273. [CrossRef] [PubMed]

26. Fry, B. *Stable Isotope Ecology*; Springer: New York, NY, USA, 2006; Volume 521.

27. Shrestha, H.L.; Bhandari, T.S.; Karky, B.S.; Kotru, R. Linking soil properties to climate change mitigation and food security in Nepal. *Environments* **2017**, *4*, 29. [CrossRef]

28. Acharya, K.P.; Nirmal, B.K.; Poudel, B.; Bastola, S.; Mahato, M.K.; Yadav, G.P.; Kaphle, K. Study on yak husbandry in Mustang district of Nepal. *J. Hill Agric.* **2014**, *5*, 100–105. [CrossRef]

29. Craig, S. Pasture management, indigenous veterinary care and the role of the horse in Mustang, Nepal. In *Rangelands and Pastoral Development in the Hindu Kush–Himalayas*; ICIMOD: Kathmandu, Nepal, 1996; pp. 147–170.

30. Balla, M.K.; Tiwari, K.R.; Kafle, G.; Gautam, S.; Thapa, S.; Basnet, B. Farmers dependency on forests for nutrients transfer to farmlands in mid-hills and high mountain regions in Nepal (case studies in Hemja, Kaski, Lete and Kunjo, Mustang district). *Int. J. Biodivers. Conserv.* **2014**, *6*, 222–229.

31. NTNC. *Sustainable Development Plan of Mustang*; National Trust for Nature Conservation (NTNC), Government of Nepal/United Nations Environment Programme: Kathmandu, Nepal, 2008.

32. Wickham, H. *ggplot2: Elegant Graphics for Data Analysis*; Springer International Publishing: Berlin, Germany, 2016.

33. Team, R.C. *R: A Language and Environment for Statistical Computing*; R Foundation for Statistical Computing: Vienna, Austria, 2013.

34. Holz, M.; Augustin, J. Erosion effects on soil carbon and nitrogen dynamics on cultivated slopes: A meta-analysis. *Geoderma* **2021**, *397*, 115045. [CrossRef]

35. Lü, Y.; Fu, B.; Chen, L.; Liu, G.; Wei, W. Nutrient transport associated with water erosion: Progress and prospect. *Prog. Phys. Geogr.* **2007**, *31*, 607–620.

36. Andersen, D.C.; Nelson, S.M. Flood pattern and weather determine Populus leaf litter breakdown and nitrogen dynamics on a cold desert floodplain. *J. Arid Environ.* **2006**, *64*, 626–650. [CrossRef]

37. Rees, R.; Chang, S.C.; Wang, C.P.; Matzner, E. Release of nutrients and dissolved organic carbon during decomposition of *Chamaecyparis obtusa* var. formosana leaves in a mountain forest in Taiwan. *J. Plant Nutr. Soil Sci.* **2006**, *169*, 792–798. [CrossRef]

38. Cogle, A.L.; Rao, K.P.C.; Yule, D.F.; Smith, G.D.; George, P.J.; Srinivasan, S.T.; Jangawad, L. Soil management for Alfisols in the semiarid tropics: Erosion, enrichment ratios and runoff. *Soil Use Manag.* **2002**, *18*, 10–17. [CrossRef]

39. Sharpley, A.N. Selective erosion of plant nutrients in runoff. *Soil Sci. Soc. Am. J.* **1985**, *49*, 1527–1534. [CrossRef]

40. Zheng, F.; He, X.; Gao, X.; Zhang, C.; Tang, K. Effects of erosion patterns on nutrient loss following deforestation on the Loess Plateau of China. *Agric. Ecosyst. Environ.* **2005**, *108*, 85–97. [CrossRef]

41. Nasir, A.; Lukman, M.; Tuwo, A.; Hatta, M.; Tambaru, R. The use of C/N ratio in assessing the influence of land-based material in Coastal Water of South Sulawesi and Spermonde Archipelago, Indonesia. *Front. Mar. Sci.* **2016**, *3*, 266. [CrossRef]

42. Jafari, M.; Kohandel, A.; Baghbani, S.; Tavili, A.; Chahouki, M.A.Z. Comparison of chemical characteristics of shoot, root and litter in three range species of *Salsola rigida*, *Artemisia sieberi* and *Stipa barbata*. *Casp. J. Environ. Sci.* **2011**, *9*, 37–46.

43. Wan, X.; Huang, Z.; He, Z.; Yu, Z.; Wang, M.; Davis, M.R.; Yang, Y. Soil C:N ratio is the major determinant of soil microbial community structure in subtropical coniferous and broadleaf forest plantations. *Plant Soil* **2015**, *387*, 103–116. [CrossRef]

44. Weil, R.R.; Weil, R.R. *The Nature and Properties of Soils*; Prentice Hall: Upper Saddle River, NJ, USA, 2017; Volume 15, pp. 662–710.

45. Staddon, P.L. Carbon isotopes in functional soil ecology. *Trends Ecol. Evol.* **2004**, *19*, 148–154. [CrossRef]

46. Garzione, C.N.; Dettman, D.L.; Quade, J.; DeCelles, P.G.; Butler, R.F. High times on the Tibetan Plateau: Paleoelevation of the Thakkhola graben, Nepal. *Geology* **2000**, *28*, 339–342. [CrossRef]

47. Cerling, T.E.; Wang, Y.; Quade, J. Expansion of C4 ecosystems as an indicator of global ecological change in the late Miocene. *Nature* **1993**, *361*, 344–345. [CrossRef]
48. Galy, A.; France-Lanord, C.; Derry, L.A. The strontium isotopic budget of Himalayan rivers in Nepal and Bangladesh. *Geochim. Cosmochim. Acta* **1999**, *63*, 1905–1925. [CrossRef]
49. France-Lanord, C.; Sheppard, S.M.F.; Fort, P.L. Hydrogen and oxygen isotope variations in the high himalaya peraluminous Manaslu leucogranite: Evidence for heterogeneous sedimentary source. *Geochim. Cosmochim. Acta* **1988**, *52*, 513–526. [CrossRef]
50. Szpak, P.; White, C.D.; Longstaffe, F.J.; Millaire, J.F.; Vasquez Sanchez, V.F. Carbon and nitrogen isotopic survey of northern peruvian plants: Baselines for paleodietary and paleoecological studies. *PLoS ONE* **2013**, *8*, e53763. [CrossRef] [PubMed]
51. Craine, J.M.; Elmore, A.J.; Aidar, M.P.; Bustamante, M.; Dawson, T.E.; Hobbie, E.A.; Kahmen, A.; Mack, M.C.; McLauchlan, K.K.; Michelsen, A.; et al. Global patterns of foliar nitrogen isotopes and their relationships with climate, mycorrhizal fungi, foliar nutrient concentrations, and nitrogen availability. *New Phytol.* **2009**, *183*, 980–992. [CrossRef]
52. Handley, L.L.; Austin, A.T.; Stewart, G.R.; Robinson, D.; Scrimgeour, C.M.; Raven, J.A.; Heaton, T.H.E.; Schmidt, S. The ^{15}N natural abundance (δ^{15}N) of ecosystem samples reflects measures of water availability. *Funct. Plant Biol.* **1999**, *26*, 185–199. [CrossRef]
53. Zhou, L.; Song, M.H.; Wang, S.Q.; Fan, J.W.; Liu, J.Y.; Zhong, H.P.; Yu, G.R.; Gao, L.P.; Hu, Z.M.; Chen, B.; et al. Patterns of Soil ^{15}N and Total N and Their Relationships with Environmental Factors on the Qinghai-Tibetan Plateau. *Pedosphere* **2014**, *24*, 232–242. [CrossRef]
54. Ruiz-Navarro, A.; Barbera, G.G.; Albaladejo, J.; Querejeta, J.I. Plant delta (15) N reflects the high landscape-scale heterogeneity of soil fertility and vegetation productivity in a Mediterranean semiarid ecosystem. *New Phytol.* **2016**, *212*, 1030–1043. [CrossRef] [PubMed]
55. Bateman, A.S.; Kelly, S.D. Fertilizer nitrogen isotope signatures. *Isot. Environ. Health Stud.* **2007**, *43*, 237–247. [CrossRef]
56. Regmi, P.P.; KC, G.B.; Bhattarai, H.P. *Impact Assessment of National Integrated Pest Management (NIMP) Program in Nepal—Final Report*; submitted to the FAO representative in Nepal, FAO country office, UN House, Pulchowk Lalitpur; Nepal Development Research Institute: Pulchwok Lalitpur, Nepal, 2014; p. 57.
57. Fogg, G.E.; Rolston, D.E.; Decker, D.L.; Louie, D.T.; Grismer, M.E. Spatial variation in nitrogen isotope values beneath nitrate contamination sources. *Groundwater* **1998**, *36*, 418–426. [CrossRef]
58. Huon, S.; Grousset, F.E.; Burdloff, D.; Bardoux, G.; Mariotti, A. Sources of one-sized organic matter in North Atlantic Heinrich Layers: δ^{13}C and δ^{15}N tracers. *Geochim. Cosmochim. Acta* **2002**, *66*, 223–239. [CrossRef]

sustainability

Article

Impact of Land Configuration and Strip-Intercropping on Runoff, Soil Loss and Crop Yields under Rainfed Conditions in the *Shivalik* Foothills of North-West, India

Anil Khokhar [1], Abrar Yousuf [1,*], Manmohanjit Singh [1], Vivek Sharma [1,2], Parminder Singh Sandhu [1] and Gajjala Ravindra Chary [3]

[1] Regional Research Station, Punjab Agricultural University, Ballowal Saunkhri, SBS Nagar, Punjab 144521, India; anilkhokhar@pau.edu (A.K.); mmjsingh@pau.edu (M.S.); sharmavivek@pau.edu (V.S.); parminderagron@pau.edu (P.S.S.)
[2] Department of Soil Science, Punjab Agricultural University, Ludhiana, Punjab 141004, India
[3] Indian Council of Agricultural Research-Central Research Institute of Dryland Agriculture, Hyderabad, Telangana 500059, India; gcravindra@gmail.com
* Correspondence: er.aywani@pau.edu; Tel.: +91-9501404847

Citation: Khokhar, A.; Yousuf, A.; Singh, M.; Sharma, V.; Sandhu, P.S.; Chary, G.R. Impact of Land Configuration and Strip-Intercropping on Runoff, Soil Loss and Crop Yields under Rainfed Conditions in the *Shivalik* Foothills of North-West, India. *Sustainability* **2021**, *13*, 6282. https://doi.org/10.3390/su13116282

Academic Editors: Bharat Sharma Acharya and Rajan Ghimire

Received: 7 April 2021
Accepted: 27 May 2021
Published: 2 June 2021

Publisher's Note: MDPI stays neutral with regard to jurisdictional claims in published maps and institutional affiliations.

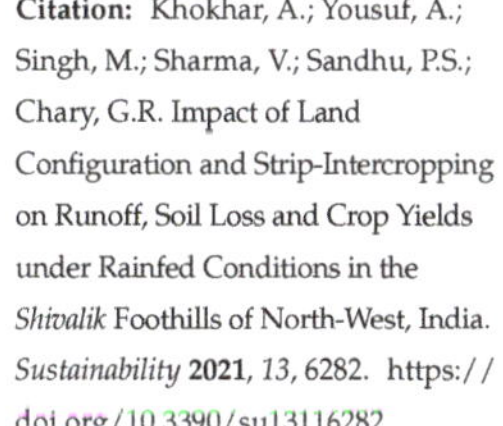

Abstract: Maintaining sustainable crop production on undulating, sloppy, and erodible soils in *Shivalik* foothills of North-west India is a challenging task. Intercropping is accepted as a highly sustainable system to reduce soil erosion and ensure sustainable production by making efficient use of resources. Field experiments were conducted in the rainy season (July to September) during 2015, 2016, and 2017 to evaluate the effect of land slopes and maize and cowpea strip-intercropping on productivity and resource conservation at the Regional Research Station, Ballowal Saunkhri located in the *Shivalik* foothills. During three years of experimentation, a total of 23–26 runoff events were observed in the maize crop grown in the rainy season. The results from this 3-year field study indicate that maize grain yield was significantly higher on a 1% slope and cowpea on a 2% slope. This accounted for significantly higher net returns (US$ 428 ha^{-1}) with a benefit-cost (BC) ratio of 2.0 on a 1% slope. Runoff, soil, and nutrient losses were higher on a 3% slope as compared to 1% and 2% slopes. N, P, and K loss on a 3% slope were 3.80, 1.82, and 4.10 kg ha^{-1} higher, respectively than a 1% slope. The adoption of a strip-intercropping system with a 4.8 m maize strip width and 1.2 m cowpea strip width resulted in significantly higher maize equivalent yield than sole maize and other strip-intercropping systems. This system showed the highest land equivalent ratio value (1.24) indicating a 24% yield advantage over sole cropping systems of maize and cowpea, and fetched the highest net returns (US$ 530 ha^{-1}) with a benefit-cost ratio (BC ratio) of 2.09. This system also reduced runoff and soil loss by 10.9% and 8.3%, respectively than sole maize crop. On all the land slopes, maize and cowpea strip-intercropping systems showed a significant reduction in N, P, K, and organic carbon loss as compared to sole maize. Thus, on sloping land, the maize and cowpea strip-intercropping system decreases surface runoff, soil, and nutrient loss, and increases yield and income of the farmers as compared to a sole maize crop.

Keywords: maize equivalent yield; nutrient loss; runoff; soil loss; slope; strip-intercropping; water use efficiency

1. Introduction

Water and wind erosion are the main soil degradation processes in drylands of the world [1]. Soil erosion not only affects the soil's physical properties but transports huge quantities of nutrients from agricultural land thereby deteriorating the soil health [2]. Soil erosion has detrimental impacts on global agricultural production as more than half (52%) of the world's arable soils have been categorized as degraded or severely degraded [3,4]. Soil erosion rate from agricultural land is about 10–100 times greater than rates of soil production, erosion under natural vegetation, and long-term geological erosion [5]. Over the last two decades, estimates for agricultural land degradation have been highly variable with figures varying from as low as 15% to as high as 80% [6–11]. Later assessments have revealed that 25%, 44%, and 10% of the present agricultural land are highly degraded, slightly to moderately degraded, and recovering from degradation, respectively [7,8,12]. The global average annual value of soil erosion is about 10.2 Mg ha^{-1}, of which 60% is anthropogenic. Soil erosion results in an \$8 billion loss to global GDP annually which has reduced the crop yields by 33.7 Tg (Teragrams $\approx$ million tonnes), and increased water abstraction by 48 billion m^3 [13]. In India, water erosion has affected about 145 million hectares of the total geographical area (328.81 million hectares). In the north-western part of India, the sub-mountainous area in the *Shivalik* foothills popularly known as the *Kandi* region is deemed as one of the most fragile ecosystems of the country. In this region, a large portion of the rainfall goes as runoff from the cultivated fields along with fertile topsoil [14,15]. The average annual erosion rate in this region is 16 Mg ha^{-1} and in some watersheds, it is more than 80 Mg ha^{-1} [16,17], which is much higher than the soil tolerance limit of 10–11.2 Mg ha^{-1} [18]. Due to the severe erosion problems, huge quantities of nutrients are lost from fertile agricultural land, rendering them barren over a period of time [18–21]. Therefore, there is a need to identify a suitable cropping system that can provide adequate soil cover to reduce the rate of soil loss to a tolerable limit and ensure sustainable crop production while maintaining the proper soil health. Maize-wheat is the principal cropping system of the *Shivalik* foot-hills. Maize, being the erosion permitting crop, augments the problem of soil erosion which is further aggravated by the high-intensity rainstorms. Hence, maize must be grown with erosion-resistant crops so as to minimize soil erosion to an acceptable level. Leguminous crops are considered to be the best soil conservators, and legumes, when grown as intercrops, impart sustainability in the cropping system [21]. Among the various factors affecting the runoff and soil erosion process, the vegetation cover in terms of structure and density plays an important role in runoff generation and hence, the soil erosion process. The vegetation cover not only reduces the amount of rainfall reaching the ground surface but also reduces the impact of raindrops on the soil surface [22]. Due to less impact velocity, the splash erosion is reduced to a larger extent. The amount of soil detached is reduced considerably due to the presence of surface/crop cover and hence the gross soil erosion is reduced as well [23]. A number of studies have been conducted wherein it has been observed that adding vegetation cover by mulching or intercropping has reduced runoff and soil loss [24–27].

Intercropping, the practice of growing two or more crops simultaneously on the same field is an age-old cropping system that aims to utilize the growth resources more efficiently than sole cropping [28]. Intercropping is identified as a cropping system that not only controls the soil loss from the agricultural fields but also increases the crop yields and enhances the soil moisture in the crop root zone [29,30]. The main advantage of intercropping is the more efficient utilization of the available resources and the increased productivity compared with each sole crop of the mixture [31,32]. Intercropping results in better pest control and reduced soil erosion [28]. Sharaiha and Ziadat [33] reported that the intercropping of vetch and barley resulted in significantly higher yields due to a reduction in runoff and soil losses. Sharma and Arora [21] conducted a study in the *Shivalik* foothills to evaluate the different intercropping systems and concluded that the intercropping not only increased yields but also conserved soil moisture. Ghosh et al. [34] reported that intercropping results in higher grain yields, net returns, and water use efficiency. Wang et al. [35] stated that hedgerow intercropping reduced the runoff, soil loss, and nutrient loss significantly as compared to sole maize crop. Sharma et al. [36] reported intercropping conserved the soil moisture in the *kharif* season, which resulted in higher crop yields during the *rabi* season, and it also reduced the runoff and soil loss by 26% and 43%, respectively.

In the rainfed tracts of Northwest India, maize is the dominant crop in the rainy (kharif) season. However, maize is an erosion permitting crop and must be grown in combinations with erosion-resistant crops preferably legumes, which due to its thick canopy cover reduces the soil and nutrient losses [37]. Few researchers have reported that crop covers proved effective in controlling runoff and soil loss in the region. Yet, limited information is available on runoff, soil, and nutrient losses under different vegetative covers in the *Shivalik* foothills region of Punjab, India. We hypothesized that the strip intercropping would result in reduced soil and nutrient loss and increase crop productivity under the different land configurations prevalent in the *Shivalik* foothills of India. Keeping this in view, the present study was conducted to evaluate the different combinations of maize and cowpea strip widths on different land slopes and their impact on soil and nutrient losses, productivity, and profitability. The objectives of the present study were (i) to determine the changes in maize and cowpea productivity from 2015 to 2017; (ii) to determine the runoff and soil loss on different land slopes and maize and cowpea strip widths, and; (iii) to identify the best combination of land slope and maize and cowpea strip width in terms of profitability and soil conservation.

2. Materials and Methods

2.1. Experimental Site

Field experiments were conducted in the rainy season (July to September) during 2015, 2016, and 2017 at the Research Farm of Punjab Agricultural University Regional Research Station, Ballowal Saunkhri, Shaheed Bhagat Singh Nagar, Punjab, India. The experimental site is geographically located between 31°6′5″ N latitude, 76°23′26″ E longitude, and at an altitude of 346 m in the *Shivalik* foothills. The climate of the region is sub-humid with hot and dry summer and extremely cold winter. The average annual rainfall of the region is 1060 mm of which 80% is received in a span of three months from mid-June to mid-September. In the past 35 years, annual rainfall has declined from over 1200 mm to 1060 mm. However, the extreme rainfall events with high intensity have increased with a decrease in the number of rainy days in the *kharif* season [38]. During the crop period 2015, 2016, and 2017, the rainfall received were 414 mm, 508.7 mm, and 726.9 mm, respectively with 6, 8, and 12 runoff events. The mean monthly minimum and maximum temperature ranged from 5.9 to 25.2 and 18.8 to 38.6 °C during the year with their respective plateau in December and May, respectively (Figure 1).

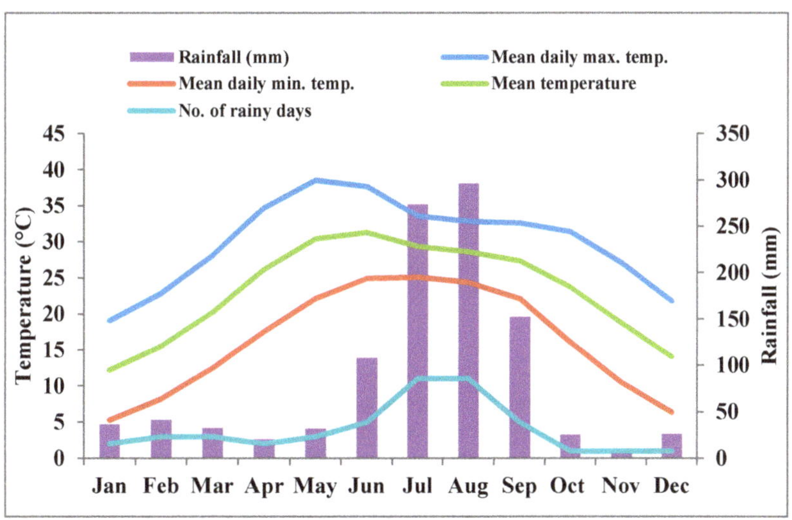

Figure 1. Ombroclimatic diagram of the study site (Mean data from 1984 to 2017).

The soil of the experimental field was loamy in texture, low in organic carbon (0.40%), available N (224 kg ha^{-1}), and medium in P (13 kg ha^{-1}) and K (192 kg ha^{-1}). The bulk density of soil was 1.54 Mg m^{-3} with pH 8.12 and electrical conductivity 0.25 dS m^{-1}. The soil moisture retention at field capacity (0.3 bars) was 19.4% and at permanent wilting point (15 bars) was 7.2% on a weight basis in a 0–180 cm soil profile. Soil pH was determined by the method of Jackson [39] in a 1:2.5 soil: water suspension, organic carbon by the method of Walkley and Black [40], available nitrogen by the alkaline-KMnO$_4$ method [41], available phosphorus by the Olsen method [42], and exchangeable potassium by the NH$_4$OAc method [39].

2.2. Experimental Details and Crop Management Practices

The experiment on strip-intercropping of maize and cowpea with different land slopes and variable strip width laid out in a randomized block design with three replications was conducted under rainfed conditions during all the years of experimentation. The treatments consisted of three land slopes, viz. (i) 1% slope (S1), (ii) 2% slope (S2), and (iii) 3% slope (S3); and five strip widths of maize and cowpea viz. (i) sole maize 6 m wide (W1), (ii) maize strip 4.8 m and cowpea strip 1.2 m wide (W2), (iii) maize strip 3 m and cowpea strip 3 m wide (W3), (iv) maize strip 1.2 m and cowpea strip 4.8 m wide (W4), and (v) sole cowpea 6 m (W5). The plot size was 12 m × 6 m for W1 and W5 treatments, 12 m × 10.8 m for W2 treatment, 12 m × 9 m for W3 treatment, and 12 m × 7.2 m for W4 treatment. The maize variety PMH-2 was sown at a spacing of 60 cm × 20 cm and cowpea dual-purpose variety CL 367 was sown at 30 cm × 10 cm spacing using 20 kg and 30 kg seed per hectare, respectively. Both the crops were sown across the slope. The recommended dose of N, P, and K for maize was 80, 40, and 30 kg ha^{-1}, respectively, and in cowpea N and P were applied @ 18.75 and 55 kg ha^{-1}, respectively. In cowpea, a full dose of N and P was applied at the time of sowing while in maize, half N and a full dose of P and K were applied as basal, and the remaining half of N was applied at knee height stage about one month after sowing. Nitrogen, phosphate, and potassium were applied through urea, di-ammonium phosphate, and muriate of potash, respectively. Maize and cowpea were sown simultaneously in the 1st week of July. Weeds were managed in the sole and intercropping system with pre-emergence application of herbicide pendimethalin @ 1.5 kg a.i. ha^{-1} followed by one hoeing in maize 25–30 DAS. No insect pest or disease attack was observed in either of the crops.

2.3. Measurement of Leaf Area and Yield

The leaf area of maize and cowpea was measured indirectly at 30 DAS with Sunscan Ceptometer, Delta-T Devices, UK. In the maize and cowpea strip cropping system, the leaf area index (LAI) of both maize and cowpea were added for comparison with sole crops and strip-cropping systems. Maize crop was harvested in the 1st week of October 2015, 2016, and 2017. The maize crop rows from the inner and border strips were harvested separately to record the grain and straw yield. In cowpea, pods were picked from time to time as they matured in 2–3 pickings and harvested in the 2nd week of October to record the fodder yield.

2.4. Measurement of Runoff, Soil Loss, Nutrient Loss, and Water Conservation Efficiency

The total surface runoff and sediments generated from each plot were measured by collecting the runoff in the collection tank (capacity 200 L) installed at the downstream end of each plot. Two drums were arranged in such a way that if the first drum is full; the excess runoff is collected in the second drum. The runoff volumes were measured immediately after the rainfall event. The plots were separated by well-constructed bunds. To measure the soil loss, the runoff water collected in the collection tanks was thoroughly stirred. The runoff samples were then collected from each tank in triplicates in the sediment sampling bottles of capacity 1.0 L. Each sample was carefully transferred to a separate beaker, evaporated in an oven at 40 °C till constant weight to determine the average soil loss per sample. The soil loss per plot was calculated by multiplying the average soil loss per sample and the total volume of runoff generated from the plot. Soil loss was calculated as per the equations given below:

$$Soil\ loss\ (kg\ ha^{-1}) = \frac{Sediment\ weight\ (g\ litre^{-1}) \times Run\ off\ (litres\ ha^{-1})}{10^3} \tag{1}$$

A separate runoff sample was also collected from each tank for nutrient analysis. The total nitrogen content of the dry soil was determined using the Kjeldahl digestion and distillation procedure [43]. Available phosphorus was determined using Olsen's method [42] and exchangeable potassium was determined by the flame photometry procedure [39].

The nutrient loss was calculated as per the equations given below:

$$Nutrient\ loss\ (kg\ ha^{-1}) = \frac{Nutrient\ content(mg\ kg^{-1}) \times Soil\ loss\ \left(kgha^{-1}\right)}{10^6} \tag{2}$$

The rainfall conserved and the conservation efficiency of different crops were assessed based on the rainfall of each storm and runoff produced. The crop water productivity was estimated from the total water conserved under the respective crop and the soybean equivalent yield of each crop cover.

The water conservation efficiency (WCE) [44,45] was determined using the following formula:

$$WCE\ (\%) = \frac{Total\ rainfall\ (mm) - Runoff\ (mm)}{Rainfall\ (mm)} \times 100 \tag{3}$$

Water use efficiency (WUE) of treatments was determined by using the formula:

$$WUE\ \left(kgha^{-1}mm^{-1}\right) = \frac{Crop\ yield\ \left(kgha^{-1}\right)}{Crop\ evapotranspiration(mm)} \tag{4}$$

2.5. Intercropping Indices

The advantage of intercropping as compared with sole cropping was evaluated using the maize equivalent yield (MEY) and land equivalent ratio (LER) equation:

MEY was calculated as per the formula given below:

$$MEY\ \left(kgha^{-1}\right) = \frac{Yield\ of\ cowpea\ \left(kgha^{-1}\right)\ x\ Price\ of\ cowpea\ \left(Rskg^{-1}\right)}{Price\ of\ maize\ \left(Rskg^{-1}\right)} + Yield\ of\ maize\ \left(kgha^{-1}\right) \tag{5}$$

LER was calculated as suggested by Willey and Rao [46] by the following formula:

$$LER = \frac{Yield\ of\ intercropped\ maize\ \left(kgha^{-1}\right)}{Yield\ of\ sole\ maize\ \left(kgha^{-1}\right)} + \frac{Yield\ of\ intercropped\ cowpea\ \left(kgha^{-1}\right)}{Yield\ of\ sole\ cowpea\ \left(kgha^{-1}\right)} \tag{6}$$

2.6. Economic Analysis

The cost of cultivation under different treatments was estimated on the basis of market prices of different inputs and outputs in the study area. The input costs include costs of seed, pesticide, fertilizers, hiring charges of human labor, field preparation, fertilizer application, plant protection, harvesting, and threshing. Gross returns were calculated on the basis of market price provided by the market committee, Balachaur, Punjab, India. Net income was calculated as the difference between gross income and total cost. The benefit-cost (B:C) ratio was calculated as gross return divided by the cost of cultivation.

2.7. Statistical Analysis

Analysis of variance (ANOVA) technique in factorial randomized block design was carried to analyze the data statistically using SAS software, version 9.4 [SAS Institute Inc., Cary, NC, USA]. The differences between the treatments were tested by the least significant difference (LSD) at a $p < 0.05$ level of significance.

3. Results

3.1. Productivity

3.1.1. Maize

Maize grain yield obtained on 1% slope was significantly higher than on 2% slope and 3% slope (Figure 2). The yield of maize declined with the increase in slope except in 2015 wherein the yield on the 2% slope was slightly less than the 3% slope. On a mean basis, the grain yield of maize on the 1% slope was 12% and 18% higher than on the 2% slope and 3% slope, respectively. Maize grain yield responded significantly to strip width. Sole maize yield varied from 2137 kg ha^{-1} in 2015 to 3645 kg ha^{-1} in 2017. The yield from a strip-intercropping combination of a 4.8 m maize strip width and 1.2 m cowpea strip width (W2) produced grain yield similar to that from sole maize (W1) but significantly higher than a 3 m maize strip width and 3 m cowpea strip width system (W3) and a 1.2 m maize strip width and 4.8 m cowpea strip width system (W4) in all the years. The average reduction in maize yields due to strip-intercropping of maize and cowpea in the W2, W3, and W4 strip-intercropping system was 4.3%, 57%, and 226%, respectively as compared to sole maize. Similar trends were also recorded for the straw yield of maize with respect to slope and strip widths.

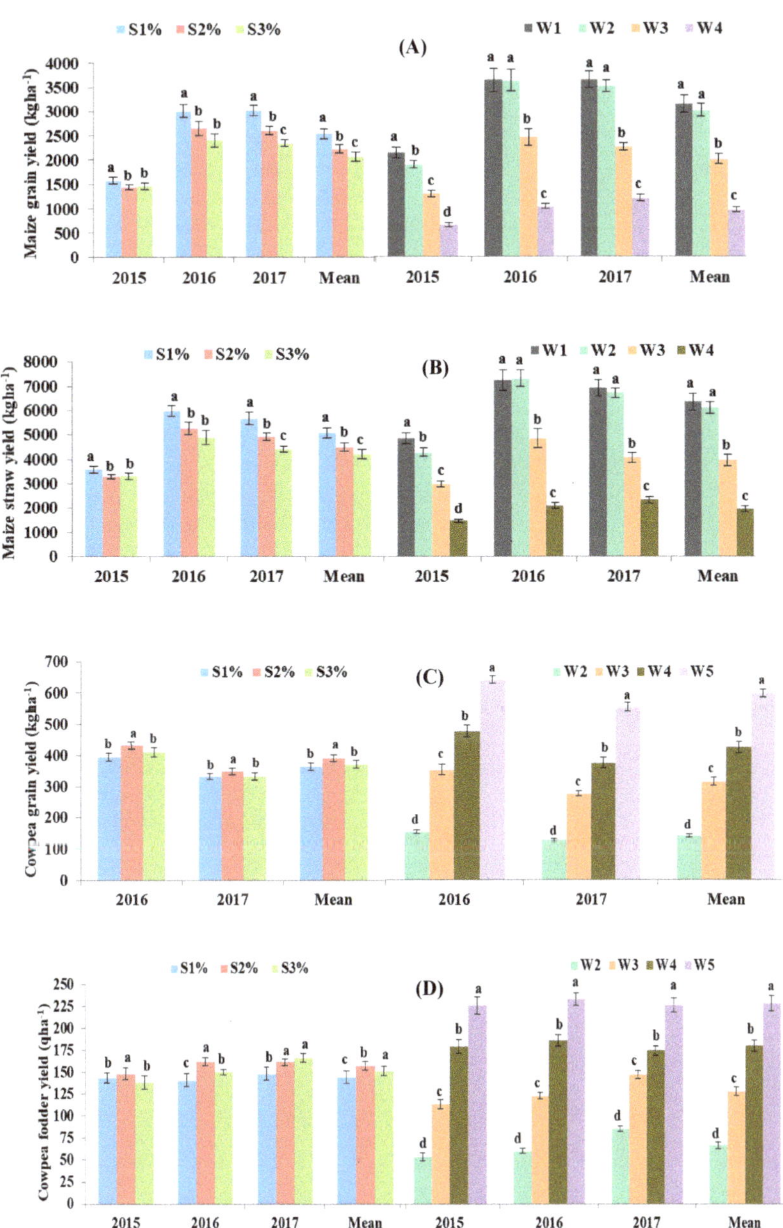

Figure 2. Effect of land slopes and strip width on (**A**) maize grain yield; (**B**) maize straw yield; (**C**) cowpea grain yield and (**D**) cowpea fodder yield. Different letters on standard error bars indicate significant differences at $p < 0.05$ between different treatments within a year according to the least significant difference (LSD) test. S1%: 1% Slope, S2%: 2% Slope, S3%: 3% Slope, W1: Sole maize (60 cm × 20 cm), W2: Maize strip 4.8 m + Cowpea strip 1.2 m, W3: Maize strip 3 m + cowpea strip 3 m, W4: Maize strip 1.2 m + cowpea strip 4.8 m and W5: Sole cowpea (at 30 cm row to row spacing).

3.1.2. Cowpea

In 2015, due to prolonged dry spell, cowpea grain yield was not obtained and the crop was harvested as fodder. The land slope had a significant influence on the cowpea grain yields during 2016 but not in 2017. The mean grain yield of cowpea was highest on the 2% slope (391 kg ha^{-1}) and it was 7% and 5% higher than the grain yield recorded on the 1% slope (364 kg ha^{-1}) and 3% slope (372 kg ha^{-1}), respectively. Cowpea grain yield from W5 plots was significantly higher than the maize and cowpea strip intercropping systems. However, the cowpea grain yield on a proportionate basis in the W2 and W3 strip-intercropping system was higher than the W4 and sole cowpea (W5). Growing of cowpea on the 2% land slope showed higher fodder yield than the 1% and 3% land slope in 2015 and 2016 but in 2017, cowpea fodder yield on the 3% land slope was the highest. The mean cowpea fodder yield on the 2% slope was 8.2% and 3.8% higher over the 1% slope and 3% slope, respectively. Fodder yield of sole cowpea was significantly higher in the maize and cowpea strip intercropping systems in all the years.

3.2. Maize Equivalent Yield (MEY) and Land Equivalent Ratio (LER)

3.2.1. Maize Equivalent Yield

A better representation of yields of component crops in intercropping/strip-intercropping system is by estimating the equivalent yield of the dominating crop, i.e., maize equivalent yield (MEY). Maize equivalent yield on the 1% slope was significantly higher than the 2% and 3% slope in all the years except in 2015 (Figure 3). The mean value of MEY on the 1% slope was 5.4 and 11.2% higher yield over the 2% slope and 3% slope, respectively. Among maize and cowpea sole cropping and maize and cowpea strip-intercropping systems, the W2 system resulted in significantly higher MEY than sole maize and other maize and cowpea strip-intercropping systems in 2016. In 2015 and 2017, MEY in the W2 system was at par with sole maize (W1) but significantly higher yield over all sole and maize and cowpea strip-intercropping systems. On a mean basis, the W2 system resulted in 11.6, 12.1, 28.6, and 41.2% higher MEY over sole maize, W3, W4, and sole cowpea, respectively. On all the land slopes, the highest MEY was recorded under maize and cowpea strip-intercropping system having 4.8 m maize strip width and 1.2 m cowpea strip width. (Figure 4).

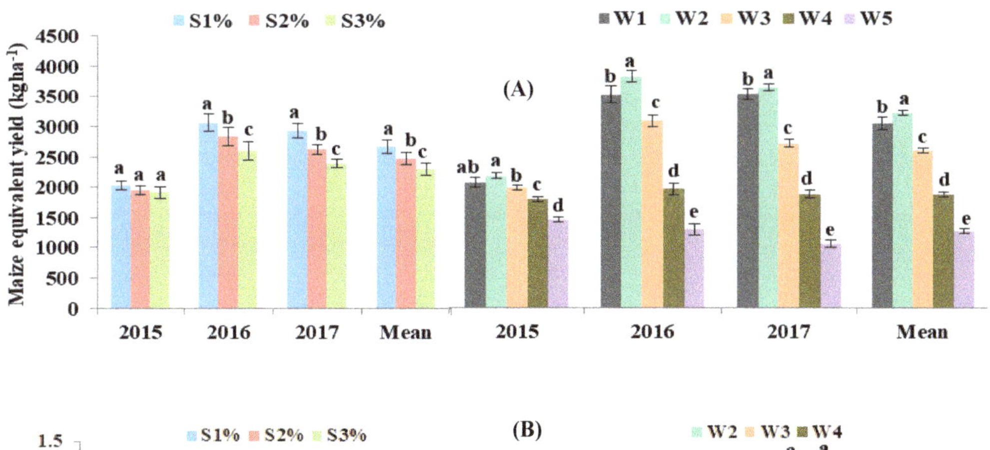

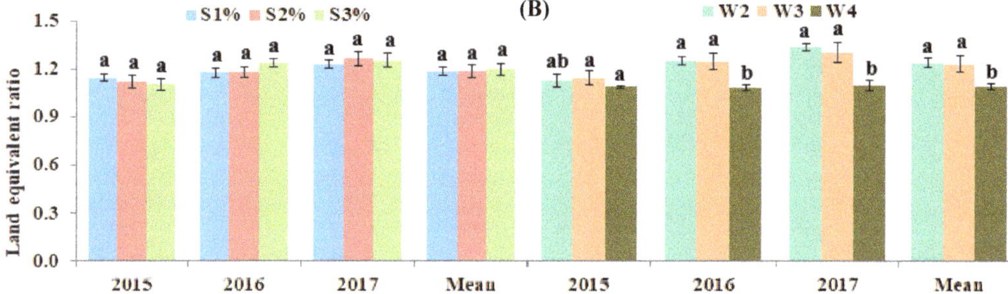

Figure 3. Effect of land slopes and strip width on (**A**) maize equivalent yield and (**B**) land equivalent ratio. Different letters on standard error bars indicate significant differences at $p < 0.05$ between different treatments within a year according to the LSD test.

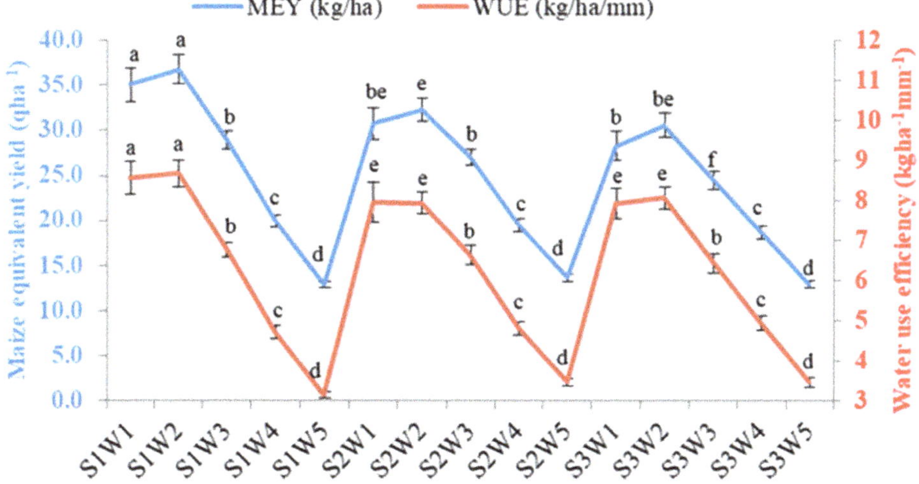

Figure 4. Combined effect of land slopes and strip width on maize equivalent yield and water use efficiency. Different letters on standard error bars indicate significant differences at $p < 0.05$ between different treatments within a year as per the ANOVA procedures of RBD analysis.

3.2.2. Land Equivalent Ratio (LER)

LER indicates the relative land area under the sole crop required to produce the yield as obtained from the intercropping system. LER is the most important parameter to evaluate the intercropping system. If the LER value is >1, then intercropping is beneficial. The effect of land slopes on the LER values was not significant in all the years. In the maize and cowpea strip-intercropping system, the LER values varied from 1.13 to 1.34 for W2, from 1.15 to 1.31 for W3, and from 1.09 to 1.10 for W3. The LER value exceeded unity in all the systems, indicating yield advantage in strip-intercropping. The 3-year mean value of the LER was the highest (1.24) in the W3 strip-intercropping system, indicating a 24% yield advantage over sole cropping systems of maize and cowpea.

3.3. Leaf Area Index (LAI)

The effect of land slopes on the leaf area index of maize and cowpea was significant. The highest LAI of maize and cowpea was measured on the 1% slope at 30 DAS and it was significantly higher than S2 and S3 (Figure 5). Among the maize and cowpea strip-intercropping systems, the W4 strip-intercropping system recorded the maximum leaf area index of maize and cowpea at 30 DAS than W1, W2, W3, and W4 systems. Among sole crops, sole cowpea provided maximum ground cover which was significantly higher than sole maize and maize and cowpea strip-intercropping systems.

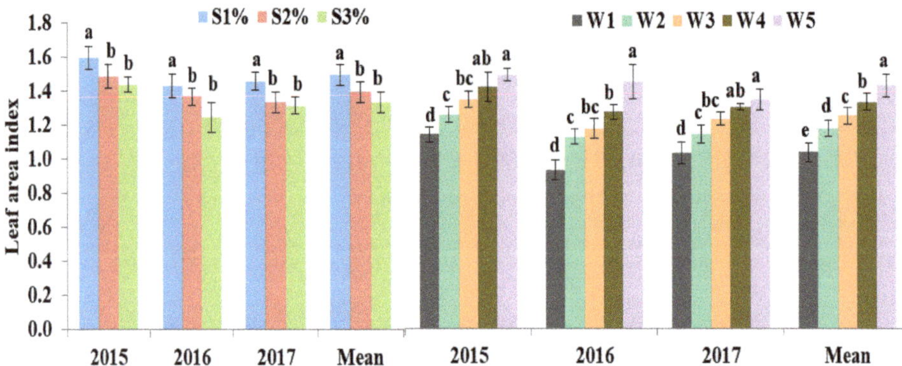

Figure 5. Effect of land slopes and strip width on leaf area index 30DAS. Different letters on standard error bars indicate significant differences at $p < 0.05$ between different treatments within a year according to the LSD test.

3.4. Resource Conservation

3.4.1. Runoff and Soil Loss

During three years of experimentation, 23–26 runoff events (6 in 2015, 8 in 2016, and 12 in 2017) were observed. A higher number (26) of runoff events were recorded from the 3% slope, 2% slope, and from sole maize plots and less (23) from the 1% slope, sole cowpea, and maize and cowpea strip-intercropping systems. Among all the land slopes, higher runoff (27.8–31.5%) and soil loss (7.01–9.79 Mg ha^{-1}) were recorded on the 3% slope followed by the 2% slope (23.1–26.5 % and 5.6–8.0 Mg ha^{-1}) and the 1% slope (19.8–23.0% and 3.75–5.74 Mg ha^{-1}). The average soil loss of 3 years on the 3% slope was 3.60 Mg ha^{-1} and 1.51 Mg ha^{-1} more than on the 1% slope and 2% slope (Figure 6). Strip-intercropping of maize and cowpea resulted in a significant reduction in runoff and soil loss as compared to sole maize. The W4 strip-intercropping system showed the highest reduction in runoff and soil loss which was at par with that under the W3 system but significantly higher than the W2 system. Among sole crops, higher runoff (28.1–31.0%) and soil loss (6.21–9.25 Mg ha^{-1}) were recorded in sole maize and lowest (21.3–24.3% and 4.92 to 6.89 Mg ha^{-1}) in the sole cowpea. This indicates the need for devising suitable conservation measures to check soil loss from undulating sloppy fields.

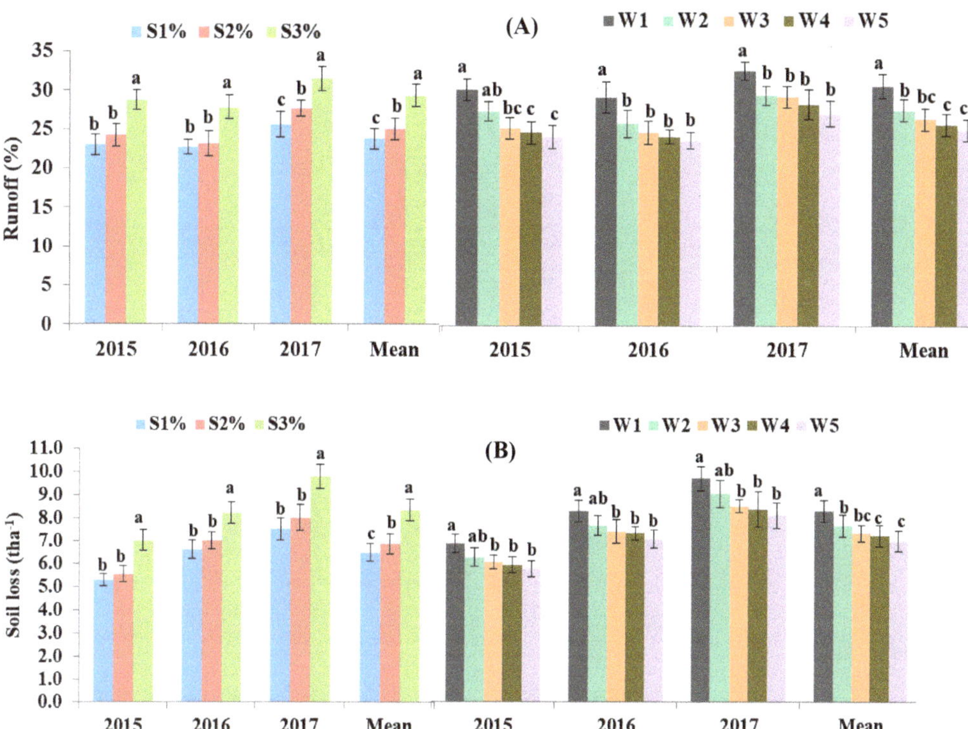

Figure 6. Effect of land slopes and strip width on (**A**) run-off and (**B**) soil loss. Different letters on standard error bars indicate significant differences at *p* < 0.05 between different treatments within a year according to the LSD test.

3.4.2. Nutrient Loss

During three years of experimentation, higher N, P, K, and organic carbon (OC) loss through runoff was observed in 2017 and lower in 2015. Among the slopes, average N, P, K, and OC loss in 3 years were lowest on the 1% slope and highest on the 3% slope (Figure 7). The nutrient loss on the 3% slope was significantly higher than on the 1% and 2% slope during all the years. On the 3% slope, the average loss of N, P, and K was 3.80, 1.82, and 4.10 kg ha^{-1} respectively more than that on the 1% slope. The average loss of N on the 3% slope was 3.8 and 2.8 kg ha^{-1} more than that on the 1% and 2% slope, respectively. Similarly, average P and K loss on the 3% slope was 6.15 kg ha^{-1} and 17.6 kg ha^{-1} respectively with corresponding values of 4.33 kg ha^{-1} and 17.1 kg ha^{-1} on the 1% slope and 4.85 kg ha^{-1} and 18.5 kg ha^{-1} on the 2% slope. The average organic carbon loss on the 3% slope was 21% and 19% more than on the 1% and 2% slope. Nutrient losses were significantly affected by different strip-intercropping systems. Among the sole crops, loss of N, P, K, and OC in runoff was lowest in W5 and highest in W1. In strip-intercropping systems, loss of N, P, K, and OC decreased with the increase in cowpea strip width. The highest loss of N, P, K, and OC were observed in the W2 system followed by W3 and W4. The W4 strip-intercropping system resulted in a significant reduction in N, P, K, and OC loss as compared to the W2 and W3 systems. The organic carbon loss in sole maize was significantly higher than W2, W3, W4, and W5 by 16%, 32%, 43%, and 56%, respectively. Interaction between slope and maize and cowpea strip width was significant for N, K, and organic carbon loss. Sole cowpea resulted in a minimum loss of N, K, and organic carbon on the 1%, 2%, and 3% slopes (Figure 8). On all the land slopes, maize and cowpea strip-intercropping systems showed a significant reduction in N, K, and organic carbon loss as compared to sole maize.

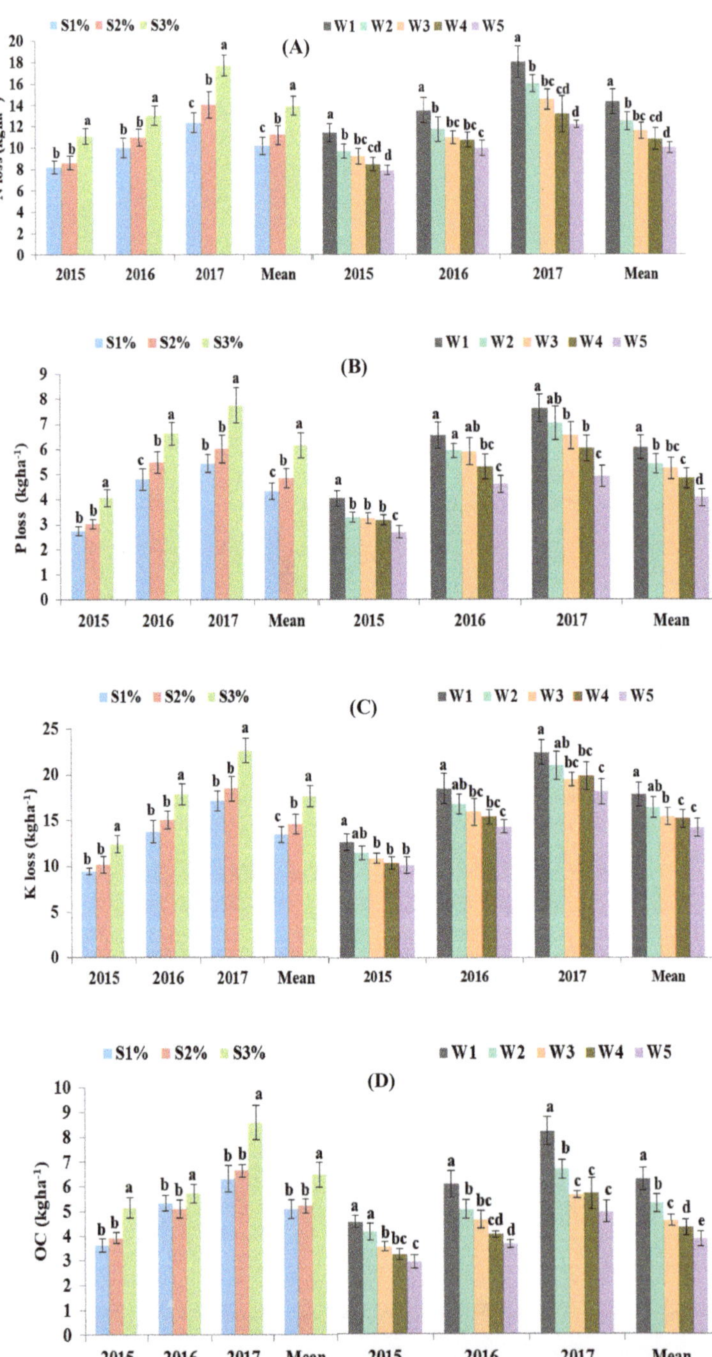

Figure 7. Effect of land slopes and strip width on (**A**) N loss; (**B**) P loss; (**C**) K loss and (**D**) organic carbon loss. Different letters on standard error bars indicate significant difference at $p < 0.05$ between different treatments within a year according to the LSD test.

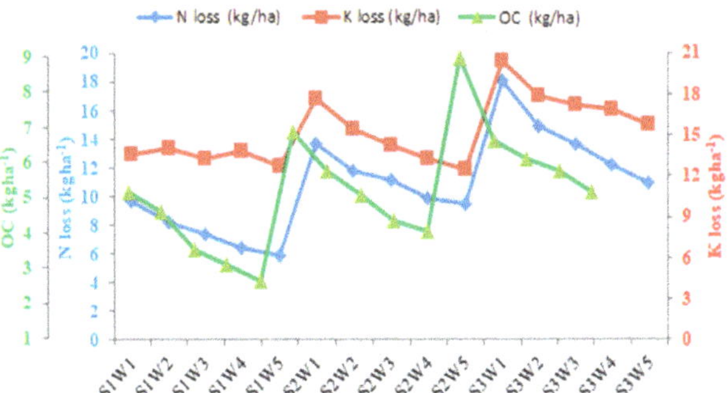

Figure 8. Combined effect of land slopes and strip width on N loss, organic carbon loss, and K loss.

3.4.3. Water Conservation Efficiency and Water Use Efficiency

The water conservation efficiency (WCE) varied significantly among the slopes. The highest water was conserved on the 1% slope (77.0–80.2%) which was significantly higher than the 2% slope (73.5–76.9%) and 3% slope (68.5–72.2%) (Figure 9). Among the sole crops and strip-intercropping systems, the highest WCE (75.7–78.7%) was recorded under sole cowpea and the lowest in sole maize (69.0–71.0%). While the W3 and W4 strip-intercropping systems showed the WCE at par with sole cowpea and significantly higher than sole maize indicating better water conservation under these treatments. Crop WUE on the 1% slope was 3.5% and 3.4% higher than on the 2% and 3% slopes. The WUE in the W2 strip-intercropping systems was 24% and 70% greater than the W3 and W4 strip-intercropping systems.

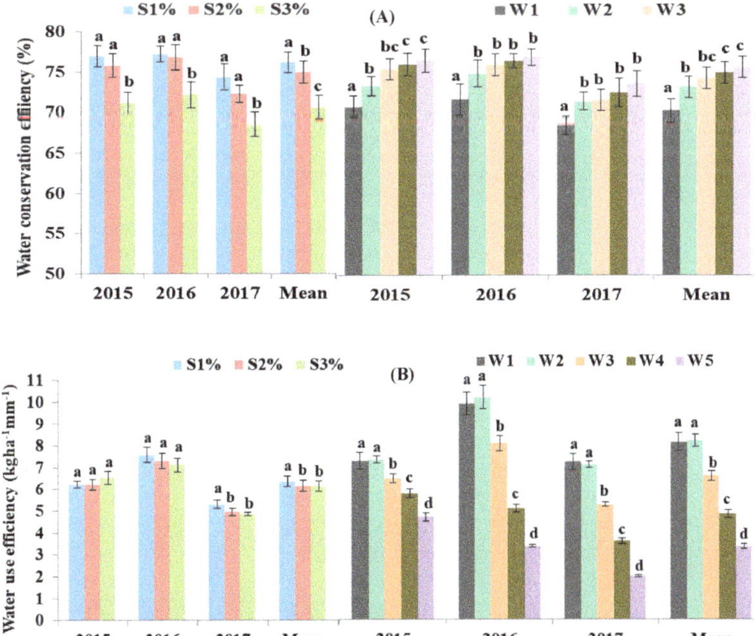

Figure 9. Effect of land slopes and strip width on (**A**) water conservation efficiency and (**B**) water use efficiency. Different letters on standard error bars indicate significant differences at $p < 0.05$ between different treatments within a year according to the LSD test.

3.5. Profitability

Among all the slopes, the average cost of cultivation was highest on the 1% slope than the 2% and 3% slopes. The highest gross returns (US$ 858 ha^{-1}) obtained on the 1% slope were higher than the gross returns obtained on the 2% and 3% slopes by US$ 809 ha^{-1} and US$ 764 ha^{-1}, respectively. Similarly, the highest net returns (US$ 428 ha^{-1}) recorded on the 1% slope were higher by US$ 45 ha^{-1} and US$ 88 ha^{-1} over the 2% and 3% slopes. The highest BC ratio of 2.00 was estimated on the 1% slope followed by 1.90 on the 2% slope and the lowest (1.80) on the 3% slope. During all the 3 years of experimentation, among the sole cropping and maize and cowpea strip-intercropping systems, the highest expenditure and net returns were in the W2 system followed by the W1, W3, W4, and the lowest in W5. The W2 system also fetched the highest net returns of US$ 530 ha^{-1} which was higher over W1, W3, W4, and W5 treatments by US$ 98.7, US$ 91.9, US$ 224, and US$ 317 ha^{-1}, respectively. The maize and cowpea strip-intercropping in 4.8 m:1.2 m (W2) resulted in the highest mean BC ratio of 2.09 followed by 1.99 in W3, 1.99 in W1, 1.77 in W4, and 1.64 in W5 (Table 1).

Table 1. Economics (mean of three years) of strip-cropping maize + cowpea on different slopes.

Slope	Cost of Cultivation (US$ ha^{-1}) *	Gross Returns (US$ ha^{-1}) **	Net Returns (US$ ha^{-1})	B:C
S_1: 1%	429	858	428	2.00
S_2: 2%	426	809	384	1.90
S_3: 3%	424	764	340	1.80
Maize + Cowpea strip-intercropping				
W_1: 6:00	470	901	432	1.92
W_2: 4.8:1.2	488	1018	530	2.09
W_3: 3:3	443	881	439	1.99
W_4: 1.2:4.8	397	704	306	1.77
W_5: 0:6	333	546	214	1.64

* 1 US $ = 66 INR. Prices (US $) of inputs (tonne or unit): urea 95; Single super phosphate 124; muriate of potash 255; maize seed 2652; cowpea seed 909; atrazine 5606; decis 9091; man labour (8 h) 4.4; land preparation ha^{-1} (one ploughing and two harrowing) 95.5. ** Prices (US $) of outputs (tonne): maize grain 253; cowpea grain 530; maize straw 15; Cowpea fodder 15.

4. Discussion

In general, during three years of experimentation (2015–2017), the highest maize grain yield (Figure 2) was recorded in the second year (2016) which can be due to the optimum and well-distributed rainfall (509 mm rainfall in 25 rainy days with only two rainy days of more than 50 mm rainfall) during the entire crop duration. The lowest grain yield of maize was observed in the first year (2015) due to the prolonged dry spell of 28 days from 21 August to 19th September starting from silking to dough stage. In 2015, due to the poor pod formation in cowpea, it was used as fodder, and grain yield was not recorded. The large variation in productivity of maize and cowpea in different years indicates that rainfall amount and its distribution have a considerable effect on crop performance. The findings of our study indicate that on the 1% and 2% slope, sole maize can be cultivated but on the 3% slope, cultivation of sole maize may cause significant runoff, soil loss, and nutrient loss. Therefore, on the sloppy field with the 3% slope, strip-intercropping of maize and cowpea should be practiced. This information on strip-intercropping of maize and cowpea on undulating land is beneficial to develop management strategies for sustaining the crop productivity in the *Shivalik* foothills of India, and other regions of the world with undulating topography [36,47–51].

4.1. Maize and Cowpea Productivity from 2015 to 2017

Maize grain yield obtained on the 1% slope was significantly higher than on the 2% slope and 3% slope and yield depicted a declining trend with increase in slope except in 2015. This might be due to the fact that on steep slopes there is a greater runoff and soil loss, which is characterized by a thinner surface horizon and lower infiltration rate resulting in lower soil productivity [52]. Soil erosion reduces crop growth and yield as it influences many soil properties such as thinning of topsoil, reducing water-stable aggregates, increasing soil bulk density, reducing water holding capacity, and reducing soil organic matter and nutrient content [53]. Wezel et al. [54] and Hoang et al. [55] reported a decrease in maize yield on sloppy fields as a result of the decrease in soil fertility due to soil erosion. Maize grain yield responded significantly to strip width. Sole maize yield varied from 2137 kg ha^{-1} in 2015 to 3645 kg ha^{-1} in 2017. The strip-intercropping combination of 4.8 m maize and 1.2 m cowpea (W2) produced grain yield similar to that from sole maize (W1) grown in 6 m strip but significantly higher than maize and cowpea strip-intercropping in 3:3 m (W3) and maize and cowpea strip-intercropping in 1.2:4.8 m (W4) in all the years. In spite of 20% less plant population, the W4 system produced only 4% less yield than sole maize. These benefits were attributed to reduced intraspecific competition among maize plants due to an increase in the number of border rows which benefits from more light interception and increased availability of nitrogen from the companion legume crop [56,57].

The grain yield of cowpea was highest when raised on the 2% slope followed by the 3% slope and the 1% slope. This might be due to the fact that on 2% slope and 3% slope, good drainage conditions ensured adequate plant growth resulting in higher seed yield [58]. Moreover, cowpea plants may have been suppressed due to the more vigorous growth of maize plants on the 1% slope [33]. Cowpea grain yield from sole cowpea (W5) plots was significantly higher than maize and cowpea strip intercropping systems. This is due to the 20%, 50%, and 80% less plant population in W4, W3, and W2 strip intercropping systems, respectively as compared to sole cowpea [59].

4.2. Runoff and Soil Loss

On sloppy arable lands, soil disturbance for agriculture operations accelerates the runoff and soil erosion [60,61]. In our study, higher runoff (27.8–31.5%) and soil loss (7.01–9.79 Mg ha^{-1}) was recorded on the 3% slope followed by the 2% slope (23.1–26.5% and 5.55–8.00 Mg ha^{-1}) and 1% slope (19.8–23.0% and 3.75–5.74 Mg ha^{-1}). This is due to the fact that on sloppy fields, water runs off quickly as the infiltration opportunity time is less. As runoff velocity increases, so does its ability to detach and transport the sediments. On flat or gently sloping land, a film of water forms on the surface during intense storms which helps to dissipate raindrop energy. Therefore, the slope is one of the most important factors in water erosion as it affects both the amount, as well as the velocity, of runoff [62]. Strip-intercropping of maize and cowpea resulted in a significant reduction in runoff and soil loss as compared to sole maize. The highest reduction in runoff and soil loss was achieved in a maize strip 1.2 m wide and cowpea strip 4.8 m wide and it increased with the reduction in the cowpea strip width. Strip-intercropping of maize with cowpea decreased runoff in all the strip-intercropping systems which can be attributed to the fact that cowpea quickly establishes a very good canopy cover which dissipates the rainfall impact thereby resulting in lower runoff [63]. The results are in agreement with the published literature [37,51], which reported that lower soil erosion losses were observed under soybean than those under widely spaced maize.

4.3. Nutrient Losses

Higher loss of N, P, K, and OC was observed on the 3% slope than on the 2% slope and 1% slope. This could have resulted from higher runoff and soil loss on the 3% slope. N which is generally soluble in water has been lost in runoff water while P, K, and OM which are adsorbed on the soil particles might have been carried away with soil aggregates in the runoff water [60]. Maximum losses of N, P, K, and OC were recorded in sole maize cultivation and the lowest in sole cowpea during 2015, 2016, and 2017. The reason ascribed is that maize is an erosion permitting crop due to wide row spacing and low leaf area per unit ground area as compared to cowpea which covers the soil surface quickly and maintains a high leaf area index throughout its growth period. Similarly, Singh et al. [51], Prasad et al. [64], and Lakaria et al. [65] reported higher losses of organic carbon, N, P, and K total nitrogen in maize and castor, whereas losses were less in groundnut.

4.4. Regression and Correlation Analysis

Regression relationships between runoff and soil and nutrient loss are depicted in Figure 10. Relatively higher R^2 values suggest that the coefficient for loss of soil and nutrients is greater for higher surface runoff. The plots with greater slope and maximum maize strip width resulted in greater soil nutrient loss as depicted from regression analysis.

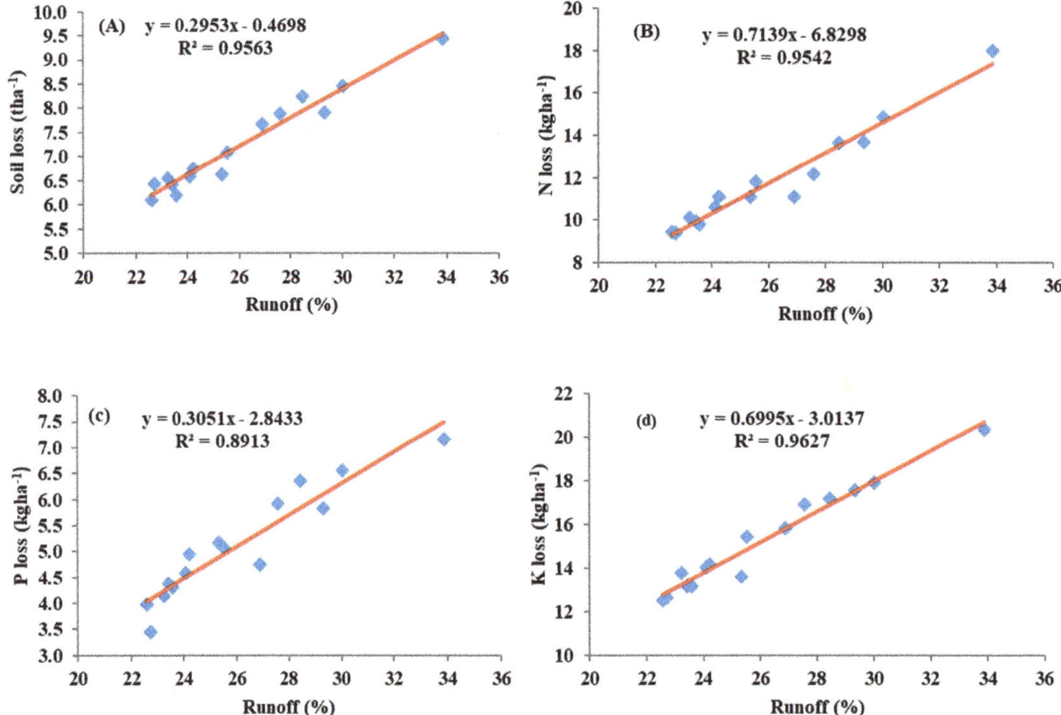

Figure 10. Regression relationship between (**A**)runoff and soil loss; (**B**) runoff and N loss; (**C**) runoff and P loss; (**D**) runoff and K loss.

A positive and strongly significant correlation (Table 2) was observed between runoff and soil and nutrient loss (NPK) suggesting that greater runoff not only erodes the soil but also results in higher dissolution of nutrients which has a perilous impact on soil fertility. Yao et al. [66] also confirmed that the relationship between runoff and nutrient loss was positively significant. A negative and significant correlation was observed between cowpea yield and nutrient loss (Table 2) indicating that a decrease in soil nutrients due to erosion impairs crop performance through reduced underground and aboveground biomass.

Table 2. Pearson's correlation among studied parameters.

	Rainfall	Runoff	Soil Loss	N Loss	P Loss	K Loss	Maize Grain Yield	Cowpea Grain Yield	Maize Straw Yield	Cowpea Straw Yield
Rainfall	1									
Runoff	0.05	1								
Soil loss	0.01	0.97 **	1							
N loss	0.06	0.97 **	0.95 **	1						
P loss	0.16	0.94 **	0.92 **	0.94 **	1					
K loss	0.08	0.98 **	0.98 **	0.96 **	0.93 **	1				
Maize grain yield	0.49	0.40	0.28	0.47	0.51	0.36	1			
Cowpea grain yield	−0.37	−0.57 *	−0.44	−0.63 *	−0.63 *	−0.52 *	−0.96 **	1		
Maize straw yield	0.49	0.41	0.28	0.48	0.52 *	0.37	0.99 **	−0.96 **	1	
Cowpea fodder yield	−0.34	−0.57 *	−0.43	−0.62 *	−0.61 *	−0.51 *	−0.95 **	0.99 **	−0.95 **	1

** Significant at 0.01 probability level, * significant at 0.05 probability level.

5. Conclusions

The results from this 3-year field study indicate that the cultivation of maize and cowpea on the 1% slope and 2% slopes, respectively resulted in the highest yield. Runoff, soil and nutrient losses were higher on the 3% slope as compared to the 1% and 2% slopes. The adoption of the strip-intercropping system with a 4.8 m maize strip width and 1.2 m cowpea strip width showed significantly higher maize equivalent yield than sole maize and other maize and cowpea strip-intercropping systems and also has the highest LER value (1.24) indicating a 24% yield advantage over sole cropping systems of maize and cowpea. This system also reduced runoff and soil loss by 10.9% and 8.3%, respectively than sole maize crop. On sloping land, the maize and cowpea strip-intercropping system enhanced water use efficiency by 12.3% and increased maize equivalent yield by 19.4% as compared to sole cropping of maize and cowpea. Hence, the strip-intercropping of maize and cowpea should be practiced in the *Shivalik* foothills of Northwest India in order to ensure sustainable crop production vis-à-vis resource conservation.

Author Contributions: Conceptualization, A.K., M.S. and G.R.C.; methodology, A.Y. and A.K.; software, A.Y.; validation, V.S. and P.S.S.; formal analysis, A.K. and V.S.; investigation, V.S. and P.S.S.; resources, M.S. and G.R.C.; data curation, A.Y.; writing—original draft preparation, A.K.; writing—review and editing, A.Y. and M.S.; visualization, A.Y. and P.S.S.; supervision, A.K., M.S. and G.R.C.; project administration, G.R.C. and M.S.; funding acquisition, G.R.C. and M.S. All authors have read and agreed to the published version of the manuscript.

Funding: The research was funded by Indian Council of Agricultural Research, New Delhi through All India Coordinated Research Project on Dryland Agriculture, Central Research Institute for Dryland Agriculture, Hyderabad, India.

Institutional Review Board Statement: Not applicable.

Informed Consent Statement: Not applicable.

Data Availability Statement: Not available.

Acknowledgments: Authors express sincere thanks to the All India Coordinated Research Project for Dryland Agriculture, Central Research Institute for Dryland Agriculture, Indian Council of Agricultural Research, Hyderabad, India for providing financial assistance to conduct this experiment. The views expressed in this paper are those of individual scientists and do not necessarily reflect the views of the donor or the authors' institution.

Conflicts of Interest: The authors declare no conflict of interest.

References

1. Dregne, H.E. Land degradation in the drylands. *Arid Land Res. Manag.* **2002**, *16*, 99–132. [CrossRef]
2. Oyedele, D.; Aina, P. A study of soil factors in relation to erosion and yield of maize on a Nigerian soil. *Soil Tillage Res.* **1998**, *48*, 115–125. [CrossRef]
3. Tiruwa, D.B.; Khanal, B.R.; Lamichhane, S.; Acharya, B.S. Soil erosion estimation using Geographic Information System (GIS) and Revised Universal Soil Loss Equation (RUSLE) in the Siwalik Hills of Nawalparasi, Nepal. *J. Water Clim. Chang.* **2021**, in press. [CrossRef]
4. UNCCD. Desertification, Land Degradation & Drought (DLDD): Some Global Facts and Figures—United Nations Conventions to Combat Desertification. 2015. Available online: http://www.unccd.int/Lists/SiteDocumentLibrary/WDCD/DLDD%20Facts.pdf (accessed on 17 February 2021).
5. Montgomery, D.R. Soil erosion and agricultural sustainability. *Proc. Natl. Acad. Sci. USA* **2007**, *104*, 13268–13272. [CrossRef]
6. Oldeman, L.R.; Hakkeling, R.T.A.; Sombroek, W.G. World Map of the Status of Human-induced Soil Degradation: An Explanatory Note, Second Revised Edition: ISRIC/UNEP. 1991. Available online: http://www.isric.org/sites/default/files/ExplanNote_1.pdf (accessed on 15 January 2021).
7. Bai, Z.G.; Dent, D.L.; Olsson, L.; Schaepman, M.E. Proxy global assessment of land degradation. *Soil Use Manag.* **2008**, *24*, 223–234. [CrossRef]
8. Bindraban, P.S.; Van der Velde, M.; Ye, L.; Van den Berg, M.; Materechera, S.; Innocent Kiba, D.; Tamene, L.; Vala Ragnarsdottir, K.; Jongschaap, R.; Hoogmoed, M.; et al. Assessing the impact of soil degradation on food production. *Curr. Opin. Environ. Sustain.* **2012**, *4*, 478–488. [CrossRef]
9. Pimentel, P.; Burgess, M. Soil erosion threatens food production. *Agriculture* **2013**, *3*, 443–463. [CrossRef]
10. Hurni, H.; Giger, M.; Liniger, H.; Studer, R.M.; Messerli, P.; Portner, B.; Schwilch, G.; Wolfgramm, B.; Breu, T. Soils, agriculture and food security: The interplay between ecosystem functioning and human well-being. *Curr. Opin. Environ. Sustain.* **2015**, *15*, 25–34. [CrossRef]
11. Gibbs, H.K.; Salmon, J.M. Mapping the world's degraded lands. *Appl. Geogr.* **2015**, *57*, 12–21. [CrossRef]
12. FAO. *The State of the World's Land and Water Resources for Food and Agriculture (SOLAW)—Managing Systems at Risk*; Food and Agriculture Organization of the United Nations: Rome, Italy; Earthscan: London, UK, 2011; Available online: http://www.fao.org/docrep/017/i1688e/i1688e.pdf (accessed on 15 January 2021).
13. McCool, D.K.; Williams, J.D. Soil erosion by water. In *Encyclopedia of Ecology*; Jorgensen, S.E., Fath, B.D., Eds.; Elsevier: Amsterdam, The Netherlands, 2008; pp. 3284–3290.
14. Bhardwaj, A.; Rana, D.S. Torrent control measures in Kandi area of Punjab-A case study. *J. Water Manag.* **2008**, *16*, 55–63.
15. Yousuf, A.; Bhardwaj, A.; Prasad, V. Simulating impact of conservation interventions on runoff and sediment yield in a degraded watershed using WEPP model. *Ecopersia* **2021**, *9*, 191–205.
16. Bhardwaj, A.; Kaushal, M.P. Two dimensional physically based finite element runoff model for small agricultural watershed. *Hydrol. Process.* **2009**, *23*, 397–407. [CrossRef]
17. Lenz, J.; Yousuf, A.; Schindewolf, M.; Werner, M.V.; Hartsch, K.; Singh, M.J.; Schmidt, J. Parameterization for EROSION-3D model under simulated rainfall conditions in lower Shivaliks of India. *Geosciences* **2018**, *8*, 396. [CrossRef]
18. Mandal, M.; Sharda, V.N. Assessment of permissible soil loss in India employing a quantitative bio-physical model. *Curr. Sci.* **2011**, *100*, 383–389.
19. Hadda, M.S.; Arora, S. Soil and nutrient management practices for sustaining crop yields under maize-wheat cropping sequence in sub-mountain Punjab, India. *Soil Environ.* **2006**, *25*, 1–5.
20. Singh, M.J.; Khera, K.L. Soil erodibility indices under different land uses in lower Shiwaliks. *Trop. Ecol.* **2005**, *49*, 113–119.
21. Sharma, K.R.; Arora, S. Runoff & soil loss as affected by different intercrops with maize in relation to productivity in the Kandi region of Jammu. In *Emerging Trends in Watershed Management*; Yadav, R.P., Tiwari, A.K., Sharma, P., Singh, P., Arya, S.L., Bhatt, V.K., Prasad, R., Sharda, V.N., Eds.; Satish Serial Publishing House: New Delhi, India, 2010; Volume 1, pp. 355–361.
22. Ma, B.; Yu, X.; Ma, F.; Li, Z.; Wu, F. Effects of crop canopies on rain splash detachment. *PLoS ONE* **2014**, *9*, e99717. [CrossRef] [PubMed]
23. Li, G.; Wan, L.; Cui, M.; Wu, B.; Zhou, J. Influence of canopy interception and rainfall kinetic energy on soil erosion under forests. *Forests* **2019**, *10*, 506. [CrossRef]
24. Barton, A.P.; Fullen, M.A.; Mitchell, D.J.; Hocking, T.J.; Liu, L.; Bo, Z.W.; Zheng, Y.; Xia, Z.Y. Effects of soil conservation measures on erosion rates and crop productivity on sub-tropical Ultisols in Yunnan Province, China. *Agric. Ecosyst. Environ.* **2004**, *104*, 343–357. [CrossRef]
25. Dunj'o, G.; Pardini, G.; Gispert, M. The role of land use land cover on runoff generation and sediment yield at a microplot scale in a small Mediterranean catchment. *J. Arid Environ.* **2004**, *57*, 99–116. [CrossRef]
26. Adekalu, K.O.; Okunade, D.A.; Osunbitan, J.A. Compaction and mulching effects on soil loss and runoff from two south-western Nigeria agricultural soils. *Geoderma* **2006**, *137*, 226–230. [CrossRef]
27. Ali, I.; Khan, F.; Bhatti, A.U. Soil and nutrient losses by water erosion under mono-cropping and legume inter-cropping on sloping land. *Pak. J. Agric. Res.* **2007**, *20*, 161–166.
28. Machado, S. Does intercropping have a role in modern agriculture? *J. Soil Water Conserv.* **2009**, *64*, 55A–57A. [CrossRef]

29. Subudhi, C.R.; Senapati, S.C. Relationship between rainfall, runoff, soil loss and productivity in north eastern ghat zone of Odisha. *Biom. Biostat. Int. J.* **2016**, *4*, 84–89.
30. Nyawade, S.O.; Gachene, C.K.K.; Karanja, N.N.; Gitari, H.I.; Schulte-Geldermann, E.; Parker, M.L. Controlling soil erosion in smallholder potato farming systems using legume intercrops. *Geoderma Reg.* **2019**, *17*, e00225. [CrossRef]
31. Lithourgidis, A.S.; Dordas, C.A.; Damalas, C.A.; Vlachostergios, D.N. Annual intercrops: An alternative pathway for sustainable agriculture. *Aust. J. Crop Sci.* **2011**, *5*, 396–410.
32. Verdelli, D.; Acciaresi, H.A.; Leguizamon, E.S. Corn and Soybeans in a Strip Intercropping System: Crop Growth Rates, Radiation Interception, and Grain Yield Components. *Int. J. Agron.* **2012**, 980284. [CrossRef]
33. Sharaiha, R.K.; Ziadat, F.M. Alternative cropping systems to control soil erosion in arid to semi-arid areas of Jordan. *Afr. Crop Sci. Conf. Proc.* **2007**, *8*, 1559–1565. [CrossRef]
34. Ghosh, B.N.; Sharma, N.K.; Dadhwal, K.S. Integrated nutrient management and intercropping/cropping system impact on yield, water productivity and net return in valley soils of north-west Himalayas. *Ind. J. Soil Conserv.* **2011**, *39*, 236–242.
35. Wang, T.; Zhu, B.; Xia, L. Effects of contour hedgerow intercropping on nutrient losses from the sloping farmland in the three Gorges Area, China. *J. Mt. Sci.* **2012**, *9*, 105–114. [CrossRef]
36. Sharma, N.K.; Singh, R.J.; Mandal, D.; Kumar, A.; Alam, N.M.; Keesstra, S. Increasing farmers income and reducing soil erosion using intercropping in rainfed maize-wheat rotation of Himalaya, India. *Agric. Ecosyst. Environ.* **2017**, *247*, 43–53. [CrossRef]
37. Jakhar, P.; Ahikary, P.P.; Naik, B.S.; Madhu, M. Finger millet (*Eleusine coracana*)–groundnut (*Arachis hypogaea*) strip-intercropping for enhanced productivity and resource conservation in uplands of Eastern Ghats of Odisha. *Ind. J. Agron.* **2015**, *60*, 365–371.
38. Sharma, V.; Kumar, V.; Sharma, S.C.; Sharma, R.K.; Khokhar, A.; Singh, M. Climatic Variability analysis at Ballowal Saunkhri in submontane Punjab (India). *Clim. Chang. Environ. Sustain.* **2017**, *5*, 83–91. [CrossRef]
39. Jackson, M.L. *Soil Chemical Analysis*, 1st ed.; Prentice Hall of India Pvt Ltd.: New Delhi, India, 1973; Volume 1, pp. 38–204.
40. Walkley, A.; Black, I.A. An examination of the Degtjareff method for determining soil organic matter, and a proposed modification of the chromic acid titration method. *Soil Sci.* **1934**, *37*, 29–38. [CrossRef]
41. Subbiah, B.V.; Asija, G.L. A rapid procedure for the determination of available nitrogen in soil. *Curr. Sci.* **1956**, *25*, 259–260.
42. Olsen, S.R.; Cole, C.V.; Watanabe, F.S.; Dean, L. *Estimation of Available Phosphorus in Soil by Extraction with Sodium Carbonate*; Cir. 939; USDA: Washington, DC, USA, 1954.
43. Nelson, D.W.; Sommers, L.E. A simple digestion procedure for estimation of total nitrogen in soils and sediments. *J. Environ. Qual.* **1972**, *1*, 423–425. [CrossRef]
44. Kassam, A.; Smith, M. FAO methodologies on crop water use and crop water productivity. In Proceedings of the Expert Meeting on Crop Water Productivity, Rome, Italy, 3–5 December 2001; Paper no. CWP-M07. pp. 1–18.
45. Lakaria, B.L.; Narayan, D.; Biswas, H.; Raj, S.; Jha, P.; Somasundaram, J. Water conservation efficiency of prominent kharif crops in Bundelkhand region. *Ind. J. Soil Conserv.* **2012**, *40*, 231–235.
46. Willey, R.W.; Rao, M.R. A competitive ratio for quantifying competition between intercrops. *Exp. Agric.* **1980**, *16*, 117–125. [CrossRef]
47. Nadal-Romero, E.; Cammeraat, E.; Pérez-Cardiel, E.; Lasanta, T. Effects of secondary succession and afforestation practices on soil properties after cropland abandonment in humid Mediterranean mountain areas. *Agric. Ecosyst. Environ.* **2016**, *228*, 91–100. [CrossRef]
48. Egarter-Vigl, L.; Depellegrin, D.; Pereira, P.; De Groot, D.; Tappeiner, U. Mapping the ecosystem service delivery chain: Capacity, flow, and demand pertaining to aesthetic experiences in mountain landscapes. *Sci. Total Environ.* **2017**, *574*, 436–442. [CrossRef]
49. Masvaya, E.N.; Nyamangara, J.; Descheemaeker, K.; Giller, K.E. Is maize and cowpea intercropping a viable option for smallholder farms in the risky environments of semiarid southern Africa? *Field Crops Res.* **2017**, *209*, 73–87. [CrossRef]
50. Romero-Díaz, A.; Ruiz-Sinoga, J.D.; Robledano-Aymerich, F.; Brevik, E.C.; Cerdà, A. Ecosystem responses to land abandonment in western Mediterranean mountains. *Catena* **2017**, *149*, 824–835. [CrossRef]
51. Singh, R.K.; Chaudhary, R.S.; Somasundaram, J.; Sinha, N.K.; Mohanty, M.; Hati, K.M.; Rashmi, I.; Patra, A.K.; Chaudhari, S.K.; Rattan, L. Soil and nutrients losses under different crop covers in vertisols of Central India. *J. Soils Sediments* **2019**, *20*, 609–620. [CrossRef]
52. Jiang, P.; Thelen, K.D. Effect of soil and topographic properties on crop yield in a north-central corn–soybean cropping system. *Agron. J.* **2004**, *96*, 252–258. [CrossRef]
53. Yang, W.; Zhang, X.; Gong, W.; Ye, Y.; Yang, Y. Soil erosion and corn yield in a cultivated catchment of the Chinese Mollisol region. *PLoS ONE* **2019**, *14*, e0221553. [CrossRef]
54. Wezel, A.; Steinmüller, N.; Friederichsen, J.R. Slope position effects on soil fertility and crop productivity and implications for soil conservation in upland northwest Vietnam. *Agric. Ecosyst. Environ.* **2002**, *91*, 113–126. [CrossRef]
55. Hoang, T.L.; Simelton, E.; Ha, V.T.; Vu, D.T.; Nguyen, T.H.; Nguyen, V.C.; Phung, Q.T.A. *Diagnosis of Farming Systems in the Agroforestry for Livelihoods of Smallholder Farmers in Northwestern Viet Nam Project*; Working Paper no. 161; World Agroforestry Centre (ICRAF) Southeast Asia Regional Program: Hanoi, Vietnam, 2013; 24p.
56. Andersen, M.K.; Hauggaard-Nielsen, H.; Ambus, P.; Jensen, E.S. Biomass production, symbiotic nitrogen fixation and inorganic N use in dual and tricomponent annual intercrops. *Plant Soil* **2004**, *266*, 273–287. [CrossRef]
57. Yang, F.; Huang, S.; Gao, R.; Liu, W.; Yong, T.; Wang, X.; Wu, X.; Yang, W. Growth of soybean seedlings in relay strip intercropping systems in relation to light quantity and red:far-red ratio. *Field Crops Res.* **2014**, *155*, 245–253. [CrossRef]

58. Hodgson, A.S.; Holland, J.F.; Rayner, P. Effects of field slope and duration of furrow irrigation on growth and yield of six grain legumes on a waterlogging-prone vertisol. *Field Crops Res.* **1989**, *22*, 165–180. [CrossRef]

59. Maruthi, V.; Reddy, K.S.; Pankaj, P.K. Strip-intercropping system as a climate adaptation strategy in semi-arid Alfisols of South-Central India. *Ind. J. Agric. Sci.* **2017**, *87*, 1238–1245.

60. Siswanto, S.Y.; Sule, M.I.S. The Impact of slope steepness and land use type on soil properties in Cirandu sub-sub catchment, Citarum watershed. *IOP Conf. Ser. Earth Environ. Sci.* **2019**, *393*, 012059. [CrossRef]

61. Zhang, Z.; Sheng, L.; Yang, J.; Chen, X.; Kong, L.; Wagan, B. Effects of land use and slope gradient on soil erosion in a red soil hilly watershed of Southern China. *Sustainability* **2015**, *7*, 14309–14325. [CrossRef]

62. Sanders, D.W. Sloping land: Soil erosion problems and soil conservation requirements, 1986. In *Land Evaluation for Land-Use Planning and Conservation in Sloping Areas*; Siderius, W., Ed.; International Institute for Land Reclamation and Improvement: Enschede, The Netherlands, 1984; pp. 40–50.

63. Guo, M.; Zhang, T.; Li, Z.; Xu, G. Investigation of runoff and sediment yields under different crop and tillage conditions by field artificial rainfall experiments. *Water* **2019**, *11*, 1019. [CrossRef]

64. Prasad, S.N.; Singh, R.; Prakash, C. Erosion studies in castor at 1% slope in vertisols of south-eastern Rajasthan. *Ind. J. Soil Conserv.* **1993**, *21*, 25–30.

65. Lakaria, B.L.; Narayan, D.; Katiyar, V.S.; Biswas, H. Evaluation of different kharif crops for minimizing runoff and soil loss in Bundelkhand region. *J. Ind. Soc. Soil Sci.* **2010**, *58*, 252–255.

66. Yao, Y.; Dai, Q.; Gao, R.; Gan, Y.; Yi, X. Effects of rainfall intensity on runoff and nutrient loss of gently sloping farmland in a karst area of SW China. *PLoS ONE* **2021**, *16*, e0246505. [CrossRef] [PubMed]

 *sustainability*

Article

How Tillage and Fertilization Influence Soil N_2O Emissions after Forestland Conversion to Cropland

Xiao Ren [1,2], Bo Zhu [1,*], Hamidou Bah [1,2] and Syed Turab Raza [1,2]

[1] Key Laboratory of Mountain Surface Processes and Ecological Regulation, Institute of Mountain Hazards and Environment, Chinese Academy of Sciences, Chengdu 610041, China; renxiao@imde.ac.cn (X.R.); bahamidou2004@imde.ac.cn (H.B.); s.turabkazmi@hotmail.com (S.T.R.)

[2] University of Chinese Academy of Sciences, Beijing 100049, China

* Correspondence: bzhu@imde.ac.cn

Received: 22 August 2020; Accepted: 22 September 2020; Published: 25 September 2020

Abstract: Soil nitrous oxide (N_2O) emissions are influenced by land use adjustment and management practices. To meet the increasing socioeconomic development and sustainable demands for food supply, forestland conversion to cropland occurs around the world. However, the effects of forestland conversion to cropland as well as of tillage and fertilization practices on soil N_2O emissions are still not well understood, especially in subtropical regions. Therefore, field experiments were carried out to continuously monitor soil N_2O emissions after the conversion of forestland to cropland in a subtropical region in Southwest China. One forestland site and four cropland sites were selected: forestland (CK), short-term croplands (tillage with and without fertilization, NC-TF and NC-T), and long-term croplands (tillage with and without fertilization, LC-TF and LC-T). The annual cumulative N_2O flux was 0.21 kg N ha^{-1} yr^{-1} in forestland. After forestland conversion to cropland, the annual cumulative N_2O flux significantly increased by 76-491%. In the short-term and long-term croplands, tillage with fertilization induced cumulative soil N_2O emissions that were 94% and 235% higher than those from tillage without fertilization. Fertilization contributed 63% and 84% to increased N_2O emissions in the short-term and long-term croplands, respectively. A stepwise regression analysis showed that soil N_2O emissions from croplands were mainly influenced by soil NO_3^- and NH_4^+ availability and WFPS (water-filled pore space). Fertilization led to higher soil NH_4^+ and NO_3^- concentrations, which thus resulted in larger N_2O fluxes. Thus, to reduce soil N_2O emissions and promote the sustainable development of the eco-environment, we recommend limiting the conversion of forestland to cropland, and meanwhile intensifying the shift from grain to green or applying advanced agricultural management practices as much as possible.

Keywords: land use change; tillage; fertilization; N_2O fluxes; subtropical region

---

1. Introduction

Nitrous oxide (N_2O) has been recognized as an important non-CO_2 greenhouse gas, with 298 times greater global warming potential than that of CO_2 based on a 100-year time horizon [1]. In the past 150 years, atmospheric N_2O concentrations have greatly increased from 270 to 324 ppb [1], and the terrestrial biosphere is still a net source of atmospheric N_2O [2]. Agricultural soils are the largest N_2O source, contributing about 60% of global anthropogenic N_2O emissions [1,3]. To guarantee a secure food supply for global human population growth, agriculture intensification might cause more N_2O emission increases in the future [4,5]. Soil N_2O is mainly produced by microbial nitrification and denitrification [6,7], and strongly affected by substrate availability (e.g., NH_4^+ and NO_3^- concentrations) and environmental conditions (e.g., temperature and moisture) [6,8–11].

Many previous studies have proven that N_2O emissions are greatly impacted by different land uses and management strategies [12,13]. In particular, the adjustment of land use is viewed as a crucial anthropogenic N_2O emission source via its significant influences on N substrates and environmental conditions [1,9,12,13]. The existing studies focusing on the effects of land use change on N_2O emissions mainly address rice paddy conversion to vegetable fields or citrus orchards [3,9,14], forest conversion to tea plantations [11,15], forest conversion to pasture or cropland [12], and cropland conversion to forestland [16,17]. Forestland converting to agricultural land has generally led to significant N_2O emissions owing to deforestation and management practices (e.g., fertilizer application and tillage) [11–13,18]. In response to increasing socioeconomic development and demands for food, the specific land use change of forestland to cropland still occurs often around the world [11,12]. Although there have been a few N_2O flux measurements since this type of conversion was published [12], the influence of forest conversion to cropland on N_2O fluxes and their driving mechanisms are still not fully understood in many regions.

Tillage and fertilization are the most important management practices after the conversion of forestland to cropland; they significantly affect soil properties as well as soil carbon (C) and nitrogen (N) availability and ultimately regulate the production and emission of N_2O [9–11,19,20]. However, tillage and fertilization have different driving mechanisms that affect N_2O emissions. Tillage can influence the soil's physical structure (e.g., aggregation, bulk density and aeration), moisture, temperature, and C and N availability [19,21], which likely influence the microbial community and activity and thus N_2O emissions [9,22]. Grandy and Robertson reported that in initial cultivation, tillage practices significantly disturbed the soil structure and released large amounts of organic C and N from macroaggregates, thus accelerating the soil N_2O fluxes [18]. Fertilization has been widely understood to significantly increase N_2O emissions, mainly through supplying more inorganic N concentrations and sufficient substrates (NH_4^+ and NO_3^-) for nitrification and denitrification [8,11,23–25]. Bouwman et al. analyzed 139 N_2O studies from agricultural regions and found that N_2O emissions generally increase with N application rates, especially above the application rates of 100 kg N ha^{-1} [26]. Recently, Chen et al. reported that forest conversion to tea fields significantly triggered substantial N_2O emissions in the first year due to a high basal fertilizer input and intense tillage [11]. Previous studies have made great progress in explaining how tillage and fertilization practices drive N_2O emissions in agricultural lands [10,19,24,26]. More studies are still needed to quantify the relative contributions of tillage and fertilization to increased N_2O emissions when specifically converting forestland to cropland, which would be helpful for developing suggestions on how to reduce N_2O emissions in the transformed croplands.

The Sichuan Basin is the largest agricultural region in Southwest China, accounting for 7% of the total cropland and supplying 10% of the total agricultural products of China [27]. In recent years, many studies have paid attention to the N_2O fluxes from agricultural soils and obtained some significant observations in the Sichuan Basin [24,25,28,29]. However, previous studies mainly focused on the influence of fertilizer application (e.g., type and rate), which has resulted in agricultural lands having more N_2O emissions compared to other land uses. Driven by afforestation policies and increasing food demand, land use conversion between forestland and cropland often occurs in this region. A recent study investigated the effects of afforestation on soil N_2O emissions and found that afforestation significantly decreased N_2O fluxes compared to those in cropland [17]. However, the effect of forestland conversion to cropland on soil N_2O emissions and its underlying mechanisms are still uncertain in this region. Understanding the effects of tillage and fertilization practices on soil N_2O emissions after forestland conversion to cropland will be beneficial for evaluating the environmental impacts of land use change on soil N_2O emissions in the Sichuan Basin, and thus for suggesting appropriate technological approaches to mitigate these effects.

In this study, we measured N_2O emissions and environmental variables from forest soils (as control), new croplands converted from forest (tillage with and without fertilization), and long-term croplands (tillage with and without fertilization) in the Sichuan Basin of Southwest China. The specific aims were (1) to quantify the influences of tillage and fertilization on soil N_2O emissions after land use conversion of forestland to cropland, (2) to evaluate the short-term and long-term land use change impacts on soil N_2O emissions, and (3) to identify the potential mechanisms driving the increased soil N_2O emissions induced by tillage and fertilization after land use conversion.

2. Materials and Methods

2.1. Study Area

This study was carried out at the Yanting Agro-Ecological Station of Purple Soil of the Chinese Academy of Sciences (31°16′ N, 105°27′ E), located in the central Sichuan Basin of Southwest China with an altitude of 400 to 600 m [27]. It exhibits a moderate subtropical monsoon climate, with a mean annual precipitation and temperature of 836 mm and 17.3 °C (30-year mean). The widely distributed soil in the study region is called purple soil locally and is classified by the FAO soil classification as a Eutric Regosol and as a Pup-Orthic Entisol by the Chinese soil taxonomy [27]. The study area is an intensive agricultural production region in China. The sloping croplands have relatively thin soil thicknesses of 30–80 cm and slopes of 3–15%, and wheat (*Triticum aestivum* L.)-maize (*Zea mays* L.) rotation is the main cropping pattern in this region [17].

2.2. Experimental Design

The selected forestland is dominated by *Cupressus funebris* with a mean diameter of 13.2 cm at breast height, a mean height of 16 m, and a density of 1595 stems ha^{-1} [30]. In late July 2016, a portion of the forestland was cleared of trees and roots and then converted to cropland. To assess the short-term effects of management practices on soil N_2O emissions, the newly converted croplands were cultivated from November 2016. In new croplands, two treatments of tillage—without fertilization (NC-T) and with fertilization (NC-TF)—were established with three replicates in a randomized block design (size 3 m × 3 m). The long-term croplands (tillage without fertilization, LC-T, and tillage with fertilization, LC-TF), which were adjacent to the newly converted croplands, were also established with three replicates in a randomized block design (size 4 m × 6 m). The long-term croplands converted from forestland have been cultivated since 2003. Moreover, the selected forestland was used as the control (CK) and had three replicate plots (size 3 m × 3 m). All the treatments had the same soil type (Regosols) and slope (5%).

Following the local cropping regimes, croplands were conventionally cultivated under a wheat-maize rotation system (winter wheat from November to May rotating with summer maize from June to October). Tillage practice involved conventional tillage with harrowing (approximately 20 cm deep) twice a year before sowing. The mineral N fertilization (urea) rates were 130 kg N ha^{-1} and 150 kg N ha^{-1} for wheat and maize with fertilization treatments, respectively [23]. All the fertilizers were manually applied and incorporated into the topsoil (0–20 cm) together with harrowing. Then, wheat and maize seeds were directly drilled into the soil. No irrigation was applied during the growth of either wheat or maize.

2.3. Measurements of N_2O Emissions

Soil N_2O emissions were continuously monitored using the static chamber-gas chromatography technique as described by Zhou et al. [5] and Zheng et al. [31] from November 2016 to October 2017 (a whole wheat-maize rotation season). Briefly, prior to the measurements, stainless-steel base collars with a uniform area of 0.25 m^2 were inserted into topsoil (10 cm in depth) and kept in place throughout the whole measurement period. The equipped chambers, with a circulating fan and an adjustable height according to the crop growth, can guarantee the chamber headspace uniformly mixed and minimize temperature changes when conducting the measurements. After tillage or fertilization practices, soil N_2O emissions were continuously observed for 7 days, and then were measured every other day of the following week. For the remaining experimental period, the measurements were conducted twice a week. For each measurement, five gas samples were collected using 50-mL volume plastic syringes after the chamber closure. Considering the low N_2O flux in the forest, the sampling intervals were 7 min in the cropland treatments (0, 7, 14, 21 and 28 min) and 15 min in the forest treatment (0, 15, 30, 45 and 60 min). The measurements were uniformly performed between 9:00 and 10:00 am local time to calculate a daily average N_2O flux. To minimize the enclosure effects of the chambers on plant growth and environmental conditions, they were immediately removed after gas sampling.

Immediately after sampling at each site, the collected gas samples were analyzed to obtain N_2O concentrations using a gas chromatograph (Agilent-7890A; Agilent Technologies, Palo Alto, CA, USA) rigged with an electron capture detector (ECD) at the research station. The soil N_2O fluxes were determined by the linear or nonlinear relationships between the gas concentration and the chamber closure time, as described in detail by Wang et al. [32]. Seasonal and annual cumulative N_2O fluxes were calculated by linear interpolation of the daily fluxes between the gas sampling dates [25]. Yield-scaled N_2O emissions (kg N Mg^{-1} grain) were calculated using annual cumulative N_2O emissions (kg N ha^{-1}) divided by the mean grain yield (Mg grain ha^{-1}) [10].

2.4. Crop Yield Measurements

For each cropland plot, three quadrats, $0.5 \times 0.5 \text{ m}^2$ for wheat and $1 \times 1 \text{ m}^2$ for maize, were randomly selected to measure crop yields. After the crop harvesting, the grains were collected separately and then oven dried at 70 °C for 48 h to constant weight to calculate the grain yield (Mg ha^{-1}).

2.5. Auxiliary Measurements of Soil Parameters

Throughout the experimental period, soil moisture, temperature, inorganic N (NO_3^- and NH_4^+), and dissolved organic C (DOC) concentrations were simultaneously measured for all the plots when gas samples were collected. The measurement procedures strictly followed the previous study of Zhou et al. [17]. For each plot, the topsoil moisture and temperature (5 cm in depth) were measured by a portable frequency domain reflector probe (RDS Technology Co. Ltd., Nanjing, Jiangsu, China) and a manual thermocouple thermometer (JM624, Tianjin Jinming Instrument Co. Ltd., Tianjin, China) with three replicates, respectively. Then the water-filled pore space (WFPS) was calculated based on the measured soil volumetric water content, bulk density and particle density (2.65 g cm^{-3}). At each plot, three soil cores (0–20 cm) were also randomly collected and completely mixed into one bulk sample. Then a 20 g fresh soil sample and 100 mL of 0.5 M K_2SO_4 were used to extract soil NH_4^+, NO_3^-, and DOC, and an AA3 continuous flow analyzer (Bran + Lubbe, Norderstedt, Germany) was employed to colorimetrically analyze the filtered extracts. During the entire experiment period, the daily rainfall and mean air temperature were automatically observed using a meteorological station.

After the maize season, for each plot, topsoil samples (0–20 cm) were also collected to measure soil properties (soil pH, total N content [TN], soil organic carbon content [SOC], soil bulk density [BD], soil particle composition), following soil agro-chemical analysis procedures [33]. In detail, soil pH was measured in a 1:2.5 (soil-to-water [w/v]) water suspension using a DMP-2 mV/pH detector (Quark Ltd., Nanjing, China). SOC content was determined by wet digestion with H_2SO_4–$K_2Cr_2O_7$, and TN content was determined by semi-micro Kjeldahl digestion using Se, $CuSO_4$ and K_2SO_4 as catalysts. Soil BD was determined by the volumetric ring method. The pipette method was used to determine soil texture. Furthermore, soil aggregates were measured according to the methods reported by Six et al. [34], in which soils were separated into four aggregate size classes (<0.053, 0.053–0.25, 0.25–2, and >2 mm) by wet sieving, and the aggregate stability was quantified by the mean weight diameter (MWD) [35].

2.6. Contribution Rates of Tillage and Fertilization to Increased Soil N_2O Emissions

To quantify the effects of tillage and fertilization on increasing soil N_2O emissions after land use conversion, the contribution rates of tillage and fertilization to increased soil N_2O emissions were calculated as follows.

$$CR_{\text{tillage}} = \left(\frac{NE_T - NE_0}{NE_{TF} - NE_0}\right) \times 100\% \tag{1}$$

where CR_{tillage} and $CR_{\text{fertilization}}$ are the relative contributions of tillage and fertilization to increased soil N_2O emissions (%), respectively. NE_0, NE_T, and NE_{TF} are the measured soil N_2O emissions from the baseline forestland (CK) and the tillage without fertilization (NC-T and LC-T) and with fertilization (NC-TF and LC-TF) treatments, respectively.

2.7. Statistical Analysis

The differences in soil N_2O fluxes and soil environmental factors (i.e., soil moisture and temperature, NO_3^-, NH_4^+ and DOC concentrations) between the different treatments were detected using one-way ANOVA analysis, followed by Duncan's range test ($p < 0.05$). The potential relationships between soil N_2O fluxes and environmental factors were evaluated using Pearson's correlation analysis. However, before the correlation analysis, the soil N_2O fluxes and environmental factors were primarily normalized by the ranked cases approach due to the original datasets not being normally distributed [17]. Moreover, multiple stepwise regression analysis was conducted to identify the key factors controlling soil N_2O emissions from croplands after land use conversion. All statistical analyses were performed using the SPSS 20.0 software (SPSS Inc., Chicago, IL, USA) and OriginPro 2015 software (OriginLab Corp., Northampton, MA, USA).

3. Results

3.1. Environmental Conditions and Soil Properties

During the whole experimental period, the total precipitation was 639.6 mm, and 67% occurred in the summer maize season (from June to October) (Figure 1). The daily mean air temperature changed from 4 to 31.2 °C with a mean of 16.9 °C.

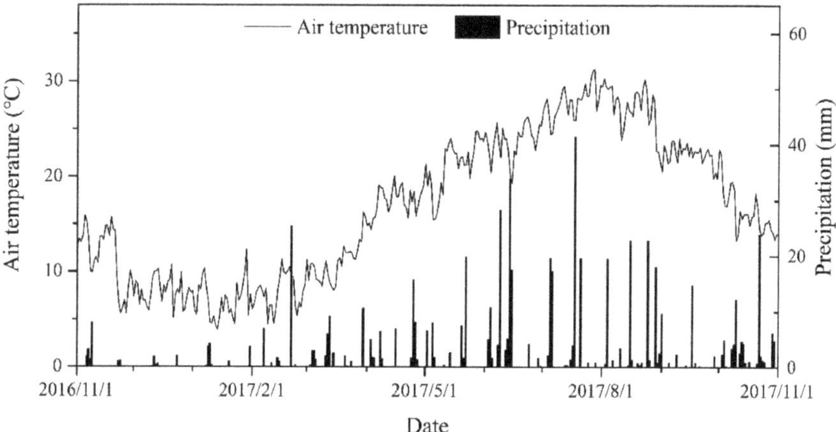

Figure 1. Air temperature and precipitation from November 2016 to October 2017.

Forestland conversion to cropland significantly influenced soil WFPS but not soil temperature at 5 cm depth (Figure 2). The WFPS values in the cropland sites (average 41.9%, 42.3%, 47.3% and 47.6%

for NC-T, NC-TF, LC-T and LC-TF, respectively) were significantly lower than that in the forestland (average 52.6%) ($p < 0.05$).

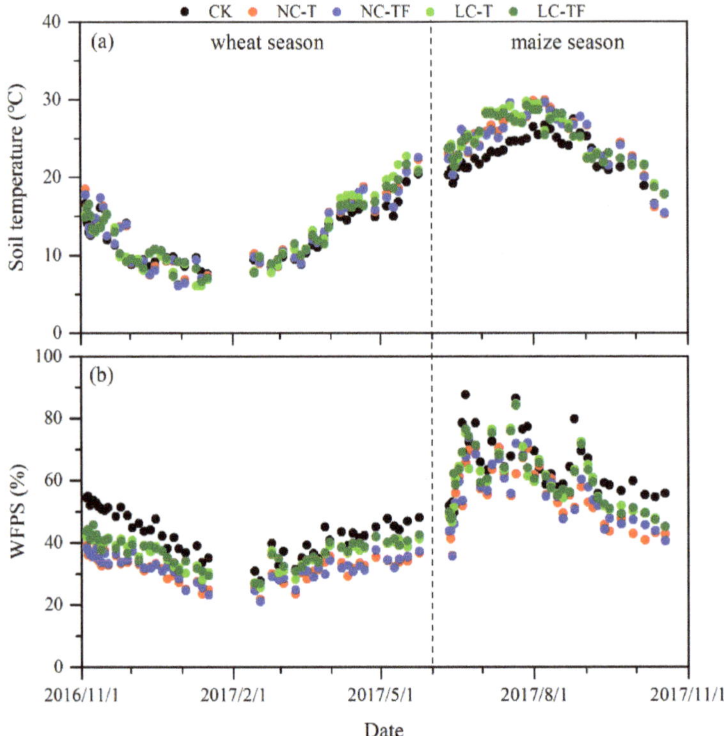

Figure 2. Temporal variations in soil temperature (**a**) and WFPS (**b**) at 5 cm depth. Abbreviations: WFPS, water-filled pore space; CK, control forestland; NC-TF and NC-T, newly converted cropland under tillage with and without fertilization, respectively; LC-TF and LC-T, long-term cropland under tillage with and without fertilization, respectively.

Compared to forestland, tillage in the croplands significantly decreased the average soil NH_4^+ concentration (average 1.79 and 1.52 mg N kg^{-1} for NC-T and LC-T, respectively) but significantly increased the average soil NO_3^- concentration (average 7.23 and 5.54 mg N kg^{-1} for NC-T and LC-T, respectively) ($p < 0.05$; Figure 3a,b). However, tillage with fertilization not only significantly increased the average soil NH_4^+ concentration (average 16.15 and 12.55 mg N kg^{-1} for NC-TF and LC-TF, respectively) but also significantly increased the average soil NO_3^- concentration (average 27.74 and 31.10 mg N kg^{-1} for NC-TF and LC-TF, respectively) compared to forestland ($p < 0.05$; Figure 3a,b). Following mineral N fertilizer application, the soil NH_4^+ concentration in the NC-TF and LC-TF treatments quickly reached peaks of 203.14 and 146.16 mg N kg^{-1} in the wheat season and 23.51 and 27.64 mg N kg^{-1} in the maize season (Figure 3a), while the soil NO_3^- concentration in the NC-TF and LC-TF treatments quickly reached peaks of 165.80 and 154.45 mg N kg^{-1} in the wheat season and 55.19 and 45.70 mg N kg^{-1} in the maize season (Figure 3b).

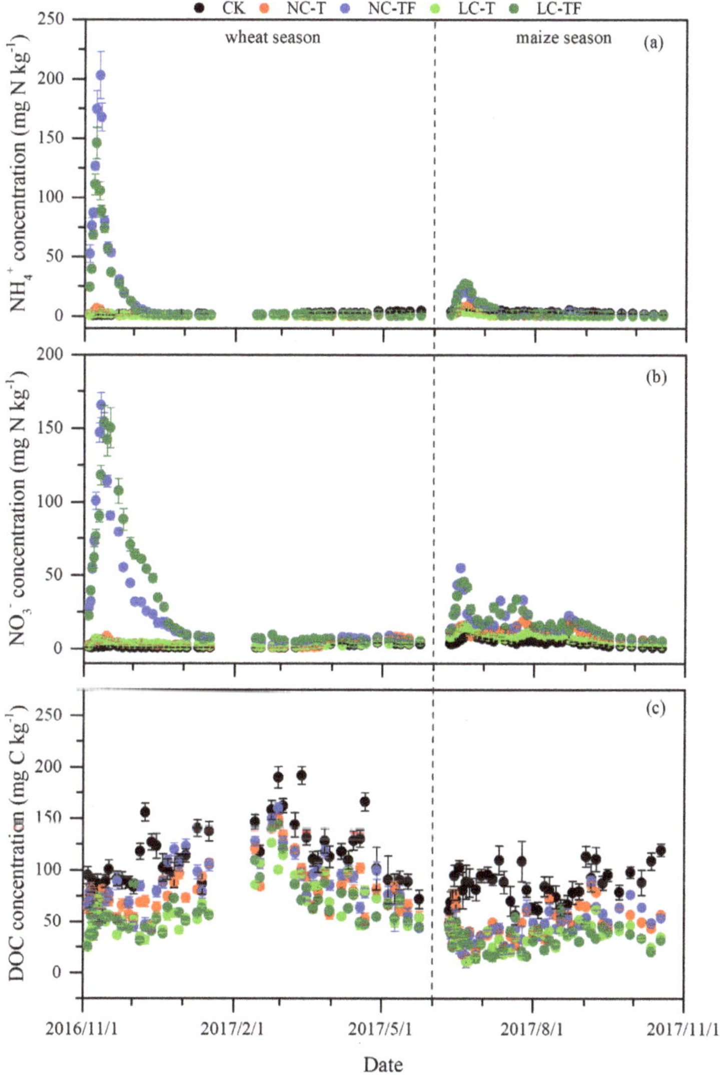

Figure 3. Temporal variations in soil NH_4^+ (**a**), NO_3^- (**b**) and DOC (**c**) concentrations. Vertical bars represent standard errors. Abbreviations: CK, control forestland; NC-TF and NC-T, newly converted cropland under tillage with and without fertilization, respectively; LC-TF and LC-T, long-term cropland under tillage with and without fertilization, respectively.

The soil DOC concentration significantly decreased after the conversion of forestland to cropland ($p < 0.05$, Figure 3c). In particular, long-term cultivation resulted in much lower soil DOC concentration (mean 51.30 and 50.54 mg C kg^{-1} for LC-T and LC-TF) than that in the short-term croplands (mean 66.37 and 68.30 mg C kg^{-1} for NC-T and NC-TF) ($p < 0.05$).

Compared to forestland, land use conversion significantly decreased soil SOC, TN, C/N ratio, and bulk density ($p < 0.05$, Table 1). Moreover, compared to the newly converted croplands, the long-term croplands had significantly lower SOC and TN contents and C/N ratio but a higher bulk density ($p < 0.05$, Table 1). Conversion did not induce significant changes in soil texture in the short term; however, long-term conversion increased the clay and silt contents and decreased the sand content ($p < 0.05$, Table 1). Furthermore, forestland conversion to cropland significantly decreased the proportion of soil macroaggregates (0.25–2 mm and 2–8 mm) and the mean weight diameter (MWD) ($p < 0.05$, Table 2).

Table 1. Topsoil properties (mean ± SE) determined after maize harvest.

Soil Properties	CK	NC-T	NC-TF	LC-T	LC-TF
pH	8.16 ± 0.04 [a]	8.13 ± 0.01 [a]	8.14 ± 0.02 [a]	8.11 ± 0.04 [a]	8.16 ± 0.02 [a]
SOC (g kg^{-1})	22.20 ± 0.42 [a]	14.83 ± 0.24 [b]	14.54 ± 0.37 [b]	7.85 ± 0.23 [c]	7.72 ± 0.27 [c]
TN (g kg^{-1})	1.62 ± 0.03 [a]	1.32 ± 0.03 [b]	1.29 ± 0.01 [b]	0.83 ± 0.02 [c]	0.79 ± 0.02 [c]
C/N ratio	13.70 ± 0.04 [a]	11.22 ± 0.22 [b]	11.30 ± 0.21 [b]	9.51 ± 0.37 [c]	9.72 ± 0.20 [c]
BD (g cm^{-3})	1.34 ± 0.01 [a]	1.16 ± 0.01 [c]	1.16 ± 0.01 [c]	1.24 ± 0.03 [b]	1.20 ± 0.01 [b]
Clay (%)	18.7 ± 0.3 [b]	19.4 ± 0.2 [b]	19.4 ± 0.2 [b]	20.9 ± 0.3 [a]	21.4 ± 0.2 [a]
Silt (%)	39.5 ± 0.5 [b]	40.4 ± 0.6 [b]	40.5 ± 0.8 [b]	42.0 ± 0.3 [a]	42.5 ± 0.3 [a]
Sand (%)	41.8 ± 0.2 [a]	40.2 ± 0.4 [a]	40.1 ± 0.8 [a]	37.1 ± 0.2 [b]	36.2 ± 0.2 [b]

BD, bulk density; CK, control forestland; NC-TF and NC-T, newly converted cropland under tillage with and without fertilization, respectively; LC-TF and LC-T, long-term cropland under tillage with and without fertilization, respectively. [a,b,c] A different letter in the same row indicates a significant difference among different treatments ($p < 0.05$).

Table 2. Aggregate size distribution and mean weight diameter (MWD) (mean ± SE) determined after maize harvest.

Treatment	Aggregate Size Distribution (%)				MWD (mm)
	2–8mm	0.25–2 mm	0.053–0.25 mm	<0.053 mm	
CK	52.77 ± 0.43 [a]	36.39 ± 0.16 [b]	6.83 ± 0.32 [c]	4.01 ± 0.29 [c]	3.06 ± 0.02 [a]
NC-T	27.93 ± 0.36 [b]	42.44 ± 0.59 [a]	16.50 ± 0.53 [b]	13.13 ± 0.27 [b]	1.90 ± 0.01 [b]
NC-TF	26.78 ± 0.41 [b]	42.78 ± 0.34 [a]	16.57 ± 0.78 [b]	13.87 ± 0.40 [b]	1.85 ± 0.02 [b]
LC-T	6.26 ± 0.14 [c]	14.59 ± 0.16 [c]	34.49 ± 0.30 [a]	44.65 ± 0.21 [a]	0.54 ± 0.01 [c]
LC-TF	7.09 ± 0.09 [c]	17.09 ± 0.29 [c]	33.86 ± 0.11 [a]	41.96 ± 0.24 [a]	0.61 ± 0.01 [c]

CK, control forestland; NC-TF and NC-T, newly converted cropland under tillage with and without fertilization, respectively; LC-TF and LC-T, long-term cropland under tillage with and without fertilization, respectively. [a,b,c] A different letter in the same column indicates a significant difference among different treatments ($p < 0.05$).

3.2. Soil N_2O Emissions

Distinct temporal variations in N_2O fluxes were observed in both forestland and cropland (Figure 4a). During the summer season, the N_2O fluxes were higher than those during the other seasons. Compared to forestland, the N_2O fluxes showed much greater temporal variations in the croplands. In the initial period of the wheat and maize season, tillage and fertilization practices induced pulse emissions of N_2O that lasted several weeks and then decreased to base levels. For the croplands with only tillage, the peak N_2O fluxes were 8.72 and 8.40 µg N m^{-2} h^{-1} in the winter wheat season and 43.58 and 33.10 µg N m^{-2} h^{-1} in the summer maize season for the NC-T and LC-T treatments, respectively. For the croplands with tillage and fertilization, the peak N_2O fluxes were 21.56 and 56.99 µg N m^{-2} h^{-1} in the winter wheat season and 185.61 and 152.00 µg N m^{-2} h^{-1} in the summer maize season for the NC-TF and LC-TF treatments, respectively. This result indicates that fertilization could induce much higher N_2O pulse emissions than tillage after land use conversion from forestland to cropland.

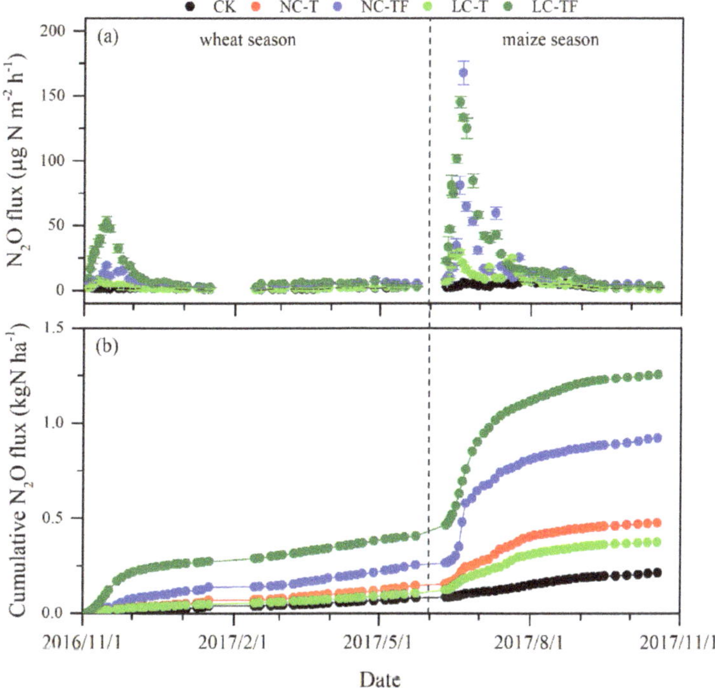

Figure 4. Temporal variations in soil N_2O emissions (**a**) and cumulative N_2O fluxes (**b**). Vertical bars represent standard errors. Abbreviations: CK, control forestland; NC-TF and NC-T, newly converted cropland under tillage with and without fertilization, respectively; LC-TF and LC-T, long-term cropland under tillage with and without fertilization, respectively.

The N_2O fluxes from forestland changed from 0.80 to 7.70 µg N m^{-2} h^{-1} over the whole experimental period, with a mean of 2.70 µg N m^{-2} h^{-1} (Figure 4a). For the croplands, the mean N_2O fluxes were 6.49 and 12.44 µg N m^{-2} h^{-1} for NC-T and NC-TF, and 5.76 and 21.59 µg N m^{-2} h^{-1} for LC-T and LC-TF, respectively. Forestland conversion to cropland significantly increased the mean N_2O fluxes ($p < 0.05$). Moreover, the average N_2O fluxes were significantly greater from tillage with fertilization treatments than those from tillage without fertilization treatments ($p < 0.05$). This result shows again that tillage with fertilization had a much greater effect on increasing N_2O emissions than tillage alone after land use conversion.

The annual cumulative N_2O emissions significantly increased after forestland conversion to cropland ($p < 0.05$, Table 3 and Figure 4b). Compared to forestland, the annual cumulative N_2O emissions increased by 124% and 334% in the short-term croplands (NC-T and NC-TF) and by 76% and 491% in the long-term croplands (LC-T and LC-TF), respectively. Moreover, tillage with fertilization (NC-TF and LC-TF) significantly increased cumulative soil N_2O emissions by 94% and 235%, compared to those from tillage without fertilization (NC-T and LC-T), respectively ($p < 0.05$, Table 3). Compared to the short-term conversion (NC-T and NC-TF), long-term tillage (LC-T) significantly decreased cumulative soil N_2O emissions by 21%, while long-term tillage with fertilization (LC-TF) greatly increased cumulative soil N_2O emissions by 36% ($p < 0.05$, Table 3).

Table 3. Cumulative N_2O emissions, grain yield, and yield-scaled N2O emissions (mean ± SE).

Treatment	Cumulative N_2O Emissions (kg N ha^{-1})			Grain Yield (Mg ha^{-1})	Yield-Scaled N_2O Emission (kg N Mg^{-1} Grain)
	Wheat Season	**Maize Season**	**Whole Year**		
CK	0.08 ± 0.001 [d]	0.13 ± 0.003 [e]	0.21 ± 0.004 [e]		
NC-T	0.15 ± 0.001 [c]	0.33 ± 0.003 [c]	0.48 ± 0.004 [c]	1.45 ± 0.08 [d]	0.33 ± 0.02 [a]
NC-TF	0.25 ± 0.013 [b]	0.67 ± 0.005 [b]	0.92 ± 0.010 [b]	4.09 ± 0.32 [b]	0.23 ± 0.01 [b]
LC-T	0.10 ± 0.001 [cd]	0.27 ± 0.003 [d]	0.38 ± 0.004 [d]	2.40 ± 0.17 [c]	0.16 ± 0.01 [d]
LC-TF	0.49 ± 0.007 [a]	0.85 ± 0.006 [a]	1.26 ± 0.007 [a]	6.72 ± 0.44 [a]	0.19 ± 0.01 [c]

CK, control forestland; NC-TF and NC-T, newly converted cropland under tillage with and without fertilization, respectively; LC-TF and LC-T, long-term cropland under tillage with and without fertilization, respectively. [a, b, c d, e] A different letter in the same column indicates a significant difference among different treatments ($p < 0.05$).

Yield-scaled N_2O emissions from the short-term converted croplands (NC-T and NC-TF) were significantly higher than those from the long-term converted croplands (LC-T and LC-TF) ($p < 0.05$, Table 3). After long-term plantation, the yield-scaled N_2O emissions under tillage with fertilization treatment (LC-TF) were significantly greater than that from only tillage practice (LC-T) ($p < 0.05$, Table 3).

3.3. Relationships between N_2O Fluxes and Soil Environmental Variables

The correlations between N_2O fluxes and soil environmental variables are presented in Figure 5. The soil N_2O fluxes were significantly positively related to soil temperature, WFPS, NH_4^+ and NO_3^- concentrations but significantly negatively related to DOC concentration for both forestland and cropland ($p < 0.05$). The further stepwise regression analysis indicated that variations in N_2O fluxes from forestland were mainly regulated by soil WFPS and temperature (82%) (Table 4). However, after land use conversion, soil NO_3^- and NH_4^+ availability and soil WFPS were the main factors influencing soil N_2O emissions from croplands with only tillage, which explained 78% and 90% of the variations in N_2O fluxes from the NC-T and LC-T treatments, respectively (Table 4). For the short-term tillage with fertilization treatment (NC-TF), soil DOC and NH_4^+ availability and soil WFPS explained 74% of the variation in N_2O fluxes (Table 4). While for the long-term tillage with fertilization treatment (LC-TF), N_2O emissions were mainly regulated by soil NH_4^+ and NO_3^- availability and WFPS, which explained 81% of the variation in N_2O fluxes (Table 4).

Table 4. Stepwise multiple linear regressions between the soil N_2O emissions and environmental factors.

Treatment	Parameter	Coefficient	*p*-Value	Adjust R^2	*p*-Value
	Intercept	0.00003	0.999		
CK	WFPS	0.49	<0.001		
	ST	0.47	<0.001	0.82	<0.001
	Intercept	0.0001	1.000		
NC-T	NO_3^-	0.38	<0.001		
	NH_4^+	0.45	<0.001		
	WFPS	0.27	<0.005	0.79	<0.001
	Intercept	−0.000004	1.000		
NC-TF	DOC	−0.38	<0.001		
	NH_4^+	0.44	<0.001		
	WFPS	0.28	<0.001	0.74	<0.001
	Intercept	−0.00005	0.999		
LC-T	NO_3^-	0.35	<0.01		
	NH_4^+	0.36	<0.001		
	WFPS	0.42	<0.001	0.90	<0.001
	Intercept	−0.0004	0.994		
LC-TF	NH_4^+	0.63	<0.001		
	WFPS	0.4	<0.001		
	NO_3^-	0.16	<0.001	0.79	<0.001

ST, soil temperature; CK, control forestland; NC-TF and NC-T, newly converted cropland under tillage with and without fertilization, respectively; LC-TF and LC-T, long-term cropland under tillage with and without fertilization, respectively. Due to the original data not normally distributed, all the data sets were normalized before analysis.

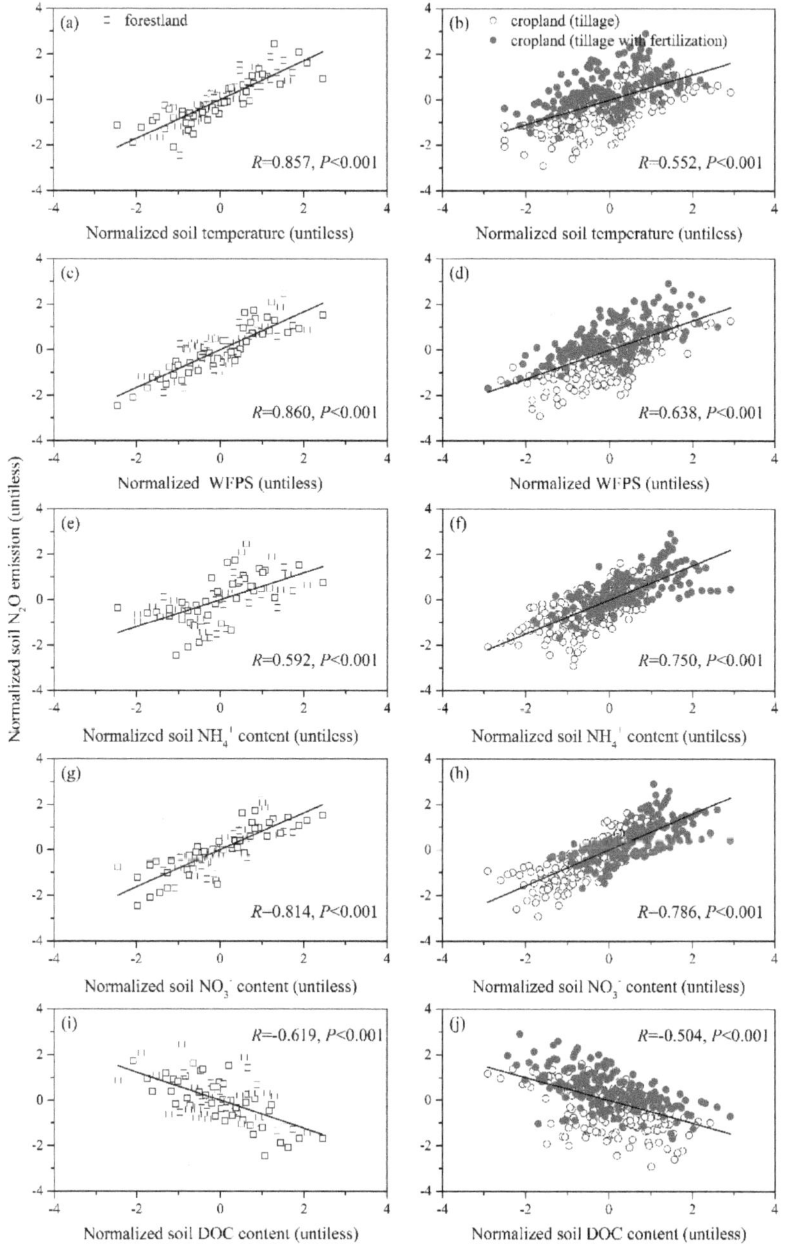

Figure 5. The correlations between soil N_2O emissions and soil temperature (**a,b**), soil WFPS (**c,d**), soil NH_4^+ (**e,f**), NO_3^- (**g,h**) and DOC (**i,j**) concentrations for forestland (n = 87) and cropland (n = 348).

4. Discussion

4.1. Effects of Land Use Conversion on Soil N_2O Emissions and Yield-Scaled N_2O Emissions

In this study, the annual cumulative N_2O flux from the forest was 0.21 kg N ha^{-1} yr^{-1}, which was significantly lower than the average annual N_2O flux of the global forest (1.429 kg N ha^{-1} yr^{-1}) [36]. Land use conversion from forestland to cropland significantly increased the annual cumulative N_2O emissions by 76–491% in this study (Table 3). van Lent et al. reviewed the literature and reported that land use conversion from forest to cropland greatly increased the annual cumulative N_2O emissions by 330% on average in the tropics and subtropics [13]. These results confirmed that land use change has a profound impact on soil N_2O fluxes [9,11–14], specifically conversion of forestland to cropland. Some studies have proved that tillage and fertilization are the most important management practices increasing soil N_2O emissions after land use conversion from forestland to cropland [11,19]. Tillage practices could significantly increase soil aeration conditions. On the one hand, good soil aeration condition enhanced soil organic matter mineralization [34], and subsequent nitrification [3], inducing higher N_2O production. On the other hand, good soil aeration condition accelerated the gas exchange between soil and atmosphere, which could promote the diffusivity of N_2O from soil into atmosphere [3,14]. Application of mineral N fertilizer could directly increase soil inorganic N concentrations (Figure 3), providing substrates for the two main N_2O production processes of nitrification and denitrification [8,9,28]. In this study, the N_2O flux from the croplands all showed pulse emissions following tillage and N fertilization events (Figure 4a), thus resulted in much greater cumulative N_2O emissions compared to the forestland (Table 3).

Previous studies regarding the effects of forestland conversion to cropland on soil N_2O emissions mainly focused on long-term cultivation croplands [12], but few studies have measured N_2O emissions from recently transitioned croplands. Our study indicated that forestland conversion to cropland in the first year significantly increased annual cumulative N_2O emissions by 124% and 334% (Table 3). Comparably, after two years of conversion of tropical forest to agriculture in French Guiana, the annual cumulative N_2O emissions from fertilized croplands increased by 90% [12]. Obviously, the increased extent of N_2O emissions in our study was higher than that in the study by Petitjean et al. [12]. This difference might be attributed to the differences in climate, soil properties and amount of N fertilizer [26]. It is remarkable that the high N_2O emissions in the initial stage after land use conversion should not be ignored.

In agricultural systems, yield-scaled N_2O emissions were widely used as a metric of the important global challenge for guaranteeing food security and reducing N_2O emissions [37,38]. In our study, the yield-scaled N_2O emissions were 0.16 and 0.19 kg N Mg^{-1} grain in croplands after long-term cultivation (Table 3), which were similar to the results reported by Bayer et al. [20] and Tang et al. [39] in other subtropical regions. However, the yield-scaled N_2O emissions in the new croplands were 21% and 106% higher than that in the long-term converted croplands (Table 3). These results further highlighted the importance of monitoring N_2O emissions at the initial stage after land use conversion [3,11]. In the newly converted croplands, the lower crop yields mean a lower uptake of available N by plant, leaving more available N for nitrification and denitrification processes, which promoted the production of N_2O. After long-term cultivation, the capacity of plant N uptake was enhanced, inducing a relative lower ratio of available N for N_2O production [10,40]. Furthermore, after long-term cultivation in the present study, the increase extent of crop yield was 64% and 66% compared to the new croplands, which were significantly higher than the increase extent of N_2O emissions (−21% and 37%), inducing significant decreases of the yield-scaled N_2O emissions. Tillage with fertilization practices resulted in yield-scaled N_2O emissions 19% higher than only tillage practice in the long-term converted croplands (Table 3). It is obvious that fertilization has induced a much higher increase extent of N_2O emissions than that of crop yield. Previous studies showed that the N fertilizer application rate was one of the key factors influencing soil N_2O production [26]. Meanwhile, N fertilizer application rate also had considerable impacts on crop yield [41]. Therefore, it is very important to determine a reasonable rate of N fertilizer application in the Sichuan Basin after forestland conversion to cropland, which can simultaneously achieve high yield and mitigate soil N_2O emission. In addition, the N fertilizer type (e.g., mineral, organic, or mix of mineral and organic), application time and proportion (e.g., all as base fertilizer, or part as base and other as topdressing), as well as application depth (surface or deep) would also influence N fertilizer efficiency and N_2O emission [41].

Overall, land use conversion from forestland to cropland and subsequent tillage and fertilization practices significantly increased the cumulative soil N_2O emissions. In particular, the yield-scaled N_2O emissions were significantly higher in the newly converted croplands than in the long-term converted croplands. These results further indicate that the effect of land use conversion on soil N_2O emissions should not be ignored in the initial years after conversion. Therefore, we strongly recommend limiting the conversion of forestland to cropland as much as possible.

4.2. Factors Regulating the Increased Soil N_2O Emissions Induced by Tillage and Fertilization

In this study, tillage in the croplands increased the cumulative N_2O emissions by 124% and 76% compared to those from forestland in the short-term and long-term after conversion, respectively (Table 3). This result indicates that tillage could induce significant increases in soil N_2O emissions after land use conversion. The statistical analyses showed that tillage significantly decreased the soil bulk density (Table 1), and thus increased soil porosity, while also decreasing the proportion of soil water-stable macroaggregates and MWD by 49% and 61% on average compared to those in forestland (Table 2). The breakage of soil macroaggregates releases more soil organic matter and increases N availability [18,21,34]. This phenomenon was observed through the decrease in the average SOC and TN contents by 49% and 34%, respectively, in croplands under only tillage compared to the contents in forestland (Table 1). Tillage physically disturbs the soil structure and increases soil aeration [34,42], and therefore enhances soil organic N mineralization and microbial activities [3,11,19]. In this study, cropland tillage induced small NH_4^+ peaks in the initial period of each cropping season that lasted for approximately three weeks (Figure 3a), thus likely increasing the NH_4^+ to NO_3^- transformation rate. This possibility was further supported by the higher NO_3^- concentrations in the tillage treatments compared to those in the forestland (Figure 3b). The increase in soil NO_3^- would further enhance the denitrification rate [17,23]. Consequently, tillage in the croplands increased soil aeration and promoted soil organic N mineralization, thus inducing N_2O pulse emissions that did not occur in the forestland and accounted for approximately 52% of the annual cumulative N_2O flux during the experimental period (Figure 4).

Tillage with fertilization significantly increased the cumulative soil N_2O emissions by 94% and 235% compared to tillage without fertilization in this study (Table 3). In another subtropical region of China, a previous study found that soil N_2O emissions were 405% higher on average in a conventional fertilizer treatment than in a tillage alone treatment after land use conversion [11]. In the current study, tillage with fertilization induced much higher N_2O pulse emissions, mainly due to the rapid increases in soil NH_4^+ and NO_3^- concentrations after mineral N fertilizer application to the croplands (Figure 3a,b). The high NH_4^+ and NO_3^- concentrations following fertilization directly provide sufficient substrates for nitrification and denitrification and thus significantly stimulate soil N_2O emissions [8,11,20,23]. This phenomenon has been reported in many previous studies on different croplands [5,10,43]. On average, pulse N_2O emissions following fertilization accounted for 67% of the annual cumulative N_2O fluxes (Figure 4). Previous study reported that N_2O emission pulses following N fertilization contributed to approximately 70% of the annual cumulative N_2O fluxes from croplands in the same study area [17].

In the newly converted cropland, 37% and 63% of the increased soil N_2O emissions could be attributed to tillage and fertilization, respectively, while in the long-term cropland, the corresponding rates were 16% and 84%. This result indicates that fertilization had a much greater effect on increasing soil N_2O emissions than tillage after forestland conversion to cropland in the Sichuan Basin, and the effect tended to be larger several years after conversion. Long-term repeated tillage significantly reduced soil structural stability and decreased organic matter (Tables 1 and 2), resulting in a lower rate of soil organic N mineralization [18–20,22]. Long-term mineral N fertilization stimulated the activity and abundance of soil ammonia-oxidizing bacteria in the study area [8], which could promote nitrification and subsequent denitrification [23].

The stepwise regression analysis results indicated that soil WFPS and temperature could explain 82% of the variations in N_2O fluxes from forestland (Table 4). Soil temperature and moisture are important factors that influence soil N_2O emissions by affecting microbial activities [8,23,44]. The low soil moisture and temperature during the winter wheat season (Figure 2) generally inhibited microbial activities, resulting in low N_2O emissions [3,11]. However, the stepwise regression analysis showed that soil N_2O emissions from the croplands were mainly influenced by soil NO_3^- and NH_4^+ availability and WFPS, which explained 78-90% of the variations in N_2O fluxes (Table 4). This result implies that after the conversion of forestland to cropland, the primary factors regulating soil N_2O emissions were soil NO_3^- and NH_4^+ availability. In the present study, we found the N_2O emissions significantly correlated to soil NH_4^+ and NO_3^- concentrations in the croplands (Figure 5f,h). The mean soil NH_4^+ and NO_3^- concentrations in the tillage with fertilization treatments were 7.7 and 3.7 times greater, on average, than those in the tillage without fertilization treatments. Therefore, after the conversion of forestland to cropland, fertilization only increased soil N_2O emissions to a greater degree than tillage, which was mainly attributed to the higher soil inorganic N levels caused by fertilization.

Additionally, the soil DOC concentration significantly decreased after the conversion of forestland to cropland (Figure 3c). The lower level of soil DOC with higher N_2O fluxes in the croplands compared to the forestland indicated the negative relationship between soil DOC concentration and N_2O fluxes induced by land use conversion (Figure 5j). Several studies have shown that soil DOC is an important factor influencing N_2O emissions [9,11,17,44]. The substantial soil DOC availability in the forestland may favor complete denitrification and thus decrease N_2O fluxes [17]. Following land use conversion, the decrease in soil DOC concentration could enhance incomplete denitrification, thereby increasing N_2O fluxes [44]. In the short-term after conversion, the combination of soil DOC and NH_4^+ availability and soil WFPS explained 74% of the variation in N_2O fluxes from the tillage and fertilization treatment (Table 4).

Applying crop residues with a high C/N ratio (such as maize) combined with synthetic N fertilizer could be an optimal strategy for mitigating N_2O emissions in the study area [28,29]. Moreover, in recent years, researchers have developed many new technological approaches to maintain the sustainable development of the agricultural ecosystem. For example, a new Biogeosystem Technique methodology reported by Kalinitchenko et al. [45,46] could improve soil aggregate structure, promote soil organic matter synthesis and reservation, and thus likely reduce greenhouse gas production. Therefore, after the conversion of forestland to cropland in the Sichuan Basin, to guarantee the sustainable food supply and ecosystem development, we also recommend application of mineral fertilizer and crop residues (such as maize), or the adoption of advanced management technologies as potential measures for conserving N in destroyed forest and mitigating N_2O emissions in the transformed croplands.

5. Conclusions

This study found that the annual cumulative N_2O flux was 0.21 kg N ha^{-1} yr^{-1} in the forestland, which significantly increased by 76–491% after land use conversion of forestland to cropland in the Sichuan Basin. In the short-term and long-term croplands, fertilization contributed 63% and 84%, while tillage contributed 37% and 16% to the increased soil N_2O emissions, respectively. Fertilization exhibited a much greater effect on increasing soil N_2O emissions than tillage after the conversion of forestland to cropland, and the effect tended to be stronger several years after conversion. After forestland conversion to cropland, the soil N_2O emissions were mainly regulated by soil NO_3^- and NH_4^+ availability. The direct land use conversion without any scientific management practices significantly influenced soil properties and thus stimulated N_2O emissions. Tillage disturbed soil structure, decreased bulk density and increased soil aeration in the croplands, thus enhancing soil organic N mineralization, while the application of mineral N fertilizer directly led to rapid increases in soil NH_4^+ and NO_3^- concentrations. Tillage and fertilization induced increases in the inorganic N concentration to different extents, which thus resulted in different magnitudes of N_2O fluxes. This study primarily suggests limiting the conversion of forestland to cropland in the Sichuan Basin as much as possible. Moreover, we recommend intensifying the shift from grain to green on the premise of ensuring the supply of grain production, and also recommend the adoption of technological approaches to mitigate N_2O emissions in both the destroyed forest and transformed croplands for the sustainable development of the ecosystem.

Author Contributions: Conceptualization, X.R. and B.Z.; data curation, X.R.; methodology, X.R., H.B., S.T.R., and B.Z.; writing—original draft, X.R.; writing—review and editing, X.R. and B.Z. All authors have read and agreed to the published version of the manuscript.

Funding: This research was funded by the National Key Research and Development Program of China (Grant No. 2017YFD0200105).

Acknowledgments: The members of the Yanting Agro-Ecological Station of Purple Soil of the Chinese Academy of Sciences are kindly thanked for providing weather data and technical help. The authors greatly appreciated the constructive comments from the reviewers, which greatly helped improve this manuscript.

Conflicts of Interest: The authors declare no conflict of interest.

References

1. IPCC. Contribution of Working Group I to the Fifth Assessment Report of the Intergovernmental Panel on Climate Change. In *Climate Change 2013: The Physical Science Basis*; Stocker, T.F., Qin, D.H., Plattner, G.K., Tignor, M., Allen, S.K., Boschung, J., Nauels, A., Xia, Y., Bex, V., Midgley, P.M., Eds.; Cambridge University Press: Cambridge, UK, 2013; p. 1535.
2. Tian, H.; Lu, C.; Ciais, P.; Michalak, A.M.; Canadell, J.G.; Saikawa, E.; Huntzinger, D.N.; Gurney, K.R.; Sitch, S.; Zhang, B.; et al. The terrestrial biosphere as a net source of greenhouse gases to the atmosphere. *Nature* **2016**, *531*, 225–228. [CrossRef] [PubMed]
3. Wu, L.; Tang, S.; He, D.; Wu, X.; Shaaban, M.; Wang, M.; Zhao, J.; Khan, I.; Zheng, X.; Hu, R.; et al. Conversion from rice to vegetable production increases N_2O emission via increased soil organic matter mineralization. *Sci. Total Environ.* **2017**, *583*, 190–201. [CrossRef] [PubMed]
4. FAO. An FAO Perspective. In *World Agriculture: Towards 2015/2030*; FAO: Rome, Italy, 2003.
5. Zhou, M.; Zhu, B.; Wang, S.; Zhu, X.; Vereecken, H.; Brüggemann, N. Stimulation of N_2O emission by manure application to agricultural soils may largely offset carbon benefits: A global meta-analysis. *Glob. Chang. Biol.* **2017**, *23*, 4068–4083. [CrossRef] [PubMed]
6. Bateman, E.J.; Baggs, E.M. Contributions of nitrification and denitrification to N_2O emissions from soils at different water-filled pore space. *Biol. Fert. Soil.* **2005**, *41*, 379–388. [CrossRef]
7. Müller, C.; Laughlin, R.J.; Spott, O.; Rütting, T. Quantification of N_2O emission pathways via a ^{15}N tracing model. *Soil Biol. Biochem.* **2014**, *72*, 44–54. [CrossRef]
8. Dong, Z.; Zhu, B.; Hua, K.; Jiang, Y. Linkage of N_2O emissions to the abundance of soil ammonia oxidizers and denitrifiers in purple soil under long-term fertilization. *Soil Sci. Plant. Nutr.* **2015**, *61*, 799–807. [CrossRef]
9. Wu, X.; Liu, H.; Fu, B.; Wang, Q.; Xu, M.; Wang, H.; Yang, F.; Liu, G. Effects of land-use change and fertilization on N_2O and NO fluxes, the abundance of nitrifying and denitrifying microbial communities in a hilly red soil region of southern China. *Appl. Soil Ecol.* **2017**, *120*, 111–120. [CrossRef]
10. Campanha, M.M.; de Oliveira, A.D.; Marriel, I.E.; Neto, M.M.G.; Malaquias, J.V.; Landau, E.C.; de Albuquerque Filho, M.R.; Ribeiro, F.P.; de Carvalho, A.M. Effect of soil tillage and N fertilization on N_2O mitigation in maize in the Brazilian Cerrado. *Sci. Total Environ.* **2019**, *692*, 1165–1174. [CrossRef]
11. Chen, D.; Li, Y.; Wang, C.; Liu, X.; Wang, Y.; Shen, J.; Qin, J.; Wu, J. Dynamics and underlying mechanisms of N_2O and NO emissions in response to a transient land-use conversion of Masson pine forest to tea field. *Sci. Total Environ.* **2019**, *693*, 133549. [CrossRef]
12. Petitjean, C.; Hénault, C.; Perrin, A.S.; Pontet, C.; Metay, A.; Bernoux, M.; Jehanno, T.; Viard, A.; Roggy, J.C. Soil N_2O emissions in French Guiana after the conversion of tropical forest to agriculture with the chop-and-mulch method. *Agr. Ecosyst. Environ.* **2015**, *208*, 64–74. [CrossRef]

13. van Lent, J.; Hergoualc'h, K.; Verchot, L.V. Reviews and syntheses: Soil N_2O and NO emissions from land use and land-use change in the tropics and subtropics: A meta-analysis. *Biogeosciences* **2015**, *12*, 7299–7313. [CrossRef]

14. Liu, H.; Liu, G.; Li, Y.; Wu, X.; Liu, D.; Dai, X.; Xu, M.; Yang, F. Effects of land use conversion and fertilization on CH_4 and N_2O fluxes from typical hilly red soil. Environ. *Sci. Pollut. Res.* **2016**, *23*, 20269–20280. [CrossRef] [PubMed]

15. Zhou, Z.; Liu, Y.; Zhu, Q.; Lai, X.; Liao, K. Comparing the variations and controlling factors of soil N_2O emissions and $NO_3{}^-$-N leaching on tea and bamboo hillslopes. *Catena* **2020**, *188*, 104463. [CrossRef]

16. Sosulski, T.; Szara, E.; Szymańska, M.; Stępień, W.; Rutkowska, B.; Szulc, W. Soil N_2O emissions under conventional tillage conditions and from forest soil. *Soil Till. Res.* **2019**, *190*, 86–91. [CrossRef]

17. Zhou, M.; Wang, X.; Ke, Y.; Zhu, B. Effects of afforestation on soil nitrous oxide emissions in a subtropical montane agricultural landscape: A 3-year field experiment. *Agric. For. Meteorol.* **2019**, *266*, 221–230. [CrossRef]

18. Grandy, A.S.; Robertson, G.P. Initial cultivation of a temperate-region soil immediately accelerates aggregate turnover and CO_2 and N_2O fluxes. *Glob. Chang. Biol.* **2006**, *12*, 1507–1520. [CrossRef]

19. Plaza-Bonilla, D.; Álvaro-Fuentes, J.; Arrúe, J.L.; Cantero-Martínez, C. Tillage and nitrogen fertilization effects on nitrous oxide yield-scaled emissions in a rainfed Mediterranean area. *Agric. Ecosyst. Environ.* **2014**, *189*, 43–52. [CrossRef]

20. Bayer, C.; Gomes, J.; Zanatta, J.A.; Vieira, F.C.B.; de Cássia Piccolo, M.; Dieckow, J.; Six, J. Soil nitrous oxide emissions as affected by long-term tillage, cropping systems and nitrogen fertilization in Southern Brazil. *Soil Till. Res.* **2015**, *146*, 213–222. [CrossRef]

21. Ayoubi, S.; Karchegani, P.M.; Mosaddeghi, M.R.; Honarjoo, N. Soil aggregation and organic carbon as affected by topography and land use change in western Iran. *Soil Till. Res.* **2012**, *121*, 18–26. [CrossRef]

22. Badagliacca, G.; Benítez, E.; Amato, G.; Badalucco, L.; Giambalvo, D.; Laudicina, V.A.; Ruisi, P. Long-term effects of contrasting tillage on soil organic carbon, nitrous oxide and ammonia emissions in a Mediterranean Vertisol under different crop sequences. *Sci. Total Environ.* **2018**, *619*, 18–27. [CrossRef]

23. Dong, Z.; Zhu, B.; Jiang, Y.; Tang, J.; Liu, W.; Hu, L. Seasonal N_2O emissions respond differently to environmental and microbial factors after fertilization in wheat-maize agroecosystem. *Nutr. Cycl. Agroecosys.* **2018**, *112*, 215–229. [CrossRef]

24. Zhou, M.; Zhu, B.; Butterbach-Bahl, K.; Zheng, X.; Wang, T.; Wang, Y. Nitrous oxide emissions and nitrate leaching from a rain-fed wheat-maize rotation in the Sichuan Basin, China. *Plant. Soil* **2013**, *362*, 149–159. [CrossRef]

25. Zhou, M.; Wang, X.; Wang, Y.; Zhu, B. A three-year experiment of annual methane and nitrous oxide emissions from the subtropical permanently flooded rice paddy fields of China: Emission factor, temperature sensitivity and fertilizer nitrogen effect. *Agric. For. Meteorol.* **2018**, *250*, 299–307. [CrossRef]

26. Bouwman, A.F.; Boumans, L.J.M.; Batjes, N.H. Emissions of N_2O and NO from fertilized fields: Summary of available measurement data. *Glob. Biogeochem. Cy.* **2002**, *16*, 1058–1071. [CrossRef]

27. Zhu, B.; Wang, T.; Kuang, F.; Luo, Z.; Tang, J.; Xu, T. Measurements of nitrate leaching from a hillslope cropland in the central Sichuan basin, China. *Soil Sci. Soc. Am. J.* **2009**, *73*, 1419–1426. [CrossRef]

28. Dong, Z.; Zhu, B.; Zeng, Z. The influence of N-fertilization regimes on N_2O emissions and denitrification in rain-fed cropland during the rainy season. *Environ. Sci. Proc. Imp.* **2014**, *16*, 2545–2553. [CrossRef]

29. Zhou, M.; Zhu, B.; Brüggemann, N.; Bergmann, J.; Wang, Y.; Butterbach-Bahl, K. N_2O and CH_4 emissions, and $NO_3{}^-$ leaching on a crop-yield basis from a subtropical rain-fed wheat-maize rotation in response to different types of nitrogen fertilizer. *Ecosystems* **2014**, *17*, 286–301. [CrossRef]

30. Wang, X.; Zhou, M.; Li, T.; Ke, Y.; Zhu, B. Land use change effects on ecosystem carbon budget in the Sichuan Basin of Southwest China: Conversion of cropland to forest ecosystem. *Sci. Total Environ.* **2017**, *609*, 556–562. [CrossRef]

31. Zheng, X.; Mei, B.; Wang, Y.; Xie, B.; Wang, Y.; Dong, H.; Xu, H.; Chen, G.; Cai, Z.; Yue, J.; et al. Quantification of N_2O fluxes from soil–plant systems may be biased by the applied gas chromatograph methodology. *Plant. Soil* **2008**, *311*, 211–234. [CrossRef]

32. Wang, K.; Zheng, X.; Pihlatie, M.; Vesala, T.; Liu, C.; Haapanala, S.; Mammarella, I.; Rannik, Ü.; Liu, H. Comparison between static chamber and tunable diode laser-based eddy covariance techniques for measuring nitrous oxide fluxes from a cotton field. *Agric. For. Meteorol.* **2013**, *171*, 9–19. [CrossRef]

33. Lu, R. *Soil Agro-Chemical Analyses*; Agricultural Technical Press of China: Beijing, China, 2000. (In Chinese)

34. Six, J.; Paustian, K.; Elliott, E.T.; Combrink, C. Soil structure and organic matter I. Distribution of aggregate-size classes and aggregate-associated carbon. *Soil Sci. Soc. Am. J.* **2000**, *64*, 681–689.

35. Whalen, J.K.; Hu, Q.; Liu, A. Compost applications increase water-stable aggregates in conventional and no-tillage systems. *Soil Sci. Soc. Am. J.* **2003**, *67*, 1842–1847.

36. Zhang, K.; Wu, H.; Li, M.; Yan, Z.; Li, Z.; Wang, J.; Zhang, X.; Yan, L.; Kang, X. Magnitude and Edaphic Controls of Nitrous Oxide Fluxes in Natural Forests at Different Scales. *Forests* **2020**, *11*, 251.

37. van Groenigen, J.W.; Velthof, G.L.; Oenema, O.; van Groenigen, K.J.; van Kessel, C. Towards an agronomic assessment of N_2O emissions: A case study for arable crops. *Eur. J. Soil Sci.* **2010**, *61*, 903–913.

38. Grassini, P.; Cassman, K.G. High-yield maize with large net energy yield and small global warming intensity. *Proc. Natl. Acad. Sci. USA* **2012**, *109*, 1074–1079.

39. Tang, Y.; Yu, L.; Guan, A.; Zhou, X.; Wang, Z.; Gou, Y.; Wang, J. Soil mineral nitrogen and yield-scaled soil N_2O emissions lowered by reducing nitrogen application and intercropping with soybean for sweet maize production in southern China. *J. Integr. Agric.* **2017**, *16*, 2586–2596.

40. Ussiri, D.A.; Lal, R.; Jarecki, M.K. Nitrous oxide and methane emissions from long-term tillage under a continuous corn cropping system in Ohio. *Soil Till. Res.* **2009**, *104*, 247–255.

41. Gao, W.; Bian, X. Evaluation of the agronomic impacts on yield-scaled N_2O emission from wheat and maize fields in China. *Sustainability* **2017**, *9*, 1201.

42. Chen, G.; Kolb, L.; Cavigelli, M.A.; Weil, R.R.; Hooks, C.R.R. Can conservation tillage reduce N_2O emissions on cropland transitioning to organic vegetable production? *Sci. Total Environ.* **2018**, *618*, 927–940.

43. Pareja-Sánchez, E.; Cantero-Martínez, C.; Álvaro-Fuentes, J.; Plaza-Bonilla, D. Impact of tillage and N fertilization rate on soil N_2O emissions in irrigated maize in a Mediterranean agroecosystem. *Agric. Ecosyst. Environ.* **2020**, *287*, 106687. [CrossRef]

44. Grave, R.A.; da Silveira Nicoloso, R.; Cassol, P.C.; da Silva, M.L.B.; Mezzari, M.P.; Aita, C.; Wuaden, C.R. Determining the effects of tillage and nitrogen sources on soil N_2O emission. *Soil Till. Res.* **2018**, *175*, 1–12. [CrossRef]

45. Kalinitchenko, V.P.; Glinushkin, A.P.; Sokolov, M.; Batukaev, A.; Minkina, T.M.; Zinchenko, V.; Chernenko, V.; Startsev, V.; Mandzhieva, S.; Sushkova, S.; et al. Biogeosystem Technique for Healthy Soil, Water and Environment. In Proceedings of the ACS Fall 2019 National Meeting & Exposition, Chemistry & Water, San Diego, CA, USA, 25–29 August 2019.

46. Kalinitchenko, V.P.; Glinushkin, A.P.; Minkina, T.M.; Mandzhieva, S.S.; Sushkova, S.N.; Sukovatov, V.A.; Il'ina, L.P.; Makarenkov, D.A. Chemical soil-biological engineering theoretical foundations, technical means, and technology for environmentally safe intra-soil waste recycling and long-term higher soil productivity. *ACS Omega* **2020**, *5*, 17553–17564. [CrossRef] [PubMed]

 *sustainability*

Review

Nitrogen Losses and Potential Mitigation Strategies for a Sustainable Agroecosystem

Kishan Mahmud [1], Dinesh Panday [2], Anaas Mergoum [3] and Ali Missaoui [1,4,5,*]

1 Center for Applied Genetic Technologies, University of Georgia, Athens, GA 30602, USA; kishan.mahmud25@uga.edu
2 Department of Biosystems Engineering & Soil Science, The University of Tennessee, Knoxville, TN 37996, USA; dpanday@utk.edu
3 Department of Internal Medicine, School of Medicine and Health Sciences, University of North Dakota, Grand Forks, ND 58102, USA; anaas.mergoum@und.edu
4 Department of Crop and Soil Sciences, University of Georgia, Athens, GA 30602, USA
5 Institute of Plant Breeding, Genetics and Genomics, University of Georgia, Athens, GA 30602, USA
* Correspondence: cssamm@uga.edu; Tel.: +1-706-542-8847

Abstract: Nitrogen (N) in the agricultural production system influences many aspects of agroecosystems and several critical ecosystem services widely depend on the N availability in the soil. Cumulative changes in regional ecosystem services may lead to global environmental changes. Thus, the soil N status in agriculture is of critical importance to strategize its most efficient use. Nitrogen is also one of the most susceptible macronutrients to environmental loss, such as ammonia volatilization (NH_3), nitrous oxide (N_2O) emissions, nitrate leaching (NO_3), etc. Any form of N losses from agricultural systems can be major limitations for crop production, soil sustainability, and environmental safeguard. There is a need to focus on mitigation strategies to minimize global N pollution and implement agricultural management practices that encourage regenerative and sustainable agriculture. In this review, we identified the avenues of N loss into the environment caused by current agronomic practices and discussed the potential practices that can be adapted to prevent this N loss in production agriculture. This review also explored the N status in agriculture during the COVID-19 pandemic and the existing knowledge gaps and questions that need to be addressed.

Keywords: nitrogen; nitrate leaching; nitrous oxide; soil resilience; soil microbiome; regenerative agriculture; ecological ditch

Citation: Mahmud, K.; Panday, D.; Mergoum, A.; Missaoui, A. Nitrogen Losses and Potential Mitigation Strategies for a Sustainable Agroecosystem. *Sustainability* **2021**, *13*, 2400. https://doi.org/10.3390/su13042400

Academic Editor: Christopher Robin Bryant

Received: 20 January 2021
Accepted: 18 February 2021
Published: 23 February 2021

Publisher's Note: MDPI stays neutral with regard to jurisdictional claims in published maps and institutional affiliations.

1. Introduction

The current global population of 7.8 billion is projected to reach over 9 billion by 2050 [1]. This projected boom in population would mandate an approximately 70% increase in global agricultural production to ensure food security in the developed and nearly 100% in the developing countries [2]. To keep up with this demand, global agriculture will continue to consume more amendments in both inorganic and organic forms that can support agricultural production and simultaneously battle the food waste crisis where one-third of the annual produced food goes to waste (1.3 billion tons) [3]. Nitrogen (N) is a critical element for all living organisms and assimilation of N by both terrestrial and aquatic plants is limited by its forms in the ecosystem [4].

Over the last five decades, the global N cycle has changed significantly due to the incessant input of nitrogen fertilizers. Nitrogen cycling involves five major steps; biological N fixation, ammonification, nitrification, N assimilation into microbial biomass pool, and finally denitrification [5]. However, both chemical N fertilizers and organic manure are often applied to soil in exceeding amounts for crop growth requirements lead to N loss. Of the applied N for crop growth, only 45–50% is being incorporated into the agricultural products [6] and the remainder is subjected to substantial loss [7]. A significant amount of N in soil and may be lost to the environment as NO_3, NH_3, or N_2O [8,9]. NO_3 may also continue to undergo recycling in the soil–water–air system and convert to N_2O and N_2 through the denitrification process and released back to the atmosphere [10]. Particularly, N_2O emission is substantial from production agriculture at the beginning stage of N fertilizer application during the cropping season [11]. Furthermore, the potential warming effect of N_2O and the short-term cooling effect of NH_3 and NO_x both has major consequences on global human and environmental health [12,13]. Additional complications in estimating N_2O arises from the nonlinear nature of N_2O emission in response to N fertilizer application along with other soil and environmental controlling factors [14]. Intergovernmental Panel on Climate Change (IPCC) linear model, one of the approaches to estimating N loss in terms of N_2O, which may overestimate N loss at lower N application rates while underestimating higher rates [15–17] (Figure 1).

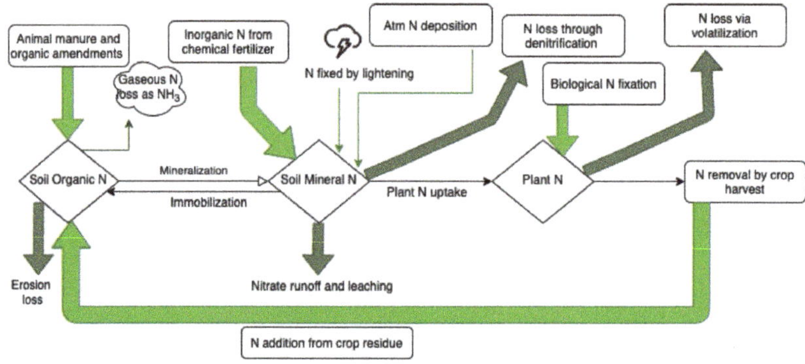

Figure 1. The nitrogen cycle in the soil.

Agricultural contribution to global N pollution requires constant monitoring and adoption of necessary efforts at the farm, regional, and national levels to mitigate and regulate environmental degradation. Nitrogen footprint, under agricultural context, is the amount of N released from resource use in agricultural production at every step of the production line both upstream or downstream [18]. Nitrogen footprint can serve as an indicator of the usage and losses of N in the production and consumption of food and energy, thereby, an ecosystem service. The objective of this review was to explore the agricultural status of N in the current agronomic production system and possible mitigation strategies to address and minimize N loss for sustainable agriculture.

2. Nitrogen Footprint in the World

Nitrogen plays such a cardinal role in ecosystem productions and the Earth's energy balance that any changes in the N cycle will bring about profound impacts on the global ecosystem and human health [19–21]. It is necessary to track the gains and losses in the N cycle and there are several tools to do just that. For instance, a quantifying tool that serves as an indicator of N losses to the environment from different stages of the N cycle ranging from production to consumption levels is called N footprint [22]. Globally, agriculture dominates N footprints [23]. Furthermore, global trades connecting different countries in importing and exporting agricultural commodities also leave a major mark on the N footprint. Nitrogen footprint can be a useful tool to help resolve the critical dilemma between the optimized N use (nitrogen use efficiency; NUE) and minimize negative impacts associated with N use. The food N footprint is the dominant component of the per capita N footprint [24,25]. On average, the per capita N footprint of ten countries from different regions of the world ranges from 15 to 47 kg N per capita per year and the principal reason behind this is the difference in the protein consumption rates and N losses during agronomic production [22]. In Asia, China has a big per capita food N footprint and it has increased almost 50% over 30 years (i.e., 1980 to 2008), however, it was still close or lower than that of countries in North America or Europe [24,26]. Of this per capita food N consumption, almost 40% is from protein, which is much lower than in other countries [26,27]. In the case of fossil fuel consumption, China leaves a lower N footprint than the USA [18] but since it has limited to no regulatory measures in place to reduce NO_x emission [28], hence, China has a higher energy N footprint compared to some of the European countries like UK and Germany [24–26]. Australia, a country in the Asia–Pacific region, due to a high protein diet and affordable food prices has a high food N footprint and a high energy N footprint due to higher N emission from electricity generation [22]. In North America, the USA has the largest N footprint for food production where the N footprint per capita is 39 kg N. This is followed by transportation where a relatively higher N footprint is observed compared to other countries [18]. In Europe, agricultural innovation was pioneered by the industrial revolution primarily at the turn of the 18th century, and again in the early 20th century with the arrival of the Haber–Bosch process. This nitrogen-based synthetic fertilizer production dramatically increased the productivity in agricultural sector, thus, countries like UK, Netherlands, Austria, and Germany have a higher food N footprint than many parts of Asia, typically ranging from 24 to 29 kg N capita^{-1} year^{-1} [18,29]. Table 1 details per-capita nitrogen footprints from different countries adopted from Oita et al. (2016) of nations [30].

Table 1. Ranked per-capita nitrogen footprints of nations (NFP = nitrogen footprint).

Country (High-Ranked in NFP)	NFP (kg N cap^{-1} yr^{-1})	Region	Country (Low Ranked in NFP)	NFP (kg N cap^{-1} yr^{-1})	Region
Hong Kong	225	Southeast Asia	Liberia	2	West Africa
Luxemburg	145	Western Europe	Moldova	3	Eastern Europe
Kuwait	102	Western Asia	Côte d'Ivoire	5	West Africa
Singapore	98	Southeast Asia	Papua New Guinea	7	Oceania
Uruguay	90	South America	Tajikistan	9	Central Asia
Australia	88	Oceania	Malawi	10	South-Eastern Africa
UAE	70	Western Asia	Sri Lanka	11	South Asia
Paraguay	67	South America	North Korea	12	East Asia
Canada	65	North America	Mozambique	15	Southern Africa
USA	62	North America	Burundi	17	East Africa

3. Nitrogen Losses in the Environment

Nitrogen contains readily converted reactive chemical (reactive nitrogen; Nr) [24] species causing a cascading effect on the environment and impacting the global ecosystems [31]. These reactive species are N_2O, NO_3^-, nitrite (NO_2^-), NH_3, and ammonium (NH_4^+) and mostly of anthropogenic origins such as fossil fuel combustion and agriculture (both legume cultivation and use of industrial fertilizers) (Figure 2).

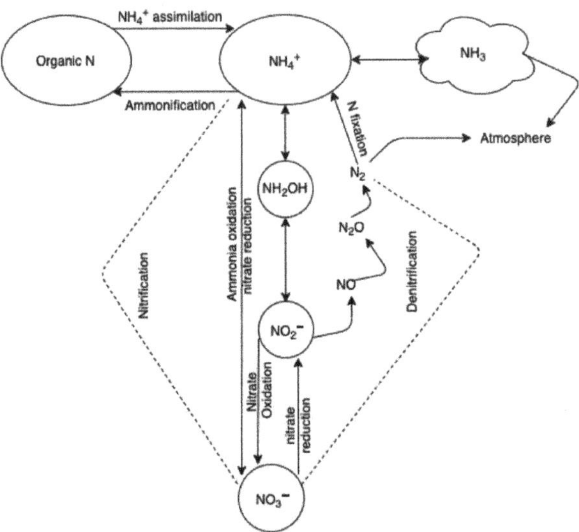

Figure 2. The major processes of the nitrogen cycle in soil.

3.1. Ammonia Volatilization

Ammonia volatilization is one of the major sources of N loss from arable farms worldwide. Different soil conditions affect the volatilization rate of NH_3 from the soil. Soils with high pH are generally prone to lose significant amounts of NH_3, however, neutral or acid soil may also lose NH_3, especially after inorganic fertilizer application like urea or organic amendments such as urine [32,33]. Moreover, soil and atmospheric temperature greatly affect the urea hydrolysis, thereby, the NH_3 (aqueous) transfer rate from the soil solution to atmospheric NH_3 (gas) [34]. Furthermore, low soil moisture tends to stimulate high soil solution concentration, thus, higher loss of NH_3 [35]. When NH_4 containing fertilizers such as ammonium nitrate (NH_4NO_3) or ammonium sulfate (NH_4SO_4) or urea (CH_4N_2O) are applied to soil during different stages of crop growth [36], they are typically subjected to immediate NH_3 volatilization. For instance, upon application, urea undergoes hydrolysis ensuing higher soil pH in the microsites of soil causing transformation of NH_4^+ to NH_3 [37,38]. The greatest amount of NH_3 is released through anthropogenic activities [39], which releases around 7.6 million kg ha^{-1} of NH_3 emission is from crop and animal husbandry accounting for more than 90% of the total emission [40]. Ammonia volatilization is a major problem to human health and the environment because it can react with acidic components of the environment such as sulfate (SO_4^{2-}) or NO_3^- and form a secondary inorganic aerosol. NH_3 emissions, moreover, significantly contribute to the formation of acid rain in the atmosphere and can be an indirect source of N_2O emissions, and promote eutrophication of surface water bodies [35,41]. The economic implications of NH_3 emission are particularly threatening to developing countries, for instance, India, as an earlier simulation study showed that almost one-third of the applied N fertilizers or manure is lost to the atmosphere as NH_3 [42,43].

3.2. Nitrous Oxide and Oxides of Nitrogen (NO_x) Emissions

The simultaneous process of nitrification (aerobic condition) and denitrification (anaerobic condition) produces N_2O as the common byproduct in soil [44]. During nitrification, NH_4-N is microbially converted to an intermediate product, hydroxylamine (NH_2OH) followed by the production of NOH, and finally NO_2. During nitrification, N_2O is produced both at NH_2OH and NO steps, while, on the contrary, denitrification, transforms NO_3 or NO to N_2 or N_2O [22]. Nitrous oxide is considered one of the most critical greenhouse gasses due to its prolonged atmospheric lifetime (120 years) and more potent in trapping heat than CO_2. Moreover, N_2O is responsible for ozone depletion by reacting with stratospheric O_2 and forming nitric acid [45]. Agricultural activities are the largest anthropogenic sources of atmospheric N_2O emissions [46]. An increase in fertilizer application rates, N sources, are the dominant factors of N_2O production in an agroecosystem [47]. Livestock, especially, traditionally grazed ruminants on poor quality fodder in pastures also plays a significant role in N_2O and methane (CH_4) emissions [48].

NO_2 is released to the stratosphere mostly through fossil fuel combustion from vehicles and industrial plants in addition to natural processes of nitrogen in the soil. It has been highlighted by the WHO as a potential health risk and that exposure to this pollutant should be minimized [49]. Several pulmonary and cardiovascular diseases have been linked to long-term exposure to elevated levels of NO_2. Some of the most documented cases include lung function growth deficit in children and compromise lung function in adults [50,51].

3.3. Nitrate Leaching

Apart from NH_4^+, NO_3 produced during nitrification in the aerobic condition is a preferable form of N for plant uptake. However, it is highly mobile in soil due to its anionic nature and in well-drained soil, NO_3 is the first nutrient to be washed off from the soil profile. The amount and distribution of rain and irrigation affect the NO_3 loss below the root zone in the soil profile [52]. Studies done in the early 90s' suggested that depending on the soil type, almost 80% of the applied N can be lost as NO_3 runoff [53]. Colder temperature contributes considerably to nitrate leaching due to greater precipitation and slower plant uptake of nitrate [54]. Earlier studies on nitrate loss in soil suggested that more than $100\,kg\,N\,ha^{-1}\,year^{-1}$ of nitrate was lost from grazeland after tillage through the soil profile over the next winter [55]. Soil structure and texture may also affect nitrate movement through the soil profile. Nitrate leaching is greatest in sandy soil with poor structure and slowest in clay soil [35]. Additionally, soil macrofauna movement and plant root growth often allow rapid nitrate movement in soil [56]. Nitrate leaching and runoff lead to severely deteriorated both ground and surface water quality levels and resulting in eutrophication and algal bloom. Severe cases of NO_3 contamination in human is evident by Blue-Baby Syndrome [57] where NO_3 molecules in drinking water combine with blood hemoglobin and hinder blood oxygen transportation [58]. Table 2 shows the major sources, amounts, and pathways of global reactive nitrogen (Nr) emission [30].

Table 2. Major sources, amounts, and pathways of global reactive nitrogen (Nr) emission.

Reactive Nitrogen (Nr)	Amount (Tg)	Source	Pathway
Nitrate (NO_3^-)	161	Industries and Agriculture	Leaching and surface runoff
Ammonia (NH_3)	45	Agriculture	Volatilized
Nitrous Oxide (N_2O)	6.2	Agriculture	Gaseous emission
NO_x	35	Transportation and emission	Gaseous emission
Nitrogen emissions potentials to water mainly as NO_3^-	28	Consumer	Mainly Sewage

4. Mitigation Strategies

4.1. Farming System Design

To establish an effective and productive resource partitioning system, a collaborative organization of the agricultural production systems, which involves the adoption and eventual adaptation of a combination of a less expansive, minimal resource extraction crop-livestock production of high-value commodities, is required. The integrated production system (IPS) is not essentially a location restrained system, rather a synergistic approach towards agricultural resource utilization with minimal external input compared to intensive systems, and IPS helps materialize the concept of integrating several on-site farming components such as crop residues and animal waste [59,60]. A recent case study from Nigeria where herdsmen and farmers exchange resources, for instance, manure (collected by farmers) and crop stubble (used by herdsmen as cattle feed) exchange helped ease societal tension and offering a limited but promising solution to an on-site N loss solution [61]. Another example of IPS has been demonstrated in the Benin Republic, West Africa where a local farm (Songhai Farm) adopted all the IPS components and successfully collaborated between the different stages of agriculture such as production, selective breeding, harvesting, product processing, product placement in the local market, and finally on-site waste recycling to reuse nutrients [62]. Thus, IPS allows all participating parties alike to place a value on the on-site waste and help in sharing spatially separated resources.

4.2. On-Farm Best Available Techniques (BATs)

Preventing the loss of ammonia can be achieved by a combination of BATs such as placing N-fertilizers subsurface or injecting, applying urea fertilizer before rainfall events and after application irrigate the crops, applying N-fertilizers with acidifying agents such as elemental sulfur (not gypsum) or urease inhibitor, etc. [35,63]. Reducing nitrate loss through the soil profile, however, requires the adoption of a wide range of management strategies, for instance, split application of N-fertilizers according to crop nitrogen requirements and if applying animal slurries as N source then according to the crop demand, spraying nitrification inhibitor, maintain a constant vegetative cover, especially over the drainage period, cultivating soil in spring rather than in fall, etc. [64–71]. Many of the mitigation strategies adopted for N_2O emission overlap with the strategies for mitigating NH_3 volatilization and nitrate loss with a few exceptions, for instance, using lime to increase soil pH, avoiding anaerobic pockets in soil by maintaining optimum irrigation rate, and reduce animal traffic to avoid soil compaction [35,72].

4.3. Improved Nitrogen Use Efficiency (NUE)

Achieving nitrogen use efficiency (NUE) in crop and livestock production will involve implementing the best management practices (BMPs) leading to increase recycling within the system [73,74]. Some BMPs of NUE are discussed hereunder.

4.3.1. Enhanced Efficiency of Fertilizer Material

The use of enhanced efficiency fertilizers composed of coatings of low permeable materials attached to an inhibitor (nitrification or urease inhibitor) as an additive may be used to regulate processes such as nitrification or urea hydrolysis to simultaneously reduce the N loss and increase N uptake by plant and soil microbial population [75–77] under both laboratory and field setup. Such examples of inhibitors are N-(n-butyl) thiophosphoric triamide (NBPT), phenylphosphorodiamidate, dicyandiamide (DCD), and 3,4-dimethyl pyrazole phosphate (DMPP). For instance, the use of NBPT coated urea compared to uncoated-urea reduced NH3 emission by 42% for sunflower in Spain [78]; from 9.5% to 1.0% of applied N for winter wheat in Australia [79], and finally in the UK, for multiple grassland and winter cereal species [80]. In New Zealand, the volatilization of NH_3 in grazing pastures was reduced through the application of NBPT (18%–28%) [81]. Moreover, the application of nitrification inhibitors such as DCD and DMPP in conjunction with urea significantly reduced N_2O emission from agricultural soils of Louisiana, USA by more than 76% and 67%, respectively [82].

4.3.2. Site-Specific Nutrient Management (SSNM) and Real-Time Nitrogen Management (RTNM)

Crops typically respond to applied N in varying degrees and the fate of that applied N depends vastly on the soil conditions. The traditional broad-spectrum blanket-recommendation or "one size fits all" often fails to address crop N requirements and cannot surveil N availability from all possible sources leading to severe economic and environmental drawbacks. A possible solution could be the site-specific nutrient management strategy (SSNM). The SSNM considers several factors in production agriculture, for instance, yield potential of the crop, plant nutrition, inherent soil capability to supply N, calibrated N dosage, and subsequent N recovery calculations [22]. The SSNM addresses the crop-specific nutritional needs, utilizes maximum available resources to obtain N, calculates the gap of N required to fill the nutrient deficit gap, and finally recommends optimum N application recommendation [83]. Contrary to the blanket application of N, split applications according to the crop requirements, based on the growth stage, could prove to be an important strategy to enhance N recycling in soil, minimize loss of N as NH_3 and N_2O. Without the destructive collection of samples, sensor-based tools have the potential to identify and correct N stress, which has already occurred during the growing season for plant production [84]. Normalized difference vegetative index (NDVI) data of the greenness of plant leaf and previous crop yield and N application data can be used to implement the splits and may lead to an increase in nitrogen use efficiency and N recovery by the plant [22]. Studies on wheat (*Triticum aestivum* L.) produced primarily on the Central Great Plains of the United States showed that sensor data converted to NDVI was used to formulate a response index and showed a 15% greater NUE compared to whole-field techniques [85]. An emerging technology for site-specific N application for maize is real-time fluoro-sensing for variable-rate nutrient management, especially at earlier growth stage (V2) of the plants [86]. Moreover, proximal sensor-based variable rate N management strategy was reported to achieved greater grain yield and improved NUE for maize in Colorado, USA [87]. Similarly, in the UK, applications based on real-time sensor data saved 15 kg N ha^{-1} that increased the NUE without any reduction in wheat grain yield [88]. In a five-year study on N recovery, approximately 25–50 greater N was recovered with RTNM in maize cropping over-scheduled N application in Punjab, India [89]. In Guangxi Zhuang, China, NUE in rice (*Oryza sativa*) was 75% higher in densely planted reduced and delayed N applied rice plants compared to conventional farmers practice (broadcast) [90].

4.3.3. Deep Placement of Urea Super Granules

If there is an anoxic layer overlying a reduced zone in the soil, which is very typical of low-lying flooded rice-production systems, N loss can be encouraged by nitrification and denitrification, simultaneously [22]. Deep placement of nitrogenous fertilizers, for instance, large urea granules in rice-fields has successfully prevented the conversion of NH_4 to NO_3 and subsequent losses. Early studies in the 90s' showed that deep placement of urea reduced N loss by 65% and resulted in a greater rice yield by 50% in grain yield compared to split application of granulated urea [91]. Recent studies also showed that NH_3 and N_2O loss were significantly suppressed by 94% with the deep placement of large urea granules ($\geq$0.7 g) in rice-growing areas [92,93]. Furthermore, policymakers across Europe have indicated a relatively new method termed as "Closed-Slot Injection Method" to reduce NH_3 emissions with inorganic fertilizers or organic inputs and this technique has proven promising due to the wide-spread availability among European farmers. For instance, NH_3 loss was reduced in maize by 75% for mineral fertilizer and 96% for organic amendments compared to surface broadcast [94].

4.4. Pasture and Livestock Management

Reducing N_2O emissions while keeping the ruminant population at the higher end of the production spectrum if not decreasing, requires either a top-quality ruminant diet or improved yield, thus, making this a livestock nutrition issue. Ruminant excretion quality is largely dependent upon ingested feed quality and depending on soil factors such as soil moisture and temperature, the N-component of the excretion may be subject to significant losses, thus, causing environmental hazards, such as increased nitrous oxide emission and greater nitrate loss [95,96]. Apart from reducing P loss to runoff [97], to facilitate enhanced nitrogen mineralization and to reduce the loss of NO_3 in the soil, strategic cattle grazing in pastures along with overseeding of susceptible areas with annual grasses may offer a sustainable solution in livestock management as opposed to traditional grazing [72]. Additional approaches, such as supplementation of animal feed and anaerobic digestion may help increasing nutrient use efficiency by recycling nutrients on-site [98]. These management strategies are particularly important for developing countries like China, India, and Brazil that are major players in the beef and dairy industries [99] where ruminant feed is mostly associated with inferior quality of feed leading to a low-efficiency animal diet [100]. On the other hand, adopting alternative approaches, such as rearing mono-gastric non-ruminants, for instance, poultry, rabbit, swine, and horses (which are typically on a better and balanced diet than beef and dairy cattle) may potentially lead to efficient nutrient use and lower nitrogen loss [98]. Additionally, supplementing animal feed with amino acids such as lysine has the benefits of reduced N loss from swine and poultry by 30% compared to traditional feeding routine [101]. In managing pasture soil health, alternative ways of adding N have been investigated. In addition to adding organic fertilizers such as poultry [102] in managed pastures, using N-fixing legumes to supply soil N is an ecologically safe practice [103]. For instance, in long term pastures in Australia and New Zealand, and organic farming in the UK, white clover (*Trifolium repens*) is grown with other grass species and the fixed N is released slowly to the grass once released into the soil through root exudates and dead legume tissue [104–106]. In more recent studies on drought-tolerant forages, inter-seeding alfalfa into established Old World bluestem (OWB) grass helped restore soil health and enhance soil microbial community complexities [107].

4.5. Managing Livestock Wastewater

On-site livestock wastewater may be managed with the use of microalgae in the wastewater where livestock feces and urine (high N content). Due to its high concentration of N content, this animal waste tends to produce pollutants such as NH_3 and N_2O, which are essential for microalgal growth [108]. In addition to inorganic P, microalgae are known to assimilate inorganic N species and transform them into organic nitrogenous compounds [109]. Nitrogen oxide emissions from the wastewater, for instance, were reduced by 80% with *Chlorella* sp. [110] and entirely with *Gracilaria birdiae* (red seaweed) [111]. Seaweed can reduce N_2O emissions by assimilating and storing N in high concentration [112] and decreasing available NOx in the wastewater system [113].

4.6. Carbon-Rich Sources

In terms of regulating N loss or impacting the N cycle, manure or litter broadcast leads to enhanced N_2O emission [114], whereas, the sole application of biochar or a combination of lime and biochar (livestock slurry) showed a significant reduction in NH_3 emissions [115] and cumulative N_2O loss [116]. In production agriculture, especially, where litter or manure is used, ammonia is the precursor of nitrous oxide emission and volatilized ammonia can travel up to 5 km from its source of origin and a minor portion (1% proportion) has the potential to be re-emitted as nitrous oxide upon redeposition [117]. The application of biochar has been shown to reduce the emission of nitrous oxide from redeposited ammonia by 69% and this was attributed to the increased aeration caused by the inherent porosity of the biochar [118]. Similar high C sources, obtained as an agro-industrial byproduct, known as char (contains 30% of total C) have shown a reduction in NH_3 volatilization up to 37% under fertilized soil compared to control [119]. In current agronomic practices, sewage sludge is commonly used, however, its use is associated with a greater risk of toxic substance accumulation, for instance, heavy metals, polycyclic aromatic hydrocarbons, or polychlorinated biphenyls, phenanthrene, and pyrene [120,121]. Furthermore, upon application, sewage sludge may phyto-accumulate and can be trophic transferred in agroecosystem food webs [122]. However, in recent years, the use of sewage sludge in conjunction with carbon adsorbents like activated carbon or biochar is gaining popularity both in terms of environmental safety and soil health [123–125]. In a recent study, conjunctive use of sewage sludge and biochar as soil amendments showed a dramatic decrease in sludge toxicity shortly after application due to nutrient immobilization and a significant reduction in nitrate loss from the soil profile [126]. Therefore, in addition to the phyto-stabilization of heavy metals [127], the use of high C sources in regulating N loss is a potentially promising avenue for further research.

4.7. Engineering Cereal Crops for Nitrogen Fixation

For the last five decades, scientists and agronomists have been studying the prospects of N fixation in cereal crops and evidently, there has been tremendous progress in areas like the expression of nitrogenase gene in eukaryotes and the nodule formation workflow in plants [128]. These efforts have led to investigate the scope of engineering cereal crops to perform symbiotic or autonomous N fixation, however, several factors, namely, population growth and a high rate of N application in production agriculture can outwit the constant efforts of developing N fixing transgenic cereals [129]. The primary tenet behind expressing nitrogenase in cereals is to re-envisage the legume-rhizobia symbiosis leading to nodule formation. The presence of building blocks for nodule formation in cereals is indicated by the presence of several common plant hormones [130,131].

Furthermore, *Nod* factors (legume-rhizobia symbiosis) and *Myc* factors (cereals and arbuscular mycorrhizal symbiosis) are structurally similar, indicating the possibilities, although abstract but promising, to engineer cereal crops to express Nod factors and initiate the first step toward nodulation [130,132,133]. On the contrary, considering the technical challenges of engineering cereals crops for nitrogen fixation, a sustainable and potential alternative approach such as root-associated diazotrophs in fixing and supplying N to cereals may offer a solution in N management with a short turnover time [134,135]. Although less sensitive than the legume-rhizobia symbiosis, nevertheless, the association between cereal crops and the rhizosphere shares a sophisticated signal route between microbes and plant host [136–139]. A recent ground-breaking study from the N depleted areas of Oaxaca, Mexico, showed mucilage associated with the aerial roots of Sierra Mixe maize can colonize free-living diazotrophic bacteria and the estimated up to 82% of the nitrogen content of maize [140]. The known association between plant and diazotrophs may improve the growth and yield of cereals in low N soils but the performance of these microbial strains is often not reproducible in the field [141,142].

4.8. Plant Growth Promoting Microbial Consortia

Soil microbes are key players in organic matter decomposition, macro, and micronutrient cycling, and facilitating nutrient availability for plant uptake [143,144]. Microbial communities associated with plants are also capable of abating environmental pollution [145]. Rhizosphere dwelling microbes are also known as plant growth-promoting microorganisms (PGPMs) [146] because they encourage plant growth and foster soil health by N fixation, P solubilization, mineralization of macro (calcium, magnesium, and potassium)/micronutrients (zinc, iodine, and nickel) and secreting phytohormones, and finally suppressing pathogens [147,148]. Additionally, animal wastes like poultry litter or composted poultry litter is a major source of organic N in production agriculture but poultry manure can cause a high loss of nitrogen via ammonia volatilization [149]. Therefore, farmers, industries, and researchers are giving considerable attention to the formulation of these microbial communities on-farm or off-site and using them to fortify the resilience of the agroecosystems. However, the function and propagation of these concocted exogenous microbes may be limited to their ability to survive under a highly diverse, competitive, and constantly changing medium such as soil [150,151]. In Southeastern USA, recent research on the production and application of locally derived exogenous microorganisms in managing nutrient availability from animal waste and utilizing them in nutrient cycling, especially in N-cycling, has shown higher potential nitrogen mineralization in soil 0–5 and 0–15 cm depth indicating a more robust and complex microbial community composition [152–154]. Additionally, microbially mediated decomposition of nitrogenous compounds in soil may enhance the N availability and consequently be intercepted by plant roots and taken up [155].

4.9. Phytogenic Approach and Fungal Utilization

As discussed above, nutrients excreted by livestock reflect the diet consumed and are an indicator of N, P, and CH_4 release into the environment. Tannins and saponin-rich plants such as *Acacia mearnsii*, *Delonix regia*, *Enterolobium cyclocarpum*, and *Musa paradisiac* may be used as animal feed supplementation to enhance N retention [156–159]. For instance, alfalfa silage has a higher crude protein and nitrate content, which upon feeding may lead to higher N_2O emission as opposed to feeding a combination of corn silage and grass hay [160]. In some countries in Asia, for instance, China and India, rice and wheat straws are burned, thus, causing environmental havoc like, smog. However, inoculating the straws with fungal species like *Aspergillus terreus* may reduce the lignocellulose content and enhance the decomposition process by soil microbes, which in turn enhance the nitrogen mineralization process [98].

4.10. Organic Agriculture as a Tool in Nitrogen Pollution Remediation

The history of organic agriculture is somewhat contentious and at the turn of the 21st century, some critics portrayed organic farming as an ideologically driven inefficient food production system [161–163]. Nonetheless, globally, the organic farming community experienced a continuous increase in the number of organic farms, acreage of land, the consumer market for organic foods, and organic agriculture-focused research funding [164]. As of now, more than 160 countries practice organic farming and more countries continue to join the community [164]. Worldwide, nearly 2.3 million organic farmers are growing organic produce in 0.99% of the total cultivable land [165] (Table 3).

Table 3. Regional distribution of land area (Mha and percent) under organic agricultural land.

Region	Area (Mha)	Percent (%) of Total Organic Agriculture Land
Oceania	17.3	40
Europe	11.6	27
Latin America	6.8	15
North America	3.7	8
Asia	3.6	7
Africa	1.3	3

There are some intrinsic weaknesses in organic farming practices, especially, in the timely release of nutrients that coincide with plant N demand from typical organic amendments such as compost [166,167] and consequent potential higher nutrient loss mainly as nitrate [168]. Nitrate leaching losses have been reported to be similar or minutely lower in conventional agriculture compared to organic practices in several studies [169]. Nevertheless, organically managed agricultural systems can potentially contribute to the mitigation of climate change through efficient nutrient management techniques leading to reduced emission of nitrogenous gases from the production system and sequester carbon in the soil. Probably the most powerful aspect of organic agriculture, especially in developing countries, is its capacity to compete and often attain equal or higher yields as compared to traditional farming practices [170]. Organic farming practices have been shown to safeguard water quality in rivers as well. In a simulation study, a researcher has shown that a combined effect of higher precipitation and ethanol production from common biofuel crops could cause the river N level to rise by 24%, whereas, simple practices that are common to organic farming such as cover cropping, use of legumes in crop rotations, etc., could decrease the river N level by 7% [171]. Organic cropping systems to succeed, a few factors need to be considered, for instance, the synchronicity between crop N demand and N delivery from animal waste, greater flexibility in designing crop rotations, pretreating the compost with microbial inoculum to facilitate greater and rapid N availability from compost for plant uptake, etc. [154,169].

4.10.1. Limited External Input

In organic agriculture, inputs such as synthetic fertilizers, chemical pesticides, and herbicides are strictly restricted, hence, external energy for the chemical synthesis of nitrogen or phosphorus fertilizer is not required. In a conventional wheat-growing system, typically, 56% of the energy burden falls on chemical fertilizers and 11% on pesticides [172], thereby, increased the chances of nutrient loss. Although, organic agriculture avoids this requirement of energy but often highly dependent on the use of fossil fuels, especially in mechanical weed management. A study in the UK compared crops grown on seven conventional and organic farms and realized that although there is a higher energy demand of machinery to produce foods, the energy balance still tilted toward energy savings in organic farms (indicated by a 15% lower energy demand) gained by waiving the use of synthetic fertilizers and pesticides [173].

4.10.2. Crop Diversification

Cropping of diverse assemblages of local plant varieties fosters resilience in agroecosystems to counter sudden environmental stresses such as droughts and economic volatility such as price variations [174]. Additionally, cropping diversity encourages the efficient use of soil nutrients and optimum yield [175].

4.11. Ecological Ditch

In addition to N_2O emission and NH_3 volatilization from agricultural fields, a major source of nitrogen pollution is agricultural drainage [176,177]. Typically, the compositional nature of agricultural runoff is complex due to the sheer number and types of nutrients, for instance, runoff often contains nitrate, ammonium, inorganic phosphorous, organic pollutants, and heavy metals [178]. These N and phosphorus (P) nutrients play a crucial role in the growth of aquatic plants, which upon a lack of regulation can lead to eutrophication in the downstream receiving aquatic systems [179]. Heavy nutrient loads from agricultural lands can potentially cause eutrophication, hypoxia, and ecological damages in nearby water bodies [180].

The "ecological ditch" (eco-ditch) is an effective component in alleviating non-point agricultural pollution. Eco-ditches are examples of best management practices and a stark contrast to the traditional agricultural drainage ditches. Eco-ditches create an exclusive ecosystem where the participating parties are aquatic plants and associated microbial communities fueled by the constant nutritional substrates [181]. Eco-ditches are designed to absorb nutrients that are otherwise lost through surface runoff and make those nutrients available for root uptake or be incorporated into microbial metabolites [182,183]. Eco-ditches designed with *Leersia oryzoides* and *Typha latifolia* reduced the load of inorganic N from 2.5% to 1.5%, accounting for more than 50% of the total reduction over 2 years in Northern Mississippi [184]. However, in designing eco-ditches, plant population diversity needs greater attention and highly efficient ditch plants should be selected. One possible constraint of these ditches could be the variability in nutrient removal capacity by plants, which is strictly dependent upon the growth stages of plants. For instance, during the growing period, plants tend to uptake more nutrients as opposed to the senesce period [185]. Plant harvesting continuously in the eco-ditches may offer a possible solution [186].

4.12. Genetic Improvement

4.12.1. Identifying Candidate Gene in Plants for Improved NUE

Nitrogen use efficiency of plants is genetically regulated; thus, nutrient use varies vastly among plant species. These variations lead to differences in several aspects in different plants including N assimilation, uptake, and remobilization capability, hence, alludes to the need to screen for potential genetic traits across these genotypes.

a. Differentially expressed genes (DFEs) to validate their roles in NUE of different genotypes of crop species can be profiled globally for different genotypes under different N treatments. DFEs have four components and these are:

    i. Hybridization based transcriptome analysis to identify differentially expressed traits with low abundance [187];

    ii. Analyzing short sequence tags of individual mRNA and then linked to form long sequences and finally cloning them [188];

    iii. Probe-targeted hybridization of immobilized cDNA molecules to generate a large amount of data and analysis of the whole genome [189];

    iv. RNA sequencing involves the sequencing of every RNA molecule and subsequently profiling a particular gene expression [190].

b. Functional validation of genes by mutation and transgenic studies.

Both the mutant population and natural variants can be studied to identify genes of interest in crop NUE. The steps involve propagation of the mutated population and screening for mutated phenotypes, finally followed by gene recovery through map-based cloning strategies.

4.12.2. Discovery of Genes by Mapping Studies

Biomarker-based mapping studies by biparental linkage analysis and association studies in naturally existing genotypes are two possible strategies in identifying the position of the NUE genes in crops. In low and optimum N systems, a meta-analysis of the quantitative trait loci (QTLs) for yield was mapped to discover linked markers with the gene that controls the specific trait, which in this case was N use and the study revealed a total of 22 meta-QTLs under low N [191]. Additionally, another association study in 196 accessions of wheat for yield components expressed 23 N-responsive regions, which can be exploited by breeders to develop highly N responsive varieties of wheat [192].

In the coming decades, one of the greatest challenges humanity faces is climate change, and agriculture is both a key contributor to crisis and will be immensely impacted by this problem [193]. Thus, minimizing the loss and emission of reactive N is crucial in slowing down the rate of climate change [194,195] (Table 4).

Table 4. Mitigation strategies to prevent potential N loss.

Mitigation Strategies	Approach
Farming System Design	Agronomic
On-Farm Best Available Techniques (BATs)	Agronomic
Improved Nitrogen Use Efficiency (NUE)	Agronomic
Pasture and Livestock Management	Agronomic and landscape
Managing Livestock Wastewater	Hydrologic
Carbon-rich sources	Agronomic
Engineering Cereal Crops for Nitrogen Fixation	Molecular
Plant Growth Promoting Microbial Consortia	Agronomic and molecular
Phytogenic Approach and Fungal Utilization	Agronomic and molecular
Organic Agriculture	Agronomic
Ecological Ditch	Landscape
Genetic Improvement	Molecular

Adapting to a combination of these mitigation strategies will enable individual growers and the farming community at large. For instance, agronomic approaches such as formulating microbial inoculum from a local source (discussed in Section 4.8) are affordable and particularly is very important for agriculture in developing countries, while molecular techniques such as N-fixing cereal crops, NUE gene identification, and mapping (discussed in Sections 4.7 and 4.12) are time-consuming and requires a long-term research investment. Nonetheless, policy-driven and ecosystem service-oriented mitigation strategies will help combat future N losses from agriculture and offer ecological safeguard.

5. Nitrogen Status in Agriculture during the COVID-19 Pandemic

On 11 March 2020, the World Health Organization (WHO) declared that the world faces a pandemic by the novel coronavirus (COVID-19) [196,197]. The tremendous disruptions across the globe caused by the COVID-19 pandemic are affecting the entire realm of human activities. A recent study investigating the relationship between long-term exposure to NO_2 and coronavirus mortality through the analysis of tropospheric NO_2 mapping and distribution data generated by the Sentinel-5P satellite showed that 78% of death cases were in five regions located in Northern Italy and Central Spain that displayed the highest NO_2 concentration levels and low circulation of air to disperse the pollution. These findings suggest that long-term exposure to this pollutant may be a major contributor to COVID-19 death in these regions and possibly in other parts of the world [198]. However, the COVID-19 outbreak and consequent social distancing activities led to an extensive decline in traffic and allowed comparisons of air quality during and before the decline to document the impacts of COVID-19 on NO_2 concentration in Florida Counties through March 2020. The results indicated a 54.07% decrease in NO_2 in the atmosphere [199]. In the context of agricultural safeguard, According to World Economic Outlook, the emerging and developing nations will face extreme severity of negative growth [200], and according to the Food and Agriculture Organization (FAO) and World Food Program (WFP), there will be food insecurity at an unprecedented level [201]. In this time of global crisis, now more than ever, soil plays a pivotal role as production agriculture and bedrock of resilience in food security [202]. The impact of this global pandemic on agriculture will likely be learned in waves in the coming years. Assessing the current soil N, carbon (C), and P content as impacted by the pandemic for devising future strategies to recover from the pandemic and establish long-term sustainable goals to maintain soil health for future needs is very crucial. Likely lower livestock production and with lower fertilizer application due to the COVID-19 restrictions, global agriculture may see improved farm management scenarios resulting in reduced GHGs, like, N_2O, lower NH_3 emissions, and low nutrient loss to surface water. For instance, there have been limited agricultural activities in livestock production that are being reflected in the reduced greenhouse gas emission such as NO_x in countries such as China and Italy; however, the same study also concluded that agricultural pollution via NH_3 emission has not changed significantly compared to the pre-pandemic era [203]. This may consequently result in improved ecosystem services and higher food quality. A closer look at the examination of macronutrients such as N and P is crucial, where N limits crop growth and P fertilization plays a crucial role in crop yield. However, during this pandemic, due to lockdown, food supply chains have been massively disrupted on a regional and global scale, thus, raising the most imminent threat from the huge addition of organic waste as mass disposal from the dairy and vegetable industry. Additionally, reduced meat consumption in the USA during the earlier months of 2020 has led to the massive burial of swine and poultry in many parts the country [202]. Such an influx of surplus organic matter to the soil, especially, P in the organic matter, may result in an imbalanced soil nutrient status. Therefore, for future crop fertilization strategies, application based on N requirements of the plant may result in over-application of P, which in turn may be fixed in the soil–mineral complex, making the soil depleted in plant-available P. In this context, the long-term consequences of the massive burial of organic waste may introduce additional complication is land use, surface, and groundwater quality, soil macro and microfauna and flora diversity, critical ecosystems services, and human well-being.

6. Knowledge Gaps and Questions

The agricultural system is complex and often the feedbacks and processes are nonlinear. Nitrogen loss mitigation strategies require sustainable, resilient, and redundant ways of producing and consuming that can also be adopted in climate change strategies in a broader sense [204]. Immediate attention is required to explore and interpret data from low-income countries in this regard. Although, the greatest emission of GHGs such as

N$_2$O is estimated to be higher in the low to middle-income parts of the world from food waste there exists close to no empirical studies on how to tackle these challenges [205,206]. The problem is further complicated as the mean income in these countries increases, the dietary habits change. High protein and carbohydrate diets lead to intense livestock and cereal production systems ensuing in greater GHGs emissions [207]. However, climate implication is yet to be explored in those countries as compared to developed parts (high-income) of the globe [208]. Another aspect of identifying and abating N loss is investing resources into research on post-harvest management of crops and crop residues [209]. Inclusive public policies and equitable funding mechanisms that are equitable and resilient are the two major catalysts that will lead to lower agricultural system N footprints. A major shortfall in global climate change strategies is the lack of financial allocation for tackling emission problems, especially in developing countries, and uncertainties shrouded by the lack of research to indicate whether the allocated money is being used for climate-smart agriculture. For instance, in 2015, out of $391(US) billion only $8 (US) billion was issued to address and adopt strategies to mitigate climate strategies like nitrogen use efficiency, soil health, and GHGs emission mitigation [210]. Furthermore, assigning sustainability indexes to agricultural commodities and assimilating sustainability benchmarks in dietary intake guidelines may lead to change in dietary regimens fostering healthy and low N footprint foods and low emission diets [211,212]. Countries such as Brazil, Germany, Qatar, and Sweden have taken steps towards this goal already, and the USA has thus far failed to adopt [213]. Government policies now must realize that voluntary measures are not adequate in lowering N footprints at farm level production, rather strict regulation, financial incentives, and subsidies for regenerative agriculture are obligated [214]. Now, the world is faced with an entirely new challenge of the COVID-19 pandemic, and the existing soil and agricultural management strategies must include the complex task of including the COVID-19 pandemic as a variable along with the existing ones, for instance, climate change, food insecurity, freshwater crisis, and continuously endangered and fading biodiversity. A radical shift from the dependencies on industrialized mono-agriculture to more diverse agroecology may be required in the coming days as the global population traverse through this pandemic.

7. Conclusions and Future Directions

The nitrogen challenge at its core requires societal recognition and indicates a potential opportunity to steer away from a fragmented policy approach, rather than toward rapid solutions [215]. Addressing the loss of N in agricultural production and proposing possible mitigation strategies should aim to be inclusive, realize the current shortcomings, and most importantly should be realistic to attain. From simple measures such as, the inclusion of nitrogen-fixing legumes in production agriculture as cover crops [216] to more complicated efforts that aim to materialize artificial symbioses or associative nitrogen fixation in non-legume plants, especially in cereals [128] are some of the possible ventures that may be undertaken. Moreover, undertaking efficient N management measures, for instance, controlled release of N or drip N fertilization in rice and maize, respectively, can direct the cost–benefit balance to lean toward profit [217,218]. Additionally, 4R (Right Source, Right Rate, Right Time, and Right Placement) N guidelines for corn are extremely profitable (40% increase) while decreasing the N application rate (21% reduction) [219]. Thus, connecting socioeconomic requirements with landscape potentials should be the central aspect in future N loss-related plans and policies [220]. Another key element to be explored is soil resilience and should be implemented wherever possible. Resilience will enable degraded or depleted soil to recover and possibly stop the soil from being a sink but a source. In the coming days, the soil should not be discussed as an isolated component rather it should be thought of as an essential tool of regenerative agriculture where possible answers to pollution questions can be sought after. Including soil as a resource in strategizing grass-root policies may reduce the risk of market vulnerabilities and associated risks. Finally, to summarize, while policy and funding apparatuses have been proposed amply but few have been implemented to address and mitigate the global N pollution status for regenerative and sustainable agriculture.

Author Contributions: Conceptualization, K.M.; writin—original draft preparation, K.M., D.P., A.M. (Anaas Mergoum) and A.M. (Ali Missaoui); writing—review and editing, A.M. (Anaas Mergoum), K.M., D.P. A.M. (Ali Missaoui); supervision, A.M. (Ali Missaoui); All authors have read and agreed to the published version of the manuscript.

Funding: Partial funding was provided by The Center for Bioenergy Innovation, a US Department of Energy Research Center supported by the Office of Biological and Environmental Research in the DOE Office of Science, for salary support of K.M.

Institutional Review Board Statement: Not applicable.

Informed Consent Statement: Not applicable.

Data Availability Statement: Not applicable.

Conflicts of Interest: The authors declare no conflict of interest.

References

1.	Grafton, R.Q.; Daugbjerg, C.; Qureshi, M.E. Towards food security by 2050. *Food Secur.* **2015**, *7*, 179–183. [CrossRef]
2.	Rodriguez, A.; Sanders, I.R. The role of community and population ecology in applying mycorrhizal fungi for improved food security. *ISME J.* **2015**, *9*, 1053–1061. [CrossRef]
3.	FAO. *Food Wastage Footprint: Impacts on Natural Resources*; Summary Report; FAO: Rome, Italy, 2013.
4.	Gruber, N.; Galloway, J.N. An Earth-system perspective of the global nitrogen cycle. *Nature* **2008**, *451*, 293–296. [CrossRef]
5.	Pathak, H.; Jain, N.; Bhatia, A.; Kumar, A.; Chatterjee, D. Improved nitrogen management: A key to climate change adaptation and mitigation. *Indian J. Fertil.* **2016**, *12*, 151–162.
6.	Houlton, B.Z.; Almaraz, M.; Aneja, V.; Austin, A.T.; Bai, E.; Cassman, K.G.; Compton, J.E.; Davidson, E.A.; Erisman, J.W.; Galloway, J.N. A world of cobenefits: Solving the global nitrogen challenge. *Earth's Future* **2019**, *7*, 865–872. [CrossRef]
7.	Xu, P.; Chen, A.; Houlton, B.Z.; Zeng, Z.; Wei, S.; Zhao, C.; Lu, H.; Liao, Y.; Zheng, Z.; Luan, S. Spatial Variation of Reactive Nitrogen Emissions from China's Croplands Codetermined by Regional Urbanization and Its Feedback to Global Climate Change. *Geophys. Res. Lett.* **2020**, *47*, e2019GL086551. [CrossRef]
8.	Fageria, N.; Baligar, V. Enhancing nitrogen use efficiency in crop plants. *Adv. Agron.* **2005**, *88*, 97–185.
9.	Maharjan, B.; Venterea, R.T.; Rosen, C. Fertilizer and irrigation management effects on nitrous oxide emissions and nitrate leaching. *Agron. J.* **2014**, *106*, 703–714. [CrossRef]
10.	Mosier, A.; Kroeze, C.; Nevison, C.; Oenema, O.; Seitzinger, S.; Van Cleemput, O. Closing the global N_2O budget: Nitrous oxide emissions through the agricultural nitrogen cycle. *Nutr. Cycl. Agroecosyst.* **1998**, *52*, 225–248. [CrossRef]
11.	Gao, W.; Yang, H.; Kou, L.; Li, S. Effects of nitrogen deposition and fertilization on N transformations in forest soils: A review. *J. Soils Sediments* **2015**, *15*, 863–879. [CrossRef]
12.	Fesenfeld, L.P.; Schmidt, T.S.; Schrode, A. Climate policy for short-and long-lived pollutants. *Nat. Clim. Chang.* **2018**, *8*, 933–936. [CrossRef]
13.	Liu, M.; Huang, X.; Song, Y.; Tang, J.; Cao, J.; Zhang, X.; Zhang, Q.; Wang, S.; Xu, T.; Kang, L. Ammonia emission control in China would mitigate haze pollution and nitrogen deposition, but worsen acid rain. *Proc. Natl. Acad. Sci. USA* **2019**, *116*, 7760–7765. [CrossRef]
14.	Venterea, R.T.; Halvorson, A.D.; Kitchen, N.; Liebig, M.A.; Cavigelli, M.A.; Grosso, S.J.D.; Motavalli, P.P.; Nelson, K.A.; Spokas, K.A.; Singh, B.P. Challenges and opportunities for mitigating nitrous oxide emissions from fertilized cropping systems. *Front. Ecol. Environ.* **2012**, *10*, 562–570. [CrossRef]
15.	Hoben, J.; Gehl, R.; Millar, N.; Grace, P.; Robertson, G. Nonlinear nitrous oxide (N_2O) response to nitrogen fertilizer in on-farm corn crops of the US Midwest. *Glob. Chang. Biol.* **2011**, *17*, 1140–1152. [CrossRef]
16.	Shcherbak, I.; Millar, N.; Robertson, G.P. Global metaanalysis of the nonlinear response of soil nitrous oxide (N_2O) emissions to fertilizer nitrogen. *Proc. Natl. Acad. Sci. USA* **2014**, *111*, 9199–9204. [CrossRef]
17.	Huddell, A.M.; Galford, G.L.; Tully, K.L.; Crowley, C.; Palm, C.A.; Neill, C.; Hickman, J.E.; Menge, D.N. Meta-analysis on the potential for increasing nitrogen losses from intensifying tropical agriculture. *Glob. Chang. Biol.* **2020**, *26*, 1668–1680. [CrossRef] [PubMed]
18.	Leach, A.M.; Galloway, J.N.; Bleeker, A.; Erisman, J.W.; Kohn, R.; Kitzes, J. A nitrogen footprint model to help consumers understand their role in nitrogen losses to the environment. *Environ. Dev.* **2012**, *1*, 40–66. [CrossRef]
19.	Erisman, J.W.; Galloway, J.N.; Seitzinger, S.; Bleeker, A.; Dise, N.B.; Petrescu, A.R.; Leach, A.M.; de Vries, W. Consequences of human modification of the global nitrogen cycle. *Philos. Trans. R. Soc. B Biol. Sci.* **2013**, *368*, 20130116. [CrossRef]

20. Westhoek, H.; Lesschen, J.P.; Leip, A.; Rood, T.; Wagner, S.; De Marco, A.; Murphy-Bokern, D.; Pallière, C.; Howard, C.M.; Oenema, O. *Nitrogen on the Table: The Influence of Food Choices on Nitrogen Emissions and the European Environment*; NERC/Centre for Ecology & Hydrology: Edinburgh, UK, 2015.
21. Westhoek, H.; Lesschen, J.P.; Rood, T.; Wagner, S.; de Marco, A.; Murphy-Bokern, D.; Leip, A.; van Grinsven, H.; Sutton, M.A.; Oenema, O. Food choices, health and environment: Effects of cutting Europe's meat and dairy intake. *Glob. Environ. Chang.* **2014**, *26*, 196–205. [CrossRef]
22. Mohanty, S.; Swain, C.K.; Kumar, A.; Nayak, A. Nitrogen Footprint: A Useful Indicator of Agricultural Sustainability. In *Nutrient Dynamics for Sustainable Crop Production*; Springer: Singapore, 2020; pp. 135–156.
23. Oita, A.; Nagano, I.; Matsuda, H. An improved methodology for calculating the nitrogen footprint of seafood. *Ecol. Indic.* **2016**, *60*, 1091–1103. [CrossRef]
24. Galloway, J.N.; Winiwarter, W.; Leip, A.; Leach, A.M.; Bleeker, A.; Erisman, J.W. Nitrogen footprints: Past, present and future. *Environ. Res. Lett.* **2014**, *9*, 115003. [CrossRef]
25. Shibata, H.; Cattaneo, L.R.; Leach, A.M.; Galloway, J.N. First approach to the Japanese nitrogen footprint model to predict the loss of nitrogen to the environment. *Environ. Res. Lett.* **2014**, *9*, 115013. [CrossRef]
26. Gu, B.; Leach, A.M.; Ma, L.; Galloway, J.N.; Chang, S.X.; Ge, Y.; Chang, J. Nitrogen footprint in China: Food, energy, and nonfood goods. *Environ. Sci. Technol.* **2013**, *47*, 9217–9224. [CrossRef] [PubMed]
27. Gu, B.; Ju, X.; Chang, J.; Ge, Y.; Vitousek, P.M. Integrated reactive nitrogen budgets and future trends in China. *Proc. Natl. Acad. Sci. USA* **2015**, *112*, 8792–8797. [CrossRef]
28. Gu, B.; Ge, Y.; Ren, Y.; Xu, B.; Luo, W.; Jiang, H.; Gu, B.; Chang, J. Atmospheric reactive nitrogen in China: Sources, recent trends, and damage costs. *Environ. Sci. Technol.* **2012**, *46*, 9420–9427. [CrossRef]
29. Smil, V. Detonator of the population explosion. *Nature* **1999**, *400*, 415. [CrossRef]
30. Oita, A.; Malik, A.; Kanemoto, K.; Geschke, A.; Nishijima, S.; Lenzen, M. Substantial nitrogen pollution embedded in international trade. *Nat. Geosci.* **2016**, *9*, 111–115. [CrossRef]
31. San-Martín, W. Global Nitrogen in Sustainable Development: Four Challenges at the Interface of Science and Policy. *Buildings* **2012**, *2*, 300–325.
32. Black, A.; Sherlock, R.; Cameron, K.; Smith, N.; Goh, K. Comparison of three field methods for measuring ammonia volatilization from urea granules broadcast on to pasture. *J. Soil Sci.* **1985**, *36*, 271–280. [CrossRef]
33. Black, A.; Sherlock, R.; Smith, N.; Cameron, K.; Goh, K. Effects of form of nitrogen, season, and urea application rate on ammonia volatilisation from pastures. *N. Z. J. Agric. Res.* **1985**, *28*, 469–474. [CrossRef]
34. McGarry, S.; O'Toole, P.; Morgan, M. Effects of soil temperature and moisture content on ammonia volatilization from urea-treated pasture and tillage soils. *Ir. J. Agric. Res.* **1987**, *26*, 173–182.
35. Cameron, K.; Di, H.J.; Moir, J. Nitrogen losses from the soil/plant system: A review. *Ann. Appl. Biol.* **2013**, *162*, 145–173. [CrossRef]
36. Acton, S.D. *The Effect of Fertiliser Application Rate and Soil PH on Methane Oxidation and Nitrous Oxide Production*; University of Aberdeen: Aberdeen, UK, 2007.
37. Mundepi, A.; Cabrera, M.; Norton, J.; Habteselassie, M. Ammonia Oxidizers as Biological Health Indicators of Elevated Zn and Cu in Poultry Litter Amended Soil. *Water Air Soil Pollut.* **2019**, *230*, 239. [CrossRef]
38. Cassity-Duffey, K.; Cabrera, M.; Franklin, D.; Gaskin, J.; Kissel, D. Effect of soil texture on nitrogen mineralization from organic fertilizers in four common southeastern soils. *Soil Sci. Soc. Am. J.* **2020**, *84*, 534–542. [CrossRef]
39. Bouwman, A.; Boumans, L.; Batjes, N. Modeling global annual N_2O and NO emissions from fertilized fields. *Glob. Biogeochem. Cycles* **2002**, *16*, 28-1–28-9. [CrossRef]
40. Paulot, F.; Jacob, D.J.; Pinder, R.; Bash, J.; Travis, K.; Henze, D. Ammonia emissions in the United States, European Union, and China derived by high-resolution inversion of ammonium wet deposition data: Interpretation with a new agricultural emissions inventory (MASAGE_NH3). *J. Geophys. Res. Atmos.* **2014**, *119*, 4343–4364. [CrossRef]
41. Sutton, M.A.; Erisman, J.W.; Dentener, F.; Möller, D. Ammonia in the environment: From ancient times to the present. *Environ. Pollut.* **2008**, *156*, 583–604. [CrossRef]
42. Carlson, J.; Daehler, K.R. The refined consensus model of pedagogical content knowledge in science education. In *Repositioning Pedagogical Content Knowledge in Teachers' Knowledge for Teaching Science*; Springer: Singapore, 2019; pp. 77–92.
43. Pathak, H.; Bhatia, A.; Prasad, S.; Singh, S.; Kumar, S.; Jain, M.; Kumar, U. Emission of nitrous oxide from rice-wheat systems of Indo-Gangetic plains of India. *Environ. Monit. Assess.* **2002**, *77*, 163–178. [CrossRef] [PubMed]
44. Caranto, J.D.; Lancaster, K.M. Nitric oxide is an obligate bacterial nitrification intermediate produced by hydroxylamine oxidoreductase. *Proc. Natl. Acad. Sci. USA* **2017**, *114*, 8217–8222. [CrossRef] [PubMed]
45. Intergovernmental Panel On Climate Change. Climate change 2007: The physical science basis. *Agenda* **2007**, *6*, 333.
46. Butterbach-Bahl, K.; Baggs, E.M.; Dannenmann, M.; Kiese, R.; Zechmeister-Boltenstern, S. Nitrous oxide emissions from soils: How well do we understand the processes and their controls? *Philos. Trans. R. Soc. B Biol. Sci.* **2013**, *368*, 20130122. [CrossRef]
47. Syakila, A.; Kroeze, C. The global nitrous oxide budget revisited. *Greenh. Gas Meas. Manag.* **2011**, *1*, 17–26. [CrossRef]
48. Sarabia, L.; Solorio, F.J.; Ramírez, L.; Ayala, A.; Aguilar, C.; Ku, J.; Almeida, C.; Cassador, R.; Alves, B.J.; Boddey, R.M. Improving the Nitrogen Cycling in Livestock Systems Through Silvopastoral Systems. In *Nutrient Dynamics for Sustainable Crop Production*; Springer: Singapore, 2020; pp. 189–213.

49. World Health Organization. *Air Quality Guidelines: Global Update 2005: Particulate Matter, Ozone, Nitrogen Dioxide, and Sulfur Dioxide*; World Health Organization: Copenhagen, Denmark, 2006.

50. Avol, E.L.; Gauderman, W.J.; Tan, S.M.; London, S.J.; Peters, J.M. Respiratory effects of relocating to areas of differing air pollution levels. *Am. J. Respir. Crit. Med.* **2001**, *164*, 2067–2072. [CrossRef] [PubMed]

51. Bowatte, G.; Erbas, B.; Lodge, C.J.; Knibbs, L.D.; Gurrin, L.C.; Marks, G.B.; Thomas, P.S.; Johns, D.P.; Giles, G.G.; Hui, J. Traffic-related air pollution exposure over a 5-year period is associated with increased risk of asthma and poor lung function in middle age. *Eur. Respir. J.* **2017**, *50*, 1602357. [PubMed]

52. Singh, B.; Sekhon, G. Nitrate pollution of groundwater from farm use of nitrogen fertilizers—A review. *Agric. Environ.* **1979**, *4*, 207–225. [CrossRef]

53. Watt, W.; Tulinsky, A.; Swenson, R.P.; Watenpaugh, K.D. Comparison of the crystal structures of a flavodoxin in its three oxidation states at cryogenic temperatures. *J. Mol. Biol.* **1991**, *218*, 195–208. [CrossRef]

54. Wild, A.; Cameron, K. Soil nitrogen and nitrate leaching. In *Soils and Agriculture*; Tinker, P.B., Ed.; Blackwell: Oxford, UK, 1980; pp. 35–70.

55. Cameron, K.; Wild, A. Potential aquifer pollution from nitrate leaching following the plowing of temporary grassland. *J. Environ. Qual.* **1984**, *13*, 274–278.

56. Silva, R.; Cameron, K.; Di, H.; Smith, N.; Buchan, G. Effect of macropore flow on the transport of surface-applied cow urine through a soil profile. *Soil Res.* **2000**, *38*, 13–24. [CrossRef]

57. Brender, J.D.; Weyer, P.J.; Romitti, P.A.; Mohanty, B.P.; Shinde, M.U.; Vuong, A.M.; Sharkey, J.R.; Dwivedi, D.; Horel, S.A.; Kantamneni, J. Prenatal nitrate intake from drinking water and selected birth defects in offspring of participants in the national birth defects prevention study. *Environ. Health Perspect.* **2013**, *121*, 1083–1089. [CrossRef]

58. Knobeloch, L.; Salna, B.; Hogan, A.; Postle, J.; Anderson, H. Blue babies and nitrate-contaminated well water. *Environ. Health Perspect.* **2000**, *108*, 675–678. [CrossRef]

59. Folke, C.; Kautsky, N.; Berg, H.; Jansson, Å.; Troell, M. The ecological footprint concept for sustainable seafood production: A review. *Ecol. Appl.* **1998**, *8*, S63–S71. [CrossRef]

60. Little, D.; Edwards, P. *Integrated Livestock-Fish Farming Systems*; Food & Agriculture Organization: Rome, Italy, 2003.

61. Sutton, M.A.; Bleeker, A.; Howard, C.; Erisman, J.; Abrol, Y.; Bekunda, M.; Datta, A.; Davidson, E.; De Vries, W.; Oenema, O. *Our Nutrient World. The Challenge to Produce More Food & Energy with Less Pollution*; Centre for Ecology & Hydrology: Edinburgh, UK, 2013.

62. FAO. *FAOSTAT Online Statistical Service*; Food and Agriculture Organization of the United Nations (FAO): Rome, Italy, 2009.

63. Horneck, D.A.; Sullivan, D.M.; Owen, J.S.; Hart, J.M. *Soil Test Interpretation Guide*; Oregon State University: Corvallis, OR, USA, 2011.

64. Di, H.; Cameron, K. Nitrate leaching in temperate agroecosystems: Sources, factors and mitigating strategies. *Nutr. Cycl. Agroecosystems* **2002**, *64*, 237–256. [CrossRef]

65. Dungait, J.A.; Cardenas, L.M.; Blackwell, M.S.; Wu, L.; Withers, P.J.; Chadwick, D.R.; Bol, R.; Murray, P.J.; Macdonald, A.J.; Whitmore, A.P. Advances in the understanding of nutrient dynamics and management in UK agriculture. *Sci. Total Environ.* **2012**, *434*, 39–50. [CrossRef] [PubMed]

66. Goulding, K.; Jarvis, S.; Whitmore, A. Optimizing nutrient management for farm systems. *Philos. Trans. R. Soc. B Biol. Sci.* **2008**, *363*, 667–680. [CrossRef] [PubMed]

67. Haynes, R. *Mineral Nitrogen in the Plant-Soil System*; Elsevier: Orlando, FL, USA, 2012.

68. Kay, P.; Edwards, A.C.; Foulger, M. A review of the efficacy of contemporary agricultural stewardship measures for ameliorating water pollution problems of key concern to the UK water industry. *Agric. Syst.* **2009**, *99*, 67–75. [CrossRef]

69. Monaghan, R.; De Klein, C.A.; Muirhead, R.W. Prioritisation of farm scale remediation efforts for reducing losses of nutrients and faecal indicator organisms to waterways: A case study of New Zealand dairy farming. *J. Environ. Manag.* **2008**, *87*, 609–622. [CrossRef] [PubMed]

70. Monaghan, R.; Hedley, M.; Di, H.; McDowell, R.; Cameron, K.; Ledgard, S. Nutrient management in New Zealand pastures—recent developments and future issues. *N. Z. J. Agric. Res.* **2007**, *50*, 181–201. [CrossRef]

71. Munoz, F.; Mylavarapu, R.; Hutchinson, C. Environmentally responsible potato production systems: A review. *J. Plant Nutr.* **2005**, *28*, 1287–1309. [CrossRef]

72. Dahal, S.; Franklin, D.; Subedi, A.; Cabrera, M.; Hancock, D.; Mahmud, K.; Ney, L.; Park, C.; Mishra, D.J.S. Strategic Grazing in Beef-Pastures for Improved Soil Health and Reduced Runoff-Nitrate-A Step towards. *Sustainability* **2020**, *12*, 558. [CrossRef]

73. Giller, K.E.; Chalk, P.; Dobermann, A.; Hammond, L.; Heffer, P.; Ladha, J.K.; Nyamudeza, P.; Maene, L.; Ssali, H.; Freney, J. Emerging technologies to increase the efficiency of use of fertilizer nitrogen. *Agric. Nitrogen Cycle Assess. Impacts Fertil. Food Prod. Environ.* **2004**, *65*, 35–51.

74. Freney, J.R. Management practices to increase efficiency of fertilizer and animal nitrogen and minimize nitrogen loss to the atmosphere and groundwater. *Tech. Bull. Food Fertil. Technol. Cent.* **2011**, *186*, 22.

75. Soares, J.R.; Cantarella, H.; de Campos Menegale, M.L. Ammonia volatilization losses from surface-applied urea with urease and nitrification inhibitors. *Soil Biol. Biochem.* **2012**, *52*, 82–89. [CrossRef]

76. Verma, J.P.; Jaiswal, D.K.; Meena, V.S.; Kumar, A.; Meena, R. Issues and challenges about sustainable agriculture production for management of natural resources to sustain soil fertility and health. *J. Clean. Prod.* **2015**, *107*, 793–794. [CrossRef]

77. Mitran, T.; Meena, R.S.; Lal, R.; Layek, J.; Kumar, S.; Datta, R. Role of soil phosphorus on legume production. In *Legumes for Soil Health and Sustainable Management*; Springer: Singapore, 2018; pp. 487–510.

78. Sanz-Cobena, A.; Misselbrook, T.H.; Arce, A.; Mingot, J.I.; Diez, J.A.; Vallejo, A. An inhibitor of urease activity effectively reduces ammonia emissions from soil treated with urea under Mediterranean conditions. *Agric. Ecosyst. Environ.* **2008**, *126*, 243–249. [CrossRef]

79. Turner, D.; Edis, R.; Chen, D.; Freney, J.; Denmead, O.; Christie, R. Determination and mitigation of ammonia loss from urea applied to winter wheat with N-(n-butyl) thiophosphorictriamide. *Agric. Ecosyst. Environ.* **2010**, *137*, 261–266. [CrossRef]

80. Chambers, B.; Dampney, P. Nitrogen efficiency and ammonia emissions from urea-based and ammonium nitrate fertilisers. In *Proceedings of Proceedings-International Fertiliser Society*; Internationl Fertiliser Society: York, UK, 2009.

81. Rodriguez, M.J.; Saggar, S.; Berben, P.; Palmada, T.; Lopez-Villalobos, N.; Pal, P. Use of a urease inhibitor to mitigate ammonia emissions from urine patches. *Environ. Technol.* **2021**, *42*, 20–31. [CrossRef]

82. Meng, Y.; Wang, J.J.; Wei, Z.; Dodla, S.K.; Fultz, L.M.; Gaston, L.A.; Xiao, R.; Park, J.-h.; Scaglia, G. Nitrification inhibitors reduce nitrogen losses and improve soil health in a subtropical pastureland. *Geoderma* **2021**, *388*, 114947. [CrossRef]

83. Dobermann, A.; Witt, C.; Dawe, D.; Abdulrachman, S.; Gines, H.; Nagarajan, R.; Satawathananont, S.; Son, T.; Tan, P.; Wang, G. Site-specific nutrient management for intensive rice cropping systems in Asia. *Field Crops Res.* **2002**, *74*, 37–66. [CrossRef]

84. Tagarakis, A.C.; Ketterings, Q.M. In-season estimation of corn yield potential using proximal sensing. *Agron. J.* **2017**, *109*, 1323–1330. [CrossRef]

85. Shanahan, J.; Kitchen, N.; Raun, W.; Schepers, J.S. Responsive in-season nitrogen management for cereals. *Comput. Electron. Agric.* **2008**, *61*, 51–62. [CrossRef]

86. Siqueira, R.; Longchamps, L.; Dahal, S.; Khosla, R. Use of fluorescence sensing to detect nitrogen and potassium variability in maize. *Remote Sens.* **2020**, *12*, 1752. [CrossRef]

87. Dahal, S.; Phillippi, E.; Longchamps, L.; Khosla, R.; Andales, A. Variable Rate Nitrogen and Water Management for Irrigated Maize in the Western US. *Agronomy* **2020**, *10*, 1533. [CrossRef]

88. Havránková, J. The evaluation of ground based remote sensing systems for canopy nitrogen management in winter wheat. *CIGR J.* **2007**. Available online: https://dspace.lib.cranfield.ac.uk/handle/1826/1711 (accessed on 23 February 2021).

89. Thind, H.; Kumar, A.; Vashistha, M. Calibrating the leaf colour chart for need based fertilizer nitrogen management in different maize (Zea mays L.) genotypes. *Field Crop. Res.* **2011**, *120*, 276–282.

90. Fu, Y.-Q.; Zhong, X.-H.; Zeng, J.-H.; Liang, K.-M.; Pan, J.-F.; Xin, Y.-F.; Liu, Y.-Z.; Hu, X.-Y.; Peng, B.-L.; Chen, R.-B. Improving grain yield, nitrogen use efficiency and radiation use efficiency by dense planting, with delayed and reduced nitrogen application, in double cropping rice in South China. *J. Integr. Agric.* **2021**, *20*, 565–580. [CrossRef]

91. Chen, Y.; Fan, P.; Mo, Z.; Kong, L.; Tian, H.; Duan, M.; Li, L.; Wu, L.; Wang, Z.; Tang, X. Deep Placement of Nitrogen Fertilizer Affects Grain Yield, Nitrogen Recovery Efficiency, and Root Characteristics in Direct-Seeded Rice in South China. *J. Plant. Growth Regul.* **2020**, *40*, 379–387. [CrossRef]

92. KhalilA, M. Physical and chemical manipulation of urea fertiliser for reducing the emission of gaseous nitrogen species. In Proceedings of the 19th World Congress of Soil Science: Soil Solutions for a Changing World, Brisbane, Australia, 1–6 August 2010; Congress Symposium 4: Greenhouse Gases from Soils; pp. 195–198.

93. Chatterjee, D.; Mohanty, S.; Guru, P.K.; Swain, C.K.; Tripathi, R.; Shahid, M.; Kumar, U.; Kumar, A.; Bhattacharyya, P.; Gautam, P. Comparative assessment of urea briquette applicators on greenhouse gas emission, nitrogen loss and soil enzymatic activities in tropical lowland rice. *Agric. Ecosyst. Environ.* **2018**, *252*, 178–190. [CrossRef]

94. Mencaroni, M.; Dal Ferro, N.; Furlanetto, J.; Longo, M.; Lazzaro, B.; Sartori, L.; Grant, B.; Smith, W.; Morari, F. Identifying N fertilizer management strategies to reduce ammonia volatilization: Towards a site-specific approach. *J. Environ. Manag.* **2021**, *277*, 111445. [CrossRef] [PubMed]

95. Lessa, A.C.R.; Madari, B.E.; Paredes, D.S.; Boddey, R.M.; Urquiaga, S.; Jantalia, C.P.; Alves, B.J.J.A. Ecosystems; Environment. Bovine urine and dung deposited on Brazilian savannah pastures contribute differently to direct and indirect soil nitrous oxide emissions. *Agric. Ecosyst. Environ.* **2014**, *190*, 104–111. [CrossRef]

96. Datta, R.; Kelkar, A.; Baraniya, D.; Molaei, A.; Moulick, A.; Meena, R.; Formanek, P. Enzymatic degradation of lignin in soil: A review. *Sustainability* **2017**, *7*, 1163. [CrossRef]

97. Subedi, A.; Franklin, D.; Cabrera, M.; McPherson, A.; Dahal, S. Grazing Systems to Retain and Redistribute Soil Phosphorus and to Reduce Phosphorus Losses in Runoff. *Soil Syst.* **2020**, *4*, 66. [CrossRef]

98. Adegbeye, M.; Reddy, P.R.K.; Obaisi, A.; Elghandour, M.; Oyebamiji, K.; Salem, A.; Morakinyo-Fasipe, O.; Cipriano-Salazar, M.; Camacho-Díaz, L. Sustainable agriculture options for production, greenhouse gasses and pollution alleviation, and nutrient recycling in emerging and transitional nations-An overview. *J. Clean. Prod.* **2020**, *242*, 118319. [CrossRef]

99. Frank, S.; Havlík, P.; Stehfest, E.; van Meijl, H.; Witzke, P.; Pérez-Domínguez, I.; van Dijk, M.; Doelman, J.C.; Fellmann, T.; Koopman, J.F. Agricultural non-CO$_2$ emission reduction potential in the context of the 1.5 °C target. *Nat. Clim. Chang.* **2019**, *9*, 66–72. [CrossRef]

100. Herrero, M.; Havlík, P.; Valin, H.; Notenbaert, A.; Rufino, M.C.; Thornton, P.K.; Blümmel, M.; Weiss, F.; Grace, D.; Obersteiner, M. Biomass use, production, feed efficiencies, and greenhouse gas emissions from global livestock systems. *Proc. Natl. Acad. Sci. USA* **2013**, *110*, 20888–20893. [CrossRef]

101. Takemasa, M.J.K. Nutritional strategies to reduce nutrient waste in livestock and poultry production. *Food* **1998**, *36*, 720–726.

102. Stephenson, G.T. The Effects of Agricultural Waste-Based Compost Amendments in Organic Pest Management. Master's Thesis, California Polytechnics University, San Luis Obispo, CA, USA, 2019.
103. Bhandari, K.B.; West, C.P.; Acosta-Martinez, V.; Cotton, J.; Cano, A. Soil health indicators as affected by diverse forage species and mixtures in semi-arid pastures. *Appl. Soil Ecol.* **2018**, *132*, 179–186. [CrossRef]
104. Peoples, M.; Baldock, J. Nitrogen dynamics of pastures: Nitrogen fixation inputs, the impact of legumes on soil nitrogen fertility, and the contributions of fixed nitrogen to Australian farming systems. *Aust. J. Exp. Agric.* **2001**, *41*, 327–346. [CrossRef]
105. Stockdale, E.; Lampkin, N.; Hovi, M.; Keatinge, R.; Lennartsson, E.; Macdonald, D.; Padel, S.; Tattersall, F.; Wolfe, M.; Watson, C. Agronomic and environmental implications of organic farming systems. *Agronomy* **2001**, *70*, 261–327.
106. White, J.; Hodgson, J.G. *New Zealand Pasture and Crop Science*; Oxford University Press: Oxford, UK, 1999.
107. Bhandari, K.B.; West, C.P.; Acosta-Martinez, V. Assessing the role of interseeding alfalfa into grass on improving pasture soil health in semi-arid Texas High Plains. *Appl. Soil Ecol.* **2020**, *147*, 103399. [CrossRef]
108. Mobin, S.; Alam, F. Biofuel production from algae utilizing wastewater. In Proceedings of the 19th Australasian Fluid Mechanics Conference, Melbourne, Australia, 8–11 December 2014.
109. Dang, N.M.; Lee, K. Recent trends of using alternative nutrient sources for microalgae cultivation as a feedstock of biodiesel production. *Appl. Chem. Eng.* **2018**, *29*, 1–9.
110. Aslan, S.; Kapdan, I.K. Batch kinetics of nitrogen and phosphorus removal from synthetic wastewater by algae. *Ecol. Eng.* **2006**, *28*, 64–70. [CrossRef]
111. Marinho-Soriano, E.; Nunes, S.; Carneiro, M.; Pereira, D. Nutrients' removal from aquaculture wastewater using the macroalgae Gracilaria birdiae. *Biomass Bioenergy* **2009**, *33*, 327–331. [CrossRef]
112. He, P.; Xu, S.; Zhang, H.; Wen, S.; Dai, Y.; Lin, S.; Yarish, C. Bioremediation efficiency in the removal of dissolved inorganic nutrients by the red seaweed, Porphyra yezoensis, cultivated in the open sea. *Water Res.* **2008**, *42*, 1281–1289. [CrossRef]
113. Webb, J.R.; Hayes, N.M.; Simpson, G.L.; Leavitt, P.R.; Baulch, H.M.; Finlay, K. Widespread nitrous oxide undersaturation in farm waterbodies creates an unexpected greenhouse gas sink. *Proc. Natl. Acad. Sci. USA* **2019**, *116*, 9814–9819. [CrossRef]
114. Meade, G.; Pierce, K.; O'Doherty, J.; Mueller, C.; Lanigan, G.; Mc Cabe, T. Ammonia and nitrous oxide emissions following land application of high and low nitrogen pig manures to winter wheat at three growth stages. *Agric. Ecosyst. Environ.* **2011**, *140*, 208–217. [CrossRef]
115. Mahmud, K.; Chowhdhury, M.; Noor, N.; Huq, S.I. Effects of different sources of biochar application on the emission of a number of gases from soil. *Can. J. Pure Appl. Sci.* **2014**, *8*, 2813–2824.
116. Brennan, R.B.; Healy, M.G.; Fenton, O.; Lanigan, G.J. The effect of chemical amendments used for phosphorus abatement on greenhouse gas and ammonia emissions from dairy cattle slurry: Synergies and pollution swapping. *PLoS ONE* **2015**, *10*, e0111965. [CrossRef] [PubMed]
117. Eggleston, S.; Buendia, L.; Miwa, K.; Ngara, T.; Tanabe, K. *2006 IPCC Guidelines for National Greenhouse Gas Inventories*; Institute for Global Environmental Strategies: Wageningen, The Netherlands, 2006; Volume 5.
118. Zhu, X.; Burger, M.; Doane, T.A.; Horwath, W.R. Ammonia oxidation pathways and nitrifier denitrification are significant sources of N_2O and NO under low oxygen availability. *Proc. Natl. Acad. Sci. USA* **2013**, *110*, 6328–6333. [CrossRef]
119. Panday, D.; Mikha, M.M.; Collins, H.P.; Jin, V.L.; Kaiser, M.; Cooper, J.; Malakar, A.; Maharjan, B. Optimum rates of surface-applied coal char decreased soil ammonia volatilization loss. *J. Environ. Qual.* **2020**, *49*, 256–267. [CrossRef] [PubMed]
120. Oleszczuk, P.; Hollert, H. Comparison of sewage sludge toxicity to plants and invertebrates in three different soils. *Chemosphere* **2011**, *83*, 502–509. [CrossRef] [PubMed]
121. Zielińska, A.; Oleszczuk, P. Evaluation of sewage sludge and slow pyrolyzed sewage sludge-derived biochar for adsorption of phenanthrene and pyrene. *Bioresour. Technol.* **2015**, *192*, 618–626. [CrossRef]
122. Ramakrishnan, B.; Maddela, N.R.; Venkateswarlu, K.; Megharaj, M. Organic farming: Does it contribute to contaminant-free produce and ensure food safety? *Sci. Total. Environ.* **2021**, *769*, 145079. [CrossRef]
123. Oleszczuk, P.; Hale, S.E.; Lehmann, J.; Cornelissen, G. Activated carbon and biochar amendments decrease pore-water concentrations of polycyclic aromatic hydrocarbons (PAHs) in sewage sludge. *Bioresour. Technol.* **2012**, *111*, 84–91. [CrossRef]
124. Stefaniuk, M.; Oleszczuk, P. Addition of biochar to sewage sludge decreases freely dissolved PAHs content and toxicity of sewage sludge-amended soil. *Environ. Pollut.* **2016**, *218*, 242–251. [CrossRef] [PubMed]
125. Oleszczuk, P.; Zielińska, A.; Cornelissen, G. Stabilization of sewage sludge by different biochars towards reducing freely dissolved polycyclic aromatic hydrocarbons (PAHs) content. *Bioresour. Technol.* **2014**, *156*, 139–145. [CrossRef] [PubMed]
126. Kończak, M.; Oleszczuk, P. Application of biochar to sewage sludge reduces toxicity and improve organisms growth in sewage sludge-amended soil in long term field experiment. *Sci. Total Environ.* **2018**, *625*, 8–15. [CrossRef] [PubMed]
127. Noor, N.; Mahmud, K.; Chowdhury, M.T.A.; Huq, S.I. The use of biochar as ameliorator for soil arsenic. *Dhaka Univ. J. Biol. Sci.* **2015**, *24*, 111–119. [CrossRef]
128. Mahmud, K.; Makaju, S.; Ibrahim, R.; Missaoui, A.J.P. Current Progress in Nitrogen Fixing Plants and Microbiome Research. *Plants* **2020**, *9*, 97. [CrossRef] [PubMed]
129. Bloch, S.E.; Ryu, M.-H.; Ozaydin, B.; Broglie, R. Harnessing atmospheric nitrogen for cereal crop production. *Curr. Opin. Biotechnol.* **2020**, *62*, 181–188. [CrossRef]
130. Rogers, C.; Oldroyd, G.E. Synthetic biology approaches to engineering the nitrogen symbiosis in cereals. *J. Exp. Bot.* **2014**, *65*, 1939–1946. [CrossRef]

131. Madsen, L.H.; Tirichine, L.; Jurkiewicz, A.; Sullivan, J.T.; Heckmann, A.B.; Bek, A.S.; Ronson, C.W.; James, E.K.; Stougaard, J. The molecular network governing nodule organogenesis and infection in the model legume Lotus japonicus. *Nat. Commun.* **2010**, *1*, 1–12. [CrossRef]

132. Maillet, F.; Poinsot, V.; André, O.; Puech-Pagès, V.; Haouy, A.; Gueunier, M.; Cromer, L.; Giraudet, D.; Formey, D.; Niebel, A. Fungal lipochitooligosaccharide symbiotic signals in arbuscular mycorrhiza. *Nature* **2011**, *469*, 58–63. [CrossRef]

133. Oldroyd, G.E. Speak, friend, and enter: Signalling systems that promote beneficial symbiotic associations in plants. *Nat. Rev. Microbiol.* **2013**, *11*, 252–263. [CrossRef]

134. Mus, F.; Crook, M.B.; Garcia, K.; Costas, A.G.; Geddes, B.A.; Kouri, E.D.; Paramasivan, P.; Ryu, M.-H.; Oldroyd, G.E.; Poole, P.S. Symbiotic nitrogen fixation and the challenges to its extension to nonlegumes. *Appl. Environ. Microbiol.* **2016**, *82*, 3698–3710. [CrossRef] [PubMed]

135. Geddes, B.A.; Ryu, M.-H.; Mus, F.; Costas, A.G.; Peters, J.W.; Voigt, C.A.; Poole, P. Use of plant colonizing bacteria as chassis for transfer of N2-fixation to cereals. *Curr. Opin. Biotechnol.* **2015**, *32*, 216–222. [CrossRef] [PubMed]

136. Rocha, F.R.; Papini-Terzi, F.S.; Nishiyama, M.Y.; Vencio, R.Z.; Vicentini, R.; Duarte, R.D.; de Rosa, V.E.; Vinagre, F.; Barsalobres, C.; Medeiros, A.H. Signal transduction-related responses to phytohormones and environmental challenges in sugarcane. *BMC Genom.* **2007**, *8*, 71. [CrossRef]

137. Ibort, P.; Imai, H.; Uemura, M.; Aroca, R. Proteomic analysis reveals that tomato interaction with plant growth promoting bacteria is highly determined by ethylene perception. *J. Plant Physiol.* **2018**, *220*, 43–59. [CrossRef] [PubMed]

138. Brusamarello-Santos, L.C.; Gilard, F.; Brulé, L.; Quilleré, I.; Gourion, B.; Ratet, P.; Maltempi de Souza, E.; Lea, P.J.; Hirel, B. Metabolic profiling of two maize (Zea mays L.) inbred lines inoculated with the nitrogen fixing plant-interacting bacteria Herbaspirillum seropedicae and Azospirillum brasilense. *PLoS ONE* **2017**, *12*, e0174576. [CrossRef]

139. Kost, T.; Stopnisek, N.; Agnoli, K.; Eberl, L.; Weisskopf, L. Oxalotrophy, a widespread trait of plant-associated Burkholderia species, is involved in successful root colonization of lupin and maize by Burkholderia phytofirmans. *Front. Microbiol.* **2014**, *4*, 421. [CrossRef]

140. Van Deynze, A.; Zamora, P.; Delaux, P.-M.; Heitmann, C.; Jayaraman, D.; Rajasekar, S.; Graham, D.; Maeda, J.; Gibson, D.; Schwartz, K.D. Nitrogen fixation in a landrace of maize is supported by a mucilage-associated diazotrophic microbiota. *PLoS Biol.* **2018**, *16*, e2006352. [CrossRef]

141. Oliveira, A.L.; Santos, O.J.; Marcelino, P.R.; Milani, K.M.; Zuluaga, M.Y.; Zucareli, C.; Gonçalves, L.S. Maize inoculation with Azospirillum brasilense Ab-V5 cells enriched with exopolysaccharides and polyhydroxybutyrate results in high productivity under low N fertilizer input. *Front. Microbiol.* **2017**, *8*, 1873. [CrossRef] [PubMed]

142. Compant, S.; Samad, A.; Faist, H.; Sessitsch, A. A review on the plant microbiome: Ecology, functions, and emerging trends in microbial application. *J. Adv. Res.* **2019**, *19*, 29–37. [CrossRef]

143. Hartmann, M.; Frey, B.; Mayer, J.; Mäder, P.; Widmer, F. Distinct soil microbial diversity under long-term organic and conventional farming. *ISME J.* **2015**, *9*, 1177–1194. [CrossRef]

144. Larsen, E.; Grossman, J.; Edgell, J.; Hoyt, G.; Osmond, D.; Hu, S. Soil biological properties, soil losses and corn yield in long-term organic and conventional farming systems. *Soil Tillage Res.* **2014**, *139*, 37–45. [CrossRef]

145. Coelho, L.M.; Rezende, H.C.; Coelho, L.M.; de Sousa, P.; Melo, D.; Coelho, N. Bioremediation of polluted waters using microorganisms. *Adv. Bioremedia. Wastewater Polluted Soil* **2015**, *10*, 60770.

146. Kloepper, J.W. Plant growth-promoting rhizobacteria on radishes. In Proceedings of the 4th International Conference on Plant Pathogenic Bacter, Station de Pathologie Vegetale et Phytobacteriologie, INRA, Angers, France, 27 August–2 September 1978; pp. 879–882.

147. Bonkowski, M.; Villenave, C.; Griffiths, B. Rhizosphere fauna: The functional and structural diversity of intimate interactions of soil fauna with plant roots. *Plant Soil* **2009**, *321*, 213–233. [CrossRef]

148. Buee, M.; De Boer, W.; Martin, F.; Van Overbeek, L.; Jurkevitch, E. The rhizosphere zoo: An overview of plant-associated communities of microorganisms, including phages, bacteria, archaea, and fungi, and of some of their structuring factors. *Plant Soil* **2009**, *321*, 189–212. [CrossRef]

149. Cao, Y.; Bai, M.; Han, B.; Impraim, R.; Butterly, C.; Hu, H.; He, J.; Chen, D. Enhanced nitrogen retention by lignite during poultry litter composting. *J. Clean. Prod.* **2020**, *277*, 122422. [CrossRef]

150. Morriën, E. Understanding soil food web dynamics, how close do we get? *Soil Biol. Biochem.* **2016**, *102*, 10–13. [CrossRef]

151. Verbruggen, E.; van der Heijden, M.G.; Rillig, M.C.; Kiers, E.T. Mycorrhizal fungal establishment in agricultural soils: Factors determining inoculation success. *New Phytol.* **2013**, *197*, 1104–1109. [CrossRef]

152. Ney, L.; Franklin, D.; Mahmud, K.; Cabrera, M.; Hancock, D.; Habteselassie, M.; Newcomer, Q. Examining trophic-level nematode community structure and nitrogen mineralization to assess local effective microorganisms' role in nitrogen availability of swine effluent to forage crops. *Appl. Soil Ecol.* **2018**, *130*, 209–218. [CrossRef]

153. Ney, L.; Franklin, D.; Mahmud, K.; Cabrera, M.; Hancock, D.; Habteselassie, M.; Newcomer, Q.; Fatzinger, B. Rebuilding Soil Ecosystems for Improved Productivity in Biosolarized Soils. *Int. J. Agron.* **2019**, *2019*. [CrossRef]

154. Mahmud, K.; Franklin, D.; Ney, L.; Cabrera, M.; Habteselassie, M.; Hancock, D.; Newcomer, Q.; Subedi, A.; Dahal, S. Improving inorganic nitrogen in soil and nutrient density of edamame bean in three consecutive summers by utilizing a locally sourced bio-inocula. *Org. Agric.* **2021**. [CrossRef]

155. Li, X.; Guo, Q.; Wang, Y.; Xu, J.; Wei, Q.; Chen, L.; Liao, L. Enhancing Nitrogen and Phosphorus Removal by Applying Effective Microorganisms to Constructed Wetlands. *Water* **2020**, *12*, 2443. [CrossRef]
156. Polyorach, S.; Wanapat, M.; Cherdthong, A.; Kang, S. Rumen microorganisms, methane production, and microbial protein synthesis affected by mangosteen peel powder supplement in lactating dairy cows. *Trop. Anim. Health Prod.* **2016**, *48*, 593–601. [CrossRef] [PubMed]
157. Alves, T.P.; Dall-Orsoletta, A.C.; Ribeiro-Filho, H.M.N. The effects of supplementing Acacia mearnsii tannin extract on dairy cow dry matter intake, milk production, and methane emission in a tropical pasture. *Trop. Anim. Health Prod.* **2017**, *49*, 1663–1668. [CrossRef]
158. Albores-Moreno, S.; Alayón-Gamboa, J.; Ayala-Burgos, A.; Solorio-Sánchez, F.; Aguilar-Pérez, C.; Olivera-Castillo, L.; Ku-Vera, J. Effects of feeding ground pods of Enterolobium cyclocarpum Jacq. Griseb on dry matter intake, rumen fermentation, and enteric methane production by Pelibuey sheep fed tropical grass. *Trop. Anim. Health Prod.* **2017**, *49*, 857–866. [CrossRef] [PubMed]
159. Freitas, C.E.S.; Duarte, E.R.; Alves, D.D.; Martinele, I.; D'Agosto, M.; Cedrola, F.; de Moura Freitas, A.A.; dos Santos Soares, F.D.; Beltran, M. Sheep fed with banana leaf hay reduce ruminal protozoa population. *Trop. Anim. Health Prod.* **2017**, *49*, 807–812. [CrossRef]
160. Gerlach, K.; Schmithausen, A.J.; Sommer, A.C.; Trimborn, M.; Büscher, W.; Südekum, K.-H. Cattle diets strongly affect nitrous oxide in the rumen. *Sustainability* **2018**, *10*, 3679. [CrossRef]
161. Reganold, J.P.; Wachter, J.M. Organic agriculture in the twenty-first century. *Nat. Plants* **2016**, *2*, 1–8. [CrossRef]
162. Trewavas, A. Urban myths of organic farming. *Nature* **2001**, *410*, 409–410. [CrossRef] [PubMed]
163. Emsley, J. Going one better than nature? *Nature* **2001**, *410*, 633–634. [CrossRef]
164. Willer, H.; Lernoud, J. *The World of Organic Agriculture. Statistics and Emerging Trends 2019*; Research Institute of Organic Agriculture FiBL and IFOAM Organics International: Frick, Switzerland; Bonn, Switzerland 2019.
165. Saffeullah, P.; Nabi, N.; Liaqat, S.; Anjum, N.A.; Siddiqi, T.O.; Umar, S. Organic Agriculture: Principles, Current Status, and Significance. In *Microbiota and Biofertilizers*; Springer: Singapore, 2021; pp. 17–37.
166. Shi, W.; Norton, J.M. Microbial control of nitrate concentrations in an agricultural soil treated with dairy waste compost or ammonium fertilizer. *Soil Biol. Biochem.* **2000**, *32*, 1453–1457. [CrossRef]
167. Habteselassie, M.Y.; Miller, B.E.; Thacker, S.G.; Stark, J.M.; Norton, J.M. Soil nitrogen and nutrient dynamics after repeated application of treated dairy-waste. *Soil Sci. Soc. Am. J.* **2006**, *70*, 1328–1337. [CrossRef]
168. Cameron, K.C.; Rate, A.W.; Noonan, M.J.; Moore, S.; Smith, N.P.; Kerr, L.E. Lysimeter study of the fate of nutrients following subsurface injection and surface application of dairy pond sludge to pasture. *Agric. Ecosyst. Environ.* **1996**, *58*, 187–197. [CrossRef]
169. Kirchmann, H.; Bergström, L. Do organic farming practices reduce nitrate leaching? *Commun. Soil Sci. Plant Anal.* **2001**, *32*, 997–1028. [CrossRef]
170. Scialabba, N.E.-H.; Müller-Lindenlauf, M. Organic agriculture and climate change. *Renew. Agric. Food Syst.* **2010**, *25*, 158–169. [CrossRef]
171. Chatterjee, R. *Projecting the Future of Nitrogen Pollution*; ACS Publications: Washington, DC, USA, 2009. [CrossRef]
172. Williams, A.; Audsley, E.; Sandars, D. Determining the Environmental Burdens and Resource Use in the Production of Agricultural and Horticultural Commodities: Defra Project Report IS0205. Available online: http://randd.defra.gov.uk/Default.aspx (accessed on 27 January 2021).
173. Rahmann, G.; Ardakani, M.R.; Bàrberi, P.; Boehm, H.; Canali, S.; Chander, M.; David, W.; Dengel, L.; Erisman, J.W.; Galvis-Martinez, A.C. Organic Agriculture 3.0 is innovation with research. *Org. Agric.* **2017**, *7*, 169–197. [CrossRef]
174. Smith, J.B.; Lenhart, S.S. Climate change adaptation policy options. *Clim. Res.* **1996**, *6*, 193–201. [CrossRef]
175. Zhang, F.; Li, L. Using competitive and facilitative interactions in intercropping systems enhances crop productivity and nutrient-use efficiency. *Plant Soil* **2003**, *248*, 305–312. [CrossRef]
176. Tilman, D. Global environmental impacts of agricultural expansion: The need for sustainable and efficient practices. *Proc. Natl. Acad. Sci. USA* **1999**, *96*, 5995–6000. [CrossRef]
177. Vymazal, J.; Březinová, T.D. Removal of nutrients, organics and suspended solids in vegetated agricultural drainage ditch. *Ecol. Eng.* **2018**, *118*, 97–103. [CrossRef]
178. Xia, Y.; Zhang, M.; Tsang, D.C.; Geng, N.; Lu, D.; Zhu, L.; Igalavithana, A.D.; Dissanayake, P.D.; Rinklebe, J.; Yang, X. Recent advances in control technologies for non-point source pollution with nitrogen and phosphorous from agricultural runoff: Current practices and future prospects. *Appl. Biol. Chem.* **2020**, *63*, 1–13. [CrossRef]
179. Conley, D.J.; Paerl, H.W.; Howarth, R.W.; Boesch, D.F.; Seitzinger, S.P.; Havens, K.E.; Lancelot, C.; Likens, G.E. Controlling eutrophication: Nitrogen and phosphorus. *Science* **2009**, *323*, 1014–1015. [CrossRef] [PubMed]
180. Howarth, R.; Chan, F.; Conley, D.J.; Garnier, J.; Doney, S.C.; Marino, R.; Billen, G. Coupled biogeochemical cycles: Eutrophication and hypoxia in temperate estuaries and coastal marine ecosystems. *Front. Ecol. Environ.* **2011**, *9*, 18–26. [CrossRef]
181. Kumwimba, M.N.; Meng, F.; Iseyemi, O.; Moore, M.T.; Zhu, B.; Tao, W.; Liang, T.J.; Ilunga, L. Removal of non-point source pollutants from domestic sewage and agricultural runoff by vegetated drainage ditches (VDDs): Design, mechanism, management strategies, and future directions. *Sci. Total Environ.* **2018**, *639*, 742–759. [CrossRef] [PubMed]
182. Dollinger, J.; Dagès, C.; Bailly, J.-S.; Lagacherie, P.; Voltz, M. Managing ditches for agroecological engineering of landscape: A review. *Agron. Sustain. Dev.* **2015**, *35*, 999–1020. [CrossRef]

183. Nsenga Kumwimba, M.; Dzakpasu, M.; Zhu, B.; Wang, T.; Ilunga, L.; Kavidia Muyembe, D. Nutrient removal in a trapezoidal vegetated drainage ditch used to treat primary domestic sewage in a small catchment of the upper Yangtze River. *Water Environ. J.* **2017**, *31*, 72–79. [CrossRef]

184. Kröger, R.; Holland, M.; Moore, M.; Cooper, C. Hydrological variability and agricultural drainage ditch inorganic nitrogen reduction capacity. *J. Environ. Qual.* **2007**, *36*, 1646–1652. [CrossRef]

185. Menon, R.; Holland, M.M. Phosphorus release due to decomposition of wetland plants. *Wetlands* **2014**, *34*, 1191–1196. [CrossRef]

186. Yu, H.; Yang, Z.; Xiao, R.; Zhang, S.; Liu, F.; Xiang, Z. Absorption capacity of nitrogen and phosphorus of aquatic plants and harvest management research. *Acta Prataculturae Sin.* **2013**, *22*, 294–299.

187. Rounsley, S.D.; Glodek, A.; Sutton, G.; Adams, M.D.; Somerville, C.R.; Venter, J.C.; Kerlavage, A.R. The construction of Arabidopsis expressed sequence tag assemblies (a new resource to facilitate gene identification). *Plant Physiol.* **1996**, *112*, 1177–1183. [CrossRef] [PubMed]

188. Velculescu, V.E.; Zhang, L.; Vogelstein, B.; Kinzler, K.W. Serial analysis of gene expression. *Science* **1995**, *270*, 484–487. [CrossRef]

189. Katagiri, F.; Glazebrook, J. Overview of mRNA expression profiling using DNA microarrays. *Curr. Protoc. Mol. Biol.* **2009**, *85*, 22.4.1–22.4.13. [CrossRef] [PubMed]

190. Wang, Z.; Gerstein, M.; Snyder, M. RNA-Seq: A revolutionary tool for transcriptomics. *Nat. Rev. Genet.* **2009**, *10*, 57–63. [CrossRef]

191. Liu, R.; Zhang, H.; Zhao, P.; Zhang, Z.; Liang, W.; Tian, Z.; Zheng, Y. Mining of candidate maize genes for nitrogen use efficiency by integrating gene expression and QTL data. *Plant Mol. Biol. Report.* **2012**, *30*, 297–308. [CrossRef]

192. Bordes, J.; Ravel, C.; Jaubertie, J.; Duperrier, B.; Gardet, O.; Heumez, E.; Pissavy, A.; Charmet, G.; Le Gouis, J.; Balfourier, F. Genomic regions associated with the nitrogen limitation response revealed in a global wheat core collection. *Theor. Appl. Genet.* **2013**, *126*, 805–822. [CrossRef]

193. Robertson, G.P.; Bruulsema, T.W.; Gehl, R.J.; Kanter, D.; Mauzerall, D.L.; Rotz, C.A.; Williams, C.O. Nitrogen–climate interactions in US agriculture. *Biogeochemistry* **2013**, *114*, 41–70. [CrossRef]

194. Suddick, E.C.; Whitney, P.; Townsend, A.R.; Davidson, E.A. The role of nitrogen in climate change and the impacts of nitrogen–climate interactions in the United States: Foreword to thematic issue. *Biogeochemistry* **2013**, *114*, 1–10. [CrossRef]

195. Townsend, A.R.; Vitousek, P.M.; Houlton, B.Z. The climate benefits of better nitrogen and phosphorus management. *Issues Sci. Technol.* **2012**, *28*, 85–91.

196. Sohrabi, C.; Alsafi, Z.; O'Neill, N.; Khan, M.; Kerwan, A.; Al-Jabir, A.; Iosifidis, C.; Agha, R. World Health Organization declares global emergency: A review of the 2019 novel coronavirus (COVID-19). *Int. J. Surg.* **2020**, *76*, 71–76. [CrossRef]

197. Jebril, N. World Health Organization declared a pandemic public health menace: A systematic review of the coronavirus disease 2019 "COVID-19", up to 26th March 2020. *SSRN* **2020**, [CrossRef]

198. Ogen, Y. Assessing nitrogen dioxide (NO_2) levels as a contributing factor to the coronavirus (COVID-19) fatality rate. *Sci. Total Environ.* **2020**, *726*, 138605. [CrossRef]

199. Karaer, A.; Balafkan, N.; Gazzea, M.; Arghandeh, R.; Ozguven, E.E. Analyzing COVID-19 Impacts on Vehicle Travels and Daily Nitrogen Dioxide (NO_2) Levels among Florida Counties. *Energies* **2020**, *13*, 6044. [CrossRef]

200. Dasgupta, D.; Rajeev, M. Paradox of a Supply Constrained Keynesian Equilibrium. *Econ. Political Week.* **2020**, *55*, 23.

201. Sampath, P.V.; Jagadeesh, G.S.; Bahinipati, C.S. Sustainable Intensification of Agriculture in the Context of the COVID-19 Pandemic: Prospects for the Future. *Water* **2020**, *12*, 2738. [CrossRef]

202. Lal, R.; Brevik, E.C.; Dawson, L.; Field, D.; Glaser, B.; Hartemink, A.E.; Hatano, R.; Lascelles, B.; Monger, C.; Scholten, T. Managing soils for recovering from the COVID-19 pandemic. *Soil Syst.* **2020**, *4*, 46. [CrossRef]

203. Le Quéré, C.; Jackson, R.B.; Jones, M.W.; Smith, A.J.; Abernethy, S.; Andrew, R.M.; De-Gol, A.J.; Willis, D.R.; Shan, Y.; Canadell, J.G. Temporary reduction in daily global CO_2 emissions during the COVID-19 forced confinement. *Nat. Clim. Chang.* **2020**, *10*, 647–653. [CrossRef]

204. Tirado, M.C.; Crahay, P.; Mahy, L.; Zanev, C.; Neira, M.; Msangi, S.; Brown, R.; Scaramella, C.; Coitinho, D.C.; Müller, A. Climate change and nutrition: Creating a climate for nutrition security. *Food Nutr. Bull.* **2013**, *34*, 533–547. [CrossRef]

205. Porter, S.D.; Reay, D.S.; Higgins, P.; Bomberg, E. A half-century of production-phase greenhouse gas emissions from food loss & waste in the global food supply chain. *Sci. Total Environ.* **2016**, *571*, 721–729. [PubMed]

206. Nemecek, T.; Jungbluth, N.; i Canals, L.M.; Schenck, R. Environmental impacts of food consumption and nutrition: Where are we and what is next. *Int. J. Life Cycle Assess.* **2016**, *5*, 607–620. [CrossRef]

207. Clonan, A.; Roberts, K.E.; Holdsworth, M. Socioeconomic and demographic drivers of red and processed meat consumption: Implications for health and environmental sustainability. *Proc. Nutr. Soc.* **2016**, *75*, 367–373. [CrossRef]

208. Jones, A.D.; Hoey, L.; Blesh, J.; Miller, L.; Green, A.; Shapiro, L.F. A systematic review of the measurement of sustainable diets. *Adv. Nutr.* **2016**, *7*, 641–664. [CrossRef]

209. Tirado, M.; Hunnes, D.; Cohen, M.; Lartey, A. Climate change and nutrition in Africa. *J. Hunger Environ. Nutr.* **2015**, *10*, 22–46. [CrossRef]

210. Niles, M.T.; Ahuja, R.; Barker, T.; Esquivel, J.; Gutterman, S.; Heller, M.C.; Mango, N.; Portner, D.; Raimond, R.; Tirado, C. Climate change mitigation beyond agriculture: A review of food system opportunities and implications. *Renew. Agric. Food Syst.* **2018**, *33*, 297–308. [CrossRef]

211. Fischer, C.G.; Garnett, T. *Plates, Pyramids, and Planets: Developments in National Healthy and Sustainable Dietary Guidelines: A State of Play Assessment*; Food and Agriculture Organization of the United Nations: Rome, Italy, 2016.

212. Rockström, J.; Stordalen, G.A.; Horton, R. *Acting in the Anthropocene: The EAT–Lancet Commission*; Lancet: London, UK, 2016.
213. Marinangeli, C.P.; Mansilla, W.D.; Shoveller, A.-K. Navigating protein claim regulations in North America for foods containing plant-based proteins. *Cereal Foods World* **2018**, *63*, 207–216.
214. Springmann, M.; Mason-D'Croz, D.; Robinson, S.; Wiebe, K.; Godfray, H.C.J.; Rayner, M.; Scarborough, P. Mitigation potential and global health impacts from emissions pricing of food commodities. *Nat. Clim. Chang.* **2017**, *7*, 69–74. [CrossRef]
215. Sutton, M.A.; Howard, C.M.; Kanter, D.R.; Lassaletta, L.; Móring, A.; Raghuram, N.; Read, N. The nitrogen decade: Mobilizing global action on nitrogen to 2030 and beyond. *One Earth* **2021**, *4*, 10–14. [CrossRef]
216. De Notaris, C.; Olesen, J.E.; Sørensen, P.; Rasmussen, J. Input and mineralization of carbon and nitrogen in soil from legume-based cover crops. *Nutr. Cycl. Agroecosyst.* **2020**, *116*, 1–18. [CrossRef]
217. Lu, J.; Hu, T.; Zhang, B.; Wang, L.; Yang, S.; Fan, J.; Yan, S.; Zhang, F. Nitrogen fertilizer management effects on soil nitrate leaching, grain yield and economic benefit of summer maize in Northwest China. *Agric. Water Manag.* **2021**, *247*, 106739. [CrossRef]
218. Lyu, Y.; Yang, X.; Pan, H.; Zhang, X.; Cao, H.; Ulgiati, S.; Wu, J.; Zhang, Y.; Wang, G.; Xiao, Y. Impact of fertilization schemes with different ratios of urea to controlled release nitrogen fertilizer on environmental sustainability, nitrogen use efficiency and economic benefit of rice production: A study case from Southwest China. *J. Clean. Prod.* **2021**, *293*, 126198. [CrossRef]
219. De Laporte, A.; Banger, K.; Weersink, A.; Wagner-Riddle, C.; Grant, B.; Smith, W. Economic and environmental consequences of nitrogen application rates, timing and methods on corn in Ontario. *Agric. Syst.* **2021**, *188*, 103018. [CrossRef]
220. Wiggering, H.; Dalchow, C.; Glemnitz, M.; Helming, K.; Müller, K.; Schultz, A.; Stachow, U.; Zander, P. Indicators for multifunctional land use—Linking socio-economic requirements with landscape potentials. *Ecol. Indic.* **2006**, *6*, 238–249. [CrossRef]

Article

Interrelationships of Chemical, Physical and Biological Soil Health Indicators in Beef-Pastures of Southern Piedmont, Georgia

Subash Dahal, Dorcas H. Franklin *, Anish Subedi, Miguel L. Cabrera, Laura Ney, Brendan Fatzinger and Kishan Mahmud

Department of Crop and Soil Sciences, University of Georgia, Miller Plant Sciences Building, 3111 Carlton St, Athens, GA 30602, USA; subash.dahal@uga.edu (S.D.); anish.subedi@uga.edu (A.S.); mcabrera@uga.edu (M.L.C.); lney@uga.edu (L.N.); brendan.fatzinger@uga.edu (B.F.); kishan.mahmud25@uga.edu (K.M.)
* Correspondence: dfranklin@uga.edu; Tel.: +1-706-542-2449

Abstract: The study of interrelationships among soil health indicators is important for (i) achieving better understanding of nutrient cycling, (ii) making soil health assessment cost-effective by eliminating redundant indicators, and (iii) improving nitrogen (N) fertilizer recommendation models. The objectives of this study were to (i) decipher complex interrelationships of selected chemical, physical, and biological soil health indicators in pastures with history of inorganic or broiler litter fertilization, and (ii) establish associations among inorganic N, potentially mineralizable N (PMN), and soil microbial biomass (SMBC), and other soil health indicators. In situ soil respiration was measured and soil samples were collected from six beef farms in 2017 and 2018 to measure selected soil health indicators. We were able to establish associations between easy-to-measure active carbon (POXC) vs. PMN ($R^2 = 0.52$), and N ($R^2 = 0.43$). POXC had a noteworthy quadratic relationship with N and nitrate, where we found dramatic increase of N and nitrate beyond an inflection point of 500 mg kg^{-1} POXC. This point may serve as threshold for soil health assessment. The relationships of loss-on-ignition (LOI) carbon with other soil health indicators were discernable between inorganic- and broiler litter-fertilized pastures. We were able to establish association of SMBC with other soil variables ($R^2 = 0.76$) and there was detectable difference in SMBC between inorganic-fertilized and broiler litter-fertilized pastures. These results could be useful for cost-effective soil health assessment and optimization of N fertilizer recommendation models to improve N use efficiency and grazing system sustainability.

Keywords: soil health indicators; grazing systems; nitrogen; permanganate oxidizable carbon; soil microbial biomass

Citation: Dahal, S.; Franklin, D.H.; Subedi, A.; Cabrera, M.L.; Ney, L.; Fatzinger, B.; Mahmud, K. Interrelationships of Chemical, Physical and Biological Soil Health Indicators in Beef-Pastures of Southern Piedmont, Georgia. *Sustainability* **2021**, *13*, 4844. https://doi.org/10.3390/su13094844

Academic Editors: Bharat Sharma Acharya and Rajan Ghimire

Received: 2 March 2021
Accepted: 23 April 2021
Published: 26 April 2021

1. Introduction

Soil is a complex and dynamic ecosystem; hence, a deep understanding of complex interrelationships between soil health indicators is required for sustainable utilization of this non-renewable resource. There is no single indicator that can describe the overall state of soil health and productivity [1,2]; thus, several indicators are used for that purpose [3]. However, many indicators provide redundant information; thus, the study of their interrelationship is very important for cost-effective assessment of soil health. Moreover, soil health is highly affected by climate and management [4–7], and a deeper understanding of interrelationship between soil health indicators and management factors [8], such as fertilizer source and grazing system, is highly important [9].

Chemical indicators are the oldest and most studied indicators of soil health and they still remain the most important ones from farmer's perspective, although biological, physical, and biochemical indicators are equally important [3], if not more [10]. Nitrogen fertilizer remains one of the most important inputs in agricultural production [11]; hence, soils' ability to store and release nitrogen has always been a high priority research area among soil scientists. Traditional fertilizer recommendation models typically depend on the soil test value of the nutrient and plant requirement; however, advances have been made to utilize other soil health indicators for optimum fertilizer recommendation [10,12,13].

In this manuscript we focused on several chemical indicators (potentially mineralizable nitrogen, inorganic nitrogen, loss on ignition carbon, permanganate oxidizable carbon, Mehlich-I phosphorus) of soil health, along with one physical (bulk density) and two biological indicators (soil respiration and soil microbial biomass). The main goal of this research was to analyze and report the interrelationships of selected soil health indicators and to provide generalized models for predicting soil microbial biomass and N availability, by utilizing a large number of soil-samples from multiple locations and two common grazing systems in the Piedmont region of Georgia.

2. Materials and Methods

2.1. Study Sites

Soil samples were collected from six study sites (Figure 1), including twelve grazed-pastures and one hay field, between May 2017 and July 2018 (Table 1). JPC and ADS (two of the sites) are research pastures owned by University of Georgia, whereas WF, TC, TH and FC are farmers' fields in Northeastern Georgia. A detailed description of the pasture characteristics is shown in Table 1. The study area has a hot humid sub-tropical climate with mean minimum annual temperature of 10.4–11.1 °C, mean maximum annual temperature of 22.5–25.6 °C and mean annual rainfall of 1190–1230 mm.

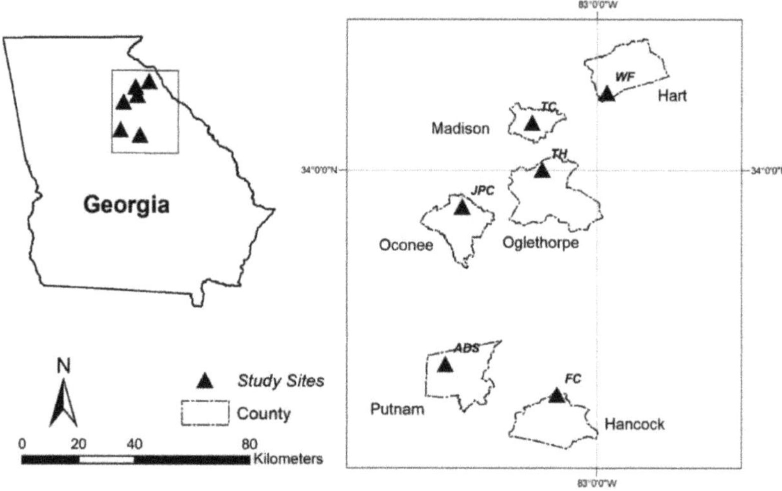

Figure 1. Map of study sites showing pasture locations and respective counties in Georgia, USA.

Table 1. Location, pasture attributes, management, soil type, and sampling time of study sites.

Site	Location	Pasture Attributes	Management/Fertilization	Soil Type	Sampling Dates	Soil Samples
JPC	J. Phil Campbell Sr. Research and Education Center Watkinsville, GA	Four Tall Fescue-Bermudagrass mixed pastures, approx. 17 ha each.	Historically (more than 10 years before 2016) conventionally grazed. In May 2016, 2 pastures were strategically grazed and other 2 were continuously grazed with rolling out of hay. Inorganic fertilizer	i. Fine, kaolinitic, thermic Typic Kanhapludults (NRCS-First Order Soil Survey)	July 2017 July 2018	492
ADS	Animal and Dairy Science Beef Cattle Farm, Eatonton, GA	Four Tall Fescue-Bermudagrass mixed pastures, approx. 17 ha each.	Same as JPC except no external fertilizer applied after May 2016.	i. Fine, kaolinitic, thermic Rhodic Kandiudults; ii. Loamy, mixed, active, thermic, shallow Typic Hapludalfs (NRCS-First Order Soil Survey)	July 2017 July 2018	528
WF	Hartwell, GA	One Tall Fescue pasture (19.71 ha) and one Hay field (4.28 ha)	Rotationally grazed; Broiler litter fertilized	i. Fine, kaolinitic, thermic Typic Kanhapludults	May 2017 June 2018	52
TC	Danielsville, GA	One Tall Fescue pasture (6.93 ha)	Rotationally grazed; Broiler litter fertilized	i. Fine, kaolinitic, thermic Rhodic Kandiudults; ii. Fine, kaolinitic, thermic Typic Kanhapludults	June 2018	25
TH	Crawford, GA	One Tall Fescue pasture (2.27 ha)	Rotationally grazed; Broiler litter fertilized	i. Fine, kaolinitic, thermic Typic Kanhapludults	June 2018	13
FC	Devereux, GA	One Tall Fescue pasture (11.26 ha)	Rotationally grazed; Broiler litter fertilized	i. Fine, kaolinitic, thermic Rhodic Kanhapludults; ii. Fine-loamy, mixed, active, nonacid, thermic Oxyaquic Udifluvents	May 2017	12

2.2. In Situ Soil Respiration and Soil Sampling

In JPC and ADS pastures, soil samples (0–5, 5–10, and 10–20 cm) were collected from 1020 locations in July 2017 and July 2018 by using a 5-cm diameter Giddings hydraulic probe (Giddings Machine Corporation, Windsor, Colorado). Two replicate soil samples were collected from each sampling location which resulted in 2040 soil samples. On the day of soil sampling, in situ alkali traps, containing 1 mol L^{-1} Sodium Hydroxide (NaOH) were installed in a static PVC chamber that was inserted to a 5-cm soil depth to measure soil respiration as described by Anderson [14]. After 24 h, alkali traps were brought to the lab and Barium Chloride (BaCl$_3$) was added to precipitate the CO$_2$ captured by NaOH, then residual NaOH was titrated with 1 N HCl (Hydrochloric acid) to calculate the amount of CO$_2$ produced by soil respiration in 24 h. Similarly, in WF, TC, TH, and FC pastures, soil samples were collected in the same manner as JPC and ADS, however only at 0–10 cm soil depth, resulting in 204 samples.

2.3. Soil Analysis

Soil samples were air-dried for two weeks, ground, and sieved (2-mm mesh) then stored in air-tight plastic bags for further analysis. Bulk density for each soil core was measured following the USDA Soil Survey Laboratory Methods Manual [15]. Samples were analyzed for Loss-on-Ignition Carbon using the combustion method as described in the USDA Soil Survey Laboratory Methods Manual [15]. Two replicate cores were then composited for further analysis, which resulted in a total of 1020 samples from JPC and ADS pastures, and a total of 102 samples from other pastures. Permanganate Oxidizable Carbon was analyzed by using the method described by Weil et al. [16]. Soil Microbial Biomass was measured using the method described by Vance et al. [17], in a smaller subset of the samples. Soil samples were extracted using 2 mol L^{-1} KCl [18] then NH$_4^+$-N was measured as described in Kempers and Zweers [19] and the NO$_3^-$-N was measured as described by Doane and Horwath [20]. Inorganic N was calculated as the sum of NH$_4^+$-N and NO$_3^-$-N fractions from 2 mol L^{-1} cold KCl extraction. Potentially mineralizable N was measured using the hot KCl extraction method [21], In this method, 20 mL of 2 mol L^{-1} KCl was added to 3 g of soil, heated at 100 °C for 4 h in a hot water bath, allowed to cool to room temperature, and filtered through Whatman #42 filter paper. Then the supernatant was analyzed for NH_4^+-N as described by Kempers and Zweers [19]. Potentially mineralizable N was calculated by subtracting cold KCL extracted NH_4^+-N from hot KCl extracted NH_4^+-N. Plant available P (Mehlich-I P) was measured using the method described by Mehlich [22]. Clay percentage in each sample was calculated from NRCS_USDA (Web Soil Survey) [23] and first-order soil survey conducted by NRCS in the pastures. Clay percentage was not used in the analysis due to coarse resolution of the data, but could be useful for interpretation of some results.

2.4. Statistical Analysis

All data processing and analysis was done using R Statistical Software [24]. Stepwise backward selection method, with minimum AIC [25], was used to identify variables. The dependent variable and a multiple regression model was fit using selected variables. The regression model was defined as

$$Y = X\beta + \varepsilon \tag{1}$$

where, Y denotes the response variable, X denotes the matrix of explanatory variable, β denotes the vectors of regression coefficients, and ε denotes the vector of random error term. Several simple linear regression models were fit between different variables to understand their interrelationships as follows:

$$y = \beta_0 + \beta_1 x + \varepsilon \tag{2}$$

where y is the response variable, x is the predictor variable, β_0 is the intercept of the model, β_1 is the slope, and ε is the error term.

For some variables, a quadratic model was more suitable:

$$y = \beta_0 + \beta_1 x + \beta_2 x^2 + \varepsilon \tag{3}$$

where y is the response variable, x is the predictor variable, β_0 is the intercept of the model, β_1 is the regression coefficient related to x, β_2 is the regression coefficient related to the quadratic term, and ε is the error term.

In addition, a factor analysis of mixed data (FAMD) was conducted to identify variables contributing most to the overall variance of the dataset as suggested by [26]. FAMD is a Principal Component Analysis (PCA) method which allows both categorical and quantitative variables. All variables were normalized in order to balance the effect of each set of variables.

3. Results and Discussion

3.1. Summary Statistics of Soil Health Indicators

A summary of all variables under study, grouped by soil-depth and fertilizer management systems, is presented in Table 2. Except bulk density, there was a general decrease in all soil parameters with soil increasing depth. Bulk density was highest in the 5–10 cm depth followed by 10–20 cm and 0–5 cm. We utilized soil samples with a wide range of soil health indicator values to increase the applicability of the developed models. For example, soils had a wide range of LOI which ranged from 1.6 to 3.6 g 100 g^{-1}, while POXC values ranged from 192 mg kg^{-1} to 1355 mg kg^{-1}. Research pastures (23%) had greater clay content compared to farmer's pastures (15.6%).

Table 2. Summary statistics of variables including number of samples, median, mean, standard deviation, minimum and maximum, grouped by type of fertilizer applied and depth of soil samples.

Soil Health Indicator	Soil Depth	N	Median	Mean	SD	Min	Max
Research Pastures (Inorganic and/or No Fertilizer)							
Loss on Ignition Carbon (LOI) ($g\ 100\ g^{-1}$)	0–5	335	73	78	27	29	211
	5–10	335	44	48	17	19	159
	10–20	323	41	45	17	16	107
Active Carbon (POXC) ($mg\ kg^{-1}$)	0–5	335	722	736	240	24.88	1355
	5–10	334	299	317	143	0.1	9996
	10–20	321	180	192	110	0.1	645
Soil Microbial Biomass (SMBC) ($mg\ CO_2\text{-}C\ kg^{-1}$)	0–5	27	168	172	85	16	409
Soil Respiration ($mg\ CO_2\text{-}C\ m^{-2}\ 24\ h^{-1}$)	0–5	316	844	1078	849	19	4136
Inorganic Nitrogen (N) ($mg\ kg^{-1}$)	0–5	329	36.6	44.9	39.1	0.7	327.1
	5–10	329	7.9	10.5	10.1	0.5	93.6
	10–20	316	4.7	6.3	7.6	0.1	79.5
Potentially mineralizable Nitrogen (PMN) ($mg\ kg^{-1}$)	0–5	330	20.7	21.6	12.2	0.1	59.2
	5–10	329	9.6	10.1	5.1	1.0	33.1
	10–20	317	6.2	6.5	3.7	0.1	31.7
Nitrate-Nitrogen ($NO_3^-\text{-}N$) ($mg\ kg^{-1}$)	0–5	330	31.7	39.3	37.8	0.00	298.2
	5–10	329	5.3	7.6	9.4	0.00	78.7
	10–20	317	2.5	4.3	7.5	0.00	77.9
Mehlich-I Phosphorus (P) ($mg\ kg^{-1}$)	0–5	302	13.3	18.3	27.5	0.2	383.5
	5–10	328	6.7	17.0	33.1	0.0	257.9
	10–20	315	3.1	14.2	34.8	0.0	321.3
Bulk Density (BD) ($g\ cm^{-3}$)	0–5	333	1.32	1.31	0.19	0.68	2.41
	5–10	330	1.51	1.51	0.15	0.82	2.41
	10 20	315	1.47	1.45	0.13	0.62	1.78
Clay%	0–5	335	23.5	21.1	7.0	10	37.5
	5–10	335	21	22.0	10.3	10	36.5
	10–20	323	21	26.6	14.9	10	50
A. Farmers' Field (Broiler Litter Fertilized)							
LOI ($g\ 100\ g^{-1}$)	0–10	102	9.4	11.8	7.8	2.5	36.1
POXC ($mg\ kg^{-1}$)	0–10	102	611	596	139	251	956
SMBC ($mg\ CO_2\text{-}C\ kg^{-1}$)	0–10	64	25.9	45.7	47.1	4.2	235.8
N ($mg\ kg^{-1}$)	0–10	102	37.5	39.9	16.3	8.5	70.1
PMN ($mg\ kg^{-1}$)	0–10	102	12.3	12.4	5.1	1.4	30.1
$NO_3^-\text{-}N$ ($mg\ kg^{-1}$)	0–10	102	28.2	28.1	12.9	5.1	53.8
P ($mg\ kg^{-1}$)	0–10	64	64.1	62.0	44.9	1.5	246.7
BD ($g\ cm^{-3}$)	0–10	102	1.41	1.41	0.62	1.13	1.67
Clay%	0–10	64	15	15.6	5.3	9.9	31

3.2. Exploratory Data Analysis

The PCA analysis showed that first two principal components accounted for 55% of total variability of the dataset. As expected, soil depth was a significant contributer in both first and second principal components. Year of sampling did not have any useful role. BD, LOI, Nitrate, PMN, Inorganic N, and POXC were important contributer on the first principal component, whereas Mehlich-1 P was an important contributor in the second principal component (Figure 2). More details on the role of qualitative and quantitative variables in overal variation on data are presented in the Supplementary Materials (Figures S1 and S2). While we realize that depth of sampling had a profound effect on the dataset, the objective of this manuscript is not to estimate the differences in soil health indicators by soil depth.

Rather, the focus is on understanding the relationships between soil health indicators in a diverse population of soil samples.

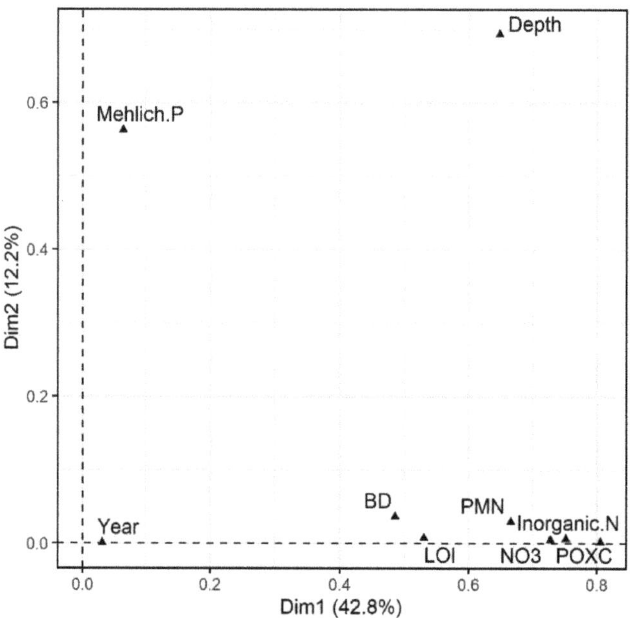

Figure 2. Principal Component Analysis of variables. Dim1 = First principal component, Dim2 = second principal component.

3.3. Relationship of Soil Nitrogen with Other Soil Health Indicators

PNM is a widely used indicator of soil health [3] and could be used for making nitrogen fertilization decisions in pastures to avoid over-application and potential loss of nitrogen via runoff, leaching, and volatilization [27]. In agreement with other researchers, we found that PMN was related to inorganic N (Figure 3A). Ros et al. [28] and Picone et al. [21] reported correlation of 0.74 and 0.68, respectively. PMN is the fraction of soil N which is yet to be mineralized and as shown here it contributes to the readily plant-available inorganic N fraction (inorganic N, Figure 3A), when mineralized [28,29].

Bulk density had a significant inverse relationship with N and PMN which was explained by a quadratic relationship (Figures 3B and 4C). Chaudhari et al. [30] also reported a significant negative correlation between bulk density and N + P + K (−0.87) content in tropical croplands, for a smaller sample size. Other studies [31,32] found a highly significant negative relationship between BD and total N. Generally, soils with BD greater than 1.6 g cm^{-3} are highly restrictive for root growth [9]; however, it has also been suggested that soils with BD as low as 1.2 might have detrimental effect on overall root and shoot growth in perennial ryegrass [31,33]. Root growth restriction coupled with low nutrient holding capacity of high BD soils might lead to over application of expensive nitrogen fertilizer and losses during runoff.

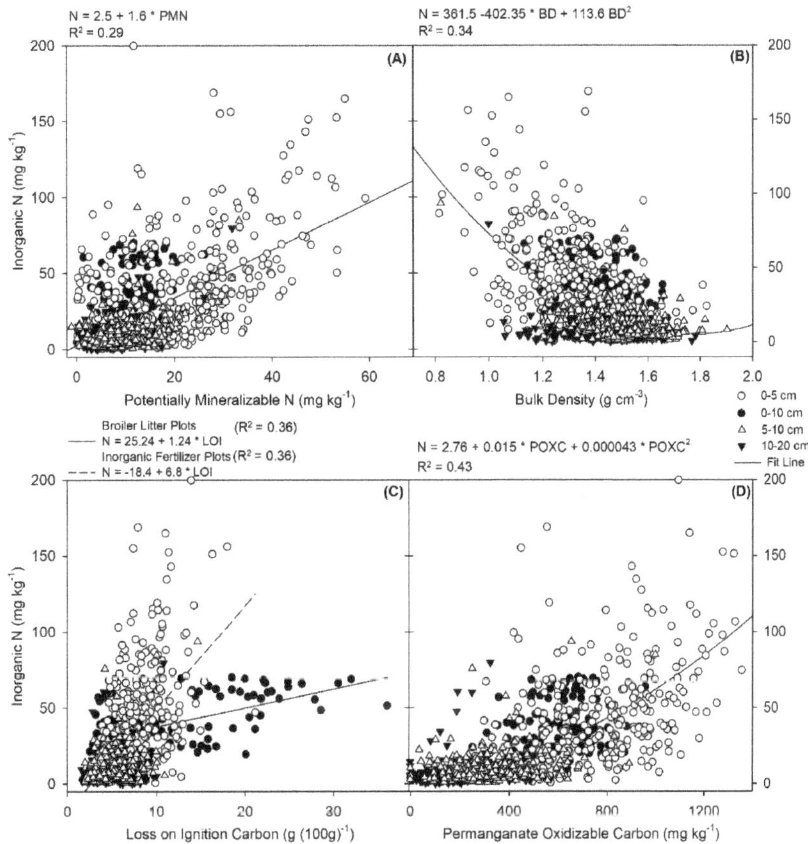

Figure 3. Relationship of inorganic N with (**A**) potentially mineralizable N, (**B**) bulk density, (**C**) loss on ignition carbon, and (**D**) active carbon. The black circles represent soils from the broiler litter applied pastures. Other symbols denote soils from inorganic fertilizer applied pastures.

LOI had positive relationship with N and PMN which were explained by simple linear models (Figures 3C and 4D). Our result is in agreement with a study by Yang et al. [34] in Tibetan grasslands, who reported that a simple linear model explains the relationship between total carbon and total nitrogen to a depth of 100 cm. Steffens et al. [35] also reported a similar relationship in arid regions of China. In this analysis, grasslands that received broiler litter as primary source of fertilizer behaved differently as compared to pastures that received inorganic fertilizers or no fertilizers (Figure 3C). Inorganically fertilized pastures had significantly ($p < 0.001$) steeper slope (6.8) compared to broiler litter fertilized pastures (1.24). Broiler litter pastures had very high (up to 40 g 100 g^{-1}) LOI; however, the rate of increase in N per unit increase in LOI was significantly lower as compared to inorganically fertilized pastures. Past studies [36,37] have reported an improved prediction of mineralizable nitrogen when organic matter was included in the model; however, our results indicate that we need to use caution as the relationships between organic matter and nitrogen can be dissimilar for different fertilizer management systems [38]. This might be attributed to high C/N ratio [39,40] of the bedding material used in poultry farms which could result in an accumulation of carbon but little corresponding increase in nitrogen. In addition to that, soils in research pastures with high in clay content could have retained more nitrogen due to greater surface area of clay particles. Upon extraction, nitrogen was released rapidly, showing a steeper slope of relation between LOI and nitrogen fractions.

As sampling depths for inorganic (0–5, 5–10 and 10–20 cm) and broiler litter pastures (0–10 cm) were different, we also calculated all soil health indicators for 0–10 cm soil depth for inorganic pastures (by averaging values for 0–5 cm and 5–10 cm) and did a confirmatory analysis. The results from this confirmatory analysis are presented in the supplementary section (Figures S3–S5). Confirmatory analysis corroborates our findings as described earlier.

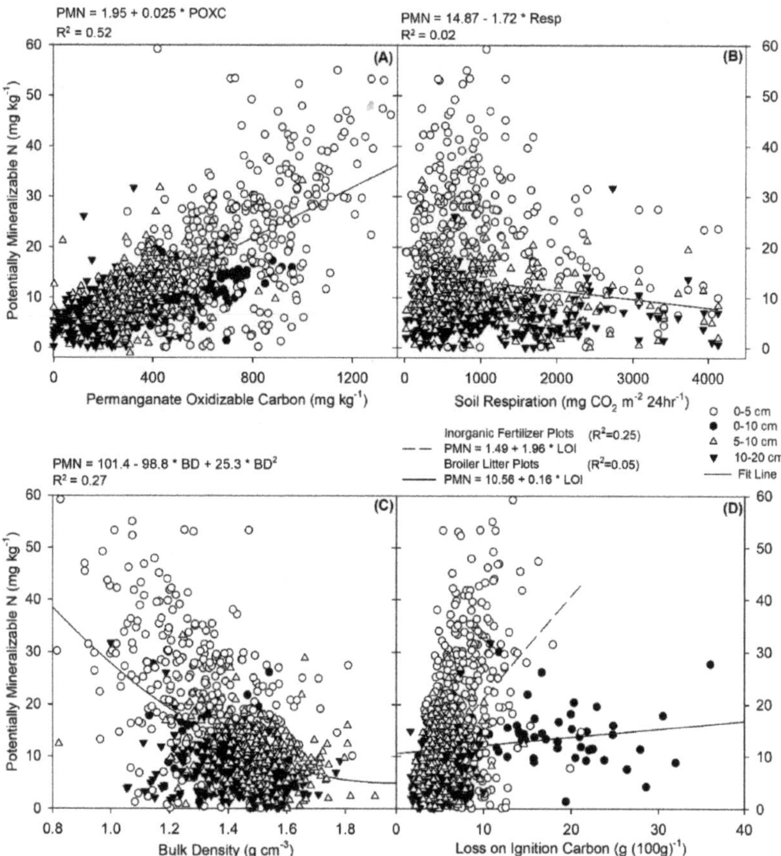

Figure 4. Relationship of potentially mineralizable N with (**A**) active carbon, (**B**) soil respiration, (**C**) bulk density, and (**D**) loss on ignition carbon. The black circles represent soils from the broiler litter applied pastures. Other symbols denote soils from inorganic fertilizer applied pastures.

POXC has been suggested as a reliable and management sensitive soil health indicator [16,41]. In these pasture soils, the POXC relationship with inorganic N was explained by a quadratic model (Figure 3D). We found that up to a value of POXC = 500 mg kg^{-1}, inorganic N was consistently low (within 50 mg kg^{-1}). When POXC was greater than 500 mg kg^{-1}, there was a dramatic increase in inorganic N. This inflection point might be utilized as a soil health criterion or a threshold for given management systems or carbon sources. The relationship of POXC and inorganic N did not have the characteristic difference (which was seen for LOI vs. N, Figure 3C) between broiler litter and inorganic pastures signifying the reliability of POXC, as a soil health indicator, because it was more stable across two fertilizer management systems. PMN also had significant strong positive relationship with POXC which was explained (R^2 = 0.52) by a simple linear model (Figure 3A). Ros et al. [28] found high correlation (r = 0.84) between hot water extractable carbon and mineralizable N for native grasslands and croplands soils with a history of mixed chemical and manure fertilization system. Since POXC is reliable and easy to measure [16] as compared to hot water extractable carbon or PMN, the POCX relationship may be a more accessible means to better understanding of nitrogen availability and nitrogen fertilizer recommendation in pastures.

3.4. Relationship of Soil Nitrogen with Other Soil Health Indicators

Soils contain more organic carbon than global vegetation and atmospheric carbon [42], and soil organic matter is central for agricultural production [43]; thus, the study of its interrelationship with other soil parameters is important to increase the productivity of agricultural systems. LOI and POXC had inverse relationship with BD and was explained by a simple linear model (Figure 5A,C). Typically, an increase in LOI is associated with reduced BD, but the two fertilizer application systems differed in the rate at which BD reduced with increasing LOI (Figure 5A). In cases of LOI, in inorganic fertilizer pastures, the slope of the equation was significantly ($p < 0.001$) steeper (-0.043) as compared to the broiler litter fertilized pastures (-0.015). In broiler litter pastures, some soil cores with LOI values as high as 30 g 100 g^{-1} had BD of 1.2. This incongruity between inorganic pastures and broiler litter pastures might also be due to differences in managerial decisions and stocking density. In addition, inherently greater bulk density of clay soils from research pastures could have added to this inconsistency between two systems by limiting the range of bulk density values. Generally our finding is in agreement with Franzluebbers [44] who reported an inverse relationship between soil organic carbon and bulk density and suggested an exponential model for predicting bulk density ($R^2 = 0.64$). The shorter-term Franzluebbers [44] study had different fertilizer sources, but there was no distinct separation between the carbon and BD relationship based on fertilizer source. Similarly, in case of POXC vs. BD, we found an inverse relationship (Figure 5C). The difference between two fertilizer systems, as was found in LOI vs. N (Figure 5A), was not observed. This again indicates that the active fraction of carbon could be a more reliable soil health indicator across pastures with different fertilizer sources.

LOI and POXC had a positive significant relationship and was explained by linear regression model (Figure 5B); however, broiler litter applied vs. inorganic pastures behaved differently. The slope of line for inorganic pastures was very steep as compared to the broiler litter applied pastures. Past studies [45–47] also reported that total organic carbon was linearly related to POXC in plantation, cropland, and pasture soils, respectively.

POXC had an interesting relationship with NO_3^--N which was explained by a quadratic model (Figure 5D). POXC values below 500 mg kg^{-1} had very low NO_3^--N (below 20 mg kg^{-1}), but beyond 500 mg kg^{-1} POXC, there was a sharp increase in NO_3^--N. This relationship might be an important consideration in fertilizing hay pastures or predicting NO_3^--N in forage to prevent nitrate toxicity in cattle.

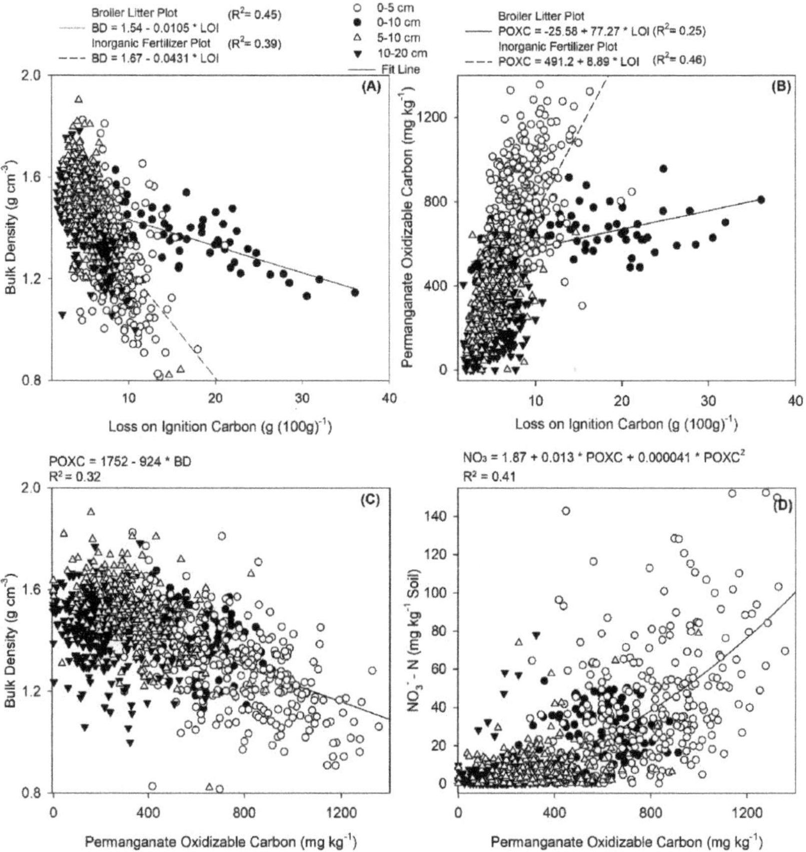

Figure 5. Relationships of loss on ignition carbon with (**A**) bulk density and (**B**) active carbon, and relationship of active carbon with (**C**) bulk density, and (**D**) nitrate-nitrogen. The black circles represent soils from the broiler litter applied pastures. Other symbols denote soils from inorganic fertilizer applied pastures.

3.5. Soil Microbial Biomass vs. Other Soil Health Indicators

Soil microbial biomass [17] has been suggested as a reliable indicator of soil health [3,44]; however, its measurement is highly time- and resource-consuming and requires hazardous chemicals. Thus, research is required in various agroecosystems to create models to assess soil microbial biomass by using other easy-to-measure soil health indicators.

The multiple regression model Equation (1) had an R^2 of 0.76 (Figure 6A), suggesting that most of the variables under consideration were useful indicators of SMBC (Table 3). The regression parameters (Table 3) show the complex nature of this biological soil health indicator. The POXC/PMN ratio had a significant positive effect, whereas the ratio of LOI/P ratio had a negative effect on SMBC. While others found a positive relationship with SMBC [16,48,49], our research showed weak though significant relationships which varied by fertilizer management systems (Figure 6B). High LOI/P ratio and high BD had negative effect, whereas LOI alone did not have any effect. The dissimilar relationship of LOI with POXC (Figure 5B) in broiler litter pastures vs. inorganic pastures may have been why we did not detect an effect of LOI in SMBC. For similar LOI values, POXC was relatively high in inorganic pastures as compared to broiler litter applier pastures. Furthermore, Franzluebbers and Stuedemann [50] suggested that larger biologically active carbon pools can be related to higher microbial carbon, whereas very high biologically resistant carbon (likely present in pastures with long history of broiler litter application) might have lower microbial biomass [51].

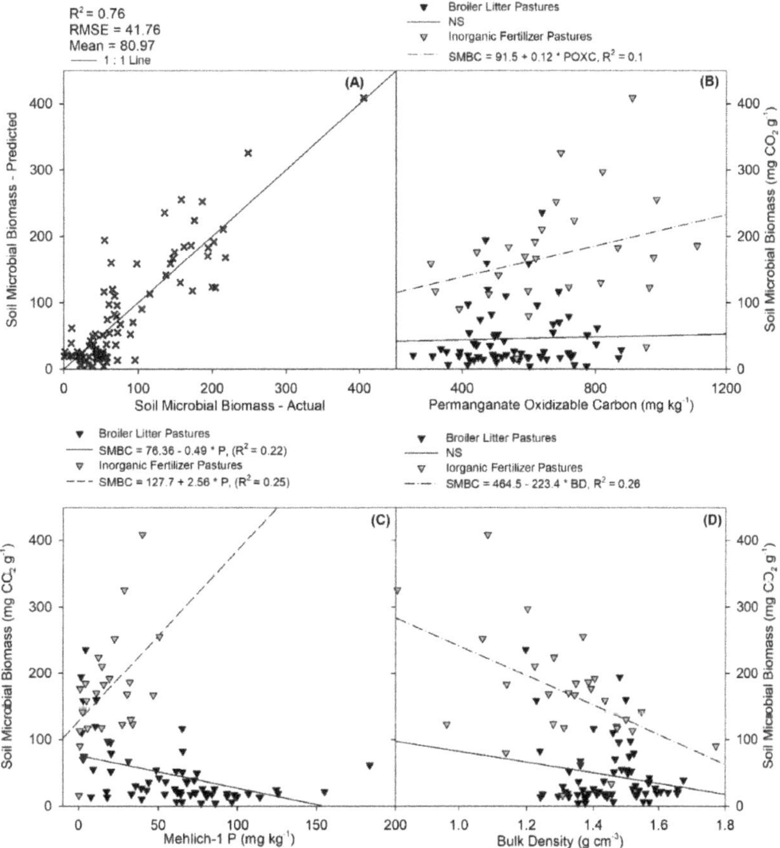

Figure 6. Graphs showing (**A**) predicted vs. actual plots for soil microbial biomass, and relationships of soil microbial biomass with (**B**) active carbon, (**C**) Mehlich-I phosphorus, and (**D**) bulk density.

Table 3. Multiple regression results for Soil Microbial Biomass prediction.

	Multiple Regression for SMBC				
Term	**Estimate**	**SE**	**t-Value**	***p*-Value**	**VIF**
Intercept	190.977	59.973	3.18	0.0021	
Inorganic-Broiler	−42.586	6.563	−6.49	<0001	1.67
POXC/PMN	0.395	0.074	5.34	<0001	1.51
P	−0.564	0.118	−4.79	<0001	1.3
PMN	3.114	0.913	3.41	0.001	1.74
LOI/P	−3.770	1.187	−3.18	0.0021	1.16
BD	−85.584	36.337	−2.36	0.021	1.59

The variables are ordered by the log-worth in descending order. SE = standard error, t-value = t score for respective term, VIF = variance inflation factor. Terms with *p*-value less than 0.05 listed are significant at $\alpha = 0.05$ significance level. POXC = active carbon, BD = bulk density, P = Mehlich-I phosphorus, Resp = soil respiration, N = inorganic nitrogen, LOI = loss-on-Ignition carbon, PMN = potentially mineralizable nitrogen.

Contrary to previous studies [50,52], broiler litter applied pastures had significantly less SMBC as compared to the inorganic pastures. However, those studies had applied poultry litter for only 4 years as compared to >20 years in our study. Typically, broiler litter applied pastures are rich in Phosphorus [53], which was the case in our study, and the negative effect of P (Table 3) might be confounded with the presence of heavy metals in broiler litter pastures. Past studies have suggested that long term application of poultry litter leads to accumulation of heavy metals such as Arsenic, Copper, Zinc, and Manganese [14,29,49,54].

3.6. Inorganic N and PMN vs. Other Soil Health Indicators

A multiple regression model was fit for Inorganic Nitrogen (Figure 7) as explained in Equation (1). Inorganic N was significantly related to all variables under consideration (Table 1). As PNM and LOI variables were correlated with POXC, they were removed from the model to address the problem of collinearity. Only variables with VIF (Variance Inflation Factor) less than 2.5 were kept in the model. Since inorganic nitrogen is the most readily plant available fraction of soil N, its accurate measurement is important from a producer's perspective. Figure 3 shows how soil health indicators under consideration relate with N. There was under-prediction below 10 mg kg^{-1} N and over-prediction above 60 mg kg^{-1} N.

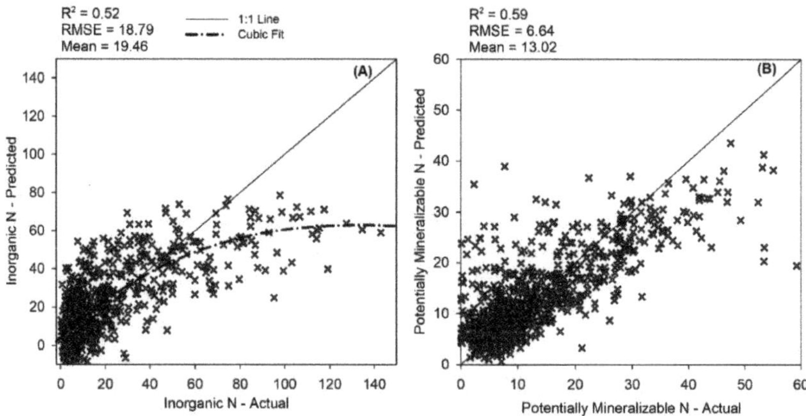

Figure 7. Predicted vs. actual plots for (**A**) Inorganic N, and (**B**) Potentially mineralizable N.

PMN was significantly affected by POXC, BD, P, and Resp (Table 4). POXC had most significant impact on PMN which was indicative of the ability of active carbon in soil to predict potentially mineralizable nitrogen in soil. Several researchers found respiration to have a significant positive effect on inorganic N, because higher soil respiration can be indicative of high microbial activity and associated N mineralization, which releases a flush of inorganic N [55,56]. Thus, soils with higher respiration rates had accumulated more inorganic N. In our work, Resp had a negative effect on PMN because actively respiring soils mineralize organic matter and release nitrogen in mineral form causing a decrease in potentially mineralizable nitrogen pool.

Table 4. Multiple regression results for Inorganic N and Potentially Mineralizable N.

Multiple Regression for N						Multiple Regression for PNM					
Term	Estimate	SE	t-Value	*p*-Value	VIF	Term	Estimate	SE	t-Value	*p*-Value	VIF
Intercept	40.134	7.598	5.28	<0001		Intercept	9.057	2.559	3.54	0.0004	
POXC	0.049	0.003	17.34	<0001	1.60	POXC	0.022	0.001	19.64	<0001	2.02
BD	−32.034	4.693	−6.83	<0001	1.58	N	0.058	0.011	5.39	<0001	1.88
P	0.136	0.021	6.5	<0001	1.02	Resp	−0.001	0.000	−5.18	<0001	1.02
Resp	0.002	0.001	2.77	0.0057	1.01	BD	−3.399	1.599	−2.13	0.0339	1.63

The variables are ordered by the log-worth in descending order. SE = standard error, t-value = t score for respective term, VIF = variance inflation factor. Terms with *p*-value less than 0.05 listed are significant at α = 0.05 significance level. POXC = active carbon, BD = bulk density, P = Mehlich-I phosphorus, Resp = soil respiration, N = inorganic nitrogen.

4. Conclusions

We documented significant relationships of active carbon (POXC) with PMN, N, LOI, BD, and SMBC, which substantiates the importance of POXC as an easily measured soil health indicator within the Southern Piedmont, USA. Of particular importance is our finding of the strong positive relationship of POXC with N and PMN, which showed the ability of active carbon-fraction to influence dynamics of nitrogen cycling in pastures. The quadratic relationship of POXC with N and nitrate-N is very interesting and needs to be studied in various grazing management systems to determine if there is one or several inflection points. These inflection points may serve as a soil health criterion/threshold or as indicator when nitrates in forages could be hazardous to grazing animals or as hay. We conclude that soil microbial biomass, a reliable and sensitive but difficult-to-measure indicator of soil health, could be assessed using other easy-to-measure variables. Fertilizer management systems significantly affect the relationship between soil health indicators and this information needs to be included in fertilizer recommendation models. The multiple regression models presented for Inorganic N, PMN and SMBC provide useful insights for developing and updating fertilizer recommendation models. Measurement of POXC and BD will help modelers, farmers, and farm managers to determine optimum nitrogen fertilizer recommendations for healthy and sustainable grazing systems.

Supplementary Materials: The following are available online at https://www.mdpi.com/article/10.3390/su13094844/s1, Figure S1: Principal Component Analysis of variables showing contribution of qualitative variables in the first and second principal components; Figure S2: Principal Component Analysis of variables showing contribution of quantitative variables in the first and second principal components; Figure S3: Relationship of inorganic N with (A) potentially mineralizable N, (B) bulk density, (C) loss on ignition carbon, and (D) active carbon; Figure S4: Relationship of potentially mineralizable N with (A) active carbon, (B) soil respiration, (C) bulk density, and (D) loss on ignition carbon; Figure S5: Relationships of loss on ignition carbon with (A) bulk density and (B) active carbon, and relationship of active carbon with (C) bulk density, and (D) nitrate-nitrogen.

Author Contributions: Conceptualization, D.H.F.; methodology, D.H.F., S.D., and A.S.; formal analysis, S.D.; investigation, D.H.F., S.D., A.S., L.N., B.F., and K.M.; resources, D.H.F.; data curation, S.D.; writing—original draft preparation, S.D.; writing—review and editing, S.D., D.H.F., M.L.C.; visualization, S.D.; supervision, D.H.F., M.L.C.; project administration, D.H.F.; funding acquisition, D.H.F. All authors have read and agreed to the published version of the manuscript.

Funding: This research was funded by USDA-NRCS.

Informed Consent Statement: Not applicable.

Data Availability Statement: Not applicable.

Acknowledgments: The authors are grateful to USDA-NRCS for their assistance with the first-order soil survey, and to the Sustainable Agriculture Laboratory team, John Rema, and Charles T. Trumbo at the University of Georgia for their endless help in the laboratory and the field.

Conflicts of Interest: The authors declare no conflict of interest.

References

1. Liebig, M.A.; Varvel, G.; Doran, J. A Simple Performance-Based Index for Assessing Multiple Agroecosystem Functions. *Agron. J.* **2001**, *93*, 313–318. [CrossRef]
2. Roper, W.R.; Osmond, D.L.; Heitman, J.L.; Wagger, M.G.; Reberg-Horton, S.C. Soil Health Indicators Do Not Differentiate among Agronomic Management Systems in North Carolina Soils. *Soil Sci. Soc. Am. J.* **2017**, *81*, 828–843. [CrossRef]
3. Moebius-Clune, B.N. *Comprehensive Assessment of Soil Health: The Cornell Framework Manual*; Cornell University: Ithaca, NY, USA, 2016. Available online: http://www.css.cornell.edu/extension/soil-health/manual.pdf (accessed on 11 February 2019).
4. Bhandari, K.B.; West, C.P.; Acosta-Martinez, V.; Cotton, J.; Cano, A. Soil health indicators as affected by diverse forage species and mixtures in semi-arid pastures. *Appl. Soil Ecol.* **2018**, *132*, 179–186. [CrossRef]
5. Bhowmik, A.; Fortuna, A.-M.; Cihacek, L.J.; Bary, A.I.; Cogger, C.G. Use of biological indicators of soil health to estimate reactive nitrogen dynamics in long-term organic vegetable and pasture systems. *Soil Biol. Biochem.* **2016**, *103*, 308–319. [CrossRef]
6. Byrnes, R.C.; Eastburn, D.J.; Tate, K.W.; Roche, L.M. A Global Meta-Analysis of Grazing Impacts on Soil Health Indicators. *J. Environ. Qual.* **2018**, *47*, 758–765. [CrossRef] [PubMed]
7. Ghimire, R.; Ghimire, B.; Mesbah, A.O.; Sainju, U.M.; Idowu, O.J. Soil Health Response of Cover Crops in Winter Wheat–Fallow System. *Agron. J.* **2019**, *111*, 2108–2115. [CrossRef]
8. Dahal, S.; Franklin, D.H.; Cabrera, M.L.; Hancock, D.W.; Stewart, L.; Ney, L.C.; Subedi, A.; Mahmud, K. Spatial Distribution of Inorganic Nitrogen in Pastures as Affected by Management, Landscape, and Cattle Locus. *J. Environ. Qual.* **2018**, *47*, 1468–1477. [CrossRef]
9. USDA. Soil Bulk Density: Soil Health Guide for Educators. Available online: https://www.nrcs.usda.gov/Internet/FSE_DOCUMENTS/nrcs142p2_050936.pdf (accessed on 10 December 2019).
10. Haney, R.L.; Haney, E.B.; Smith, D.R.; Harmel, R.D.; White, M.J. The soil health tool—Theory and initial broad-scale application. *Appl. Soil Ecol.* **2018**, *125*, 162–168. [CrossRef]
11. Parfitt, R.; Yeates, G.; Ross, D.; Mackay, A.; Budding, P. Relationships between soil biota, nitrogen and phosphorus availability, and pasture growth under organic and conventional management. *Appl. Soil Ecol.* **2005**, *28*, 1–13. [CrossRef]
12. Schomberg, H.H.; Cabrera, M.L. Modeling in situ N mineralization in conservation tillage fields: Comparison of two versions of the CERES nitrogen submodel. *Ecol. Model.* **2001**, *145*, 1–15. [CrossRef]
13. Veum, K.S.; Sudduth, K.A.; Kremer, R.J.; Kitchen, N.R. Sensor data fusion for soil health assessment. *Geoderma* **2017**, *305*, 53–61. [CrossRef]
14. Anderson, T.-H.; Domsch, K. Application of eco-physiological quotients (qCO2 and qD) on microbial biomasses from soils of different cropping histories. *Soil Biol. Biochem.* **1990**, *22*, 251–255. [CrossRef]
15. Staff, S.S.L. Soil survey laboratory methods manual. In *Soil Survey Investigation Report 42*; National Soil Survey Center: Lincoln, NE, USA, 1996.
16. Islam, K.R.; Stine, M.A.; Gruver, J.B.; Samson-Liebig, S.E.; Weil, R.R. Estimating active carbon for soil quality assessment: A simplified method for laboratory and field use. *Am. J. Altern. Agric.* **2003**, *18*, 3–17. [CrossRef]
17. Vance, E.; Brookes, P.; Jenkinson, D. An extraction method for measuring soil microbial biomass C. *Soil Biol. Biochem.* **1987**, *19*, 703–707. [CrossRef]
18. Maynard, D.; Kalra, Y.; Crumbaugh, J. *Nitrate and Exchangeable Ammonium Nitrogen*, 2nd ed.; CRC Press: Boca Raton, FL, USA; Taylor and Francis Group: Abingdon, UK, 1993; Volume 1, pp. 71–80.
19. Kempers, A.; Zweers, A. Ammonium determination in soil extracts by the salicylate method. *Commun. Soil Sci. Plant Anal.* **1986**, *17*, 715–723. [CrossRef]
20. Doane, T.A.; Horwáth, W.R. Spectrophotometric Determination of Nitrate with a Single Reagent. *Anal. Lett.* **2003**, *36*, 2713–2722. [CrossRef]
21. Picone, L.I.; Cabrera, M.L.; Franzluebbers, A.J. A Rapid Method to Estimate Potentially Mineralizable Nitrogen in Soil. *Soil Sci. Soc. Am. J.* **2002**, *66*, 1843–1847. [CrossRef]
22. Mehlich, A. Determination of P, Ca, Mg, K, Na, and NH4. In *North Carolina Soil Test Division (Mimeo 1953)*; 1953; pp. 23–89. Available online: http://www.ncagr.gov/AGRONOMI/pdffiles/mehlich53.pdf (accessed on 10 December 2019).
23. Soil Survey Staff, U. Web Soil Survey; US Department of Agriculture—Natural Resources Conservation Service. Available online: http://websoilsurvey.sc.egov.usda.gov/ (accessed on 10 December 2019).
24. R Core Team. *R: A Language and Environment for Statistical Computing*; R Foundation for Statistical Computing: Vienna, Austria, 2013.
25. Bozdogan, H. Model selection and Akaike's information criterion (AIC): The general theory and its analytical extensions. *Psychometrika* **1987**, *52*, 345–370. [CrossRef]
26. Kassambara, A. Practical guide to principal component methods in R: PCA, M (CA), FAMD, MFA, HCPC, Factoextra; Sthda. Available online: https://cloud.r-project.org/package=factoextra/factoextra.pdf (accessed on 12 August 2019).
27. Saint-Fort, R.; Frank, K.; Schepers, J. Role of nitrogen mineralization in fertilizer recommendations. *Commun. Soil Sci. Plant Anal.* **1990**, *21*, 1945–1958. [CrossRef]
28. Ros, G.H.; Hanegraaf, M.C.; Hoffland, E.; van Riemsdijk, W.H. Predicting soil N mineralization: Relevance of organic matter fractions and soil properties. *Soil Biol. Biochem.* **2011**, *43*, 1714–1722. [CrossRef]
29. Kingery, W.; Wood, C.; Delaney, D.; Williams, J.; Mullins, G. Impact of Long-Term Land Application of Broiler Litter on Environmentally Related Soil Properties. *J. Environ. Qual.* **1994**, *23*, 139–147. [CrossRef]

30. Chaudhari, P.R.; Ahire, D.V.; Ahire, V.D.; Chkravarty, M.; Maity, S. Soil bulk density as related to soil texture, organic matter content and available total nutrients of Coimbatore soil. *Int. J. Sci. Res. Publ.* **2013**, *3*, 1–8.

31. Houlbrooke, D.J.; Thom, E.R.; Chapman, R.; McLay, C.D.A. A study of the effects of soil bulk density on root and shoot growth of different ryegrass lines. *N. Z. J. Agric. Res.* **1997**, *40*, 429–435. [CrossRef]

32. Hu, C.; Li, F.; Xie, Y.-H.; Deng, Z.-M.; Chen, X.-S. Soil carbon, nitrogen, and phosphorus stoichiometry of three dominant plant communities distributed along a small-scale elevation gradient in the East Dongting Lake. *Phys. Chem. Earth Parts A/B/C* **2018**, *103*, 28–34. [CrossRef]

33. Chapman, R.; Allbrook, R. The effects of subsoiling compacted soils under grass-a progress report. Available online: https://www.agronomysociety.nz/uploads/94803/files/1987_13._Subsoiling_compacted_soils_under_grass.pdf (accessed on 2 June 2012).

34. Yang, Y.; Fang, J.; Guo, D.; Ji, C.; Ma, W. Vertical patterns of soil carbon, nitrogen and carbon: Nitrogen stoichiometry in Tibetan grasslands. *Biogeosci. Discuss.* **2010**, *7*. [CrossRef]

35. Steffens, M.; Kölbl, A.; Totsche, K.U.; Kögel-Knabner, I. Grazing effects on soil chemical and physical properties in a semiarid steppe of Inner Mongolia (P.R. China). *Geoderma* **2008**, *143*, 63–72. [CrossRef]

36. Dessureault-Rompré, J.; Zebarth, B.J.; Burton, D.L.; Sharifi, M.; Cooper, J.; Grant, C.A.; Drury, C.F. Relationships among Mineralizable Soil Nitrogen, Soil Properties, and Climatic Indices. *Soil Sci. Soc. Am. J.* **2010**, *74*, 1218–1227. [CrossRef]

37. Schomberg, H.H.; Wietholter, S.; Griffin, T.S.; Reeves, D.W.; Cabrera, M.L.; Fisher, D.S.; Endale, D.M.; Novak, J.M.; Balkcom, K.S.; Raper, R.L.; et al. Assessing Indices for Predicting Potential Nitrogen Mineralization in Soils under Different Management Systems. *Soil Sci. Soc. Am. J.* **2009**, *73*, 1575–1586. [CrossRef]

38. Wyngaard, N.; Cabrera, M.L.; Shober, A.; Kanwar, R. Fertilization strategy can affect the estimation of soil nitrogen mineralization potential with chemical methods. *Plant Soil* **2018**, *432*, 75–89. [CrossRef]

39. Franklin, D.; Bender-Özenç, D.; Özenç, N.; Cabrera, M. Nitrogen mineralization and phosphorus release from composts and soil conditioners found in the Southeastern United States. *Soil Sci. Soc. Am. J.* **2015**, *79*, 1386–1395. [CrossRef]

40. Riffaldi, R.; Saviozzi, A.; Levi-Minzi, R. Carbon mineralization kinetics as influenced by soil properties. *Biol. Fertil. Soils* **1996**, *22*, 293–298. [CrossRef]

41. Melero, S.; López-Garrido, R.; Madejón, E.; Murillo, J.M.; Vanderlinden, K.; Ordóñez, R.; Moreno, F. Long-term effects of conservation tillage on organic fractions in two soils in southwest of Spain. *Agric. Ecosyst. Environ.* **2009**, *133*, 68–74. [CrossRef]

42. Lehmann, J.; Kleber, M. The contentious nature of soil organic matter. *Nat. Cell Biol.* **2015**, *528*, 60–68. [CrossRef] [PubMed]

43. Joshi, D.R.; Clay, D.E.; Smart, A.; Clay, S.A.; Kharel, T.P.; Mishra, U. Soil and land-use change sustainability in the Northern Great Plains of the USA. In *Land Use Change and Sustainability*; IntechOpen: London, UK, 2019. [CrossRef]

44. Franzluebbers, A.J.; Pehim-Limbu, S.; Poore, M.H. Soil-Test Biological Activity with the Flush of CO2: IV. Fall-Stockpiled Tall Fescue Yield Response to Applied Nitrogen. *Agron. J.* **2018**, *110*, 2033–2049. [CrossRef]

45. Hendricks, T.; Franklin, D.; Dahal, S.; Hancock, D.; Stewart, L.; Cabrera, M.; Hawkins, G. Soil carbon and bulk density distribution within 10 Southern Piedmont grazing systems. *J. Soil Water Conserv.* **2019**, *74*, 323–333. [CrossRef]

46. Mendham, D.S.; O'Connell, A.M.; Grove, T.S. Organic matter characteristics under native forest, long-term pasture, and recent conversion to Eucalyptus plantations in Western Australia: Microbial biomass, soil respiration, and permanganate oxidation. *Soil Res.* **2002**, *40*, 859. [CrossRef]

47. Plaza-Bonilla, D.; Álvaro-Fuentes, J.; Cantero-Martínez, C. Identifying soil organic carbon fractions sensitive to agricultural management practices. *Soil Tillage Res.* **2014**, *139*, 19–22. [CrossRef]

48. Culman, S.W.; Snapp, S.S.; Freeman, M.A.; Schipanski, M.E.; Beniston, J.; Lal, R.; Drinkwater, L.E.; Franzluebbers, A.J.; Glover, J.D.; Grandy, A.S.; et al. Permanganate Oxidizable Carbon Reflects a Processed Soil Fraction that is Sensitive to Management. *Soil Sci. Soc. Am. J.* **2012**, *76*, 494–504. [CrossRef]

49. Islam, K.; Weil, R. Land use effects on soil quality in a tropical forest ecosystem of Bangladesh. *Agric. Ecosyst. Environ.* **2000**, *79*, 9–16. [CrossRef]

50. Franzluebbers, A.J.; Stuedemann, J.A. Bermudagrass management in the southern piedmont USA. III. Particulate and biologically active soil carbon. *Soil Sci. Soc. Am. J.* **2003**, *67*, 132–138. [CrossRef]

51. Franzluebbers, A.J.; Haney, R.L.; Honeycutt, C.W.; Schomberg, H.H.; Hons, F.M. Flush of Carbon Dioxide Following Rewetting of Dried Soil Relates to Active Organic Pools. *Soil Sci. Soc. Am. J.* **2000**, *64*, 613–623. [CrossRef]

52. Acosta-Martínez, V.; Harmel, R.D. Soil Microbial Communities and Enzyme Activities under Various Poultry Litter Application Rates. *J. Environ. Qual.* **2006**, *35*, 1309–1318. [CrossRef] [PubMed]

53. Franklin, D.; Truman, C.; Potter, T.; Bosch, D.; Strickland, T.; Jenkins, M.; Nuti, R. Nutrient losses in runoff from conventional and no-till pearl millet on pre-wetted Ultisols fertilized with broiler litter. *Agric. Water Manag.* **2012**, *113*, 38–44. [CrossRef]

54. Brookes, P.C. The use of microbial parameters in monitoring soil pollution by heavy metals. *Biol. Fertil. Soils* **1995**, *19*, 269–279. [CrossRef]

55. Francis, G.S.; Haynes, R.J.; Williams, P.H. Effects of the timing of ploughing-in temporary leguminous pastures and two winter cover crops on nitrogen mineralization, nitrate leaching and spring wheat growth. *J. Agric. Sci.* **1995**, *124*, 1–9. [CrossRef]

56. Rustad, L.E.; News, G.-; Campbell, J.L.; Marion, G.M.; Norby, R.J.; Mitchell, M.J.; Hartley, A.E.; Cornelissen, J.H.C.; Gurevitch, J. A meta-analysis of the response of soil respiration, net nitrogen mineralization, and aboveground plant growth to experimental ecosystem warming. *Oecologia* **2001**, *126*, 543–562. [CrossRef] [PubMed]

sustainability

Review

Biochar Role in the Sustainability of Agriculture and Environment

Muhammad Ayaz [1,*], Dalia Feizienė [1], Vita Tilvikienė [1], Kashif Akhtar [2], Urte Stulpinaitė [1] and Rashid Iqbal [3]

[1] Institute of Agriculture, Lithuanian Research Center for Agriculture and Forestry, LT-58344 Kedainiu r, Lithuania; dalia.feiziene@lammc.lt (D.F.); vita.tilvikiene@lammc.lt (V.T.); urte.stulpinaite@lammc.lt (U.S.)

[2] Institute of Nuclear Agricultural Sciences, College of Agriculture and Biotechnology, Zhejiang University, Hangzhou 310058, China; kashif@zju.edu.cn

[3] Department of Agronomy, Faculty of Agriculture and Environment, The Islamia University of Bahawalpur, Bahawalpur 63100, Punjab, Pakistan; rashid.iqbal@iub.edu.pk

* Correspondence: muhammad.ayaz@lammc.lt; Tel.: +370-654-22-196

Abstract: The exercise of biochar in agribusiness has increased proportionally in recent years. It has been indicated that biochar application could strengthen soil fertility benefits, such as improvement in soil microbial activity, abatement of bulk density, amelioration of nutrient and water-holding capacity and immutability of soil organic matter. Additionally, biochar amendment could also improve nutrient availability such as phosphorus and nitrogen in different types of soil. Most interestingly, the locally available wastes are pyrolyzed to biochar to improve the relationship among plants, soil and the environment. This can also be of higher importance to small-scale farming, and the biochar produced can be utilized in farms for the improvement of crop productivity. Thus, biochar could be a potential amendment to a soil that could help in achieving sustainable agriculture and environment. However, before mainstream formulation and renowned biochar use, several challenges must be taken into consideration, as the beneficial impacts and potential use of biochar seem highly appealing. This review is based on confined knowledge taken from different field-, laboratory- and greenhouse-based studies. It is well known that the properties of biochar vary with feedstock, pyrolysis temperature (300, 350, 400, 500, and 600 °C) and methodology of preparation. It is of high concern to further investigate the negative consequences: hydrophobicity; large scale application in farmland; production cost, primarily energy demand; and environmental threat, as well as affordability of feedstock. Nonetheless, the current literature reflects that biochar could be a significant amendment to the agroecosystem in order to tackle the challenges and threats observed in sustainable agriculture (crop production and soil fertility) and the environment (reducing greenhouse gas emission).

Keywords: biochar; food security; socio-economics benefits; sustainable agriculture; sustainable environment

Citation: Ayaz, M.; Feizienė, D.; Tilvikienė, V.; Akhtar, K.; Stulpinaitė, U.; Iqbal, R. Biochar Role in the Sustainability of Agriculture and Environment. *Sustainability* **2021**, *13*, 1330. https://doi.org/10.3390/su13031330

Received: 3 January 2021
Accepted: 20 January 2021
Published: 27 January 2021

1. Introduction

The world's population is increasing day by day and is expected to reach 9.8 billion by 2050 (United Nations Department of Economics and Social Affairs, New York, NY, USA), which will put the world's agricultural system under an increasing threat. Thus, to feed the increasing population and fulfil the constantly growing demand for grains and organic food, the farming system has become dependent on technological and chemical inputs [1]. Some parts of the world have met the needs for food through improved farming system technologies. Such farming systems have been classified as agroforestry, agroecology, sustainable agriculture, organic agriculture, etc. [2]. The objective of all these improvements in the farming systems is to reduce hankering and enhance crop yield to obtain sustainable agriculture and the environment [3]. This concept has directed the attention of the scientific community and farmers towards natural residue and organic matters instead of commercially prepared products [4]. Biochar is one of the outcomes of scientific experiments, which has an important role in achieving sustainable agriculture and the environment [5]. Biochar, a type of charcoal obtained after the combustion of feedstock under no or very limited supply of oxygen, is considered as a potential soil conditioner [3]. It is also an efficient measure to sequestrate carbon to tackle climate change and global warming. It is highly durable when applied to the soil and can remain in soil for hundreds to thousands of years [5]. Biochar has become a public interest in the framework of bio-based industries, which depends on the alteration of feedstock into value-added chemicals and energy.

The term "sustainable agriculture" is defined as the consolidation of bioprocesses, chemical processes, physical activities, ecological processes, and socio-economic sciences in a holistic manner to design new agricultural practices that are safe and environmentally friendly [6]. Sustainable agriculture is a procedure by which agrofarming can nourish itself over an extended period by preserving and maintaining all its natural resources, e.g., maintaining the fertility of the soil, safeguarding surfaces and underground resources, developing renewable sources of energy, and seeking solutions to revamp farming methods to climate change [7,8]. Agrofarming must also consider the sustainability of the vast area and social groups.

Biochar is also being examined to rehabilitate environments, to diminish pollutant mobility in contaminated soils, and to reduce alteration of perilous elements to agronomic crops [7]. Mostly, biochar is produced from waste residues such as agricultural wastes, animal manures, and forest residues. The significance of these feedstocks is to produce biochar in a way that potentially transforms waste into a useful and valuable product [9]. Its impact on soil amendment includes increased soil quality and plant growth with enhancement in crop yield. The response and behavior of biochar can be substantially influenced by its manufacturing process, soil conditions and types where applied, and as well as the kind of crop to be grown [10,11]. In keeping with the importance of biochar, many researchers have studied the adaptability of biochar for the improvement of soil and environmental health. This review highlights the production processes of various feedstocks and pyrolysis temperatures at which biochar is produced and their impact on agriculture sustainability via improving soil ecosystem functions and services. This review is intended to help researchers globally in the selection of proper biochar produced at a certain temperature to improve agriculture and environment sustainability without compromising crop yield.

2. Brief Methodology

Data and literature were collected from Web of Science eBooks Freedom Collection (ScienceDirect) https://www.sciencedirect.com/; EBSCO Publishing (elFL.net duomenų bazių paketas) http://search.epnet.com/; Emerald Management e-Journals Collection https://www.emeraldinsight.com; Science Direct; Taylor & Francis https://www.tandfonline.com/; Springer LINK https://link.springer.com/. We collected and synthesized published literature from 1997 to 2020 using keywords "biochar", biochar and soil nutrients, "biochar and environment", etc., in the database. Though more than 1000 articles were downloaded, we focused on those indicating empirical outcomes. The cited literature was based on field studies as well as greenhouse pot or laboratory studies (Figure 1a,b). The online data search was irrespective of the region, biochar type, etc.

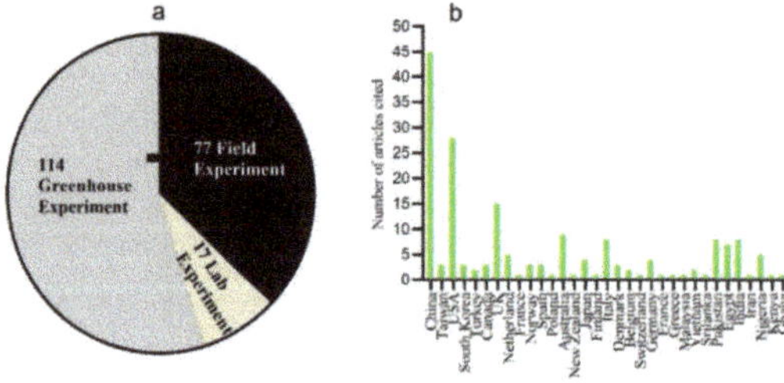

Figure 1. The details of cited information are (**a**) types of experiment and (**b**) countrywise published cited research.

3. Formulation, Morphology and Biochemistry of Biochar

Biochar is the bioproduct of thermal decomposition of renewable feedstock (forest and agricultural residues, hard woods, bamboos, livestock manure, etc.) under zero or low oxygen (O) conditions [12,13]. Feedstocks lose their mass when pyrolyzed at a minimum of 200–250 °C, as the thermal deterioration of the pulp occurs and leaves behind a spongy structure. Slow pyrolysis at low to intermediate heating (around 300 °C) and outstretched reaction times have been used for a long time to transform wood feedstock into high yields of biochar (biocarbon) [14]. The slow pyrolysis process also produces lower yields of bio-oil and gaseous by-products. In the past three decades, fast pyrolysis accomplished at medium temperatures ($\leq$500 °C) and very short processing times (couple of seconds) has received substantial interest as a method for generating higher yields with considerably higher energy density than the original feedstock, in conjunction with 20% of biochar and 15% of gas. Biochar yield and physiochemical properties considerably depend upon the pyrolysis process and feedstock used [15,16]. Biochar produced at a low temperature contains more aliphatic compounds in the pores that increase the hydrophobicity [17,18], while high temperature pyrolysis allows a much smaller number of aliphatic compounds to remain in the pores [17]. In a few studies, biochar had no significant influence on soil water repellency (SWR) and water holding capacity (WHC) of hydrophobic soil with high total organic carbon content [19]. In this study, to better understand the dynamics of biochar, thermogravimetric analysis (TGA) of swine manure (feedstock) was performed prior to its conversion to biochar for the ongoing field experiment. It was found that the digestate starts losing humidity in between 15 and 20 min at 200–300 °C of temperature followed by the loss of different volatiles, which leads to pure black carbon at 900 °C and evolution of porous structure.

The formulation of biochar was reported in [10–22] rice straw (RS), vinasse (VI), *Phyllostachys pubescens* (PP), *Arundo donax* (AD), chicken manure (CM) and sugarcane bagasse (SB) residues. The samples were cut into pieces (<5 cm) or crushed and dried for 24 h to achieve a constant weight before pyrolysis. Two kilograms (Kg) of each sample was put into the furnace and heated to the recommended temperature. The obtained biochar was passed through an 80 mm-mesh sieve to obtain finer biochar. Similarly, in [23], the same process of preparing biochar from Pine and Jarrah wood was indicated. The biochar obtained was alkaline in nature and had a large surface area, with tremendous porosity and molar ratio, e.g., C/N; H/C; and a large number of beneficial elements and organic matter, e.g., C, H, N, S, and O [20,22]. The aromaticity and the surface charge of the biochars decreased after coating with FA [15] and humic acid [12]. Antonangelo [24] reported that biochar obtained from witchgrass (*Panicum capillare*) and poultry manure (SGB and PLB, respectively) was thermally decomposed at various temperatures, e.g., 350 and 700 °C. The pH and elemental configuration of biochars were found to be alkaline and nutrient rich, and a strong correlation between accessible nutrients and ash contents was recorded [24,25]. The internal porosity of the biochar influences the surface chemistry and the bulk density of the biochar. Additionally, the source of feedstock controls the hydrophobicity and porosity of the material, along with production temperature.

4. Biochar and Nutrients

The nutrient composition of biocarbon always differs with the type of biomass and pyrolysis temperature (Table 1). The concentration of nutrients in animal-derived biochar would not necessarily be higher than in plant-derived biochar pyrolyzed at the same temperature [26]. Biochar produced from lantana camara at 300 °C was rich in available P (0.64 mg kg^{-1}), available Ca (5880 mg kg^{-1}), available Mg (1010 mg kg^{-1}) and available Na (1145 mg kg^{-1}) [24]. Dried swine manure waste-derived biochar under slow pyrolysis (300–750 °C) was found to be rich in soil micronutrients and macronutrients, such as Ca, Mg, Na, Fe, Mn, Cu, Zn, N, P and K [25]. Total N contents were significantly greater for poultry manure-derived biochar pyrolyzed at 400 and 600 °C treatments in silt loam and sandy soils; however, they were not affected by swine manure-derived biochar (400 and 600 °C) and wood chip biochar (1000 °C) in the same soils [26]. Similarly, freshly prepared biochar is a rich source of available nutrients and could discharge a significant amount of N ranging from 23 to 635 mg kg^{-1} and P ranging from 46 to 1664 mg kg^{-1} [27]. Jiang et al.

reported that old biochar was not as effective for soil organic carbon (SOC) protection as fresh biochar. The decline in SOC stability with old biochar might be associated with the attenuated sorption of SOC on aged biochar [28]. Compared with old biochar, the addition of fresh biochar in sandy loam soil increased the biomass production [29]. Major nutrients such as N, P and K could assume the role of fertilizer and be absorbed by plants and soil microbes. Therefore, these examples indicate that biochar can potentially influence soil nutrition. Several studies have assessed the availability of nutrients in biochar by carrying out transient and long-term leaching experiments in recent decades. Mallee wood-derived biochar was easily drainable with double distilled water after a day (24 h) (15–20% Ca, 10–60% of P and 2% of N) [30]. However, it is necessary to choose the suitable biochar for its long-term nutrient availability to plants.

Table 1. Nutrient composition of various biochars at different pyrolysis temperatures.

Biochar Feedstock	Pyrolysis Temp. (°C)	pH	C	N	C/N	P	K	CA	MG	References
						(%)				
Corn cob	600	10.1	79.1	4.25	19	-	-	-	-	[31]
Corn stover	600	9.95	69.8	1.01	70	0.181	2.461	0.938	0.858	[32]
Peanut hull	400	10.0	65.5	2.0	33	0.00162	0.0015	0.00044	-	[33]
Pearl millet	400	10.6	64	1.10	58	1.60	2.52	1.47	1.06	[34]
Corn stover	300	7.33	59.5	1.16	51	0.137	1.705	0.648	0.588	[32]
Dairy manure	700	9.9	56.7	1.51	38	1.69	2.31	4.48	2.06	[35]
Poultry litter	350	8.7	51.1	4.45	12	2.08	4.58	2.66	0.94	[35]
Turkey litter	700	9.9	44.8	1.94	23	3.63	5.59	5.61	1.24	[35]
Cow manure	500	9.20	33.6	0.15	22	0.814	0.005	0.042	0.034	[36]

5. Biochar and Chemical Properties

The additions of biochar with organic matter and humic substances are getting increasing attention regarding their influence on soil fertility and crop yield [37]. Cacao shell- and rice husk-derived biochar at 600 and 500 °C, respectively, increased the pH and released dissolved organic matter from the soil [38]. Straw-derived (500–600 °C) biochar enhanced the degradation of organic matter and maturity and increased soil nutrients [39]. Composting dynamics are influenced by biochar via increasing the speed of organic matter decomposition and enhancing soil porosity, therefore improving composting efficiency and humification processes [40,41]. Ten percent of poultry manure-derived biochar and cow manure-derived biochar application into a composting mixture increased carbon content in humic and fulvic acids [42]. Acacia saligna-derived biochar at 380 °C and sawdust-derived biochar at 450 °C were the potential sources of humic substances (17.7 and 16.2%, respectively) [43]. Adding biochar during the composting process to maize straw and sewage sludge increased the available water content in sandy soil [44]. Biochar with mushroom residues and with corn straw could accelerate biodegradation of polycyclic aromatic hydrocarbon [45]. The amendments of biochar help in carbon (C) abetment and improving soil quality [46–48]. However, there are several studies comparing the physicochemical and morphological characteristics of biochar obtained from various feedstock sources and at different temperatures (slow, medium and high), as their effect when used in soil acclimatization may vary, since thermal decomposition has a great impact on biochar characterization [49,50]. Biochar produced at low thermal decomposition is often rich in carbon biomass content [51,52]. Liard [53] reported that slow pyrolyzed biochar has a higher amount of available P content compared to fast pyrolyzed biochar. This could be attributed to the lower percentage of crystallized P-associated minerals in slow pyrolyzed biochar. Moreover, the total K and available K (water soluble) content increases with an increase in pyrolysis temperature [54]. C richness allied with high adsorption capacity, porous structure and high alkalinity makes biochar inclusion into soil a practicable and effectual way to enhance soil quality and fertility [55–58]. The alkaline nature of biochar and organic carbon richness also enhance cation exchange capacity (CEC), which leads to a greater heavy metal adsorption capacity [59], and thus improves soil quality [60]. Further, numerous studies have focused on the physicochemical properties of biochar and its influence on soil nutrients and crop yield (Tables 1 and 2).

Table 2. Outline of the primary literature cited about biochar dynamics and its effect on crop yields.

Biochar Types	Temperature	Country/Type of Experiment	Application Rate	Biochar Properties	Soil Type/Texture	Result	Reference
Wheat Straw	300–500 °C	China/Greenhouse	3% w/w	pH 10.60	Psammaquent and Plinthudult	Increased rice yields in both soils	[61]
Wheat Straw	300, 400 and 500 °C	China/Greenhouse	1% w/w	pH = 6.74, 7.8 and 8.0 C = 52, 62 and 66 g % N = 23.8, 19.4 and 18 g kg^{-1}	Sandy clay loam and Calcisols Yermi	Biochar prepared at 300 °C significantly increased Maize crop yield	[62]
Rice Straw and Corn Stalk	450 °C	China/Field	1, 2 and 4 ton/ha	C = 71.7 and 63.5%, H = 3.70% and 1.6, O = 16.50 and 9.2% N = 2.40 and 1.3%	Inceptisol	Increased Corn, peanut and sweet potato by 5%, 15% and 20%, respectively	[63]
Miscanthus Giganteous Straw	500–750 °C	Norway/Field	8 and 25 ton/ha	pH = 7.86 C = 80% H = 1.2% O = 0.6% N = 6.6%	Silty clay loam Albeluvisol	No effect on crop yield	[64]
Cow Manure	600 °C	Japan/Greenhouse	0, 10, 15 and 20 ton/ha	pH = 9.20 C = 33.61% N = 1.51%	Sandy soil	Significantly enhanced Maize crop yield	[65]
Rice Husks	450 °C	China/Greenhouse	0, 10, 25 and 50 tha^{-1}	pH = 9.21 C = 465.4 g kg^{-1} N = 6.2 g kg^{-1}	Upland soil and paddy soil	Increased rice and wheat yield by 12% and 17%, respectively	[66]
Maize Stover	600 °C	USA/Field	0, 1, 3, 12 and 30 tha^{-1}	pH = 10.02 C = 290 mg g^{-1} N = 3.02 mg g^{-1}	Kendaia silt loam	No significant effect on crop yield	[67]
Woodchips	290 °C	Taiwan/Greenhouse	2% w/w	pH 6.3 C = 59.1%, N = 0.35%, H = 5.73%, K = 0.78 g kg^{-1}	Clay texture and sandy loam texture	No significant effect on crop yield	[68]
Woodchips	700 °C	USA/Field	5% w/w	pH = 9.6 C = 83.0%, N = 0.34%, H = 2.57%, K = 3.90 g kg^{-1}	Clay texture and sandy loam texture	No significant effect on crop yield	[68]

Table 2. *Cont.*

Biochar Types	Temperature	Country/Type of Experiment	Application Rate	Biochar Properties	Soil Type/Texture	Result	Reference
Rice straw	300,400 and 500 °C	Taiwan/Greenhouse	1% w/w	pH = 6.74, 7.8 and 8.0 C = 52, 62 and 66 g kg^{-1} N = 23.8, 19.4, 18 g kg^{-1}	Sandy clay loam, Calcisols Yermi	No effect on Maize crop yield	[62]
Sorghum	500 °C	USA/Laboratory	200 bushels ha^{-1}	C = 750.5 g kg^{-1} N = 13.5 g kg^{-1}	Norfolk soil and Dunbar soil (fine, kaolinitic, thermic, Aeric Paleaquults)	Wheat yield increased by 31%	[69]
Maize Cobs	300–550 °C	Ghana/Field	0, 2 and 6 t ha^{-1}	pH = 7.6–9.7 C = 69–81g kg^{-1} N = 0.6–0.7%	Sand and loamy sandy soil	Has positive effect on crop yield	[70]
Wheat Straw	370 °C	Spain/Greenhouse	0, 0.5, 1 and 2.5% w/w	pH = 9.8–11 Total C = 483–894 g kg^{-1} Total N = 3.7–8.3 g kg^{-1}	Haplic Luvisol	20–30% increase in wheat grain yield	[71]
Hard Wood	500 °C	Nigeria/Field	0,10, 20 and 30 t ha^{-1}	pH = 7.5 Total N = 0.65% Organic carbon = 52%	Sandy loam	30 t ha^{-1} of biochar significantly enhanced Cocoyam crop yield	[72]
Eucalyptus Polybractea	550 °C	UK/Greenhouse	10 t ha^{-1}	pH = 9.5 Total N = 1.1% Total C = 42	Ferrosol Soil	No effect on Cauliflower, peas or broccoli crops	[73]

6. Biochar and Physical Properties

Biochar amendment reduces soil bulk density and enhances water holding capacity (WHC) and nutrient holding capacity (NHC) as a result of its large surface area which increases water and nutrient use efficiency (Figure 2) [74–77]. Biochar could decrease soil bulk density by 3 to 31% and increase porosity by 14 to 64%. It shows a promising behavior of WHC and NHC in sandy soil due to its macropores and lower surface area [78]. Biochar could have a positive impact on WHC (Figure 2) and NHC, thus increasing water and nutrient availability to plants in sandy soil [79]. Barrow [80] proposed that biochar amendment could be an effective strategy to combat desertification and promote plant growth. Straw-derived biochar at 525 and 400 °C has a long-term effect on soil physiochemical properties, as it is most efficient in enhancing plant available water and soil aggregate stability in a coarse-textured Planosol [81]. The information on the physical and chemical properties of biochar is also presented in Table 1.

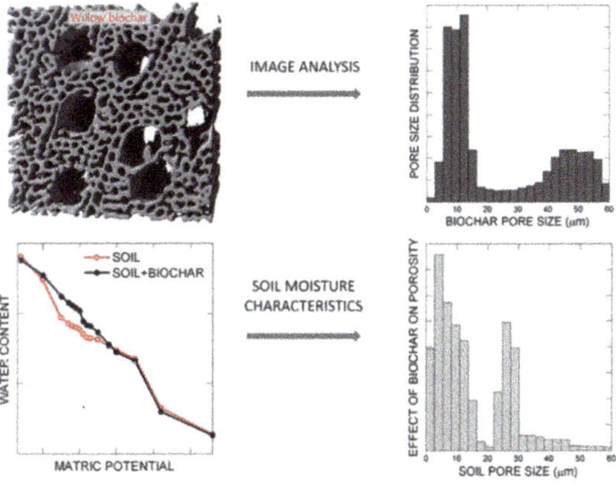

Figure 2. Determination of pore size distribution by 3D image analysis of X-ray tomography image (top panel) and the change in the pore size distribution due to biochar addition and determination of soil water content [82].

7. Effect of Biochar on Microbiota

The effect of biochar on the activity of soil microbes is dependent on types of soil and crop [83]. Wood-derived biochar application at a rate of 30 and 60 t ha^{-1} has a very short-lived positive effect on the microbial community [84]. In a recent study, Lu [85] reported that the porous structure of biochar substantially enhanced soil microbiota, due to the niche environment favorable to microbes. Otsuka [86] reported that multifariousness of the soil bacterial community expanded by 25% following biochar amendment compared to untreated soils. However, microbial biomass carbon and N mineralization were lowered with biochar amendment, thus reflecting that any boon from the liming effect of biochar is counterbalanced by a decrease in mass and community of soil microbes [87,88]. The application of biochar may increase soil pH. An increase in alkaline microsites may also alter the ammonia oxidizer population, particularly in acidic soil [89,90]. Similarly, rice straw biochar application significantly decreased Actinobacteria and Ascomycota fungi communities; however, soil microbial species diversification and copiousness may vary after biochar application [91]. Further, biochar amendment alters the soil nutrient cycling and nutrient supply, which in turn may affect the microbial community [92].

8. Biochar and Abatement of Greenhouse Gases

Climate change is usually attributed to the enhancing atmospheric abundance of greenhouse gases (GHGs) due to anthropogenic activities. Ninety percent of the anthropogenic climate warming is caused by three major GHGs, i.e., CO_2, CH_4 and N_2O [93].

Biochar has been suggested as an idle matter and beneficial soil amendment for carbon (C) sequestration to reduce CO_2 emission and its abundance from the environment [94,95]. Biochar offers a multitude of benefits for ensuring environmental safety. In view of their importance and diverse dynamics, a new terminology, "biochar culture", has been introduced to encircle the implementational and environmental gains of biochar [96]. It is an extremely valuable soil conditioner, as it changes a number of soil physicochemical characteristics (enhances soil moisture retention and increases air permeability) and impacts the soil microbial activity [97]. Biochar amendment also helps in the mitigation of greenhouse gases or setback by efficient management of agrofarming. Consequently, the "carbon foot-printing" of a specific land management has to be distinguished earlier than the targeted use for reducing emission by increasing "carbon sequestration" [96,97]. As "carbon sequestration" aims to reduce global warming by capturing the GHGs for a longer period, it can be proficient, because biochar overwhelmingly contains cyclic carbon with high aromaticity and exhausted H and O, which confer defiance to microbial strike on amendment in soil [98]. This increased obstinacy aspect, which helped in mitigating the emission of GHGs by reducing microbial decay and carbon digestion of organic biomass. Therefore, manufacturing of biochar and its amendment principally exploits the natural process of photosynthesis for crop biomass generation with the exception of removing atmospheric CO_2 by autotrophic microbes, that is, plants, mitigation of CO_2 to carbohydrate and other materials. Rondon [99] was the first scientist who reported N_2O emissions from a greenhouse experiment amended with biochar. Biochar pyrolysis at high temperatures and low N content might be more suitable to mitigate CO_2 and N_2O emission. A study from Terra Preta soil of the Amazonian region [100] reported that biochar can mitigate CO_2 emission for hundreds to thousands of years. Further, Wang [101] demonstrated from his dimensional analysis studies of the putrefaction and dressing effects on biochar stability in soil that only 3% of biochar C is bioavailable, and the rest is rendered into long-term stability. We expect a 4.0 °C rise in temperature by the end of the 21st century. Such environmental changes are the result of an increase in the atmospheric denseness of GHGs [102]. Biochar is anticipated to possess the desirable conditions for combating global warming and climate change. Biochar helps in atmospheric expulsion of CO_2, and due to its intractable nature, it captures the carbon to facilitate a huge carbon-negative economy [103]. Biochar amendment to soil curbs the emission of not only CO_2 but also about a hundred other potent GHGs, particularly nitrous oxide and methane [104]. Numerous studies advocate a substantial reduction in emissions of CO_2, N_2O and CH_4 gasses due to biochar application in land use under vegetable cultivation [105,106], but more long-term studies are needed.

9. Biochar and Soil Fertility

9.1. Effect of Biochar on Soil Nutrients

The effects of biochar on soil physicochemical properties are shown in Table 3. Biochar application is an effective practice for restoration of the functionality of degraded soils, and maintenance of long-term soil functions and fertility [107]. The addition of biochar improves degraded and low fertility soils, and thus improves crop production [108,109]. A. El-Naggar reported that biochar has the potential to be the best management practice for low fertility soils [110]. Major nutrients in biochar might not be necessarily available to plants in the desired amount [111]; the available NH^{4+}, $NO3^-$, $PO4^{-3}$ and K^+ might be associated with the amount of total N, P and K. For instance, the total N loss leads to a reduction in available N in highly thermally decomposed biochar [112]. The absorptivity of different nutrient ions on the surface of biochar and release occur due to variation in the CEC and pH of soil amended with biochar [113]. Yao [114] reported that the uptake of N and P as ammonium/nitrate and phosphate ions is significantly decreased by biochar application, decreasing their frequency in soil leachates by a high proportion. The properties of soil, e.g., texture, SOC, clay-to-sand contents and pH can change the biochar nutrient-sorption characteristics [114]. Additionally, the nutrient movement also changes biochar adsorption and succeeding release properties. The nutrient use efficiency of added organic fertilizer also increases with biochar amendment. The dynamism of nitrogen, principally engaging a decrease in nitrate transformation for subsequent reduced N loss, happens in response to biochar amendment, which can be deemed significant for optimal nitrogen use efficiency [78]. Thus, nutrient dynamics of biochar also assist in temperature and pH- reliant slow release of adsorbed nutrients by capturing nutrients from draining, runoff, leaching, microbial digestion and physical volatility processes. Therefore, plants and crops can potentially uptake nutrients, as these will be in plant-available forms in the root zone [112,113]. A sustained improvement of the physical characteristics of soil with biochar amendment involves better aggregate stability and formation, and alteration in the soil microbial community and activities imposes an indirect effect on retaining highly and reasonably mobile nutrients such as N and P [114].

Table 3. Different feedstock of biochar alters soil physicochemical properties.

Biochar Feedstock	Type of Soil	Sand	Silt	Clay	pH	TN	TC	References
		%						
Wheat straw	Sandy loam	-	-	16	5.6	0.18%	2.01%	[115]
Charcoal biochar	sandy	90.9	4.6	4.5	6.8	0.1 g kg^{-1}	1.0 g kg^{-1}	[116]
Charcoal biochar	Sandy loam	67.3	25.9	6.8	6.1	1.7 g kg^{-1}	31.0 g kg^{-1}	[116]
Oak and wood Bamboo	Clay loam	22	40	38	4.57	0.94 g kg^{-1}	5.50 g kg^{-1}	[117]
Poultry litter	Silt loam	26.6	33.7	39.7	7.99	0.13%	0.70%	[118]
Fruit tree and stem branches	Sandy loam	61.7	32.1	6.17	7.33	0.71	12.6 g kg^{-1}	[119]
Poultry litter	Sandy clay loam	52	17	31	3.95	0.25%	3.5%	[120]
Sewage sludge	Loam	71	25	4	6.50	0.04	5.48%	[121]
Wheat straw	Silt clay loam	16	52	32	8.3	1.0 g kg^{-1}	8.1 g kg^{-1}	[122]
Maize straw	Silty loam	13	72	15	7.9	0.99 g kg^{-1}	15.1 g kg^{-1}	[123]
Commercial biochar	Silt loam	16.1	64.1	19.8	6.90	0.13%	1.96%	[123]
Bamboo biochar	Sandy loam	49.2	39.2	11.6	4.72	0.17%	1.83%	[124]
Pine sawdust	Silt loam	30	56	14	5.7	2.2 g kg^{-1}	21.3 g kg^{-1}	[125]
Apple branches	Silty clay	10.7	73.0	16.8	6.23	0.47 g kg^{-1}	3.32 g kg^{-1}	[126]
Wheat straw (WSB) and miscanthus straw (MSB)	Sandy loam	73	15	12	6.46	1.28 g kg^{-1}	9.84 g kg^{-1}	[127]
Corncob	Silty loam	12.0	85.1	2.94	7.94	0.95 g kg^{-1}	8.23 g kg^{-1}	[128]
Sugarcane bagasse	Sandy loam	77.3	20.3	14.5	7.54	13.40 g kg^{-1}	4.20 g kg^{-1}	[129]
Pine sawdust	Clay loam	29	36	35	6.3	9.0 g kg^{-1}	97.2 g kg^{-1}	[125]

9.2. Effect of Biochar on Soil Organic Matter

The anticipation of enhanced soil fertility assigned to biochar amendment originates from the investigation of the terra preta that comprises a large percentage of black carbon [130]. Terra preta soil was found to be rich in organic matter content, which reflects the earlier evidence of biochar existence in the soil. Wang [131] reported that biochar enhanced C storage in macroaggregates of the fine-coarse soil and thereby increased the physical security of soil organic matter (SOM); C storage in stable microaggregates can promote the stabilization of SOM for a long period of time [132]. Biochar-stimulated physical fixation of C may be related to the existence of partially carbonized, highly degradable organic residues, often a characteristic of low thermally decomposed biochars [133]. Though SOM is usually more vulnerable to digestion in coarser rather than finer textured soils because of the lower surface area of mineral binding sites that can brace organic particles, Fang [134] indicated that Eucalyptus saligna wood biochars enhanced the mineralization of indigenous organic C in sandy soil, but not in a clayey textured soil. Moreover, biochar amendment to a grassland soil results in the arousal of mineralization of indigenous soil organic C because of the positive short-term priming effects [135].

10. Affinity of Biochar and Soil Characteristics

Biochar increases water and nutrient holding capacity (Table 2, Figure 3); however, these characteristics depend not only on biochar types but also on retention capability of soil [136]. Jien [136] reported that due to the physical structure of biochar, it improves soil porosity and structure, aggregate stabilization [137], nutrient cycle [138], penetration resistance [139] and tensile strength [140]. Asai [141] added that biochar enhances soil infiltration and lowers water runoff, thus decreasing erosion due to its bulkiness and spongy structure. Biochar amendment alters the physiochemical properties of soil, which highly influences P retention in soil. [142]. Thus, the effect could be different depending on soil properties when biochar produced from the same feedstock is added to various soil types. The release of phosphorus in sandy soil is quicker compared with that in clay soil. Therefore, biochar acts as a holding agent of P and prevents the leaching or runoff loss of P from sandy soil [143,144]. The P release characteristic of biochar-amended soil is even independent of the pyrolysis temperature at which biochar is produced [145].

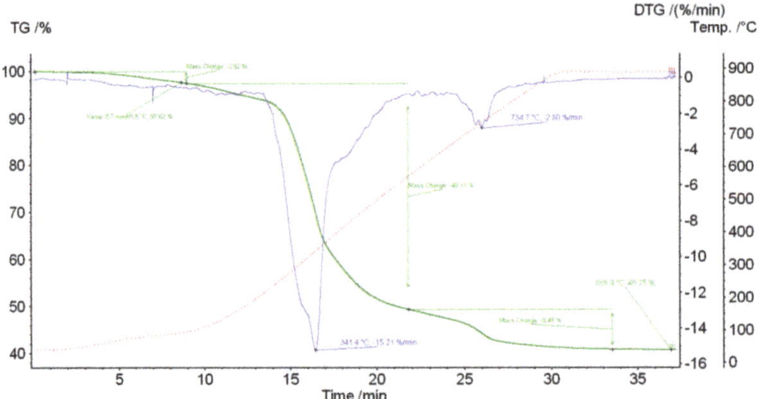

Figure 3. Thermogravity of swine digestate performed by author.

With the evolution of pore structure, the morphology of biochar also undergoes tremendous fluctuation under pyrolysis. Biochar gains the parental features of feedstock irrespective of the rate of temperature [146]. The permeable configuration in corncob [147], "beehive-like" pore structures in sugarcane [148], the symmetrical structure in wood [149] and the origin of surface morphological structure in rice husk and sawdust [150] were all retained after pyrolysis at 400–900 °C. At higher pyrolysis temperature (>1000 °C), melting with substantial deformation of biochar structure could be observed [150]. Under pyrolysis at 2000 °C, the macropores of biochar surface disappeared due to melting, while small grains emerged as the accumulation of beads on the surface of biochar [151].

The process of pyrolysis of organic feedstock into biochar brings stability to oxidized carbon fractions existing in the organic debris [152] that can persist in soils for years [153]. Therefore, biochar amendment substantially reduces greenhouses gases [154] and can be deemed as a climate change mitigation strategy. Due to these facts, biochar, the black diamond, acts as an optimistic soil conditioner of high economic and environmental value [155]. Several studies demonstrated the positive effect of biochar application on crop yield enhancement via different mechanisms. For instance, biochar amendment to soil significantly improves soil micro and macronutrients [156], despite the fact that biochar bears a higher pH value [155–158]. Nonetheless, biochar serves as a slow-release fertilizer due the strong adsorption of soil nutrients [159]. So, it is considered that biochar could be a perfect utility to acidic soil rather than alkaline or calcareous soil. Additionally, biochar, due to its high surface area and large porous structure [160], causes indirect impacts on soil physical characteristics, for instance, it significantly enhances water retention [158–162] and hydraulic conductivity [163] while decreasing soil bulk density in sandy soil [164]. However, the efficiency of biochar application to soil is not always the same. Rather, it depends on properties of applied biochar and soil conditions. Due to large surface area porosity, biochar has a significant adsorbing ability in increasing the water holding capacity (WHC) [165] and plant-available water capacity of soil (AWC) [166]. In the case of coarse sandy soil, the water and nutrient storage is generally lower under a drought condition, and thus large proportions of hydrophilic micropores (0.2–30 mm) are found in biochar, potentially retaining plant-available water and benefiting coarse sandy soils [167]. Further, gasification of biochar (GB) improves root development and thus enhances soil water retention, hence improving crop productivity [168].

11. Immutability of Soil Organic Matter and Soil Configuration

The influence of biochar on aggregate formation and organic matter stability is of high importance. Pituello [169] stated that biochar is a potential amendment for the stabilization of aggregates, especially if soil has a coarse texture and a low organic content [169]. These recommendations were further supported by Ma [170], who reported a significant increase in aggregate stability and soil organic carbon. However, Fungo [171] reported from his two year experiment that biochar had no effect on soil aggregates in tropical Ultisol. Moreover, increasing biochar amendment to fine sand and sandy loam textured soil may decrease aggregate strength [172]. Thus, the effect of biochar is multifarious and could vary with soil types and textures.

Biochar improves the structure and fertility of soil [117,118]. Glaser [173] reported that a substantial amount of biochar in terra preta was present in vulnerable fractions. However, in [174], it was indicated that biochar was associated primarily with the ultrafine, sub-50 μm soil chunk, and in [175], it was found that biochar, rather than as a free organic matter, was preponderantly available in small clumps of soil particles or soil aggregates. Brodowski [176] also found large macro-aggregate fractions with a small amount of practical biochar (>2 mm); thus, biochar might act as a binding agent for organic matter in aggregate formation and soil against degradation. Due to the interaction of biochar with soil organic matter and microorganisms and minerals, it may influence soil aggregates and its stability [177,178]. The slow oxidation properties of biochar determine the long-term effect on soil aggregation [179].

12. Biochar and Sustainability

The key obstacles with the current agrofarming systems are to enhance crop yield in a more sustainable and environmentally friendly manner [180,181]. Post-green revolution, agricultural practices enhanced their dependency on organic fertilizer for securing higher crop yield. Chemical fertilizers do increase crop yield, but they also risk the sustainability of the environment by provoking key ecological disparities, such as biodiversity loss, global warming and inclusion of heavy metals in living organisms [182,183]. Thus, adopting a more natural way of farming will reduce the reliance on organic fertilizers and sustain agricultural production and productivity.

More recently, biochar is thought to be an auspicious soil conditioner to sustain carbon and nutrients in soil, and thereby reflects the environmental problems regarding sustainable agricultural nutrient management [184,185]. Contemporary research on biochar is predominantly focused on customizing biochar properties to enhance their elimination capability for organic and inorganic pollutants [89]. Biochar has comprehensive environmental use due to its idiosyncratic properties, e.g., large surface area, microporosity, higher adsorption capacity and ion exchange capacity [99,100]. These properties have substantial consequences to its competency and potency in sustainability of the environment. The transformation of feedstock into biochar is a carbon-negative technique and has been indicated to sequestrate around 87% of carbon [186]. This not only reduces the problems of waste disposal of agricultural residues but also provides a viable and frugal method of waste transformation into value-added products. Due to its exceptional surface characteristics, biochar shows remarkable efficacy in reducing contaminants such as antibiotics, herbicides, dyes, pesticides and heavy metals and plays a key role in alleviating global climate change [187]. Biochar is thus a promising way to return lost C into the soil [188].

Many investigators have suggested biochar as an efficient soil supplement to encourage C storage [189], to augment value to agricultural products and to foster plant growth for sustainable agriculture [190]. Biochar has an exceptional function to immobilize rhizospheric heavy metals and farm chemicals on its large surface and inhibits their movement into the plants/crops, thus improving crop productivity [191]. Biochar substantially increased crop grain yield and biomass, and such favorable impacts of biochar were greater under rational P fertilization where half (50%) of P is from a natural source and the remaining 50% is from an inorganic fertilizer. The synergetic effect on nutrient accessibility and plant uptake is necessary for better crop yield and soil fertility, which can be gained by combinative use of organic and inorganic fertilizers with biochar [192]. In addition, biochar amendment increase in grain yield could signal the instrumental role of biochar in the conservation of soil nutrients and moisture, increasing nutrient uptake for potential crop yield and development [193]. Biochar can strengthen crop biomass and growth by enhancing nutrient availability [194]. Biochar application has been shown to decrease the saturated hydraulic conductivity of the soil, especially in light textured soils [195,196]. For example, Ajayi and Horn reported a decrease between 23 and 82% in the saturated hydraulic conductivity of a fine-sand soil when amended with a large application rate (2–5%, w/w) of biochar [197]. Several field trials have been conducted side by side to greenhouse pot experiments on biochar effects on soil nutrients (Table 2). While soil amendment with biochar resulted in an increase in crop production and improved soil fertility under different natural and agricultural environments [198], the immediate impact of biochar addition on soil fertility and nutrition is incoherent and weakly understood. Biochar has a consistent effect on some parameters of soil but not in all conditions [199]. While the beneficial effects and usage of biochar are widely discussed, more research is warranted to understand its perks and magnitudes, as well as the constraints of biochar amendment, in agroecosystems (Figure 4). Farming systems mostly depend upon the locally produced waste materials, e.g., crop residues and animal manures, as farmers have very limited resources to buy commercially prepared organic fertilizers [200]. The Oxisol class of soil by its very nature is poor in nutrients and organic matter content [201], which further limits microbial activities, thus leading to low crop yield. Smallholder farmers have access to bundles of local waste, so mutually rewarding benefits of crop yield and waste management can be gained, if policies associated with biochar are made for its governance in developing countries.

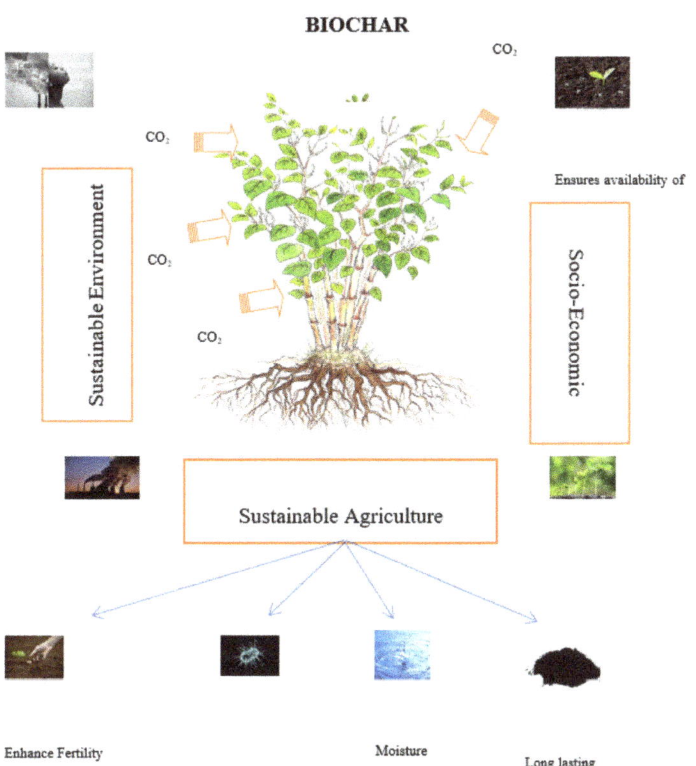

Figure 4. A diagrammatic representation of biochar dynamics and its role in agroecosystem and environmental sustainability.

13. Constraints of Biochar Application

Even though, overwhelmingly, the literature outcomes reflect the valuable prospect of biochar application, there are some constraints of biochar application which deserve attention. It is documented that weed yield increased by 200% with an increase in wheat straw-derived biochar (300–1100 °C) application rate from 15 t ha^{-1}, thus becoming a big competitor for soil nutrients to the main crop [202]. Biochar has a repressive effect on soil aging, and sporadic amendment of fresh biochar might be needed for the maximum nutrient cycling and aqua environment in soil [203]. For example, Anyanwu [204] reported that aged biochar derived from rice husk in soil has a substantially negative effect on the growth of soil earthworms and fungi. Biochar application may cause a delay in flowering in plants [205]. Additionally, Zhao [206] reported that aged biochar led to a significant reduction in the root biomass of Oryza sativa and Solanum lycopersicum in the soil. Biochar application at 14 t ha^{-1} enhanced vegetative growth but not tomato crop fruit yield, thus indicating the impact of biocarbon and crop yield dependency on plant species or the targeted part of the plant [207]. Furthermore, biochar is characterized by a selective capability to assimilate pollutants. For example, dichlorodiphenyltrichloroethane (DDT) chemical pesticide absorption was not limited by biochar application in a soil [208]. Thus, Table 2 presents the current research cited on the effect of biochar amendment on crop yield and fertility dynamics. The results fluctuate, both positively and negatively, depending on the biochar type, amount, soil type, crop type, etc.

14. Conclusions

Agroecosystems are extremely important to ensure food security and abate GHG emissions. Measures to reduce chemical fertilizer inputs and alleviate GHGs emissions include increasing soil C sequestration by addition of biochar, and thus increased crop-use efficiency of fertilizer-N. Smart choice of biochar type, rate, and affinity with agrofarming systems should not be ignored before its application. Biochar is an approach to slow the release of nutrients, and thus protect the environment without compromising crop yield. The beneficial capacity of biochar to amend agroecosystems and achieve a sustainable environment needs rational research knowledge as well as economic and social research. The practice of biochar application could enhance soil quality, increase the resilience of agroecosystems and agroforestry and support their adaptation capacity to the fluctuating climatic conditions. Nevertheless, the effects of biochar would be site dependent. Of course, biochar is not a solution to all agroecosystem problems; however, it could be a substantial strategy that deserves cognizance to achieve a sustainable agroecosystem in the future. This review has indicated many benefits, complexities and effects of biochar; however, more research is needed to provide a better understanding of biochar mechanisms and their interactive effects on plants, soil and the environment.

Author Contributions: M.A. and K.A., Collection of literature; M.A., Writing original draft; D.F., Review and editing, visualization, investigation and conceptualization; V.T., Review and editing, resource investigation and fund acquisition, U.S. Review and editing, grammar, etc.; K.A. and R.I., Review and editing, grammar, etc. All authors have read and agreed to the published version of the manuscript.

Funding: The authors received no external funding.

Institutional Review Board Statement: Not applicable.

Informed Consent Statement: Not applicable.

Data Availability Statement: No new data were created or analyzed in this study. Data sharing is not applicable to this article.

Acknowledgments: We would like to thank the scientific community of the Lithuanian Research Centre for Agriculture and Forestry for strong intellectual discussions, recommendations and changes made to the article. We are also thankful to the Lithuanian Research Council for continuous support to this study.

Conflicts of Interest: The authors declare that they have no known competing financial interests or personal relationships that could have influenced the work reported in this paper.

References

1. Mueller, N.D.; Gerber, J.S.; Johnston, M.; Ray, D.K.; Ramankutty, N.; Foley, J.A. Closing yield gaps through nutrient and water management. *Nature* **2012**, *490*, 254–257. [CrossRef]
2. Nair, V.D.; Nair, P.K.; Dari, B.; Freitas, A.M.; Chatterjee, N.; Pinheiro, F.M. Biochar in the agroecosystem–climate-change–sustainability nexus. *Front. Plant. Sci.* **2017**, *8*, 2051. [CrossRef] [PubMed]
3. Lehmann, J.; Joseph, S. Biochar for environmental management: An introduction. Biochar for environmental management. *Sci. Technol.* **2009**, *1*, 1–12.
4. Ilvo, P.; Uog, P.; Slu, P.; Inra, P.; Upm, P.; Unitus, P.; Iamo, P.; Iea-ar, P. D5.2 Participatory Impact Assessment of Sustainabil-ity and Resilience of EU Farming Systems. *2019*. No. June 2017. Available online: https://www.surefarmproject.eu/wordpress/wp-content/uploads/2019/06/D5.2-FoPIA-SURE-Farm-Cross-country-report.pdf (accessed on 26 January 2021).
5. Lu, L.; Yu, W.; Wang, Y.; Zhang, K.; Zhu, X.; Zhang, Y.; Chen, B. Application of biochar-based materials in environmental remediation: From multi-level structures to specific devices. *Biochar* **2020**, *2*, 1–31. [CrossRef]
6. Lichtfouse, E.; Navarrete, M.; Debaeke, P.; Souchère, V.; Alberola, C.; Ménassieu, J. Agronomy for sustainable agriculture: A review. In *Sustainable Agriculture*; Springer: Dordrecht, The Netherlands, 2009; pp. 1–7.
7. Jalal, F.; Arif, M.; Akhtar, K.; Khan, A.; Naz, M.; Said, F.; Ali, M. Biochar Integration with Legume Crops in Summer Gape Synergizes Nitrogen Use Efficiency and Enhance Maize Yield. *Agronomy* **2020**, *10*, 58. [CrossRef]
8. Akhtar, K.; Wang, W.; Ren, G.; Khan, A.; Nie, E.; Khan, A.; Feng, Y.; Yang, G.; Wang, H. Straw mulching with inorganic nitrogen fertilizer reduces soil CO_2 and N_2O emissions and improves wheat yield. *Sci. Total Environ.* **2020**, *741*, 140488. [CrossRef]

9. Brewer, C.E.; Chuang, V.J.; Masiello, C.A.; Gonnermann, H.; Gao, X.; Dugan, B.; Davies, C.A. New approaches to measuring biochar density and porosity. *Biomass Bioenergy* **2014**, *66*, 176–185. [CrossRef]
10. Awad, Y.M.; Lee, S.S.; Kim, K.H.; Ok, Y.S.; Kuzyakov, Y. Carbon and nitrogen mineralization and enzyme activities in soil aggregate-size classes: Effects of biochar, oyster shells, and polymers. *Chemosphere* **2018**, *198*, 40. [CrossRef]
11. Arif, M.; Ali, S.; Ilyas, M.; Riaz, M.; Akhtar, K.; Ali, K.; Wang, H. Enhancing phosphorus availability, soil organic carbon, maize productivity, and farm profitability through biochar and organic-inorganic fertilizers in an irrigated maize agroecosystem under semi-arid climate. *Soil Use Manag.* **2020**. [CrossRef]
12. Pandey, D.; Daverey, A.; Arunachalam, K. Biochar: Production, properties, and emerging role as a support for enzyme immobilization. *J. Clean. Prod.* **2020**, *255*, 120267. [CrossRef]
13. Sewu, D.D.; Tran, H.N.; Ohemeng-Boahen, G.; Woo, S.H. Facile magnetic biochar production route with new goethite nanoparticle precursor. *Sci. Total Environ.* **2020**, *717*, 137091. [CrossRef] [PubMed]
14. Dai, Z.; Meng, J.; Muhammad, N.; Liu, X.; Wang, H.; He, Y.; Xu, J. The potential feasibility for soil improvement, based on the properties of biochars pyrolyzed from different feedstocks. *J. Soils Sediments* **2013**, *13*, 989–1000. [CrossRef]
15. Demirbas, A. Combustion characteristics of different biomass fuels. *Prog. Energy Combust. Sci.* **2004**, *30*, 219–230. [CrossRef]
16. De Freitas, M.I.; Lucas, A.A.T.; Gonzaga, M.I.S. Biochar and Its Impact on Soil Properties and on The Growth of Okra Plants. *Colloquium Agrariae.* 2020, Volume 16, pp. 29–39. Available online: http://revistas.unoeste.br/index.php/ca/article/view/3302 (accessed on 26 January 2021).
17. Chen, B.; Zhou, D.; Zhu, L. Transitional adsorption and partition of nonpolar and polar aromatic contaminants by biochars of pine needles with different pyrolytic temperatures. *Environ. Sci. Technol.* **2008**, *42*, 5137–5143. [CrossRef]
18. Gray, M.; Johnson, M.G.; Dragila, M.I.; Kleber, M. Water uptake in biochars: The roles of porosity and hydrophobicity. *Biomass Bioenergy* **2014**, *61*, 196–205. [CrossRef]
19. Jiefei, M.; Zhang, K.; Chen, B. Linking hydrophobicity of biochar to the water repellency and waterholding capacity of biochar-amended soil. *Environ. Pollut.* **2019**, *253*, 779–789.
20. Jia, Y.; Hu, Z.; Mu, J.; Zhang, W.; Xie, Z.; Wang, G. Preparation of biochar as a coating material for biochar-coated urea. *Sci. Total Environ.* **2020**, *731*, 139063. [CrossRef]
21. Tang, Y.H.; Liu, S.H.; Tsang, D.C. Microwave-assisted production of CO^2-activated biochar from sugarcane bagasse for electrochemical desalination. *J. Hazard. Mater.* **2020**, *383*, 121192. [CrossRef]
22. Chen, H.; Awasthi, S.K.; Liu, T.; Duan, Y.; Ren, X.; Zhang, Z.; Awasthi, M.K. Effects of microbial culture and chicken manure biochar on compost maturity and greenhouse gas emissions during chicken manure composting. *J. Hazard. Mater.* **2020**, *389*, 121908. [CrossRef]
23. Lehmann, J.; Rillig, M.C.; Thies, J.; Masiello, C.A.; Hockaday, W.C.; Crowley, D. Biochar effects on soil biota–A review. *Soil Boil. Biochem.* **2011**, *43*, 1812–1836. [CrossRef]
24. Antonangelo, J.A.; Zhang, H. Heavy metal phytoavailability in a contaminated soil of northeastern Oklahoma as affected by biochar amendment. *Environ. Sci. Pollut. Res.* **2019**, *26*, 33582–33593. [CrossRef] [PubMed]
25. Albert, H.A.; Li, X.; Jeyakumar, P.; Wei, L.; Huang, L.; Huang, Q.; Kamran, M.; Shaheen, S.M.; Hou, D.; Rinklebe, J. Influence of Biochar and Soil Properties on Soil and Plant Tissue Concentrations of Cd and Pb: A Meta-Analysis. *Sci. Total Environ.* **2020**, *755*, 142582. [CrossRef] [PubMed]
26. Chen, H.; Yang, X.; Wang, H.; Sarkar, B.; Shaheen, S.M.; Gielen, G.; Rinklebe, J. Animal carcass-and wood-derived biochars improved nutrient bioavailability, enzyme activity, and plant growth in metal-phthalic acid ester co-contaminated soils: A trial for reclamation and improvement of degraded soils. *J. Environ. Manag.* **2020**, *261*, 110246. [CrossRef]
27. Xinyu, J.; Tan, X.; Cheng, J.; Michelle, L.; Haddix, M.; Cotrufo, F. Interactions between aged biochar, fresh low molecular weight carbon and soil organic carbon after 3.5 years soil-biochar incubations. *Geoderma* **2019**, *333*, 99–107.
28. Aller, D.; Rathke, S.; Laird, D.; Cruse, R.; Hatfield, J. Impacts of fresh and aged biochars on plant available water and water use efficiency. *Geoderma* **2017**, *307*, 114–121. [CrossRef]
29. Mukherjee, A.; Zimmerman, A.R. Organic carbon and nutrient release from a range of laboratory-produced biochars and biochar–soil mixtures. *Geoderma* **2013**, *193*, 122–130. [CrossRef]
30. Wang, K.; Peng, N.; Lu, G.; Dang, Z. Effects of pyrolysis temperature and holding time on physicochemical properties of swine-manure-derived biochar. *Waste Biomass Valoriz.* **2020**, *11*, 613–624. [CrossRef]
31. Zheng, H.; Wang, Z.; Deng, X.; Zhao, J.; Luo, Y.; Novak, J.; Herbert, S.; Xing, B. Characteristics and nutrient values of biochars produced from giant reed at different temperatures. *Bioresour. Technol.* **2013**, *130*, 463–471. [CrossRef]
32. Mandal, A.; Singh, N.; Purakayastha, T.J. Characterization of pesticide sorption behaviour of slow pyrolysis biochars as low cost adsorbent for atrazine and imidacloprid removal. *Sci. Total Environ.* **2017**, *577*, 376. [CrossRef]
33. Enders, A.; Hanley, K.; Whitman, T.; Joseph, S.; Lehmann, J. Characterization of biochars to evaluate recalcitrance and agronomic performance. *Bioresour. Technol.* **2012**, *114*, 644. [CrossRef]
34. Gaskin, J.W.; Das, K.C.; Tassistro, A.S.; Sonon, L.; Harris, K.; Hawkins, B. Characterization of char for agricultural use in the soils of the southeastern United States. In *Amazonian Dark Earths: Wim Sombroek's Vision*; Springer: Dordrecht, Germany, 2009; p. 433.
35. Purakayastha, T.J.; Kumari, S.; Pathak, H. Characterization, stability, and microbial effects of four biochars produced from crop residues. *Geoderma* **2015**, *239*, 293. [CrossRef]

36. El-Bassi, L.; Azzaz, A.A.; Jellali, S.; Akrout, H.; Marks, E.A.N.; Ghimbeu, C.M.; Jeguirim, M. Application of Olive Mill Waste-Based Biochars in Agriculture: Impact on Soil Properties, Enzymatic Activities and Tomato Growth. *Sci. Total Environ.* **2021**, *755*, 142531. [CrossRef] [PubMed]

37. Lin, H.Y.; Chang, S.T. Antioxidant potency of phenolic phytochemicals from the root extract of Acacia confusa. *Ind. Crops Prod.* **2013**, *49*, 871–878. [CrossRef]

38. Mandal, S.; Pu, S.; Adhikari, S.; Ma, H.; Kim, D.-H.; Bai, Y.; Hou, D. Progress and Future Prospects in Biochar Composites: Application and Reflection in the Soil Environment. *Crit. Rev. Environ. Sci. Technol.* **2020**, 1–53. [CrossRef]

39. Wei, L.; Shutao, W.; Jin, Z.; Tong, X. Biochar influences the microbial community structure during tomato stalk composting with chicken manure. *Bioresour. Technol.* **2014**, *154*, 148–154.

40. Smebye, A.; Alling, V.; Vogt, R.D.; Gadmar, T.C.; Mulder, J.; Cornelissen, G.; Hale, S.E. Biochar amendment to soil changes dissolved organic matter content and composition. *Chemosphere* **2016**, *142*, 100–105. [CrossRef]

41. Zhang, J.; Chen, G.; Sun, H.; Zhou, S.; Zou, G. Straw biochar hastens organic matter degradation and produces nutrient-rich compost. *Bioresour. Technol.* **2016**, *200*, 876–883. [CrossRef]

42. Jain, M.S.; Jambhulkar, R.; Kalamdhad, A.S. Biochar amendment for batch composting of nitrogen rich organic waste: Effect on degradation kinetics, composting physics, and nutritional properties. *Bioresour. Technol.* **2018**, *253*, 204–213. [CrossRef]

43. Masto, R.E.; Ansari, M.A.; George, J.; Selvi, V.; Ram, L. Co-application of biochar and lignite fly ash on soil nutrients and biological parameters at different crop growth stages of Zea mays. *Ecol. Eng.* **2013**, *58*, 314–322. [CrossRef]

44. Jindo, K.; Sonoki, T.; Matsumoto, K.; Canellas, L.; Roig, A.; Sanchez-Monedero, M.A. Influence of biochar addition on the humic substances of composting manures. *Waste Manag.* **2016**, *49*, 545–552. [CrossRef]

45. Ding, Y.; Liu, Y.; Liu, S.; Li, Z.; Tan, X.; Huang, X.; Zheng, B. Biochar to improve soil fertility. A review. *Agron. Sustain. Dev.* **2016**, *36*, 36. [CrossRef]

46. Głąb, T.; Żabiński, A.; Sadowska, U.; Gondek, K.; Kopeć, M.; Mierzwa–Hersztek, M.; Tabor, S. Effects of co-composted maize, sewage sludge, and biochar mixtures on hydrological and physical qualities of sandy soil. *Geoderma* **2018**, *315*, 27–35. [CrossRef]

47. Bao, H.; Wang, J.; Zhang, H.; Li, J.; Li, H.; Wu, F. Effects of biochar and organic substrates on biodegradation of polycyclic aromatic hydrocarbons and microbial community structure in PAHs-contaminated soils. *J. Hazard. Mater.* **2020**, *385*, 121595. [CrossRef] [PubMed]

48. Okareh, O.T.; Gbadebo, A.O. Enhancement of Soil Health Using Biochar. *Appl. Biochar Environ. Saf.* **2020**, *143*. [CrossRef]

49. Pariyar, P.; Kumari, K.; Jain, M.K.; Jadhao, P.S. Evaluation of change in biochar properties derived from different feedstock and pyrolysis temperature for environmental and agricultural application. *Sci. Total Environ.* **2020**, *713*, 136433. [CrossRef]

50. Lehmann, J.; Joseph, S. *Biochar for Environmental Management: Science, Technology, and Implementation*; Routledge: Abingdon, UK, 2015; pp. 1–33.

51. Spokas, K.A.; Cantrell, K.B.; Novak, J.M.; Archer, D.W.; Ippolito, J.A.; Collins, H.P.; Lentz, R.D. Biochar: A synthesis of its agronomic impact beyond carbon sequestration. *J. Environ. Q.* **2012**, *41*, 973–989. [CrossRef]

52. Zhang, X.; Zhang, P.; Yuan, X.; Li, Y.; Han, L. Effect of pyrolysis temperature and correlation analysis on the yield and physicochemical properties of crop residue biochar. *Bioresour. Technol.* **2020**, *296*, 122318. [CrossRef]

53. Laird, D.A.; Fleming, P.; Davis, D.D.; Horton, R.; Wang, B.; Karlen, D.L. Impact of biochar amendments on the quality of a typical Midwestern agricultural soil. *Geoderma* **2010**, *158*, 443–449. [CrossRef]

54. Mullen, C.A.; Boateng, A.A.; Goldberg, N.M.; Lima, I.M.; Laird, D.A.; Hicks, K.B. Bio-oil and bio-char production from corn cobs and stover by fast pyrolysis. *Biomass Bioenergy* **2010**, *34*, 67–74. [CrossRef]

55. Laird, D.A.; Brown, R.C.; Amonette, J.E.; Lehmann, J. Review of the pyrolysis platform for coproducing bio-oil and biochar. *Biofuels Bioprod. Biorefining* **2009**, *3*, 547–562. [CrossRef]

56. Cantrell, K.B.; Hunt, P.G.; Uchimiya, M.; Novak, J.M.; Ro, K.S. Impact of pyrolysis temperature and manure source on physico-chemical characteristics of biochar. *Bioresour. Technol.* **2012**, *107*, 419–428. [CrossRef] [PubMed]

57. Antonangelo, J.A.; Zhang, H.; Sun, X.; Kumar, A. Physicochemical properties and morphology of biochars as affected by feedstock sources and pyrolysis temperatures. *Biochar* **2019**, *1*, 325–336. [CrossRef]

58. Qian, K.; Kumar, A.; Patil, K.; Bellmer, D.; Wang, D.; Yuan, W.; Huhnke, R.L. Effects of biomass feedstocks and gasification conditions on the physiochemical properties of char. *Energies* **2013**, *6*, 3972–3986. [CrossRef]

59. Gul, S.; Whalen, J.K.; Thomas, B.W.; Sachdeva, V.; Deng, H. Physico-chemical properties and microbial responses in biochar-amended soils: Mechanisms and future directions. *Agric. Ecosyst. Environ.* **2015**, *206*, 46–59. [CrossRef]

60. Spokas, K.A. Review of the stability of biochar in soils: Predictability of O: C molar ratios. *Carbon Manag.* **2010**, *1*, 289–303. [CrossRef]

61. Muhammad, N.; Aziz, R.; Brookes, P.C.; Xu, J. Impact of wheat straw biochar on yield of rice and some properties of Psammaquent and Plinthudult. *J. Soil Sci. Plant. Nutr.* **2017**, *17*, 808–823. [CrossRef]

62. Naeem, M.A.; Khalid, M.; Aon, M.; Abbas, G.; Tahir, M.; Amjad, M.; Akhtar, S.S. Effect of wheat and rice straw biochar produced at different temperatures on maize growth and nutrient dynamics of a calcareous soil. *Arch. Agron. Soil Sci.* **2017**, *63*, 2048–2061. [CrossRef]

63. Yang, Y.; Ma, S.; Zhao, Y.; Jing, M.; Xu, Y.; Chen, J. A field experiment on enhancement of crop yield by rice straw and corn stalk-derived biochar in Northern China. *Sustainability* **2015**, *7*, 13713–13725. [CrossRef]

64. O'toole, A.; Moni, C.; Weldon, S.; Schols, A.; Carnol, M.; Bosman, B.; Rasse, D.P. Miscanthus biochar had limited effects on soil physical properties, microbial biomass, and grain yield in a four-year field experiment in Norway. *Agriculture* **2018**, *8*, 171. [CrossRef]

65. Azeem, M.; Hayat, R.; Hussain, Q.; Ahmed, M.; Pan, G.; Tahir, M.I.; Imran, M.; Irfan, M. Biochar Improves Soil Quality and N2-Fixation and Reduces Net Ecosystem CO2 Exchange in a Dryland Legume-Cereal Cropping System. *Soil Tillage Res.* **2019**, *186*, 172–182. [CrossRef]

66. Wang, J.; Pan, X.; Liu, Y.; Zhang, X.; Xiong, Z. Effects of biochar amendment in two soils on greenhouse gas emissions and crop production. *Plant. Soil* **2012**, *360*, 287–298. [CrossRef]

67. Güereña, D.; Lehmann, J.; Hanley, K.; Enders, A.; Hyland, C.; Riha, S. Nitrogen dynamics following field application of biochar in a temperate North American maize-based production system. *Plant. Soil* **2013**, *365*, 239–254. [CrossRef]

68. Lai, W.Y.; Lai, C.M.; Ke, G.R.; Chung, R.S.; Chen, C.T.; Cheng, C.H.; Chen, C.C. The effects of woodchip biochar application on crop yield, carbon sequestration and greenhouse gas emissions from soils planted with rice or leaf beet. *J. Taiwan Inst. Chem. Eng.* **2013**, *44*, 1039–1044. [CrossRef]

69. Sigua, G.C.; Stone, K.C.; Hunt, P.G.; Cantrell, K.B.; Novak, J.M. Increasing biomass of winter wheat using sorghum biochars. *Agron. Sustain. Dev.* **2015**, *35*, 739–748. [CrossRef]

70. Martinsen, V.; Mulder, J.; Shitumbanuma, V.; Sparrevik, M.; Børresen, T.; Cornelissen, G. Farmer-led maize biochar trials: Effect on crop yield and soil nutrients under conservation farming. *J. Plant. Nutr. Soil Sci.* **2014**, *177*, 681–695. [CrossRef]

71. Alburquerque, J.A.; Salazar, P.; Barrón, V.; Torrent, J.; del Campillo, M.D.C.; Gallardo, A.; Villar, R. Enhanced wheat yield by biochar addition under different mineral fertilization levels. *Agron. Sustain. Dev.* **2013**, *33*, 475–484. [CrossRef]

72. Adekiya, A.O.; Agbede, T.M.; Olayanju, A.; Ejue, W.S.; Adekanye, T.A.; Adenusi, T.T.; Ayeni, J.F. Effect of Biochar on Soil Properties, Soil Loss, and Cocoyam Yield on a Tropical Sandy Loam Alfisol. *Sci. World J.* **2020**. [CrossRef]

73. Boersma, M.; Wrobel-Tobiszewska, A.; Murphy, L.; Eyles, A. Impact of biochar application on the productivity of a temperate vegetable cropping system. *N. Z. J. Crop. Horticult. Sci.* **2017**, *45*, 277–288. [CrossRef]

74. Bolan, N.; Kunhikrishnan, A.; Thangarajan, R.; Kumpiene, J.; Park, J.; Makino, T.; Scheckel, K. Remediation of heavy metal (loid) s contaminated soils–to mobilize or to immobilize? *J. Hazard. Mater.* **2014**, *266*, 141–166. [CrossRef]

75. Qian, K.; Kumar, A.; Zhang, H.; Bellmer, D.; Huhnke, R. Recent advances in utilization of biochar. *Renew. Sustain. Energy Rev.* **2015**, *42*, 1055–1064. [CrossRef]

76. Werdin, J.; Fletcher, T.D.; Rayner, J.P.; Williams, N.S.; Farrell, C. Biochar made from low density wood has greater plant available water than biochar made from high density wood. *Sci. Total Environ.* **2020**, *705*, 135856. [CrossRef]

77. Razzaghi, F.; Obour, P.B.; Arthur, E. Does biochar improve soil water retention? A systematic review and meta-analysis. *Geoderma* **2020**, *361*, 114055. [CrossRef]

78. Radwan, N.M.; Marzouk, E.R.; El-Melegy, A.M.; Hassan, M.A. Improving Soil Properties by Using Biochar Under Drainage Conditions in North Sinai. *Sinai J. Appl. Sci.* **2020**. [CrossRef]

79. Laghari, M.; Naidu, R.; Xiao, B.; Hu, Z.; Mirjat, M.S.; Hu, M.; Abudi, Z.N. Recent developments in biochar as an effective tool for agricultural soil management: A review. *J. Sci. Food Agric.* **2016**, *96*, 4840–4849. [CrossRef] [PubMed]

80. Barrow, C.J. Biochar: Potential for countering land degradation and for improving agriculture. *Appl. Geogr.* **2012**, *34*, 21–28. [CrossRef]

81. Blanco-Canqui, H. Biochar and soil physical properties. *Soil Sci. Soc. Am. J.* **2017**, *81*, 687–711. [CrossRef]

82. Rasa, K.; Heikkinen, J.; Hannula, M.; Arstila, K.; Kulju, S.; Hyväluoma, J. How and why does willow biochar increase a clay soil water retention capacity? *Biomass Bioenergy* **2018**, *119*, 346–353. [CrossRef]

83. Uzoma, K.C.; Inoue, M.; Andry, H.; Fujimaki, H.; Zahoor, A.; Nishihara, E. Effect of cow manure biochar on maize productivity under sandy soil condition. *Soil Use Manag.* **2011**, *27*, 205–212. [CrossRef]

84. Sánchez-Monedero, M.A.; Cayuela, M.L.; Sánchez-García, M.; Vandecasteele, B.; D'Hose, T.; López, G.; Mondini, C. Agronomic evaluation of biochar, compost, and biochar-blended compost across different cropping systems: Perspective from the European project FERTIPLUS. *Agronomy* **2019**, *9*, 225. [CrossRef]

85. Lu, H.; Yan, M.; Wong, M.H.; Mo, W.Y.; Wang, Y.; Chen, X.W.; Wang, J.J. Effects of biochar on soil microbial community and functional genes of a landfill cover three years after ecological restoration. *Sci. Total Environ.* **2020**, *717*, 137133. [CrossRef]

86. Otsuka, S.; Sudiana, I.; Komori, A.; Isobe, K.; Deguchi, S.; Nishiyama, M.; Shimizu, H.; Senoo, K. Community structure of soil bacteria in a tropical rainforest several years after fire. *Microbes Environ.* **2008**, *23*, 49–56. [CrossRef] [PubMed]

87. Rutigliano, F.A.; Romano, M.; Marzaioli, R.; Baglivo, I.; Baronti, S.; Miglietta, F.; Castaldi, S. Effect of biochar addition on soil microbial community in a wheat crop. *Eur. J. Soil Biol.* **2014**, *60*, 9–15. [CrossRef]

88. Dempster, D.N.; Gleeson, D.B.; Solaiman, Z.I.; Jones, D.L.; Murphy, D.V. Decreased soil microbial biomass and nitrogen mineralisation with Eucalyptus biochar addition to a coarse textured soil. *Plant. Soil* **2012**, *354*, 311–324. [CrossRef]

89. De Boer, W.; Kowalchuk, G.A. Nitrification in acid soils: Micro-organisms and mechanisms. *Soil Biol. Biochem.* **2001**, *33*, 853–866. [CrossRef]

90. Gao, L.; Wang, R.; Shen, G.; Zhang, J.; Meng, G.; Zhang, J. Effects of biochar on nutrients and the microbial community structure of tobacco-planting soils. *J. Soil Sci. Plant. Nutr.* **2017**, *17*, 884–896. [CrossRef]

91. Farrell, M.; Kuhn, T.K.; Macdonald, L.M.; Maddern, T.M.; Murphy, D.V.; Hall, P.A.; Singh, B.P.; Baumann, K.; Krull, E.S.; Baldock, J.A. Microbial utilization of biochar-derived carbon. *Sci. Total Environ.* **2013**, *465*, 288–297. [CrossRef]

92. He, Y.; Zhou, X.; Jiang, L.; Li, M.; Du, Z.; Zhou, G.; Wallace, H. Effects of biochar application on soil greenhouse gas fluxes: A meta-analysis. *GCB Bioenergy* **2017**, *9*, 743–755. [CrossRef]
93. Smith, J.L.; Collins, H.P.; Bailey, V.L. The effect of young biochar on soil respiration. *Biol. Biochem.* **2010**, *42*, 2345–2347. [CrossRef]
94. Lunde, C.F.; Hake, S. Florets & Rosettes: Meristem Genes in Maize and Arabidopsis. *Maydica* **2005**, *50*, 451–458.
95. Reddy, S.B.N. *Biochar Culture: Biochar for Environment and Development*; MetaMeta: s-Hertogenbosch, The Netherlands, 2014.
96. Jindo, K.; Mizumoto, H.; Sawada, Y.; Sanchez-Monedero, M.A.; Sonoki, T. Physical and chemical characterization of biochars derived from different agricultural residues. *Biogeoscience* **2014**, *11*, 6613–6621. [CrossRef]
97. Mclennon, E.; Solomon, J.K.Q.; Neupane, D.; Davison, J. Biochar and Nitrogen Application Rates Effect on Phosphorus Removal from a Mixed Grass Sward Irrigated with Reclaimed Wastewater. *Sci. Total Environ.* **2020**, *715*, 137012. [CrossRef] [PubMed]
98. Glaser, B.; Haumaier, L.; Guggenberger, G.; Zech, W. The 'Terra Preta' phenomenon: A model for sustainable agriculture in the humid tropics. *Naturwissenschaften* **2001**, *88*, 37–41. [CrossRef] [PubMed]
99. Rondon, M.; Ramirez, J.A.; Lehmann, J. Greenhouse gas emissions decrease with charcoal additions to tropical soils. In Proceedings of the 3rd USDA Symposium on Greenhouse Gases and Carbon Sequestration, Baltimore, MD, USA, 21–24 March 2005; Volume 208.
100. Solomon, S.; Qin, D.; Manning, M.; Chen, Z.; Marquis, M.; Averyt, K.B.; Tignor, M.; Miller, H.L. *IPCC. Climate change, the physical science basis: Contribution of Working Group I to the Fourth Assessment Report of the Intergovernmental Panel on Climate Change*; Cambridge University Press: Cambridge, UK; New York, NY, USA, 2007; p. 996.
101. Wang, J.; Xiong, Z.; Kuzyakov, Y. Biochar stability in soil: Metaanalysis of decomposition and priming effects. *GCB Bioenergy* **2016**, *8*, 512–523. [CrossRef]
102. Vanholme, B.; Desmet, T.; Ronsse, F.; Rabaey, K.; van Breusegem, F.; De Mey, M. Towards a carbon negative sustainable bio-based economy. *Front. Plant. Sci.* **2013**, *4*, 174. [CrossRef] [PubMed]
103. Downie, A.; Munroe, P.; Cowie, A.; van Zwieten, L.; Lau, D.M.S. Biochar as a geo engineering climate solution: Hazard identification and risk management. *Crit. Rev. Environ. Sci. Technol.* **2012**, *42*, 225–250. [CrossRef]
104. Jia, J.; Li, B.; Chen, Z.; Xie, Z.; Xiong, Z. Effects of biochar application on vegetable production and emissions of N_2O and CH_4. *Soil Sci. Plant. Nutr.* **2012**, *58*, 503–509. [CrossRef]
105. Sun, L.; Li, L.; Chen, Z.; Wang, J.; Xiong, Z. Combined effects of nitrogen deposition and biochar application on emissions of N2O, CO2 and NH3 from agricultural and forest soils. *Soil Sci. Plant. Nutr.* **2014**, *60*, 254–265. [CrossRef]
106. Hale, S.E.; Nurida, N.L.; Mulder, J.; Sørmo, E.; Silvani, L.; Abiven, S.; Joseph, S.; Taherymoosavi, S.; Cornelissen, G. The Effect of Biochar, Lime and Ash on Maize Yield in a Long-Term Field Trial in a Ultisol in the Humid Tropics. *Sci. Total Environ.* **2020**, *719*, 137455. [CrossRef]
107. Kuppusamy, S.; Thavamani, P.; Megharaj, M.; Venkateswarlu, K.; Naidu, R. Agronomic and remedial benefits and risks of applying biochar to soil: Current knowledge and future research directions. *Environ. Int.* **2016**, *87*, 1–12. [CrossRef]
108. Randolph, P.; Bansode, R.R.; Hassan, O.A.; Rehrah, D.; Ravella, R.; Reddy, M.R.; Watts, D.W.; Novak, J.M.; Ahmedna, M. Effect of biochars produced from solid organic municipal waste on soil quality parameters. *J. Environ. Manag.* **2017**, *192*, 271–280. [CrossRef]
109. El-Naggara, A.; Lee, S.S.; Rinklebed, J.; Farooq, M.; Song, H.; Sarmah, A.K.; Zimmerman, A.R.; Ahmad, M.; Shaheend, S.M.; Ok, Y.S. Biochar application to low fertility soils: A review of current status, and future prospects. *Geoderma* **2019**, *337*, 536–554. [CrossRef]
110. Koide, R.T. Biochar—Arbuscular mycorrhiza interaction in temperate soils. In *Mycorrhizal Mediation of Soil*; Elsevier: New York, NY, USA, 2017; pp. 461–477.
111. Mohanty, P.; Nanda, S.; Pant, K.K.; Naik, S.; Kozinski, J.A.; Dalai, A.K. Evaluation of the physiochemical development of biochars obtained from pyrolysis of wheat straw, timothy grass and pinewood: Effects of heating rate. *J. Anal. Appl. Pyrol.* **2013**, *104*, 485–493. [CrossRef]
112. Filiberto, D.M.; Gaunt, J.L. Practicality of biochar additions to enhance soil and crop productivity. *Agriculture* **2013**, *3*, 715–725. [CrossRef]
113. Oguntunde, P.G.; Fosu, M.; Ajayi, A.E.; Van De Giesen, N. Effects of charcoal production on maize yield, chemical properties, and texture of soil. *Biol. Fertil Soils* **2004**, *39*, 295–299. [CrossRef]
114. Yao, Y.N.; Gao, B.; Chen, H.; Jiang, L.; Inyang, M.; Zimmerman, A.R. Adsorption of sulfamethoxazole on biochar and its impact on reclaimed water irrigation. *J. Hazard. Mater.* **2012**, *209*, 408–413. [CrossRef]
115. Wu, F.; Jia, Z.; Wang, S.; Chang, S.X.; Startsev, A. Contrasting effects of wheat straw and its biochar on greenhouse gas emissions and enzyme activities in a Chernozemic soil. *Biol. Fert. Soils* **2013**, *49*, 555–565. [CrossRef]
116. Awad, Y.M.; Blagodatskaya, E.; Ok, Y.K.; Kuzyakov, Y. Effects of polyacrylamide, biopolymer, and biochar on decomposition of soil organic matter and plant residues as determined by C and enzyme activities. *Eur. J. Soil Biol.* **2012**, *48*, 1–10. [CrossRef]
117. Demisie, W.; Liu, Z.Y.; Zhang, M.K. Effect of biochar on carbon fractions and enzyme activity of red soil. *Catena* **2014**, *121*, 214–221. [CrossRef]
118. Akça, M.O.; Namli, A. Effects of poultry litter 1053 biochar on soil enzyme activities and tomato, pepper and lettuce plants growth. *Eur. J. Soil Sci.* **2014**, *4*, 161–168.
119. Shang, J.; Geng, Z.C.; Wang, Y.T.; Chen, X.X.; Zhao, J. Effect of biochar amendment on soil microbial biomass carbon and nitrogen and enzyme activity in tier soils. *Sci. Agric. Sin.* **2016**, *49*, 1142–1151.

120. Paz-Ferreiro, J.; Fu, S.; Mendez, A.; Gasco, G. Biochar modifies the thermodynamic parameters of soil enzyme activity in a tropical soil. *J. Soil Sediment.* **2015**, *15*, 578–583. [CrossRef]
121. Paz-Ferreiro, J.; Gascó, G.; Gutiérrez, B.; Méndez, A. Soil biochemical activities and the geometric mean of enzyme activities after application of sewage sludge and sewage sludge biochar to soil. *Biol. Fert. Soils* **2012**, *48*, 511–517. [CrossRef]
122. Zhang, M.; Cheng, G.; Feng, H.; Sun, B.; Zhao, Y.; Chen, H.; Chen, J.; Dyck, M.; Wang, X.; Zhang, J.; et al. Effects of straw and biochar amendments on aggregate stability, soil organic carbon, and enzyme activities in the Loess Plateau, China. *Environ. Sci. Pollut. Res.* **2017**, *24*, 10108–10120. [CrossRef] [PubMed]
123. Zhu, L.; Xiao, Q.; Chenga, H.; Shi, B.; Shen, Y.; Li, S. Seasonal dynamics of soil microbial activity after biochar addition in a dry land maize field in North-Western China. *Ecol. Eng.* **2017**, *104*, 141–149. [CrossRef]
124. Chen, J.; Li, S.; Lianga, C.; Xu, O.; Li, Y.; Qin, H.; Fuhrman, J.J. Response of microbial community structure and function to short-term biochar amendment in an intensively managed bamboo (Phyllostachys praecox) plantation soil: Effect of particle size and addition rate. *Sci. Total Environ.* **2017**, *574*, 24–33. [CrossRef]
125. Pokharel, P.; Kwak, J.H.; Ok, Y.S.; Chang, S.X. Pine sawdust biochar reduces GHG emission by decreasing microbial and enzyme activities in forest and grassland soils in a laboratory experiment. *Sci. Total Environ.* **2018**, *625*, 1247–1256. [CrossRef]
126. Li, S.; Liang, C.; Shangguan, Z. Effects of apple branch biochar on soil C mineralization and nutrient cycling under two levels of N. *Sci. Total Environ.* **2017**, *607–608*, 109–119. [CrossRef]
127. Mierzwa-Hersztek, M.; Gondek, K.; Klimkowicz-Pawlas, A.; Baran, A. Effect of wheat and Miscanthus straw biochars on soil enzymatic activity, ecotoxicity, and plant yield. *Int. Agrophys.* **2017**, *31*, 367–375. [CrossRef]
128. Pei, J.; Zhuang, S.; Cui, J.; Li, J.; Li, B.; Wu, J.; Fang, C. Biochar decreased the temperature sensitivity of soil carbon decomposition in a paddy field. *Agric. Ecosyst. Environ.* **2017**, *249*, 156–164. [CrossRef]
129. Bashir, S.; Hussain, Q.; Akmal, M.; Riaz, M.; Hu, H.; Ijaz, S.S.; Iqbal, M.; Abro, S.; Mehmood, S.; Ahmad, M. Sugarcane bagasse-derived biochar reduces the cadmium and chromium bioavailability to mash bean and enhances the microbial activity in contaminated soil. *J. Soil Sediment.* **2017**, *18*, 874–886. [CrossRef]
130. Wang, D.; Fonte, S.J.; Parikh, S.J.; Six, J.; Scow, K.M. Biochar additions can enhance soil structure and the physical stabilization of C in aggregates. *Geoderma* **2017**, *303*, 110–117. [CrossRef] [PubMed]
131. Six, J.; Conant, R.T.; Paul, E.A.; Paustian, K. Stabilization mechanisms of soil organic matter: Implications for C-saturation of soils. *Plant. Soil* **2002**, *241*, 155–176. [CrossRef]
132. Zimmerman, A.R. Abiotic and microbial oxidation of laboratory-produced black carbon (biochar). *Environ. Sci. Technol.* **2010**, *44*, 1295–1301. [CrossRef] [PubMed]
133. Fang, Y.; Singh, B.; Singh, B.P. Effect of temperature on biochar priming effects and its stability in soils. *Soil Biol. Biochem.* **2015**, *80*, 136–145. [CrossRef]
134. Singh, B.P.; Cowie, A.L. Long-term influence of biochar on native organic carbon mineralisation in a low-carbon clayey soil. *Sci. Rep.* **2014**, *4*, 3687. [CrossRef] [PubMed]
135. Jien, S.H.; Wang, C.S. Effects of biochar on soil properties and erosion potential in a highly weathered soil. *Catena* **2013**, *110*, 225–233. [CrossRef]
136. Kimetu, J.M.; Lehmann, J. Stability and stabilisation of biochar and green manure in soil with different organic carbon contents. *Soil Res.* **2010**, *48*, 577–585. [CrossRef]
137. Harvey, O.R.; Kuo, L.J.; Zimmerman, A.R.; Louchouarn, P.; Amonette, J.E.; Herbert, B.E. An index-based approach to assessing recalcitrance and soil carbon sequestration potential of engineered black carbons (biochars). *Environ. Sci. Technol.* **2012**, *46*, 1415–1421. [CrossRef]
138. Joseph, S.D.; Camps-Arbestain, M.; Lin, Y.; Munroe, P.; Chia, C.H.; Hook, J.; Lehmann, J. An investigation into the reactions of biochar in soil. *Soil Res.* **2010**, *48*, 501–515. [CrossRef]
139. Chan, K.Y.; Van Zwieten, L.; Meszaros, I.; Downie, A.; Joseph, S. Agronomic values of greenwaste biochar as a soil amendment. *Soil Res.* **2008**, *45*, 629–634. [CrossRef]
140. Asai, H.; Samson, B.K.; Stephan, H.M.; Songyikhangsuthor, K.; Homma, K.; Kiyono, Y.; Horie, T. Biochar amendment techniques for upland rice production in Northern Laos: 1. Soil physical properties, leaf SPAD and grain yield. *Field Crops Res.* **2009**, *111*, 81–84. [CrossRef]
141. Wu, L.; Zhang, S.; Wang, J.; Ding, X. Phosphorus retention using iron (II/III) modified biochar in saline-alkaline soils: Adsorption, column, and field tests. *Environ. Pollut.* **2020**, *261*, 114223. [CrossRef]
142. Burrell, L.D.; Zehetner, F.; Rampazzo, N.; Wimmer, B.; Soja, G. Long-term effects of biochar on soil physical properties. *Geoderma* **2016**, *282*, 96–102. [CrossRef]
143. Glaser, B.; Lehr, V.I. Biochar effects on phosphorus availability in agricultural soils: A meta-analysis. *Sci. Rep.* **2019**, *9*. [CrossRef]
144. Dharmakeerthi, R.S.; Kumaragamage, D.; Goltz, D.; Indraratne, S.P. Phosphorus release from unamended and gypsum-or biochar-amended soils under simulated snowmelt and summer flooding conditions. *J. Environ. Q.* **2019**, *48*, 822–830. [CrossRef] [PubMed]
145. Li, F.; Liang, X.; Niyungeko, C.; Sun, T.; Liu, F.; Arai, Y. Effects of biochar amendments on soil phosphorus transformation in agricultural soils. *Adv. Agron.* **2019**, *158*, 131–172.
146. Pottmaier, D.; Costa, M.; Farrow, T.; Oliveira, A.A.; Alarcon, O.; Snape, C. Comparison of rice husk and wheat straw: From slow and fast pyrolysis to char combustion. *Energy Fuels* **2013**, *27*, 7115–7125. [CrossRef]

147. Bourke, J.; Manley-Harris, M.; Fushimi, C.; Dowaki, K.; Nunoura, T.; Antal, M.J. Do all carbonized charcoals have the same chemical structure? 2. A model of the chemical structure of carbonized charcoal. *Ind. Eng. Chem. Res.* **2007**, *46*, 5954–5967. [CrossRef]

148. Kondo, Y.; Fukuzawa, Y.; Kawamitsu, Y.; Ueno, M.; Tsutsumi, J.; Takemoto, T.; Kawasaki, S. A new application of bagasse char as a solar energy absorption and accumulation material. *Earth Environ. Sci. Trans. Royal Soc. Edinburgh* **2012**, *103*, 31–38. [CrossRef]

149. Burhenne, L.; Damiani, M.; Aicher, T. Effect of feedstock water content and pyrolysis temperature on the structure and reactivity of spruce wood char produced in fixed bed pyrolysis. *Fuel* **2013**, *107*, 836–847. [CrossRef]

150. Zeng, K.; Minh, D.P.; Gauthier, D.; Weiss-Hortala, E.; Nzihou, A.; Flamant, G. The effect of temperature and heating rate on char properties obtained from solar pyrolysis of beech wood. *Bioresour. Technol.* **2015**, *182*, 114–119. [CrossRef] [PubMed]

151. Trubetskaya, A.; Jensen, P.A.; Jensen, A.D.; Steibel, M.; Spliethoff, H.; Glarborg, P. Influence of fast pyrolysis conditions on yield and structural transformation of biomass chars. *Fuel Proc. Technol.* **2015**, *140*, 205–214. [CrossRef]

152. Song, X.; Li, Y.; Yue, X.; Hussain, Q.; Zhang, J.; Liu, Q.; Cui, D. Effect of cotton straw-derived materials on native soil organic carbon. *Sci. Total Environ.* **2019**, *663*, 38–44. [CrossRef] [PubMed]

153. Giagnoni, L.; Maienza, A.; Baronti, S.; Vaccari, F.P.; Genesio, L.; Taiti, C.; Mancuso, S. Long-term soil biological fertility, volatile organic compounds and chemical properties in a vineyard soil after biochar amendment. *Geoderma* **2019**, *344*, 127–136. [CrossRef]

154. Awad, Y.M.; Wang, J.; Igalavithana, A.D.; Tsang, D.C.; Kim, K.H.; Lee, S.S.; Ok, Y.S. Biochar effects on rice paddy: Meta-analysis. *Adv. Agron.* **2018**, *148*, 1–32.

155. Abdelhafez, A.A.; Abbas, M.H.; Li, J. Biochar: The black diamond for soil sustainability, contamination control and agricultural production. *Eng. Appl. Biochar* **2017**, *2*. [CrossRef]

156. Elshony, M.; Farid, I.M.; Alkamar, F.; Abbas, M.H.; Abbas, H. Ameliorating a sandy soil using biochar and compost amendments and their implications as slow release fertilizers on plant growth. *Egypt. J. Soil Sci.* **2019**, *59*, 305–322. [CrossRef]

157. Rahman, G.M.; Rahman, M.M.; Alam, M.S.; Kamal, M.Z.; Mashuk, H.A.; Datta, R.; Meena, R.S. Biochar and organic amendments for sustainable soil carbon and soil health. In *Carbon and Nitrogen Cycling in Soil*; Springer: Singapore, 2020; pp. 45–85.

158. Zheng, H.; Wang, X.; Luo, X.; Wang, Z.; Xing, B. Biochar-induced negative carbon mineralization priming effects in a coastal wetland soil: Roles of soil aggregation and microbial modulation. *Sci. Total Environ.* **2018**, *610*, 951–960. [CrossRef]

159. Dai, Y.; Wang, W.; Lu, L.; Yan, L.; Yu, D. Utilization of biochar for the removal of nitrogen and phosphorus. *J. Clean. Prod.* **2020**, *257*, 120573. [CrossRef]

160. Brassard, P.; Godbout, S.; Raghavan, V. Soil biochar amendment as a climate change mitigation tool: Key parameters and mechanisms involved. *J. Environ. Manag.* **2016**, *181*, 484. [CrossRef]

161. Manolikaki, I.; Diamadopoulos, E. Agronomic potential of biochar prepared from brewery byproducts. *J. Environ. Manag.* **2020**, *255*, 109856. [CrossRef] [PubMed]

162. Wiersma, W.; van der Ploeg, M.J.; Sauren, I.J.; Stoof, C.R. No effect of pyrolysis temperature and feedstock type on hydraulic properties of biochar and amended sandy soil. *Geoderma* **2020**, *364*, 114209. [CrossRef]

163. Cooper, J.; Greenberg, I.; Ludwig, B.; Hippich, L.; Fischer, D.; Glaser, B.; Kaiser, M. Effect of Biochar and Compost on Soil Properties and Organic Matter in Aggregate Size Fractions under Field Conditions. *Agric. Ecosyst. Environ.* **2020**, *295*, 106882. [CrossRef]

164. Bruun, E.W.; Petersen, C.T.; Hansen, E.; Holm, J.K.; Hauggaard-Nielsen, H. Biochar amendment to coarse sandy subsoil improves root growth and increases water retention. *Soil Use Manag.* **2014**, *30*, 109–118. [CrossRef]

165. Abel, S.; Peters, A.; Trinks, S.; Schonsky, H.; Facklam, M.; Wessolek, G. Impact of biochar and hydrochar addition on water retention and water repellency of sandy soil. *Geoderma* **2013**, *202–203*, 183–191. [CrossRef]

166. Hardie, M.; Clothier, B.; Bound, S.; Oliver, G.; Close, D. Does biochar influence soil physical properties and soil water availability? *Plant. Soil* **2014**, 1–15. [CrossRef]

167. Hansen, V.; Hauggaard-Nielsen, H.; Petersen, C.T.; Mikkelsen, T.N.; Müller-Stöver, D. Effects of gasification biochar on plant-available water capacity and plant growth in two contrasting soil types. *Soil Tillage Res.* **2016**, *161*, 1–9. [CrossRef]

168. Ebenezer, A.A.; Rainer, H.O.R.N. Biochar-induced changes in soil resilience: Effects of soil texture and biochar dosage. *Pedosphere* **2017**, *27*, 236–247.

169. Pituello, C.; Dal Ferro, N.; Francioso, O.; Simonetti, G.; Berti, A.; Piccoli, I.; Morari, F. Effects of biochar on the dynamics of aggregate stability in clay and sandy loam soils. *Eur. J. Soil Sci.* **2018**, *69*, 827–842. [CrossRef]

170. Ma, N.; Zhang, L.; Zhang, Y.; Yang, L.; Yu, C.; Yin, G.; Ma, X. Biochar improves soil aggregate stability and water availability in a mollisol after three years of field application. *PLoS ONE* **2016**, *11*. [CrossRef]

171. Fungo, B.; Lehmann, J.; Kalbitz, K.; Thiongo, M.; Okeyo, I.; Tenywa, M.; Neufeldt, H. Aggregate size distribution in a biochar-amended tropical Ultisol under conventional hand-hoe tillage. *Soil Tillage Res.* **2017**, *165*, 190–197. [CrossRef] [PubMed]

172. Atkinson, C.J.; Fitzgerald, J.D.; Hipps, N.A. Potential mechanisms for achieving agricultural benefits from biochar application to temperate soils: A review. *Plant. Soil* **2010**, *337*, 1–18. [CrossRef]

173. Glaser, B.; Lehmann, J.; Zech, W. Ameliorating physical and chemical properties of highly weathered soils in the tropics with charcoal–A review. *Biol. Fertile. Soils* **2002**, *35*, 219–230. [CrossRef]

174. Glaser, B.; Balashov, E.; Haumaier, L.; Guggenberger, G.; Zech, W. Black carbon in density fractions of anthropogenic soils of the Brazilian Amazon region. *Organ. Geochem.* **2000**, *31*, 669–678. [CrossRef]

175. Liang, B.; Lehmann, J.; Solomon, D.; Sohi, S.; Thies, J.E.; Skjemstad, J.O.; Wirick, S. Stability of biomass-derived black carbon in soils. *Geochim. Cosmochim. Acta* **2008**, *72*, 6069–6078. [CrossRef]
176. Brodowski, S.; John, B.; Flessa, H.; Amelung, W. Aggregate-occluded black carbon in soil. *Eur. J. Soil Sci.* **2006**, *57*, 539–546. [CrossRef]
177. Piccolo, A.; Pietramellara, G.; Mbag Demisie, J.S.C. Use of humic substances as soil conditioners to increase aggregate stability. *Geoderma* **1997**, *75*, 267–277. [CrossRef]
178. Verheijen, F.; Jeffery, S.; Bastos, A.C.; Van der Velde, M.; Diafas, I. Biochar application to soils. A critical scientific review of effects on soil properties, processes, and functions. *EUR* **2010**, *24099*, 162.
179. Patel, J.S.; Singh, A.; Singh, H.B.; Sarma, B.K. Plant genotype, microbial recruitment and nutritional security. *Front. Plant. Sci.* **2015**, *6*, 1–3. [CrossRef]
180. Hamilton, C.E.; Bever, J.D.; Labbé, J.; Yang, X.; Yin, H. Mitigating Climate Change through Managing Constructed-Microbial Communities in Agriculture. *Agric. Ecosyst. Environ.* **2016**, *216*, 304–308. [CrossRef]
181. Srivastav, A.L. Chemical fertilizers and pesticides: Role in groundwater contamination. In: Agrochemicals detection, treatment, and remediation. *Butterworth-Heinemann* **2020**, 143–159. [CrossRef]
182. Ye, L.; Zhao, X.; Bao, E.; Li, J.; Zou, Z.; Cao, K. Bio-organic fertilizer with reduced rates of chemical fertilization improves soil fertility and enhances tomato yield and quality. *Sci. Rep.* **2020**, *10*, 1–11. [CrossRef]
183. Mandal, S.; Sarkar, B.; Bolan, N.; Novak, J.; Ok, Y.S.; Van Zwieten, L.; Singh, B.P.; Kirkham, M.B.; Choppala, G.; Spokas, K.; et al. Designing advanced biochar products for maximizing greenhouse gas mitigation potential. *Crit. Rev. Environ. Sci. Technol.* **2016**, *46*, 1367–1401. [CrossRef]
184. El-Naggar, A.; Awad, Y.M.; Tang, X.Y.; Liu, C.; Niazi, N.K.; Jien, S.H.; Tsang, D.C.W.; Song, H.; Yong, S.O.; Sang, S.L. Biochar influences soil carbon pools and facilitates interactions with soil: A field investigation. *Land Degrad. Dev.* **2018**. [CrossRef]
185. Yu, K.L.; Show, P.L.; Ong, H.C.; Ling, T.C.; Chen, W.H.; Salleh, M.A.M. Biochar production from microalgae cultivation through pyrolysis as a sustainable carbon sequestration and biorefinery approach. *Clean. Technol. Environ. Policy* **2018**, *20*, 2047–2055. [CrossRef]
186. Panwar, N.L.; Pawar, A.; Salvi, B.L. Comprehensive review on production and utilization of biochar. *SN Appl. Sci.* **2019**, *1*, 168. [CrossRef]
187. Schiermeier, Q. Putting the carbon back: The hundred billion tonne challenge. *Nature* **2006**, *442*, 620–624. [CrossRef]
188. Lehmann, J.; Gaunt, J.; Rondon, M. Bio-char sequestration in terrestrial ecosystems—A review. *Mitig Adapt. Strat Glob. Chang.* **2006**, *11*, 403–427. [CrossRef]
189. Oguntunde, P.G.; Abiodun, B.J.; Ajayi, A.E.; van de Giesen, N. Effects of charcoal production on soil physical properties in Ghana. *J. Plant. Nutr. Soil Sci.* **2008**, *171*, 591–596. [CrossRef]
190. Dang, V.M.; Joseph, S.; Van, H.T.; Mai, T.L.A.; Duong, T.M.H.; Weldon, S.; Taherymoosavi, S. Immobilization of heavy metals in contaminated soil after mining activity by using biochar and other industrial by-products: The significant role of minerals on the biochar surfaces. *Environ. Technol.* **2019**, *40*, 3200–3215. [CrossRef]
191. Bolan, N.S.; Choppala, G.; Kunhikrishnan, A.; Park, J.; Naidu, R. Microbial transformation of trace elements in soils in relation to bioavailability and remediation. In *Reviews of Environmental Contamination and Toxicology*; Springer: New York, NY, USA, 2013; pp. 1–56.
192. Aziz, T.; Ullah, S.; Sattar, A.; Nasim, M.; Farooq, M.; Khan, M.M. Nutrient availability and maize (Zea mays. L) growth in soil amended with organic manures. *Int. J. Agric. Biol.* **2010**, *12*, 621–624.
193. Alshankiti, A.; Gill, S. Integrated plant nutrient management for sandy soil using chemical fertilizers, compost, biochar and biofertilizers. *J. Arid. Land Stud.* **2016**, *26*, 101–106.
194. Bohara, H.; Dodla, S.; Wang, J.J.; Darapuneni, M.; Acharya, B.S.; Magdi, S.; Pavuluri, K. Influence of poultry litter and biochar on soil water dynamics and nutrient leaching from a very fine sandy loam soil. *Soil Tillage Res.* **2019**, *189*, 44–51. [CrossRef]
195. Brockhoff, S.R.; Christians, N.E.; Killorn, R.J.; Horton, R.; Davis, D.D. Physical and mineral-nutrition properties of sand-based turfgrass root zones amended with biochar. *Agron. J.* **2010**, *102*, 1627–1631. [CrossRef]
196. Lehman, J.; Silva, J.P.D.; Steiner, C.; Nehls, T.; Zech, W.; Glaser, B. Nutrients availability and leaching in an archaeological Anthrosol and Ferralsol of the Central Amazon basin: Fertilizer, manure, and charcoal amendments. *J. Plant. Soil* **2003**, *249*, 343–357. [CrossRef]
197. Tian, X.; Li, C.; Zhang, M.; Wan, Y.; Xie, Z.; Chen, B.; Li, W. Biochar derived from corn straw affected availability and distribution of soil nutrients and cotton yield. *PLoS ONE* **2018**, *13*, e0189924. [CrossRef]
198. Yu, H.; Zou, W.; Chen, J.; Chen, H.; Yu, Z.; Huang, J.; Gao, B. Biochar amendment improves crop production in problem soils: A review. *J. Environ. Manag.* **2019**, *232*, 8–21. [CrossRef]
199. Nartey, O.D.; Zhao, B. Biochar Preparation, Characterization, and Adsorptive Capacity and Its Effect on Bioavailability of Contaminants: An Overview. *Adv. Mater. Sci. Eng.* **2014**. [CrossRef]
200. Poole, N. *Smallholder Agriculture and Market. Participation*; Food and Agriculture Organization of the United Nations (FAO): Rome, Italy, 2017.
201. Buol, S.W.; Eswaran, H.O. *Advances in Agronomy*; Academic Press: Cambridge, MA, USA, 1999; Volume 68, pp. 151–195.
202. Safaei Khorram, M.; Fatemi, A.; Khan, M.A.; Kiefer, R.; Jafarnia, S. Potential risk of weed outbreak by increasing biochar's application rates in slow-growth legume, lentil (Lens culinaris Medik.). *J. Sci. Food Agric.* **2018**, *98*, 2080–2088. [CrossRef]

203. Kavitha, B.; Reddy, P.V.L.; Kim, B.; Lee, S.S.; Pandey, S.K.; Kim, K.H. Benefits and limitations of biochar amendment in agricultural soils: A review. *J. Environ. Manag.* **2018**, *227*, 146–154. [CrossRef]
204. Anyanwu, I.N.; Alo, M.N.; Onyekwere, A.M.; Crosse, J.D.; Nworie, O.; Chamba, E.B. Influence of biochar aged in acidic soil on ecosystem engineers and two tropical agricultural plants. *Ecotoxicol. Environ. Saf.* **2018**, *153*, 116–126. [CrossRef]
205. Hol, W.G.; Vestergård, M.; ten Hooven, F.; Duyts, H.; van de Voorde, T.F.; Bezemer, T.M. Transient negative biochar effects on plant growth are strongest after microbial species loss. *Soil Biol. Biochem.* **2017**, *115*, 442–451. [CrossRef]
206. Zhao, J.; Ren, T.; Zhang, Q.; Du, Z.; Wang, Y. Effects of biochar amendment on soil thermal properties in the North China Plain. *Soil Sci. Soc. Am. J.* **2016**, *80*, 1157–1166. [CrossRef]
207. Vaccari, F.P.; Maienza, A.; Miglietta, F.; Baronti, S.; Di Lonardo, S.; Giagnoni, L.; Valboa, G. Biochar stimulates plant growth but not fruit yield of processing tomato in a fertile soil. *Agric. Ecosyst. Environ.* **2015**, *207*, 163–170. [CrossRef]
208. Denyes, M.J.; Rutter, A.; Zeeb, B.A. Bioavailability assessments following biochar and activated carbon amendment in DDT-contaminated soil. *Chemosphere* **2016**, *144*, 1428–1434. [CrossRef]

sustainability

Article

Effect of Biochar on Soil and Water Loss on Sloping Farmland in the Black Soil Region of Northeast China during the Spring Thawing Period

Pengfei Yu [1,2,3], Tianxiao Li [1,2,3,*], Qiang Fu [1,2,3,*], Dong Liu [1,2,3], Renjie Hou [1,2,3] and Hang Zhao [1,2,3]

1   School of Water Conservancy and Civil Engineering, Northeast Agricultural University, Harbin 150030, China; yupengfei9995@126.com (P.Y.); liudong9599@yeah.net (D.L.); hourenjie888@126.com (R.H.); m15663583512@163.com (H.Z.)
2   Key Laboratory of Effective Utilization of Agricultural Water Resources of Ministry of Agriculture, Northeast Agricultural University, Harbin 150030, China
3   Heilongjiang Provincial Key Laboratory of Water Resources and Water Conservancy Engineering in Cold Region, Northeast Agricultural University, Harbin 150030, China
*   Correspondence: litianxiao@neau.edu.cn (T.L.); fuqiang0629@126.com (Q.F.)

Abstract: Biochar, as a kind of soil amendment, has attracted wide attention from scholars in various countries, and the effects of biochar on soil and water loss have been well reported. However, soil erosion is significantly affected by geographical conditions, climate, and other factors, and research on the characteristics of soil erosion and the effects of biochar application in seasonally frozen soil areas is currently unclear. The purpose of this study was to explore the effect of corn straw biochar application on soil and water conservation during the spring thawing period. Specifically, through field experiments, the addition of 0, 6, and 12 kg m^{-2} biochar on slopes of 1.8, 3.6, 5.4, and 7.2° and the effects on runoff and the soil erosion rate of farmland were analyzed. The results showed that in the 6 and 12 kg m^{-2} biochar addition treatments, the saturated water content of the soil increased by 24.17 and 42.91%, and the field capacity increased by 32.44 and 51.30%, respectively. Compared with the untreated slope, with an increase in biochar application rate, runoff decreased slightly, and soil erosion decreased significantly. This study reveals that biochar can be used as a potential measure to prevent soil and water loss on sloping farmland in cold regions.

Keywords: evaluation of soil and water conservation; simulated rainfall events; soil denudation; water and sediment process

Citation: Yu, P.; Li, T.; Fu, Q.; Liu, D.; Hou, R.; Zhao, H. Effect of Biochar on Soil and Water Loss on Sloping Farmland in the Black Soil Region of Northeast China during the Spring Thawing Period. *Sustainability* **2021**, *13*, 1460. https://doi.org/10.3390/su13031460

Academic Editor: Danilo Spasiano
Received: 3 December 2020
Accepted: 25 January 2021
Published: 30 January 2021

1. Introduction

Soil erosion has always been an environmental problem faced by humans. In modern times, the development of large-scale industry and agriculture has intensified the occurrence of soil erosion, causing sharp deterioration in the ecological environment, which severely restricts agricultural development and threatens the survival of humankind. There are many forms of soil erosion, and regional differences in climate characteristics, topography, and soil vegetation types result in different forms of soil erosion [1,2]. For example, the soil in the loess region of Northwest China tends to be loose, poorly agglomerated, and structurally unstable; thus, the soil can be susceptible to erosion due to rainfall and runoff [3]. However, in the black soil region of Northeast China, due to seasonal climate change, freeze–thaw cycling between winter and spring leads to soil accumulation in ditches caused by changes in soil structure, permeability, water conductivity, water content, strength, and aggregate water stability, which make the soil in this region vulnerable to erosion [4,5]. Soil erosion as a result of freezing and thawing is an important process and occurs mainly at high latitudes and high altitudes. When the environmental temperature changes, the water in the soil undergoes a phase change. This change results in the soil being mechanically damaged and then migrating and accumulating under the action of gravity and runoff [6–8]. The soil erosion process caused by freezing and thawing is very complex; therefore, studying soil erosion under the action of freezing and thawing has important practical significance for regional agricultural development and environmental governance [9,10].

The black soil region in Northeast China is an important base of grain commodity production [11]. The black soil region is an area with a concentrated distribution of black soil, chernozem soil, and meadow soil. Because the surface of these soils is rich in organic matter (OM), they have an obvious black color and similar properties (The Ministry of Agriculture of the People's Republic of China, 1996) [12]. Therefore, the distribution areas of black soil, chernozem, and meadow soil are collectively referred to as the northeastern black soil region [13]. Most of the three large black soil areas in China are located in the northeast. Among them, the Songnen black soil area is the largest, at approximately 0.208 million km^2, representing 65.38% of the agricultural land area in the northeast [14]. Due to the combined effects of agricultural development and seasonal changes, the soil and water losses in the region are significant. Yu et al. [15] showed that 37.9% of the cultivated land in this area was threatened by significant soil erosion. Therefore, mitigating soil erosion in this area is an urgent problem that must be solved.

In recent years, biochar has shown potential for retarding land degradation and enriching soil organic carbon (SOC), and it has received extensive attention from environmental departments in China and abroad [16,17]. Biochar is composed of a wide range of raw materials, including agricultural and forestry wastes, such as wood, straw, and fruit peels, as well as industrial and urban organic wastes [18]. The raw materials used for biochar production vary regionally [19–21]. For example, large areas of cotton are planted in the Yangtze River Basin of China [22]. In this area, cotton straw is used as the main raw material to produce biochar, while in Northeast China, where maize, rice, and soybean are the main crops, the straw remaining after harvesting provides sufficient raw material for the preparation of biochar. The total amount of agricultural waste produced in China in 2018 was 9×10^8 tons. The safe disposal and utilization of carbon-rich biomass residues are major challenges [23]. Traditional methods (e.g., incineration) not only fail to effectively recycle resources but also lead to severe atmospheric pollution (emission of greenhouse gases).

The unique and large specific surface area and multiporous structure of biochar reduce soil bulk density, increase porosity, actively improve soil hydraulic parameters, and increase the soil water-holding capacity, thus aiding in overcoming land degradation and other issues [24,25]. Atkinson et al. [26] found that the incorporation of biochar into soil can affect soil physical properties, such as structure, texture, porosity, particle size distribution (PSD), and bulk density [27,28]. Biochar application can reduce soil bulk density and increase porosity, thus affecting soil infiltration, erosion, and runoff by affecting the soil hydraulic characteristics [21,29–32]. Biochar is an organic material obtained by pyrolysis and carbonization in the complete or partial absence of oxygen. Biochar has the characteristics of low density, a high pH value, a high cation exchange capacity (CEC), and high stability [33–35]. After biochar was applied to soil by Wang et al. [36], the proportion of carbon storage in the soil increased. In the application of biochar, Shang et al. [37] and Liu et al. [38] found that biochar increased the SOC content, supplemented mineral nutrients, reduced the use of agricultural chemical fertilizer, and increased crop yield. In the black soil area of Northeast China, Wei et al. [39] applied 50 t hm^{-2} biochar to sloping farmland for two consecutive years; the generalized soil structure index (GSSI) was greatly improved, and water savings and yield increases were achieved. When biochar is mixed with soil, its porous structure reduces the soil bulk density [24], changes PSD [25], increases porosity [26], and actively improves soil hydraulic parameters and the soil water-holding capacity [29,30]. By affecting soil hydraulic characteristics, biochar application can affect soil runoff, infiltration, and erosion, thus helping to overcome land degradation [31,32].

Dong et al. [40] found that biochar is stable in dry–wet and freeze–thaw cycles and under tillage conditions in the field, and Kettunen and Saarnio [41] noted that biochar application can reduce nitrogen loss in winter soil and increase crop yield in the second year. Zhou et al. [42], Lee et al. [43], and Sadeghi et al. [44] also found that the application of biochar to soil significantly reduced soil loss and increased inorganic nitrogen and total phosphorus contents to levels higher than those in untreated soil. However, previous studies have shown that the ability of biochar to improve soil quality, slope erosion, and soil physical and chemical properties depends on the type of soil and the amount of biochar applied [45]. In addition, the relationship between the effective amount of biochar on different slopes, the soil type, and other environmental factors remains unclear. For example, Li et al. [46] found that as the biochar application rate on a 27% slope on the Loess Plateau increased, the impact on soil and water loss increased. However, Peake et al. [47] determined that the lowest biochar content best enhanced soil water saturation, and Reddy et al. [48] believed that the water conductivity and shear strength of soils increased with a decrease in the applied biochar content. Therefore, among the different soil types and different slopes, the level to which biochar decreases soil erosion differs. Moreover, soil erosion is greatly affected by climate conditions, especially under the effects of dramatic climate change, and the occurrence of soil erosion may be aggravated [2]. Compared with nonfrozen soil areas, in seasonally frozen soil areas with high latitude and high altitude, the free–thaw process may be an important process leading to soil erosion [49]. When the environmental temperature changes rapidly, the water in the soil undergoes a phase transition, which leads to mechanical damage [50], and the resistance to erosion of the soil decreases [4].

A large number of studies have reported that biochar can reduce soil erosion; however, in seasonally frozen regions such as Northeast China, after freezing and thawing, it is not clear whether biochar has a positive effect on soil and water conservation during spring thawing. Moreover, research on soil erosion following freezing and thawing has been conducted mostly in indoor simulated freezing and thawing environments, while indoor simulated and natural environments have different climatic conditions; therefore, the impact of freeze–thaw action on soil erosion requires further study. Based on the freeze–thaw interactions in the cold northeastern region, outdoor sloped land on the Songnen Plain of Northeast China was selected as the research area of this study, and the impact of biochar on sloping farmland soil erosion was explored. The thawing period of the frozen spring soil layer was selected; an artificial rainfall simulation including runoff, infiltration, and soil erosion was conducted; and the effects of biochar content on the soil erosion of different slopes under these climatic conditions were explored.

2. Materials and Methods

2.1. Test Area Overview

The experiments were conducted at the comprehensive test site of the School of Water Conservancy and Civil Engineering at Northeast Agricultural University. The geographical location is 45°44′22″ N, 126°43′6″ E, and the average elevation is 138 m. The test area location is shown in Figure 1, and the experimental area is located in the southeastern part of the Songnen Plain of Northeast China [51]. The area is mainly dominated by plains and has a mid-temperate continental monsoon climate. The average daily temperature is between −2.66 and 7.92 °C, the four seasons are distinct, and rain is concentrated in the summer. The average annual precipitation is approximately 583 mm, and summer precipitation accounts for 65% of the total annual precipitation. The rainfall duration is short and concentrated between June and September. Winters are cold and prolonged, with a regional soil freezing period of approximately 110 days and a snowfall of approximately 109 mm from November to April [52].

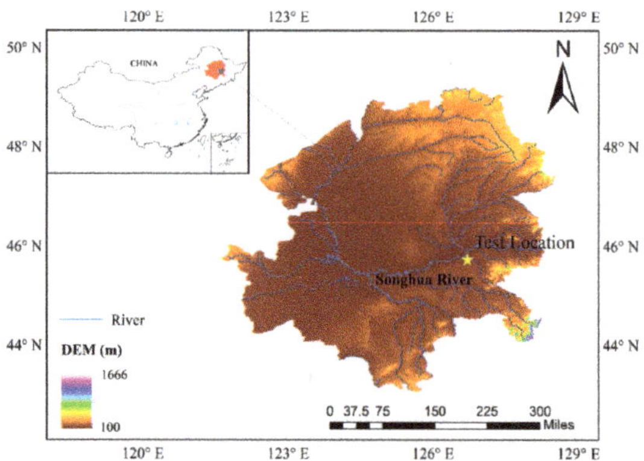

Figure 1. Test location.

Liu et al. [11] obtained slope information from 90×90 m topographic survey (Shuttle Radar Topography Mission (SRTM)) data from the US space shuttle. The areas of land with slopes of 0–1°, 1–2°, 2–3°, 3–4°, 4–5°, and 5–8° accounted for 72.25, 17.77, 6.37, 2.12, 0.79, and 0.58% of the total area, respectively, and slopes of 0–8 degrees accounted for 99.88% of the total area. According to the USDA Soil Taxonomy, soil in the area is classified as Argiborolls, Haploborolls, Cryoborolls, and Haprostolls of mollisols, and the soil in this area is also listed as phaiozem in the United Nations World Soil legend. In the Keys to Chinese Soil Taxonomy (3rd edition), the phaiozem in Northeast China can be divided into three major types: black soil, chernozem, and meadow soil. The cultivated land vegetation is mainly corn and soybean and is grown in the drylands [53–55].

2.2. Test Methods

Rainfall was simulated using a rainfall simulation system with downward sprinkling. The artificial simulated rainfall equipment was designed and manufactured by Harbin Tianyu Automation Instrument Co., Ltd (Harbin, China). The rainfall simulator includes five nozzles, and the raindrop size and rainfall intensity can be adjusted by adjusting the nozzle aperture and water pressure, which can be set at any rainfall intensity between 50 and 100 mm h^{-1}. The rainfall height can be adjusted in the range of 2.5–3.5 m, the rainfall uniformity is greater than 85%, the simulated raindrop diameter distribution is approximately 0.15–3 mm, and approximately 80%–90% of the raindrop diameters are less than 1.5 mm. The above artificial rainfall parameters can evenly cover an area of 2×7 m, which can meet the rainfall uniformity of each experimental plot; additionally, the size and distribution of simulated raindrops are similar to those of natural rainfall, and the simulated rainfall intensity was designed to be 80 mm h^{-1} given the intensity of erosion-causing rainstorms in the black soil area in Northeast China.

Twelve trapezoidal runoff test plots (5 m length, 2 m width) (No. P1–P12 in Table 1) with a slope adjusted from 0 to 15° by filling soil were used in this study. Each runoff test plot was trapezoidal box shape, the upper surface was of an inclined plane, and the lower surface was connected with the ground, as shown in Figure 2. The design gradients of this test were 1.8° (P1–P3), 3.6° (P4–P6), 5.4° (P7–P9), and 7.2° (P10–P12) because this slope range caused the most serious soil erosion in the northeastern black soil area during the last 30 years. The lower end of the test area was equipped with a collecting port, and runoff samples were collected through the connected bucket. The soil used was collected from the plow layer in Harbin, located in the black soil area of Northeast China. The soil in the experimental plot was loaded according to the soil type of the sloped farmland in the black soil area. Impurities such as gravel and straw were removed from the soil. To best retain its natural state, the soil was not sifted; thus, its aggregates maintained their original aggregation state and manually filled into the test plots. After adding the parent material layer soil, the soil was compacted artificially, then the surface layer soil was added, and the same autumn plowing treatment as the local farmland was used. The surface layer of 0–30 cm was classified as loam (particle size fractionation: 46.3% sand, 20.4% silt, and 33.3% clay), and the bulk density was approximately 1.15 g cm^{-3}. The 30–60 cm parent material was clay loam, and the particle size fractionation was 38.7% sand, 24.7% silt, and 36.6% clay. The average bulk density of the soil was 1.30 g cm^{-3}, which is the same as that of the bottom of the plow layer in the northeastern black soil area.

Table 1. Slope gradient and biochar application in the experimental plots.

Biochar Content	Slope (°)			
(kg m^{-2})	1.8	3.6	5.4	7.2
0	P1	P4	P7	P10
6	P2	P5	P8	P11
12	P3	P6	P9	P12

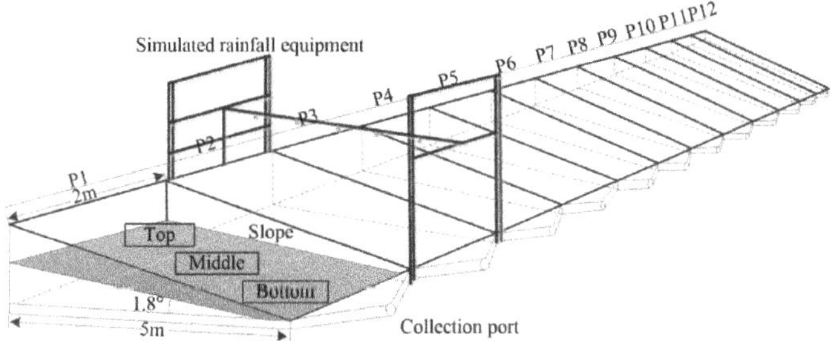

Figure 2. Sketch of the test areas.

The biochar used was corn straw biochar purchased from Liaoning Jinhefu Agricultural Development Co., LTD (Anshan, China). The preparation of this biochar involves high-temperature cracking at 450 to 500 °C under low-oxygen or limited-oxygen conditions. The basic physical and chemical properties are as follows: particle size of 1.5–2.0 mm, pH value of 9.14; nitrogen content of 1.53%, phosphorus content of 0.78%, potassium content of 1.68%, total organic carbon (TOC) content of 409.7 g kg^{-1}, and ash content of 31.8%.

From late September to early October 2017, 0, 6, and 12 kg m^{-2} biochar was applied on the surface of the test area at a depth of 0–30 cm. The biochar was evenly mixed with the cultivated soil by the traditional agricultural tillage method, and a biochar–soil mixed layer (approximately 0–30 cm) was formed. The control group (0 kg m^{-2}) was also tilled to the same extent. The volume ratio of biochar to 0–30 cm surface layer soil was approximately 3.33–6.67%. Table 1 shows the amount of biochar applied in each test area.

During the freeze–thaw period from October to April of the next year, plots P1–P12 in the test area were allowed to keep their natural snow cover without human interference. The test areas were adjacent to each other, with no differences in climate or environment. On 8 April 2018, 9 equidistant sampling points (3 × 3) were selected for each plot. Undisturbed soil samples (0–10 cm) were collected by the cutting ring method [56], the natural water content was determined by the drying method [57], the content of soil organic matter was determined by the potassium dichromate oxidation external heating method [58], and a simulated rainfall experiment was carried out. The meteorological conditions on that day were as follows: the average daily temperature was 0.3 °C, the average wind speed was 2.6 m s^{-1}, the average relative humidity was 56%, the daily evaporation rate was 2.8 mm d^{-1}, the accumulated precipitation was 0.2 mm, and there was no precipitation in the test area on that day. The meteorological data were collected from the China Meteorological Data Network (http://data.cma.cn/). On 8 April, the range of the thawing depth was 45–50 cm, and the frozen soil depth was measured by using a frozen soil device composed of a PVC pipe and rubber hose (LQX-DT, Jinzhou Licheng, Jinzhou, Liaoning, China) [51]. To ensure that the initial water content of each plot was consistent and to minimize the impact of rainfall on the thawing depth, a rainfall pretreatment with a rainfall intensity of 80 mm h^{-1} was carried out 24 h before the test. When runoff appeared on the slope, the rainfall was stopped immediately to ensure that the initial soil moisture content of all slopes reached saturation. In the rainfall-runoff experiment, the time started when runoff appeared on the slope and lasted for 40 min. Three repeated rainfall experiments were carried out in each plot. The runoff water and eroded soil material were collected through the collection port of the test chamber [59], and the time was recorded with a timer with an accuracy of 0.01 s [60]. The runoff water and eroded soil material were collected every 5 min after runoff generation, and the method of sample collection was consistent throughout the test. After the rainfall, all samples were weighed with an electronic balance with an accuracy of 0.01 g, and the runoff was measured. Then, the samples were allowed to stand for 12 h for sedimentation. After removing the supernatant, the sediment mixtures were put into an oven, dried at 105 °C for more than 24 h, and then weighed with an electronic balance with an accuracy of 0.01 g [61].

2.3. Calculation of the Average Infiltration Rate

Water infiltration into the soil is a dynamic process in which water, such as water from rainfall and runoff, migrates under the action of gravity and other potential forces and is stored as soil water. The extent of bare soil and the soil texture and slope under the same rainfall and intensity conditions are important factors affecting the amount of soil infiltration. The soil infiltration rate is an important index reflecting soil permeability characteristics because it directly reflects the soil water retention capacity. According to rainfall and runoff data, the average infiltration rate of a slope under the conditions of unchanged rainfall and rainfall intensity can be calculated according to Formula (1) [62]:

$$f_i = \frac{Ptcos\alpha - \frac{10R}{S}}{t} \tag{1}$$

where f_i represents the average infiltration rate of the slope (mm min^{-1}), P represents the rainfall intensity (mm min^{-1}), α represents the slope (°), t represents the rainfall time (min), R represents the runoff (mL) generated during rainfall time t, and S represents the actual rain-affected area (cm^2).

2.4. Grey Relational Projection Model

The main factors affecting soil and water conservation were determined. In this study, the soil water-holding capacity, runoff, and soil loss rate were selected as indicators for the grey relational projection method. The test results were made dimensionless by the extremum method [63]. Generally, the indexes of the grey relational projection model include "benefit type" and "cost type." The benefit index refers to the index with the larger attribute value, the better; the cost index refers to the index with the smaller attribute value, the better. The effect of soil and water conservation includes soil water-holding capacity, runoff rate, infiltration rate, and soil loss rate. These indexes have "benefit type" and "cost type" indexes, respectively. Therefore, 12 runoff plots are regarded as 12 soil and water conservation schemes, and the grey relational projection model is used to find the relative best decision-making scheme among the 12 schemes. In the grey relational projection model, it is necessary to determine the weighted vector of the index. The analytic hierarchy process (AHP) can decompose the relevant key index factors according to the actual problems. By constructing the judgment matrix, the weight of the key factors can be calculated through the matrix. After the consistency test, it provides an objective and scientific judgment method for comprehensive decision-making. The judgment matrix $A = (a_{ij})_{n \times n}$ is constructed by using the numbers 1–9 and their reciprocals as the scales of each criterion layer and index layer. The judgment matrix, ranking calculation, and consistency test together formulate a persuasive final result that has obvious advantages compared with other methods, making this approach more suitable for determining the weight of the evaluation index of a decision scheme. Test plots P1 to P12 were considered to correspond to 12 schemes, and the grey correlation projection value of each scheme was calculated.

The decision matrix is determined as follows. The set of runoff and soil conservation schemes in the 12 small areas is defined as A. The soil and water conservation index set of the experimental plot is V. The value matrix Y_{ij} of scheme A_i is attributed to evaluation index V_j:

$$A = \{P1, P2, \cdots, P12\} \tag{2}$$

$$V = \{moisture\ content,\ infiltration\ rate,\ soil\ loss\ rate\} \tag{3}$$

$$Y_{ij}(i = 1, 2 \cdots, 12; j = 1, 2, 3) \tag{4}$$

The optimal scheme of the three indexes is selected, comprising the maximum moisture content, the maximum infiltration rate, and the soil loss rate from the test data. The optimal residual augmented matrix is formed with the original indicators.

$$Y = \left[Y_{ij}\right]_{(12+1)\times 3}(i = 0, 1, 2, \cdots, 12; j = 1, 2, 3) \tag{5}$$

The evaluation index is dimensionless, and the initial decision matrix Y'_{ij} is obtained.

$$Y'_{ij} = Y_{ij}/Y'_{0j}, i = 0, 1, 2, \cdots, 12; j = 1, 2, \cdots, 3. \tag{6}$$

The grey correlation matrix r_{ij} is determined, and the coefficient γ is taken as 0.5.

$$r_{ij} = \frac{\min\limits_{12} \min\limits_{3}\left|Y'_{0j} - Y'_{ij}\right| + \gamma\max\limits_{12} \max\limits_{3}\left|Y'_{0j} - Y'_{ij}\right|}{\left|Y'_{0j} - Y'_{ij}\right| - \gamma\max\limits_{12} \max\limits_{3}\left|Y'_{0j} - Y'_{ij}\right|} \tag{7}$$

The grey relational judgment matrix f composed of r_{ij} is established.

$$F_{ij} = \begin{vmatrix} F_{01} & F_{02} & F_{03} \\ F_{11} & F_{12} & F_{13} \\ \vdots & \vdots & \vdots \\ F_{12\,1} & F_{12\,2} & F_{12\,3} \end{vmatrix} \tag{8}$$

If the evaluation index of weighted vector $W = [W_1, W_2, W_3]^T > 0$, the weight is normalized, and $\overline{W}_j$ is the grey correlation projection weight vector.

$$\overline{W}_j = W^2{}_j / \sqrt{\sum_{j=1}^{m} W_j^2}, j = 1, 2, 3. \tag{9}$$

On the basis of Formulas (8) and (9), the grey correlation projection value D_j is calculated.

$$D_j = F_{ij}\overline{W}_j,\ j = 1, 2, 3. \tag{10}$$

The larger the projection value D_j of a given test scheme is, the closer the test results are to the optimal result.

2.5. Analytical Methods

SPSS 22 and Origin 2017 were used for data processing, drawing, and tabulating. Data points were summarized by calculating the mean and standard deviation. The least-significant difference (LSD) method of single-factor square analysis (ANOVA) was used to test the difference in slope water content and sediment loss in different treatment modes with a significance level of $p = 0.05$.

3. Results

3.1. Effects of Biochar on the Natural Moisture Content of the Slope Surface

The basic slope soil physical indicators after biochar application are shown in Table 2. The saturated water content and field water-holding capacity of the soil were significantly higher than those of the soil without biochar application and increased as the amount of biochar applied increased. However, after the first application of biochar and freezing in winter, during the thawing period in spring, the soil OM content changed slightly.

Table 2. Basic soil physical properties after biochar application.

Biochar Application Rate (kg m^{-2})	Saturated Moisture Content (g kg^{-1})	Field Water-Holding Rate (g kg^{-1})	Dry Weight of Soil (g cm^{-3})	Organic Matter (OM) Content (g kg^{-1})
0	392.7 ± 9.73	281.7 ± 3.15	1.43 ± 0.015	38.6 ± 0.95
6	487.6 ± 8.58	373.1 ± 6.26	1.35 ± 0.018	39.1 ± 1.36
12	561.2 ± 7.21	426.2 ± 3.86	1.17 ± 0.013	38.8 ± 0.89

Note: The results are presented as the means ± standard deviations (n = 3), and the data in the table were measured before the simulated rainfall.

As shown in Figure 3, compared with those of the untreated plot (P1), the natural moisture contents of P2 and P3 were significantly higher at a slope of 1.8° ($p < 0.05$), and the difference between P2 and P3 was not significant. This result indicated that biochar played a role in increasing the natural water content on this slope, but the difference between 6 and 12 kg m^{-2} was not significant. The natural water content of P6 was significantly higher than that of P4 at a 3.6° slope ($p < 0.05$). The difference in the natural water content of P5 from those of P4 and P6 was not significant, indicating that only biochar application at 12 kg m^{-2} played a significant role in increasing the natural water content of soil at a 3.6° slope. There was no significant difference in natural water content between the biochar-treated slopes at 5.4° (P7–P9) and 7.2° (P10–P12) and the untreated slopes. The results showed that the natural water content of the slope at 1.8° significantly increased with biochar application. Only biochar application at 12 kg m^{-2} on the 3.6° slope significantly increased the natural water content. It may be that the gradients of 5.4 and 7.2° were relatively large. Under the action of gravity, the adsorption of water by the medium and large voids after applying biochar was insufficient, which led to the movement of water to the bottom of the slope and the failure to improve the water-holding capacity of the soil.

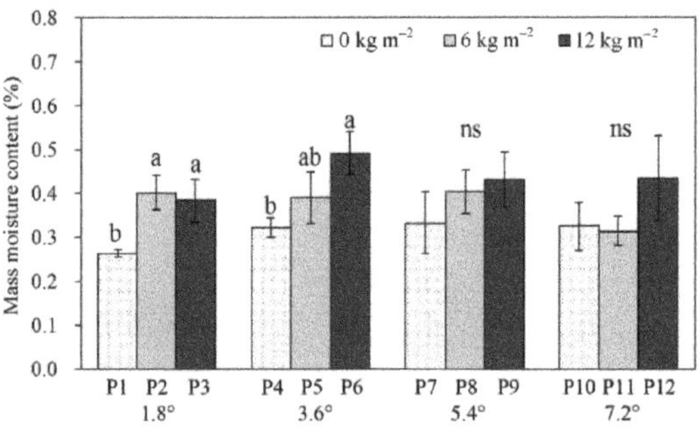

Figure 3. Mass soil moisture content of each of the twelve experimental plots before the simulated rainfall. Error bars represent the standard deviations of means (n = 3). In each gradient, the different letters (a, b) indicate significant differences between treatments at the $p < 0.05$ level according to the least-significant difference (LSD) test; ns denotes nonsignificant.

3.2. Effects of Biochar on Rainfall Runoff and Slope Infiltration during the Thawing Period

The runoff and infiltration rates of the slopes are shown in Figure 4. According to the change in runoff rate, 0–40 min was divided into a nonsteady stage from 0 to 20 min and a quasisteady stage from 20 to 40 min. The runoff and infiltration on the slope were analyzed. Overall, in the 1.8° slope test plot, the runoff rate began to increase rapidly in the nonsteady stage, and the difference between the three treatments was small and gradually stabilized in the quasisteady stage. In the quasisteady stage, the runoff rate was ranked P1 > P2 > P3 and decreased with increasing biochar content. The infiltration rate decreased rapidly in the nonsteady stage. In the quasisteady stage, the average infiltration rate of the plots treated with 6 and 12 kg m^{-2} biochar increased by 3.94% and 7.12%, respectively. In the 3.6° test plots, the runoff rates of P4, P5, and P6 were almost the same in the nonsteady stage. In the quasisteady stage, the runoff rate of P4 continued to increase, exceeded the runoff rates of P5 and P6, and tended to become stable; the average infiltration rates of the P5 and P6 plots increased by 6.68% and 11.44%, respectively. Although the runoff rates of P8 and P9 were less than that of P7 among the 5.4° test plots, biochar application had little effect on the runoff rate at this slope.

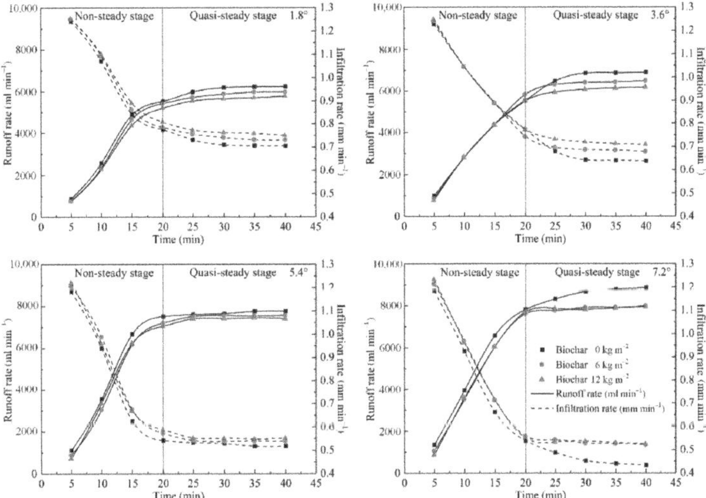

Figure 4. Runoff and infiltration rates of the four slopes from 0 to 40 min.

In the quasisteady stage, the average infiltration rates of P8 and P9 increased by 3.42% and 5.58%, respectively, compared with that of P7. In the 7.2° test plots, the runoff rates of P11 and P12 decreased significantly compared with that of P10, and the difference between P11 and P12 was very small. The runoff rates under biochar application at 6 and 12 kg m^{-2} decreased with the slope. However, the difference between the two biochar treatments was very small. Compared with that of P10, the average permeation rates of P11 and P12 in the quasisteady stage increased by 16.97% and 17.38%, respectively.

3.3. Effects of Biochar on the Soil Loss Rate of the Slopes

The soil loss rate (g m^{-2} min^{-1}) is the mass of sediment lost per unit time and unit area of soil on a slope under the interaction of rainfall and runoff [64]. The variation in the soil loss rate with rainfall duration on the four slopes is shown in Figure 5.

Within 0–20 min after the onset of runoff, the soil loss rates of plots P1–P12 rapidly increased; the soil loss rates tended to become stable from 20 to 40 min. We also found that in the quasistable stage, the effects of slope and biochar content on the soil loss rate were significant ($p < 0.001$), as shown in Table 3.

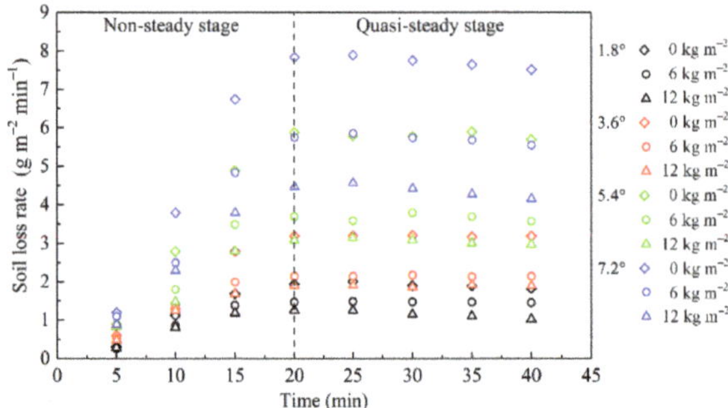

Figure 5. Soil loss rate of the four slopes from 0 to 40 min under the three treatment conditions.

Table 3. Average soil loss rate (g m^{-2} min^{-1}) of the four slopes under the three treatment conditions in the quasistable stage from 20 to 40 min. $n = 4$. Values of the soil loss rate followed by different letters (i.e., a through c and A through D) are significantly different at $p < 0.001$ according to ANOVA. Different capital letters in the same row indicate significant differences on different slopes with the same biochar content ($p < 0.001$); different lowercase letters in the same column indicate significant differences in biochar content on the same slope ($p < 0.001$).

Biochar Content	Slope Gradient (°)			
(kg m^{-2})	1.8°	3.6°	5.4°	7.2°
0	1.905 ± 0.068 Da	3.193 ± 0.015 Ca	5.795 ± 0.071 Ba	7.708 ± 0.140 Aa
6	1.465 ± 0.011 Db	2.155 ± 0.015 Cb	3.670 ± 0.088 Bb	5.708 ± 0.112 Ab
12	1.132 ± 0.086 Dc	1.895 ± 0.023 Cc	3.060 ± 0.068 Bc	4.363 ± 0.158 Ac

In the quasistable stage, the soil loss rates under biochar application at different concentrations were significantly different at the same slope ($p < 0.001$), the soil loss rates were significantly lower than those under no treatment at the same slope, and the soil loss rates under 12 kg m^{-2} biochar application were less than those under no treatment or 6 kg m^{-2} biochar application. Under the same treatment conditions, the soil loss rate significantly differed among slopes ($p < 0.001$). Under the same biochar application rate, the soil loss rate increased with increasing slope.

In the total runoff period from 0 to 40 min, compared with those of P1, the soil loss rates of P2 and P3 decreased by 21.37% and 32.75%, respectively; at 3.6°, the soil loss rates of P5 and P6 were 24.88% and 33.35% lower than that of P4, respectively; and at 7.2°, the soil loss rates of P8 and P9 were 31.54% and 41.61% lower than that of P7, and those of P11 and P12 were 25.16% and 40.71% lower than that of P10. According to the above results, the effect of 6 and 12 kg m^{-2} biochar application on reducing the soil loss rate first increased and then decreased with increasing slope gradient. Under these conditions, the rate of soil loss decreased the most at 5.4°, by 31.54% and 41.61% for 6 and 12 kg m^{-2} biochar applications, respectively. Although there were differences among previous studies in terms of the amount of biochar applied, soil type, and climatic conditions, similar results in the literature show that biochar has a significant effect on reducing soil loss on slopes.

3.4. Evaluation of the Soil and Water Conservation Ability of Test Plots P1–P12 Based on the Grey Relational Projection Model

The weighted values of the three characteristic indexes correspond to eigenvector values of 0.086, 0.258, and 0.657. The projection values (D_j) of test plots P1–P12 obtained through the grey relational projection model are shown in Figure 6. The D_j of P3 was larger than that of P2. The D_j of P2 was larger than that of P1. The D_j under 12 kg m^{-2} biochar application was higher than that under 6 kg m^{-2} biochar application for the same slope. This result was also found in the test plots with slopes of 3.6, 5.4, and 7.2°. The

D_j of the biochar-treated test plots was higher than that of the plot with no biochar at the same slope, and the D_j of the test plot under 12 kg m^{-2} biochar application was higher than that of the test plot under 6 kg m^{-2} biochar application. In other words, biochar can enhance soil and water conservation on the slope of 1.8–7.2 degrees, and the effect of B12 treatment was the most significant. In addition, the D_j values under 12 kg m^{-2} biochar application at 3.6, 5.4, and 7.2° were higher than those under no treatment at 1.8, 3.6, and 5.4°. The results show that biochar application can effectively control soil and water losses on sloping farmland with gradients between 1.8 and 7.2° after freezing and thawing and that the effect of the biochar application rate on soil and water conservation is consistent in this gradient range. In the test plots with slopes between 1.8 and 7.2°, the degree of soil and water loss after biochar application was less than that in the untreated test plot, but the loss in the untreated plot was lower at 1.8°.

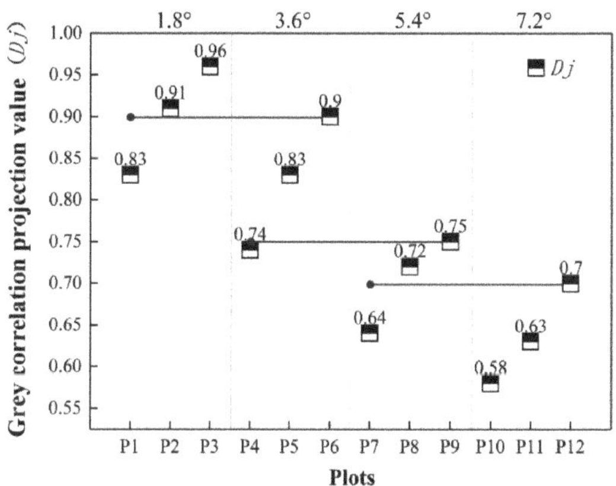

Figure 6. Grey relational projection model results for P1–P12. The function of the horizontal line in the figure is only for a more intuitive horizontal comparison.

4. Discussion

Usually, the water storage capacity of soil after biochar application increases with the number of medium and large aggregates and the soil pore size (30–75 μm) [65], and the pores in biochar itself can also store water. Fu et al. [65] considered that the effect of the combined application of biochar and freeze–thaw cycling on soil water content was mainly achieved by reducing the proportion of micropores (<0.3 μm) in the soil layer and increasing the proportion of medium pores (>30–75 μm). Therefore, under the same natural conditions, the natural water content of the soil treated with biochar was higher than that of the untreated soil, and the soil water content at 1.8 and 3.6° was significantly affected by biochar application. However, the two biochar treatments at 5.4 and 7.2° had less of an effect on the soil water content. The soil dry weight decreased as the amount of applied biochar increased, which is consistent with the effect of biochar on basic soil physical properties reported by Li et al. [66]. The soil moisture content in the spring during the thawing period is higher than that of dry unfrozen soil, which is affected by the melting snow water [67]. Brodowski et al. [68], and Sachdeva et al. [69] believed that the main mechanism of the bioenhancement of soil OM was that biochar particles are the core component and are gradually covered by clay, silt, and OM particles. The reason for the small change in the soil OM content may be because biochar application occurs in autumn, and the subsequent lower ambient temperature and the freezing effect of winter cause the biochar to freeze for most of the winter, but measurements were taken in the thawing period of the freezing and thawing process. As a result, the OM and microorganism contents in the biochar and soil particles were not fully aggregated in the following spring, and because the microorganisms in the soil had a low activity for a long time, the rate of decomposition of animal and plant residues was slow [70], resulting in a very small change in OM content.

The slope runoff rate is an important factor affecting soil erosion [46]. He et al. [71] found that when biochar was mixed with soil, the soil intergranular pore space could increase through the presence of macrovoids in the biochar, thus improving the efficiency of soil water movement, increasing the soil permeability rate, and reducing the runoff rate. Liu et al. [72] reported that biochar is an important factor in changing soil properties. Similar to the results of Jien and Wang [45], Lee et al. [73] and Sadeghi et al. [32], appropriate biochar application was used to protect the soil surface from scabbing and increase the seepage rate, thus delaying the runoff generation time and reducing runoff. In this study, the difference between the two biochar treatments was less than 1%. The results show that biochar can increase the infiltration rate and reduce rainfall runoff on slopes, which is similar to the results of Sadeghi et al. [32]. Smetanova et al. [74] showed that 10% biochar application could increase the soil infiltration rate, reduce runoff events by up to 40%, and decrease the runoff coefficient by 16%. Bissonnais et al. [75] also found that biochar could reduce the impact of runoff on slopes and enhance the resistance of soil to raindrop impacts and runoff scouring. In this study, the runoff rate of the biochar-treated plots was slightly lower than that of the untreated plots at the same slope, but the differences were less than those in previous studies due to the limited conditions of the test site, climatic conditions, and other factors. To ensure a consistent freezing depth, the test utilized limited climatic and geographical conditions. The effect of biochar on the infiltration rate in this study may be related to the thawing depth. Therefore, it can be considered that the influence of biochar on slope runoff is constrained by the amount of biochar applied, the soil type, and the slope. Similar to the results of this experiment, a previous study found that the effect of biochar application on slope runoff was not significant in the red soil sloping farmland of southern China [76]. In contrast, the results of Jien and Wang [45] and Abrol [77] showed that with an increase in the amount of biochar applied, the control effect on slope runoff decreased. We speculate that perhaps there were differences in slope and soil texture.

Raindrops detach soil particles, destroy soil structure, and finally increase runoff and soil erosion [61]. Lee et al. [78] speculated that biochar could stabilize soil aggregates and form loose biochar–soil mixtures under Coulomb and van der Waals forces; these mixtures could absorb or buffer the energy of larger raindrops to prevent the separation of soil particles and soil scabs, thus reducing soil loss. Abrol et al. [77] significantly reduced soil loss by biochar application. These authors believed that biochar increased the soil surface roughness and that a large number of biochar particles accumulated on the soil surface. Greater soil surface roughness will likely interfere with the lateral movement of separated soil particles in runoff, thus reducing soil transport. Jien and Wang [45] and Doan et al. [79] found that the application of biochar could reduce the soil loss rate by 15–78%. In our study, soil loss decreased by 21.37–41.61%, similar to the soil loss reduction reported by Li et al. [46], and biochar application had an obvious effect on reducing soil erosion. In addition, we found some black biochar particles during the collection of runoff. We speculate that biochar may migrate upward because higher-density sand particles displace lower-density biochar particles and that the influence of raindrops helps this migration. We consider another dynamic mechanism of biochar migration to the upper soil layer to be buoyancy because of the tar and oil on the biochar surface, as biochar itself is hydrophobic and its dry density is lower than that of water [80]. It may also be that the very fine fraction of biochar particles move upward toward the biochar near the surface and that water can penetrate and fill most of the voids in the soil particles. New pore spaces in shallow soils are available for deep biochar entry. Some studies have also found biochar particles in runoff [32], which means that biochar application may need to be combined with treatment measures such as straw mulching, fertilizer application, or polyacrylamide treatment to improve the viscosity of biochar and soil particles and avoid floating after rainfall.

The soil and water conservation effects of the 6 and 12 kg m^{-2} biochar treatments significantly improved in the spring thawing period after the winter freezing period, and the effect of the 12 kg m^{-2} biochar treatment was higher than that of the 6 kg m^{-2} biochar treatment. This finding is similar to the results of many scientific studies on the influence of biochar on slope runoff [17,66]. However, the conditions in Abrol et al. [77] and Wei et al. [80] did not include freeze–thaw effects after biochar treatment, as a result of which biochar had a lower effect on the slope runoff rate but a similar effect on significantly reducing the soil loss rate. The soil and water conservation effects of the 6 kg m^{-2} and 12 kg m^{-2} biochar treatments were evaluated by the grey relational projection method. The results showed that biochar treatment could improve the degree of soil and water conservation, and the effect of the 12 kg m^{-2} biochar treatment was higher than that of the 6 kg m^{-2} biochar treatment in the range of 1.8° to 7.2°. Therefore, biochar treatment can be used as a soil conditioner to improve soil and water conservation in seasonally frozen soil areas. To further improve the economic efficiency of biochar application, exploring the most appropriate biochar dosage in different slope ranges should be the focus of research related to the large-scale application of biochar.

5. Conclusions

Three different proportions of corn straw biochar were used on four different slopes in a seasonally frozen soil area. The results showed that biochar had a positive effect on the soil water-holding capacity of the 0–20 cm soil layer on plots with small slopes, which was beneficial for improving the soil and water conservation effects. However, when the slope increased, the effect of biochar treatment on the soil water-holding capacity was limited. In addition, the soil loss rates of the 6 kg m^{-2} and 12 kg m^{-2} biochar treatments decreased significantly in the spring thawing period and increased with increasing biochar content. The grey correlation results for the effects of biochar on slope soil and water conservation showed that the effects of the two biochar treatments on soil and water conservation were better than those of no treatment. The results of the field experiment with seasonal freezing and thawing show that biochar treatment has a positive effect on reducing the soil loss rate in seasonally frozen soil areas. For the extensive application of biochar, we also need to understand the matching of different biochar types and amounts with different soil types in long-term experiments, and the impact on soil organic matter is also worth further exploration.

Author Contributions: Conceptualization, Q.F. and T.L.; methodology, D.L.; software, Q.F.; validation, Q.F., T.L. and D.L.; formal analysis, R.H.; investigation, H.Z.; resources, Q.F.; data curation, T.L.; writ-ing—original draft preparation, P.Y.; writing—review and editing, P.Y.; visualization, H.Z.; su-pervision, T.L.; project administration, R.H.; funding acquisition, Q.F. All authors have read and agreed to the published version of the manuscript.

Funding: This research was funded by China National Fund for Distinguished Young Scientists, grant number 51825901; joint fund of the National Natural Science Foundation of China, grant number U20A20318; the Heilongjiang Provincial Science Fund for Distinguished Young Scholars, grant number YQ2020E002; the "Young Talents" Project of Northeast Agricultural University, grant number 18QC28; and a China Postdoctoral Science Foundation Grant [2019M651247].

Institutional Review Board Statement: Not applicable.

Informed Consent Statement: Not applicable.

Data Availability Statement: Data is contained within the article.

Acknowledgments: Thanks to the hard-working editors and valuable comments from reviewers.

Conflicts of Interest: The authors declare no conflict of interest.

References

1. Hu, G.; Wu, Y.; Liu, B.; Zhang, Y.; You, Z.; Yu, Z. The characteristics of gully erosion over rolling hilly black soil areas of Northeast China. *J. Geogr. Sci.* **2009**, *19*, 309–320. [CrossRef]
2. Mullan, D. Soil erosion under the impacts of future climate change: Assessing the statistical significance of future changes and the potential on-site and off-site problems. *Catena* **2013**, *109*, 234–246. [CrossRef]
3. Chen, H.; Shao, M.; Li, Y. Soil desiccation in the Loess Plateau of China. *Geoderma* **2008**, *143*, 91–100. [CrossRef]
4. Zhao, Y.; Wang, E.; Cruse, R.M.; Chen, X. Characterization of seasonal freeze-thaw and potential impacts on soil erosion in northeast China. *Can. J. Soil Sci.* **2012**, *92*, 567–571. [CrossRef]
5. Cerdà, A.; Keesstra, S.D.; Rodrigo-Comino, J.; Novara, A.; Pereira, P.; Brevik, E.C.; Giménez-Morera, A.; Fernández-Raga, M.; Pulido, M.; Di Prima, S.; et al. Runoff initiation, soil detachment and connectivity are enhanced as a consequence of vineyards plantations. *J. Environ. Manag.* **2017**, *202*, 268–275. [CrossRef] [PubMed]
6. Ollesch, G.; Kistner, I.; Meissner, R.; Lindenschmidt, K.-E. Modelling of snowmelt erosion and sediment yield in a small low-mountain catchment in Germany. *Catena* **2006**, *68*, 161–176. [CrossRef]
7. Su, J.; Van Bochove, E.; Thériault, G.; Novotná, B.; Khaldoune, J.; Denault, J.; Zhou, J.; Nolin, M.; Hu, C.; Bernier, M.; et al. Effects of snowmelt on phosphorus and sediment losses from agricultural watersheds in Eastern Canada. *Agric. Water Manag.* **2011**, *98*, 867–876. [CrossRef]
8. Nunes, J.; Seixas, J.; Keizer, J. Modeling the response of within-storm runoff and erosion dynamics to climate change in two Mediterranean watersheds: A multi-model, multi-scale approach to scenario design and analysis. *Catena* **2013**, *102*, 27–39. [CrossRef]
9. Behzadfar, M.; Sadeghi, S.H.; Khanjani, M.J.; Hazbavi, Z. Effects of rates and time of zeolite application on controlling runoff generation and soil loss from a soil subjected to a freeze-thaw cycle. *Int. Soil Water Conserv. Res.* **2017**, *5*, 95–101. [CrossRef]
10. Sadeghi, S.H.; Raeisi, M.B.; Hazbavi, Z. Influence of freeze-only and freezing-thawing cycles on splash erosion. *Int. Soil Water Conserv. Res.* **2018**, *6*, 275–279. [CrossRef]
11. Liu, J.; Yu, Z.; Wang, G.; Yao, Q.; Sui, Y.; Shi, Y.; Chu, H.; Tang, C.; Franks, A.E.; Jin, J.; et al. Ammonia-oxidizing archaea show more distinct biogeographic distribution patterns than ammonia-oxidizing bacteria across the black soil zone of Northeast China. *Front. Microbiol.* **2018**, *9*, 171. [CrossRef] [PubMed]
12. The Ministry of Agriculture of the People's Republic of China. *Classification of Type Regions and Fertility of Cultivated Land in China*; NY/T 309; The Ministry of Agriculture of the People's Republic of China: Beijing, China, 1996; pp. 1–3.
13. Yao, Q.; Liu, J.; Yu, Z.; Li, Y.; Jin, J.; Liu, X.; Wang, G. Three years of biochar amendment alters soil physiochemical properties and fungal community composition in a black soil of northeast China. *Soil Biol. Biochem.* **2017**, *110*, 56–67. [CrossRef]
14. Liu, B.Y.; Yan, B.X.; Shen, B.; Wang, Z.Q.; Wei, X. Current status and comprehensive control strategies of soil erosion for cultivated land in the Northeastern black soil area of China. *Sci. Soil Water Conserv.* **2008**, *6*, 1–8. [CrossRef]
15. Yu, D.; Shen, B.; Xie, J. Damages caused by soil and water erosion and their controlling approaches in the black earth area in Northeast China. *J. Soil Water Conserv.* **1992**, *12*, 25–34. [CrossRef]
16. Busch, D.; Kammann, C.; Grünhage, L.; Müller, C. Simple biotoxicity tests forevaluation of carbonaceous soil additives: Establishment and reproducibility off four test procedures. *J. Environ. Qual.* **2012**, *41*, 1023–1032. [CrossRef] [PubMed]
17. Wang, C.; Walter, M.; Parlange, J.-Y. Modeling simple experiments of biochar erosion from soil. *J. Hydrol.* **2013**, *499*, 140–145. [CrossRef]
18. Singh, B.; Singh, B.P.; Cowie, A.L. Characterisation and evaluation of biochars for their application as a soil amendment. *Aust. J. Soil Res.* **2010**, *48*, 516–525. [CrossRef]
19. Singh, R.; Srivastava, P.; Singh, P.; Sharma, A.K.; Singh, H.; Raghubanshi, A.S. Impact of rice-husk ash on the soil biophysical and agronomic parameters of wheat crop under a dry tropical ecosystem. *Ecol. Indic.* **2019**, *105*, 505–515. [CrossRef]
20. Singh, P.; Singh, R.; Borthakur, A.; Madhav, S.; Singh, V.K.; Tiwary, D.; Srivastava, V.C.; Mishra, P. Exploring temple floral refuse for biochar production as a closed loop perspective for environmental management. *Waste Manag.* **2018**, *77*, 78–86. [CrossRef]
21. Singh, R.; Singh, P.; Singh, H.; Raghubanshi, A.S. Impact of sole and combined application of biochar, organic and chemical fertilizers on wheat crop yield and water productivity in a dry tropical agro-ecosystem. *Biochar* **2019**, *1*, 229–235. [CrossRef]
22. Xu, N.Y.; Li, J. Regional distribution characteristics of cotton fiber quality in main cotton production areas in China. *Chin. J. Eco. Agric.* **2016**, *24*, 1547–1554. [CrossRef]
23. Liu, C.; Chen, L.; Ding, D.; Cai, T. From rice straw to magnetically recoverable nitrogen doped biochar: Efficient activation of peroxymonosulfate for the degradation of metolachlor. *Appl. Catal. B Environ.* **2019**, *254*, 312–320. [CrossRef]
24. Butnan, S.; Deenik, J.L.; Toomsan, B.; Antal, M.J.; Vityakon, P. Biochar characteristics and application rates affecting corn growth and properties of soils contrasting in texture and mineralogy. *Geoderma* **2015**, *237*, 105–116. [CrossRef]
25. El-Naggar, A.; Lee, S.S.; Awad, Y.M.; Yang, X.; Ryu, C.; Rizwan, M.; Rinklebe, J.; Tsang, D.C.; Ok, Y.S. Influence of soil properties and feedstocks on biochar potential for carbon mineralization and improvement of infertile soils. *Geoderma* **2018**, *332*, 100–108. [CrossRef]
26. Atkinson, C.J.; Fitzgerald, J.D.; Hipps, N.A. Potential mechanisms for achieving agricultural benefits from biochar application to temperate soils: A review. *Plant Soil* **2010**, *337*, 1–18. [CrossRef]
27. Farrell, M.; Macdonald, L.M.; Baldock, J.A. Biochar differentially affects the cycling and partitioning of low molecular weight carbon in contrasting soils. *Soil Biol. Biochem.* **2015**, *80*, 79–88. [CrossRef]

28. De La Rosa, J.M.; Paneque, M.; Hilber, I.; Blum, F.; Knicker, H.; Bucheli, T.D. Assessment of polycyclic aromatic hydrocarbons in biochar and biochar-amended agricultural soil from Southern Spain. *J. Soils Sediments* **2016**, *16*, 557–565. [CrossRef]
29. Ouyang, L.; Wang, F.; Tang, J.; Yu, L.; Zhang, R. Effects of biochar amendment on soil aggregates and hydraulic properties. *J. Soil Sci. Plant Nutr.* **2013**, *13*, 991–1002. [CrossRef]
30. Huang, Y.; Liu, D.; An, S. Effects of slope aspect on soil nitrogen and microbial properties in the Chinese Loess region. *Catena* **2015**, *125*, 135–145. [CrossRef]
31. Hazbavi, Z.; Sadeghi, S.H.; Kiani-Harchegani, M. Application of Biochar on temporal variability of runoff volume and coefficient. In Proceedings of the Third WASWAC Conference, Belgrade, Serbia, 22–26 September 2016; Conference Abstracts Books: Belgrade, Serbia, 2016; pp. 186–187.
32. Sadeghi, S.H.; Hazbavi, Z.; Harchegani, M.K. Controllability of runoff and soil lossfrom small plots treated by vinasse-produced biochar. *Sci. Total Environ.* **2016**, *541*, 483–490. [CrossRef]
33. Gul, S.; Whalen, J.K.; Thomas, B.W.; Sachdeva, V.; Deng, H. Physico-chemical properties and microbial responses in biochar-amended soils: Mechanisms and future directions. *Agric. Ecosyst. Environ.* **2015**, *206*, 46–59. [CrossRef]
34. Sika, M.P.; Hardie, A. Effect of pine wood biochar on ammonium nitrate leaching and availability in a South African sandy soil. *Eur. J. Soil Sci.* **2014**, *65*, 113–119. [CrossRef]
35. Yuan, P.; Wang, J.; Pan, Y.; Shen, B.; Wu, C. Review of biochar for the management of contaminated soil: Preparation, application and prospect. *Sci. Total Environ.* **2019**, *659*, 473–490. [CrossRef] [PubMed]
36. Wang, D.; Fonte, S.J.; Parikh, S.J.; Six, J.; Scow, K.M. Biochar additions can enhance soil structure and the physical stabilization of C in aggregates. *Geoderma* **2017**, *303*, 110–117. [CrossRef] [PubMed]
37. Shang, J.; Geng, Z.C.; Chen, X.X.; Zhao, J.; Geng, R.; Wang, S. Effects of biochar on soil organic carbon and nitrogen and their fractions in a rainfed farmland. *J. Environ. Sci. (China)* **2015**, *34*, 509–517.
38. Liu, X.; Mao, P.; Li, L.; Ma, J. Impact of biochar application on yield-scaled greenhouse gas intensity: A meta-analysis. *Sci. Total Environ.* **2019**, *656*, 969–976. [CrossRef] [PubMed]
39. Wei, Y.X.; Zhang, Y.P.; Zhang, Y.F.; Wang, R.Y.; Ma, Y.Y.; Zhang, Y. Influences of two consecutive years supply of biochar on soil improvement, water saving and yield increasing in sloping farmland of black soil region. *Trans. CSAM* **2018**, *284*, 291–312.
40. Dong, X.; Li, G.; Lin, Q.; Zhao, X. Quantity and quality changes of biochar aged for 5 years in soil under field conditions. *Catena* **2017**, *159*, 136–143. [CrossRef]
41. Kettunen, R.; Saarnio, S. Biochar can restrict N_2O emissions and the risk of nitrogen leaching from an agricultural soil during the freeze-thaw period. *Agric. Food Sci.* **2015**, *22*, 373–379. [CrossRef]
42. Zhou, Y.; Berruti, F.; Greenhalf, C.; Henry, H.A. Combined effects of biochar amendment, leguminous cover crop addition and snow removal on nitrogen leaching losses and nitrogen retention over winter and subsequent yield of a test crop (*Eruca sativa* L.). *Soil Biol. Biochem.* **2017**, *114*, 220–228. [CrossRef]
43. Lee, C.-H.; Wang, C.-C.; Lin, H.-H.; Lee, S.S.; Tsang, D.C.; Jien, S.-H.; Ok, Y.S. In-situ biochar application conserves nutrients while simultaneously mitigating runoff and erosion of an Fe-oxide-enriched tropical soil. *Sci. Total Environ.* **2018**, *619*, 665–671. [CrossRef] [PubMed]
44. Sadeghi, S.H.; Panah, M.H.G.; Younesi, H.; Kheirfam, H. Ameliorating some quality properties of an erosion-prone soil using biochar produced from dairy wastewater sludge. *Catena* **2018**, *171*, 193–198. [CrossRef]
45. Jien, S.-H.; Wang, C.-S. Effects of biochar on soil properties and erosion potential in a highly weathered soil. *Catena* **2013**, *110*, 225–233. [CrossRef]
46. Li, Y.; Zhang, F.; Yang, M.; Zhang, J.; Xie, Y. Impacts of biochar application rates and particle sizes on runoff and soil loss in small cultivated loess plots under simulated rainfall. *Sci. Total Environ.* **2019**, *649*, 1403–1413. [CrossRef] [PubMed]
47. Peake, L.R.; Reid, B.J.; Tang, X. Quantifying the influence of biochar on the physical and hydrological properties of dissimilar soils. *Geoderma* **2014**, *235*, 182–190. [CrossRef]
48. Reddy, K.R.; Yaghoubi, P.; Aksoy, Y.Y. Effects of biochar amendment on geotechnical properties of landfill cover soil. *Waste Manag. Res.* **2015**, *33*, 524–532. [CrossRef]
49. Gao, H.; Shao, M. Effects of temperature changes on soil hydraulic properties. *Soil Tillage Res.* **2015**, *153*, 145–154. [CrossRef]
50. Ma, Q.; Zhang, K.; Jabro, J.D.; Ren, L.; Liu, H. Freeze–thaw cycles effects on soil physical properties under different degraded conditions in Northeast China. *Environ. Earth Sci.* **2019**, *78*, 321. [CrossRef]
51. Hou, R.; Li, T.; Fu, Q.; Liu, D.; Cui, S.; Zhou, Z.; Yan, P.; Yan, J. Effect of snow-straw collocation on the complexity of soil water and heat variation in the Songnen Plain, China. *Catena* **2019**, *172*, 190–202. [CrossRef]
52. Wu, Y.; Ouyang, W.; Hao, Z.; Lin, C.; Liu, H.; Wang, Y. Assessment of soil erosion characteristics in response to temperature and precipitation in a freeze-thaw watershed. *Geoderma* **2018**, *328*, 56–65. [CrossRef]
53. Fang, H.-J.; Yang, X.; Zhang, X.-P.; Liang, A. Using 137 CS tracer technique to evaluate erosion and deposition of black soil in northeast china. *Pedosphere* **2006**, *16*, 201–209. [CrossRef]
54. Yu, Z.-H.; Yu, J.; Ikenaga, M.; Sakai, M.; Liu, X.-B.; Wang, G.-H. Characterization of root-associated bacterial community structures in soybean and corn using locked nucleic acid (LNA) oligonucleotide-PCR clamping and 454 pyrosequencing. *J. Integr. Agric.* **2016**, *15*, 1883–1891. [CrossRef]
55. Fu, Q.; Hou, R.; Li, T.; Wang, M.; Yan, J. The functions of soil water and heat transfer to the environment and associated response mechanisms under different snow cover conditions. *Geoderma* **2018**, *325*, 9–17. [CrossRef]

56. Tian, D.; Qu, Z.Y.; Gou, M.M.; Li, B.; Lv, Y.J. Experimental study of influence of biochar on different texture soil hydraulic characteristic parameters and moisture holding properties. *Pol. J. Environ.* **2015**, *24*, 1435–1442.
57. O'Kelly, B.C. Accurate determination of moisture content of organic soils using the oven drying method. *Dry. Technol.* **2004**, *22*, 1767–1776. [CrossRef]
58. Shen, H.; Zheng, F.; Wen, L.; Han, Y.; Hu, W. Impacts of rainfall intensity and slope gradient on rill erosion processes at loessial hillslope. *Soil Tillage Res.* **2016**, *155*, 429–436. [CrossRef]
59. Liu, G.; Hu, F.; Zheng, F.; Zhang, Q. Effects and mechanisms of erosion control techniques on stairstep cut-slopes. *Sci. Total Environ.* **2019**, *656*, 307–315. [CrossRef]
60. Lv, J.; Luo, H.; Xie, Y.-S. Effects of rock fragment content, size and cover on soil erosion dynamics of spoil heaps through multiple rainfall events. *Catena* **2019**, *172*, 179–189. [CrossRef]
61. Gholami, L.; Sadeghi, S.H.; Homaee, M. Straw mulching effect on splash erosion, runoff, and sediment yield from eroded plots. *Soil Sci. Soc. Am. J.* **2013**, *77*, 268–278. [CrossRef]
62. Li, Z.B.; Li, S.X.; Ren, Z.P.; Li, P.; Wang, T. Effects of freezing-thawing on hillslope erosion process. *J. Soil Water Conserv.* **2015**, *29*, 56–60. [CrossRef]
63. Ye, Z.Y. On the Selection of Index Forward and Dimensionless Methods in Multi-index Comprehensive Evaluation. *J. Stat. Theory Pract.* **2003**, *4*, 24–25. [CrossRef]
64. Liu, Y.; Wang, X.; Zhou, L.; Zan, X.; Sheng, S. Relationship analysis between soil detachment rate and erosion factors on freeze-thaw slope. *Trans. CSAE* **2016**, *32*, 136–141. [CrossRef]
65. Fu, Q.; Zhao, H.; Li, T.; Hou, R.; Liu, D.; Ji, Y.; Zhou, Z.; Yang, L. Effects of biochar addition on soil hydraulic properties before and after freezing-thawing. *Catena* **2019**, *176*, 112–124. [CrossRef]
66. Li, Z.-G.; Gu, C.; Zhang, R.; Ibrahim, M.; Zhang, G.-S.; Wang, L.; Chen, F.; Liu, Y. The benefic effect induced by biochar on soil erosion and nutrient loss of slopping land under natural rainfall conditions in central China. *Agric. Water Manag.* **2017**, *185*, 145–150. [CrossRef]
67. Juan, Y.; Tian, L.; Sun, W.; Qiu, W.; Curtin, D.; Gong, L.; Liu, Y. Simulation of soil freezing-thawing cycles under typical winter conditions: Implications for nitrogen mineralization. *J. Soils Sediments* **2019**, *20*, 143–152. [CrossRef]
68. Brodowski, S.; John, B.; Flessa, H.; Amelung, W. Aggregate-occluded black carbon in soil. *Eur. J. Soil Sci.* **2006**, *57*, 539–546. [CrossRef]
69. Sachdeva, V.; Hussain, N.; Husk, B.R.; Whalen, J.K. Biochar-induced soil stability influences phosphorus retention in a temperate agricultural soil. *Geoderma* **2019**, *351*, 71–75. [CrossRef]
70. Wang, G.; Jin, J.; Chen, X.; Liu, J.; Liu, X.; Herbert, S. Biomass and catabolic diversity of microbial communities with long-term restoration, bare fallow and cropping history in Chinese Mollisols. *Plant Soil Environ.* **2008**, *53*, 177–185. [CrossRef]
71. He, X.; Yin, H.; Sun, X.; Han, L.; Huang, G. Effect of different particle-size biochar on methane emissions during pig manure/wheat straw aerobic composting: Insights into pore characterization and microbial mechanisms. *Bioresour. Technol.* **2018**, *268*, 633–637. [CrossRef]
72. Liu, Z.; Dugan, B.; Masiello, C.A.; Barnes, R.T.; Gallagher, M.E.; Gonnermann, H. Impacts of biochar concentration and particle size on hydraulic conductivity and DOC leaching of biochar–sand mixtures. *J. Hydrol.* **2016**, *533*, 461–472. [CrossRef]
73. Lee, S.S.; Shah, H.S.; Awad, Y.M.; Kumar, S.; Ok, Y.S. Synergy effects of biochar and polyacrylamide on plants growth and soil erosion control. *Environ. Earth Sci.* **2015**, *74*, 2463–2473. [CrossRef]
74. Smetanová, A.; Dotterweich, M.; Diehl, D.; Ulrich, U.; Dotterweich, N.F. Influence of biochar and terra preta substrates on wettability and erodibility of soils. *Z. Geomorphol.* **2013**, *57*, 111–134. [CrossRef]
75. Le Bissonnais, Y.; Arrouays, D. Aggregate stability and assessment of soil crustability and erodibility: II. Application to humic loamy soils with various organic carbon contents. *Eur. J. Soil Sci.* **1997**, *48*, 39–48. [CrossRef]
76. Peng, X.; Zhu, Q.H.; Xie, Z.; Darboux, F.; Holden, N.M. The impact of manure, straw and biochar amendments on aggregation and erosion in a hillslope Ultisol. *Catena* **2016**, *138*, 30–37. [CrossRef]
77. Abrol, V.; Ben-Hur, M.; Verheijen, F.G.A.; Keizer, J.J.; Martins, M.A.S.; Tenaw, H.; Tchehansky, L.; Graber, E.R. Biochar effects on soil water infiltration and erosion under seal formation conditions: Rainfall simulation experiment. *J. Soils Sediments* **2016**, *16*, 2709–2719. [CrossRef]
78. Lee, S.S.; Gantzer, C.J.; Thompson, A.L.; Anderson, S.H.; Ketcham, R.A. Using High-Resolution Computed Tomography Analysis to Characterize Soil-Surface Seals. *Soil Sci. Soc. Am. J.* **2008**, *72*, 1478–1485. [CrossRef]
79. Doan, T.T.; Henry-Des-Tureaux, T.; Rumpel, C.; Janeau, J.-L.; Jouquet, P. Impact of compost, vermicompost and biochar on soil fertility, maize yield and soil erosion in Northern Vietnam: A three year mesocosm experiment. *Sci. Total Environ.* **2015**, *514*, 147–154. [CrossRef]
80. Wei, Y.X.; Shi, G.X.; Wu, Y.; Liu, H. Effect and comprehensive evaluation of biochar application mode on slope farmland in black soil region. *Chin. Soc. Agric. Mach.* **2018**, *49*, 251–259. [CrossRef]

sustainability

MDPI

Article

The Synergic Effect of Whey-Based Hydrogel Amendment on Soil Water Holding Capacity and Availability of Nutrients for More Efficient Valorization of Dairy By-Products

Jarmila Čechmánková [1,*], Jan Skála [1], Vladimír Sedlařík [2], Silvie Duřpeková [2], Jan Drbohlav [3], Alexandra Šalaková [3] and Radim Vácha [1]

[1] Department of Soil Hygiene, Research Institute for Soil and Water Conservation, Zabovreska 250, Zbraslav, 15627 Prague, Czech Republic; skala.jan@vumop.cz (J.S.); vacha.radim@vumop.cz (R.V.)
[2] Centre of Polymer Systems, University Institute, Tomas Bata University in Zlin, Tr. T. Bati 5678, 76001 Zlin, Czech Republic; sedlarik@utb.cz (V.S.); durpekova@utb.cz (S.D.)
[3] Dairy Research Institute, Ke Dvoru 12a, 16000 Prague, Czech Republic; drbohlav@milcom-as.cz (J.D.); salakova@milcom-as.cz (A.Š.)
* Correspondence: cechmankova.jarmila@vumop.cz; Tel.: +420-606737098

Abstract: Agricultural production is influenced by the water content in the soil and the availability of nutrients. Recently, changes in the quantity and seasonal water availability are expected to impact agriculture due to climate change. This study aimed to test an agricultural product with promising properties to improve soil quality and water-holding capacity during agricultural application. Most of the traditional hydrogels are low-biodegradable synthetic materials with under-researched long-term fate in field soil conditions. The novel, biodegradable hydrogel made from acid whey and cellulose derivatives cross-linked with citric acid was used. The soil-improving effects were tested under controlled experimental conditions with the sandy artificial soil consisting of 10% finely ground sphagnum peat, 20% kaolinite clay, and 70% quartz sand. Soil pH, the content of organic carbon (Cox), total nitrogen (N), available forms of the essential macronutrients (P, K, Ca, and Mg), the cation exchange capacity (CEC), the maximum water capacity (MWC) and water holding capacity (WHC) were determined. The results showed a positive effect on water retention and basic soil properties after the different levels of hydrogel had been introduced into the soil. Generally, the addition of whey-based hydrogel increases the available nutrients concentration and water retention in soil.

Keywords: hydrogel; soil quality; chemico-physical properties; sustainability

Citation: Čechmánková, J.; Skála, J.; Sedlařík, V.; Duřpeková, S.; Drbohlav, J.; Šalaková, A.; Vácha, R. The Synergic Effect of Whey-Based Hydrogel Amendment on Soil Water Holding Capacity and Availability of Nutrients for More Efficient Valorization of Dairy By-Products. *Sustainability* **2021**, *13*, 10701. https://doi.org/10.3390/su131910701

Academic Editors: Bharat Sharma Acharya and Rajan Ghimire

Received: 25 August 2021
Accepted: 19 September 2021
Published: 26 September 2021

Publisher's Note: MDPI stays neutral with regard to jurisdictional claims in published maps and institutional affiliations.

1. Introduction

The concept of agricultural sustainability spans both a way of thinking as well as of farming practices towards an environmentally sound, productive, economical, and socially needed agriculture [1]. The use of absorbent hydrogels in the field of agriculture presents several benefits for soil—conservation of water, reducing the usage of soil nutrients, and lowering the negative effects of dehydration and moisture stress in crop plants [2].

Hydrogels are a special part of gels, obtained by the chemical stabilization of hydrophilic polymers in a tridimensional network [3], having a remarkable ability to absorb water [4,5]. They show desirable physical and mechanical properties, and also economical reasonability in some cases. Polymer technology has provided a massive contribution to the development of novel cross-linking methods to design hydrogels. Various hydrogel preparation methods have been published for the formation of materials with various compositions, structures, and properties [6]. Due to their properties, they have been widely used in various fields ranging from industry and tissue engineering to pharmacy, medicine, cosmetics, and grocery [7]. Hydrogels are also widely utilized in forestry [8], reclamation [9], and agriculture where their good water retention and slow-release capacity showed promising results for the improvement of soil properties, reduction of fertilizer loss [10,11], or ameliorative advance fertilization [3].

An ideal hydrogel/superabsorbent for agricultural or industrial use must meet several requirements, including high capacity for absorption water and absorption rate, reasonable economy, stability after swelling and during storage, rewetting capability, and nontoxicity [12]. Most of the traditional hydrogels on the market are acrylate-based products, thus low-biodegradable materials with complicated fate in soils [13]. Whilst industrial utilization may profit from the high stability of (semi)-synthetic hydrogels with an effect of long service life, the same property may turn undesirable for environmentally friendly use in agriculture. With regard to the growing attention to environmental issues, biodegradable hydrogels are of particular interest for potential commercial use in agriculture [14]. Hydrogels derived from natural sources offer advantages over synthetic forms, owing to their biocompatibility, physicochemical, physico-mechanical, and environmentally friendly properties.

Numerous studies have reported the potential for increasing water retention properties after the use of synthetic polyacrylicacid [15] and polyacrylamide [16] hydrogel with a proven large capacity for water retention [8,15]. These acrylate-based products have been successfully used as soil conditioners to increase water holding capacity or nutrient retention of soils during practical agricultural testing [17–19].

Recently, various natural ingredients were effectively incorporated into hydrogels to increase their biocompatibility and biodegradability and resulted in novel hydrogels for agricultural applications—i.e., biodegradable cellulose-based superabsorbent [3,20,21] hydrogels based on pectin, cashew gum, starch, Arabic gum [22], kernel gum [23], natural and semi-synthetic chitosan [12], various proteins, polysaccharides and their combinations with different cross-linking methods [6], and casein [24]. Nevertheless, the practical agricultural testing of their behavior under field conditions has been applied to few studies only. Practical use was presented e.g., by Demitri et al. [20] or Montesano et al. [3] who conducted pilot studies of the cellulose-based hydrogel with special attention to the changes in the water absorption capacity of the soil.

Acid whey is a dairy industry waste obtained from cheese production [25] for which the industry has long tried to find a sustainable utilization [26]. The increased demand for dairy products creates an excessive supply of acid whey (approximately 9 kg of whey per kilogram of cheese) that must be either disposed of or repurposed [27]. The worldwide production of whey is approx. 115 MM tons annually which involves both sweet whey (pH cca 6.0) with numerous utilizations in the food production and less beneficial acid whey (pH cca 4.6) [28]. Applying the recent trends toward eliminating waste and keeping products in circulation, the various ways of acid whey valorization are investigated [29]. The whey is typically compound from proteins, lactose, milk fat, minerals, lactic acids, and amino acids [30]. The chemical properties of whey give it the potential to be used as a direct source of nutrients (N, P, K, S, Ca, Na, and Mg) [29,31,32], and as potential biological fertilizer for plant growth [27,33].

For our experimental testing, acid whey was used as a base together with carboxymethylcellulose sodium salt and hydroxyethylcellulose cross-linked with citric acid to develop novel biodegradable hydrogel with the aim to obtain an effective soil conditioner with a high swelling capacity enabling optimal conditions to improve the soil quality in terms of water conservation and nutrients supply. In order to minimize the effects other than the dose and duration connected to the hydrogel application, the artificial soil was used as a standard soil medium and reference matrix [34]. Especially, the advantage is that the natural soil-inherent diversity can be reduced under strictly controlled experimental conditions of a simple artificial soil system. The artificial soil mixture was prepared with sandy textures since the sandy soils are highly permeable with poor water retention capacity and nutrient deficiency [35]; hence, suitable candidates to examine the efficacy of the hydrogel application.

Since the laboratory characterization of the novel whey/cellulose hydrogel cross-linked by citric acid showed a decent capacity for water uptake together with desirable properties of thermal stability and viscoelasticity [36], the objective of this study was to explore the effects of different soil–hydrogel mixtures on several critical soil parameters under controlled experimental conditions. The investigated soil parameters were chosen to cover the basic desirable properties (water retention characteristics), synergic ameliorative effects (nutritional status of soil), and the potentially adverse effects (undue pH drop, excessive mobilization of soil macro-elements). The main aim of this research is to determine the suitability of whey-based hydrogel for application in agriculture with benefits for environment and waste management by testing soil quality improvement after hydrogel application. The purpose of presented research was to prove the synergic effect of whey-based hydrogel amendment on soil water-holding capacity and the availability of nutrients under controlled conditions of artificial soil.

2. Materials and Methods

2.1. Hydrogel Preparation and Analysis

Carboxymethylcellulose sodium (Sigma Aldrich No.C5013, St. Louis, MO, USA) salt and 2-hydroxyethylcellulose (Sigma Aldrich No. 434973) (3%) at a weight ratio of 3:1 were dissolved in the acid whey (a by-product of the milk processing) to yield the solution for hydrogel preparation. The citric acid in anhydrous form (Lach-Ner, Neratovice, Czech Republic) was added at concentrations equal to 10% wt as a cross-linking agent. The hydrogel samples were prepared according to the procedure detailed by Durpekova et al. [36]. The hydrogel samples were analyzed by accredited methods (Table 1) in the Research Institute for Soil and Water Conservation.

2.2. Artificial Soil Mixture

Soil composition was fixed by using artificial soil mixtures that cover the typical sandy soils, which are assumed to preferentially benefit from the hydrogel supplementation. It reduces both the variability of initial soil physio-chemical properties and the effects of field management effects of field soils (e.g., possible post-harvest residue input to cropland, the uncertainties of former fertilizing). The artificial soil (standard soil) was prepared according to the standard guidelines of ISO 11268-1:2012 [37]. Artificial soil consists of 10% finely ground sphagnum peat, 20% kaolinite clay, and 70% quartz sand. Quartz sand of 0.05–0.2 mm grain size from Střeleč, kaolinite clay (30% kaolinite) from Sedlec and commercially available sphagnum peat were used. The individual components were mixed with an electric mixer, dried spontaneously, and incubated for one month before use.

2.3. Hydrogel Application in a Pot Experiment

A pot experiment was set up with four variants in three replicates: the pure artificial soil as a control soil, artificial soil mixed with three variants of the hydrogel supply (Figure 1. Plastic planting containers with a volume equal to 3.2 L were filled with 3.6 kg of artificial soil with a bulk density of 1.22 g·cm^{-3}. The prepared hydrogel was weighed and mixed with artificial soil in a particular amount (Figure 1). The soil–hydrogel mixtures were homogenized in the entire volume and irrigated immediately. Pots were watered with 0.5 L of pure water three times a week for a period of four weeks. Soils were sampled 24 h, 7 days, and over the next 3 weeks after the experimental setup and basic chemical properties were determined.

Soil samples for physical soil properties were prepared for saturation (part physical properties). Presented values resulted from six independent measurements for each variant—i.e., control experiment with pure artificial soil without hydrogel and variants with a dose escalation of whey-based hydrogel in three soil–hydrogel mixtures.

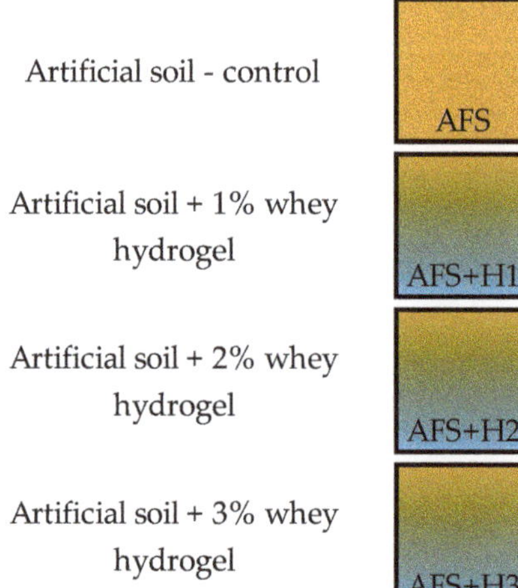

Figure 1. Pot experiment variants.

2.4. Soil Analysis

The purpose of the analyses was to evaluate the effect of the hydrogel amendment on the chemical properties of the soil. The analyses were performed in the accredited laboratory of the Research Institute for Soil and Water Conservation, Prague. The quality assurance of analytical data is guaranteed by the control process and certified methods of the analyses. For the soils from experiments, a stable set of chemical soil properties was determined, as given in Table 1.

2.4.1. Chemical Properties

Soil pH was determined by using the instrumental method [38] for the routine determination of pH using a glass electrode in a 1:5 (volume fraction) suspension of soil in water (pH_{H2O}); the content of organic carbon by sulfochromic oxidation according to ISO 14235 [39] and total nitrogen according to ISO 11261 [40] by the modified Kjeldahl method. The nutritional status of soil was explored using the available forms of the essential macronutrients (P, K, Ca, and Mg) after the common extraction in the Mehlich III solution [41]. Mehlich III is a weak acid soil extraction used worldwide for extracting bioavailable nutrients in soils and is standardized as a combination of five reagents at a dilution ratio of 1:10 (0.2 mol·L^{-1} glacial acetic acid, 0.25 mol·L^{-1} ammonium nitrate, 0.015 mol·L^{-1} ammonium fluoride, 0.013 mol·L^{-1} nitric acid, and 0.001 mol·L^{-1} ethylene diamine tetra acetic acid). The inductively coupled plasma optical emission spectrometry (ICP-OES) was used to analyze the extractions from the soil samples. As the complementary soil property for the potential of available nutrients, the cation exchange capacity (CEC) was measured by a barium chloride solution buffered at pH = 8.1 using triethanolamine [42].

Table 1. Overview of the studied chemical soil properties and used methods.

	Method	**Reference**	**Accuracy (% rel.)**
pH	Determination of pH	ISO 10390 [38]	4–5
C	Oxidimetric method	ISO 14235 [39]	10–15
N	Modified Kjeldahl method	ISO 11261 [40]	15–20
P	Mehlich III solution	Mehlich (1984) [41]	20
K	Mehlich III solution	Mehlich (1984) [41]	20
Ca	Mehlich III solution	Mehlich (1984) [41]	20
Mg	Mehlich III solution	Mehlich (1984) [41]	20
CEC	Barium chloride solution	ISO 13536 [42]	20

2.4.2. Soil Physical Properties

In this study, the maximum water capacity (MWC, full water capacity) and water holding capacity (WHC) were used as indicators for the effectiveness of hydrogel in improving water availability in soil. The soil hydraulic parameters were measured by the adjusted instrumental method according to EN 13041:2011 [43], ISO 11508:2017 [44], and ISO 11272:2017 [45]. The soil–hydrogel mixture was placed into a two-part volumetric stainless-steel ring. The prepared sample rings were placed into a water bath where the pure water gradually flowed in until it reached the height of 1 cm below the rims. The samples were then left to saturate for 24 h. The saturated rings are weighed and then moved into sand tanks. The sand tanks are the PVC boxes with sealed bottom filled with multi-layered substrates and equipped with the suction cup with height compensator, allowing to create a partial vacuum where the negative fluid pressure is set as a function of the depression of the suction cup. The rings were left for suction for 48 h by the negative pressure of 1 kPa. After the first suction period, the two-part sample rings were separated—the upper part is removed and weighed, the lower part was given back into the sand tank and left for suction for the period of 48 h by the negative pressure of 3 kPa. After the second suction period, the sample rings were oven-dried at 105 °C for the period until successive weighing agrees to ± 0.1 mg.

As soon as the sample rings were proceeded using the described instrumentation and the measured weights were gathered, the maximum water capacity (MWC) and water holding capacity (WHC) were calculated from the following equations:

$$MWC = (M1 - M3)/V1 \times 100 \ (\% \ vol.) \tag{1}$$

$$WHC = (M2 - M3)/V1 \times 100 \ (\% \ vol.) \tag{2}$$

where M1 is the mass of saturated sample including ring (g), M2 is the mass of the sample after the first suction section (-1 kPa, 48 h), M3 is the mass of the sample with the ring after the second suction section (-3 kPa, 48 h), and V1 is the measured volume of the steel ring. The first equation (Equation (1)) targets the soil state when practically all pore spaces are completely filled with fluid and there is plenty of water available to the crop at saturation. Only a portion of the available water is easily used by the crop, and hence, the second equation (Equation (2)) targets the available moisture that a given soil can hold for crop use. In the field conditions, soil texture and organic matter are the crucial determinants for soil water holding capacity, while in our artificial soil experiment these factors were fixed; hence, enabling us to observe the effects of hydrogel addition in various application rates.

2.4.3. Statistical Analysis

Soil samples were assayed in three (chemical properties) and six (physical properties) replicates for each analysis. The data were subjected to a normality analysis using the Shapiro–Wilk test. One-way ANOVA was conducted to test the significance of differences in the soil water characteristics among the different soil–hydrogel mixtures using the Statistica software (Version 10 for Windows; StatSoft, Inc., Tulsa, OK, USA). Statistical significance was considered at $p < 0.05$ unless otherwise stated.

3. Results

The whey-based hydrogel was analytically characterized before the experiment. The chemical properties of hydrogel and artificial soil are given in Table 2. Measured pH_{H2O} was 4.17. Since both the initial whey additive and cross-linking agent yielded a low pH, the possible acidification may be a matter of attention due to several adverse effects of the pH drop on the availability of soil nutrients, and the toxicity or mobility of soil microelements.

Table 2. Whey-based hydrogel characteristics.

Hydrogel Characteristics		
Properties		
pH_{H2O}		4.17
C	%	40.1
N	%	1.26
P	$mg \cdot kg^{-1}$	6683
K	$mg \cdot kg^{-1}$	13,059
Ca	$mg \cdot kg^{-1}$	6093
Mg	$mg \cdot kg^{-1}$	806
CEC	$mmol \cdot 100\ g^{-1}$	82.2

Direct chemical extraction of nutrients in whey-based hydrogel showed a shift of chemical composition towards higher contents and availability of three elements (K > P~Ca) and relative depletion of nitrogen.

The artificial soil is characterized as a sandy soil with a near-neutral pH (pH_{H2O} was 7.6). Since the quartz sand dominates the artificial soil composition, the soil can be characterized as poor in nutrients (N, P, K) (Table 3) with a low cation change capacity, typical for sand-textured soils. Nevertheless, the admixture of sphagnum peat resulted in an above-average C_{ox} content and relatively higher content of Mg or Ca.

Table 3. Artificial soil characteristic.

Artificial Soil Characteristics		
Soil Properties		
pH_{H2O}		7.65
C_{ox}	%	2.54
N	%	0.075
P	$mg \cdot kg^{-1}$	18.0
K	$mg \cdot kg^{-1}$	73.5
Ca	$mg \cdot kg^{-1}$	3204
Mg	$mg \cdot kg^{-1}$	487
CEC	$mmol \cdot 100\ g^{-1}$	9.35

3.1. Effect of Hydrogel on Chemical Soil Properties

There was observed a trend of pH drop with the changing ratios of soil to hydrogel in mixtures. Soil pH decreased after hydrogel application from 7.65 to 7.56 (H1) and 7.52 (H3) in a period of 24 h (Table 4). The lowermost pH was detected after the application of the H3 dose.

Table 4. pH_{H2O} values.

pH_{H2O}			
Variant	**24**	**7**	**3**
AFS control	7.65	7.75	7.62
AFS+H1	7.56 **	7.73	7.57
AFS+H2	7.67	7.62	7.62
AFS+H3	7.52 **	7.57	7.51

** mean values significantly different at $\alpha = 0.05$ compared to control. AFS control—artificial soil, AFS+H1—artificial soil with 1% of hydrogel, AFS+H2—artificial soil with 2% of hydrogel, AFS+H3—artificial soil with 3% of hydrogel.

The maximum pH drop was 1.7%. The addition of whey hydrogel did not cause the soil pH to drop to a point that would be injurious to plants. Within 7 days there was perceptible lower pH in all variants with the hydrogel application but without statistical significance in the linear model. After 4 weeks after the hydrogel application, differences between hydrogel variants and control soil decreased (Figure 2) and soil pH stabilized again due to the buffering capacity of higher Ca content in both artificial soil and whey-hydrogel. In the artificial soil, the addition of hydrogel in testing doses did not cause a long-term decrease of soil pH and did not decline more than 3%.

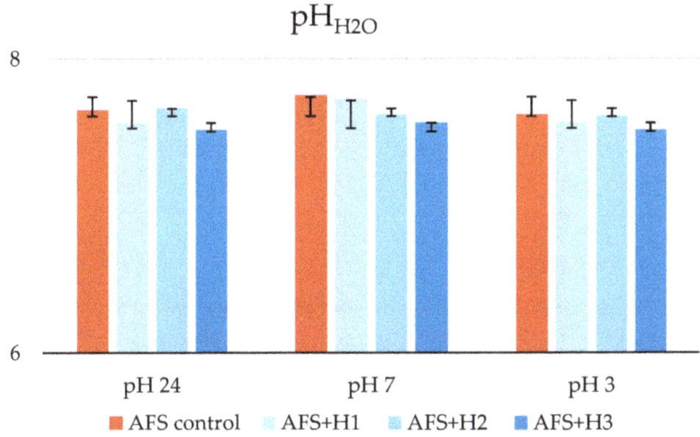

Figure 2. The effect of whey hydrogel application on soil pH$_{H2O}$. AFS control—artificial soil, AFS+H1—artificial soil with 1% of hydrogel, AFS+H2—artificial soil with 2% of hydrogel, AFS+H3—artificial soil with 3% of hydrogel, pH24—pH measured after 24 h, pH7—pH measured after 7 days, pH3—pH measured after next 3 weeks.

The nitrogen content in the artificial soil (control) at the beginning of the experiment was 755 mg·kg^{-1}. As shown in Table 5, concentrations of this element immediately increased after the hydrogel addition while the highest content after the addition was found in the H1 variant. Differences between variants were very low and were not statistically significant. The nitrogen amount in whey hydrogel was very low and soil nitrogen was not affected by the hydrogel addition. After 7 days (Figure 3) from the hydrogel application, the amount of nitrogen further increased, and the highest amount of nitrogen contained variant with the highest amount of hydrogel. After another three weeks, a decreased content of nitrogen was found in all variants.

Table 5. Nitrogen, phosphorus, and potassium mean values in variants.

	Variant	24	7	3
N	AFS control	755	800	700
	AFS+H1	785	805	745
	AFS+H2	780	885	670
	AFS+H3	760	915	740
P	AFS control	18	17	19
	AFS+H1	21	24	27 **
	AFS+H2	27 **	28 **	29 **
	AFS+H3	29 **	35 **	44 **
K	AFS control	74	82	72
	AFS+H1	102	99 **	101
	AFS+H2	114 **	114 **	113 **
	AFS+H3	135 **	140 **	160 **

** mean values significantly different at α = 0.05 compared to control. AFS control—artificial soil, AFS+H1—artificial soil with 1% of hydrogel, AFS+H2—artificial soil with 2% of hydrogel, AFS+H3—artificial soil with 3% of hydrogel, 24—measured after 24 h, 7—measured after 7 days, pH3—measured after next 3 weeks.

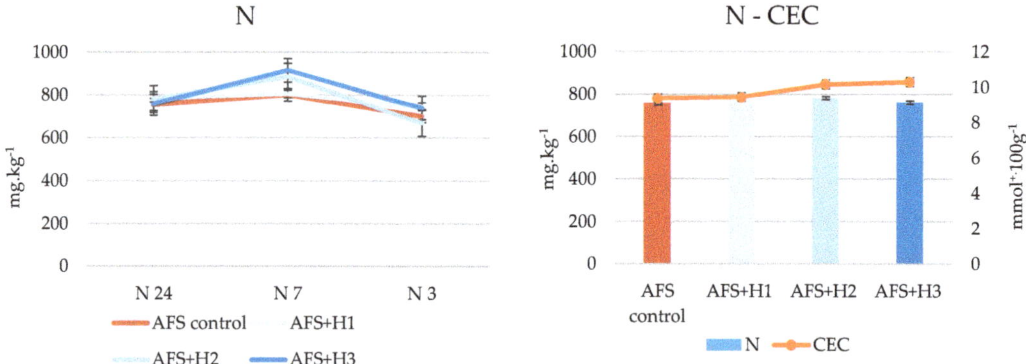

Figure 3. The effect of whey hydrogel application on nitrogen (N) and cation exchange capacity (CEC). AFS control—artificial soil, AFS+H1-artificial soil with 1% of hydrogel, AFS+H2-artificial soil with 2% of hydrogel, AFS+H3-artificial soil with 3% of hydrogel, N 24-N measured after 24 h, N 7-pH measured after 7 days, N 3-N measured after next 3 weeks.

The phosphorus content increased across all the hydrogel variants and sampling periods. After the hydrogel application, phosphorus increased immediately. The application of soil mixed with the highest hydrogel ratio (H3) increased phosphorus by about 50% compared to control, the same rise caused the mixture with the moderate ratio of hydrogel (H2). Lowest change (16%) was following the application of the soil–hydrogel mixture H1 (Figure 4). Moreover, the available content of phosphorus increased also along the period of sampling after the application, and especially a boosting effect was the most radical for the mixture with the highest hydrogel proportion (H3). Nevertheless, major differences among soil–hydrogel variants were proved statistically significant (Table 5).

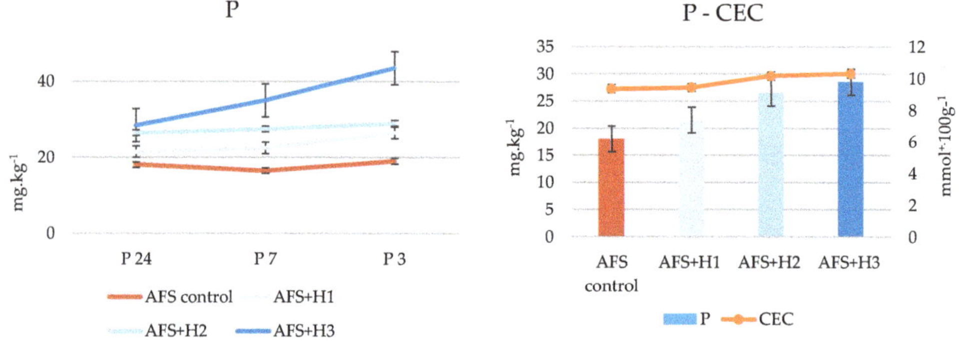

Figure 4. The effect of whey hydrogel application on phosphorus (P) content and cation exchange capacity (CEC). AFS control-artificial soil, AFS+H1-artificial soil with 1% of hydrogel, AFS+H2-artificial soil with 2% of hydrogel, AFS+H3-artificial soil with 3% of hydrogel, P 24-P measured after 24 h, P 7-P measured after 7 days, P 3-P measured after next 3 weeks.

Potassium showed a similar trend of rapid increase similar to phosphorus after the hydrogel application. Application of whey hydrogel in the lowest soil–hydrogel proportion (H1) caused an increase by about 39%, and with the increasing proportion of hydrogel to soil in the mixture, the relative enrichment of K increased up to 55% (the middle soil–hydrogel proportion—H2), resp. 84% (H3 variant). Similar to P, the ameliorative effects of hydrogel application were temporally for K contents. In the contrary, the K contents remained stable for variants with lower proportions of hydrogel to soil (H1, H2 variants)

(Figure 5). The K content significantly exceeded the control variant (up to 122%) after the addition of the highest dose of hydrogel (H3). Significant differences in K contents among the soil–hydrogel variants were found using one-way ANOVA (Table 5).

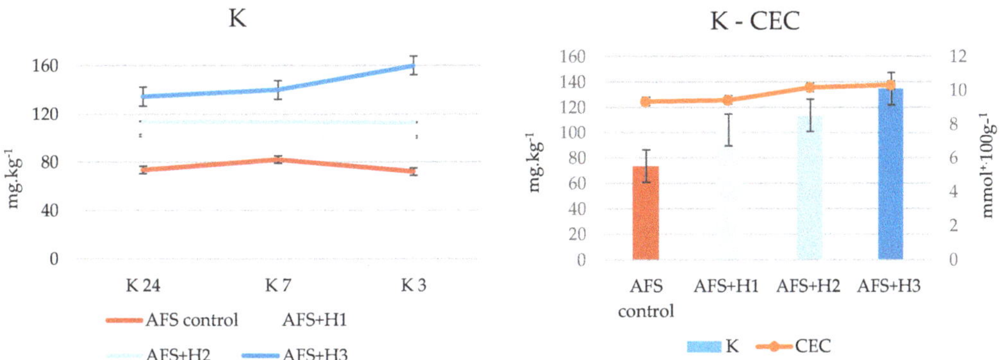

Figure 5. The effect of whey hydrogel application on potassium content in soil and cation exchange capacity. AFS control-artificial soil, AFS+H1-artificial soil with 1% of hydrogel, AFS+H2-artificial soil with 2% of hydrogel, AFS+H3-artificial soil with 3% of hydrogel, K 24-K measured after 24 h, K 7-K measured after 7 days, K 3-K measured after next 3 weeks.

Generally, the results showed the gradual increase of nutrients' available pools with the changing hydrogel dose. As shown in Figures 3–5, along with the nutrients' level, cation exchange capacity (CEC) also significantly changed after hydrogel addition. Since the cation exchange capacity measured by a pH 8.1 buffered barium chloride solution usually shows a decent correlation with exchangeable cations in soil samples, there was observed a similar dose gradient of CEC for various soil–hydrogel mixtures. Differences in soil CEC between control and hydrogel application were found evident soon after the application of the H3 dose into artificial soil and then remained relatively stable during the consecutive sampling periods.

The contents of available calcium and magnesium were also determined (Figure 6). In both cases, there was an immediate decline in the available nutrients' pools. After the application of hydrogel, the lowest values were determined after the application of the H3 dose in the first sampling period (24 h). The relative decrease of available Ca pools was proportional to the hydrogel dose in the mixture and decreased from 5.5% for the low dose (H1) up to 9 resp. 14% for higher hydrogel doses (H2 resp. H3 dose). During the consecutive sampling periods, differences between soil–hydrogel variants and the control soil were still considerable while differences among particular soil–hydrogel variants were low. A similar trend was found in the case of magnesium, for which the hydrogel dose gradient in various soil–hydrogel mixtures resulted in an immediate decline by about 5, 10, and 16%. These results confirmed the crucial role of soluble Ca, Mg for buffering the hydrogel-induced acidification of soil. Since the experimental artificial soil was relatively rich in Ca and Mg, the soil pH changes were controlled by their pools. The question about the soil acidification issues remains open for soils with very poor pH buffering capacity.

3.2. Effect of Hydrogel on Physical Soil Properties

In general, the determination of the physical properties of disturbed soils showed that the hydrogel additions increased the water retention properties in untreated soil (Figures 7 and 8); hence, the lowermost measured values for maximum water capacity (MWC, water field capacity, soil saturation) and water holding capacity (WHC) were mostly detected in control soils without the application of hydrogel. MWC values range from 44.94% (control) up to 46.75% (H2), 47.67 (H2), and 49.95% (H3). In sandy artificial soil, the addition of 1% hydrogel (H1) increased maximum water capacity by 4%, compared with untreated artificial soil (Figure 7). Amendment of 2% hydrogel (H2) had an increase of 6% compared to the control. Amending artificial soil by 3% of hydrogel (H3) led to an increase of maximum water capacity by about 11%. Differences between single doses of hydrogel were not statistically proved and the highest influence on maximum water capacity was found after H3 dose application into artificial soil. Statistically proved was the difference between artificial soil (control variant without hydrogel application) and artificial soil with 3% of whey–hydrogel.

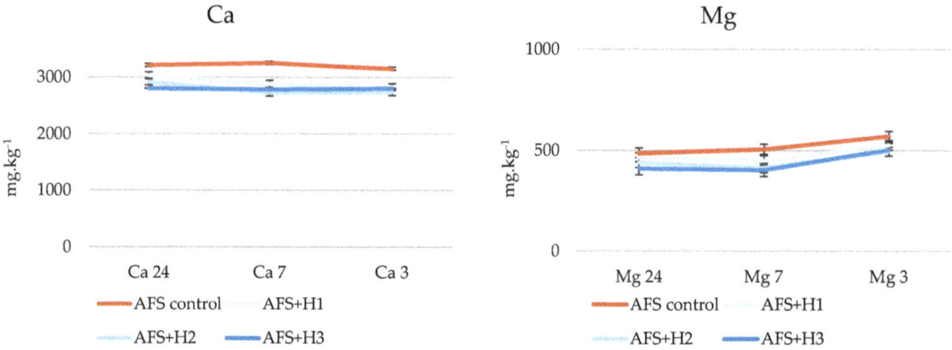

Figure 6. The effect of whey hydrogel application on calcium (Ca) and magnesium (Mg) content in soil. AFS control-artificial soil, AFS+H1artificial soil with 1% of hydrogel, AFS+H2-artificial soil with 2% of hydrogel, AFS+H3-artificial soil with 3% of hydrogel, Ca 24-Ca measured after 24 h, Ca 7-Ca measured after 7 days, Ca 3-Ca measured after next 3 weeks, Mg 24-Mg measured after 24 h, Mg 7-Mg measured after 7 days, Mg 3-Mg measured after next 3 weeks.

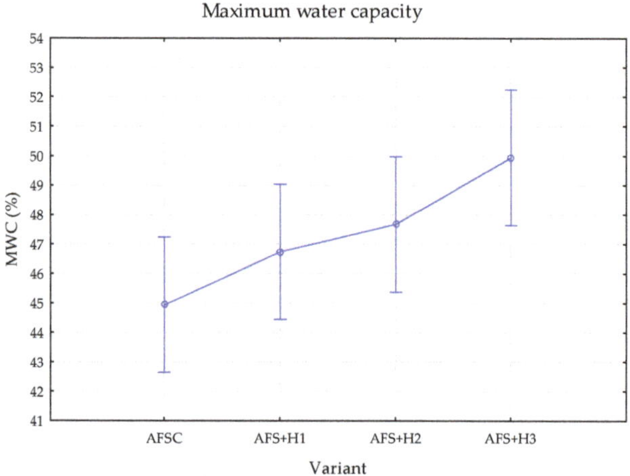

Figure 7. Maximum water capacity of single hydrogel variant, control and different doses of hydrogel, results of ANOVA. AFSC—artificial soil (control), AFS+H1—artificial soil with 1% of hydrogel, AFS+H2—artificial soil with 2% of hydrogel, AFS+H3—artificial soil with 3% of hydrogel.

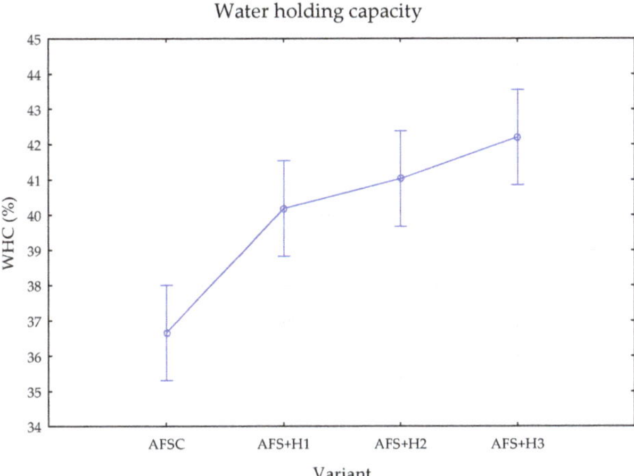

Figure 8. Water holding capacity of single hydrogel variant, control and different doses of hydrogel, results of ANOVA. AFSC—artificial soil (control), AFS+H1—artificial soil with 1% of hydrogel, AFS+H2—artificial soil with 2% of hydrogel, AFS+H3—artificial soil with 3% of hydrogel.

Water holding capacity—available water-holding capacity is a key attribute, as it quantifies the amount of water available for plants. Water retention in the soil is strongly influenced by the soil composition, as well as the practices of soil use. The sandy artificial soil used for our experiment dispose of 36.66% WHC. As in the case of maximum water capacity, values of water holding capacity escalated from control variant up to variants amended with increasing doses of hydrogel (from 40.19%, 41.03%, up to 42.21%). Our results (Figure 8) showed an unequivocally positive effect of whey-based hydrogel on WHC. The addition of 1% of hydrogel into the artificial soil increased WHC by 10% compared to control soil. Under higher hydrogel doses (H1 resp. H2 dose), the water holding capacity was measured at the level of 112 resp. 115% of the untreated variant. The mean WHC for various experimental variants were proved to significantly differ at the significance level $\alpha = 0.05$. Statistically proved differences were between artificial soil (control variant) and all variants with hydrogel amendment and between variants with 1% of hydrogel and 3% of hydrogel.

4. Discussion

Synthetic hydrogels based on acrylates and acrylamides demonstrate high mechanical strength and the potential to absorb large amounts of water [46]. Due to the problem of their lower biodegradability, alternative (semi)natural biopolymers, such as alginate, agar, cellulose, chitosan, and starch, have been attempted to replace them [3,12,20–23]. In spite there are plenty of hydrogel formulations developed in laboratory conditions, only a few materials meet the requirements for effective and safe agricultural utilizations i.e., simultaneous nontoxicity and biodegradability with favorable properties for application (efficiency, applicability in field conditions) [47].

Several different polysaccharides have been used to prepare hydrogels as slow-release fertilizer hydrogels with both water retention and slow-release properties [22,48]. Although biopolymers could be used as macro- and micro-elements carriers for fertilization [46], our research aimed to test the straight use of whey-based hydrogel as nutrients' source for soil amelioration.

Often, the hydrogel in different forms (granular, particle) is mixed with soil at the 0.1% concentration ratio [21]. Miller and Naeth [9] tested the applicability of various doses—i.e., recommended dose (488 kg·ha^{-1}), under- (one-half), and over-dosed (double) application of a synthetic acrylic hydrogel cross-linked with polyacrylamide for reclamation of mine-impacted soils. They concluded a high effect of applied co-substrates for creating hydrogel-amended soil substrates for reclamation. Similar to our results, the effects of hydrogel addition led to initial high effects with application rate dependency on water retention that faded away with time from the application. Concerning the polyacrylate hydrogel, the common concentration ratios range between 0.1 and 0.4% [49,50]. When combining chitosan with polyacrylamide hydrogel, Ritonga et al. [51] used the application rate of 0.1% for soil testing. During our study, the application rates were set to 1, 2, and 3% of whey-based hydrogel tally with amounts of cellulose-based hydrogels used for agricultural purposes [3,20]. Song et al. [52] applied 0.375, 0.650, and 0.975% of lignosulfonate and sodium alginate hydrogel to optimize the soil available water content for tobacco crop grown in loess soils under drought experimental conditions.

Since the acid whey was the source material for the synthesis of hydrogel, and pH was recognized as one of the most important factors that control the healthy plant, the direct sequential measurements of pH changes were conducted after certain time intervals. Whey source as a mild acid suspension with typical solids concentrations near 8% has a high content of soluble salts (including Ca, Na, and K salts), and hence, its direct application into soil decreased soil pH and increases Ca solubility [53]. The mild acidity of whey usually resulted in soil pH drop by neutralizing soil solution (bi)carbonates and consequently, increased the solubility of calcium carbonates [54]. Long-term practicing of the land disposal of acid whey in the field conditions, the benefits of direct whey application were demonstrated for various soil-crop systems in Scotland [55], New Zealand [56], or the US [31,32,57]. The pH changes may temporally reach a range of 2 pH units (depending on application rate, original soil pH, and soil type) [58]. Especially for acid soils, the pH drop may temporally reach the point beyond the optimal range for crop production. On the contrary, the land-disposed whey rendered decent plant nutrition for crops on calcareous soils in the 7.6 to 8.8 pH range under irrigation in an arid climate [54,59]. The pH drop may impact early plant development, and hence, it is recommended to postpone planting of sensitive crops to a few days later after whey application to eliminate soil acidity effect on germination. Analyzing the germination capacity of seeds and plant growth of three crops (soya, maize, broccoli), Grosu et al. [33] proved an inhibitory effect of whey over the germination process while the positive influence of whey addition on plants growth in the later development phase in a pot experiment. After the experimental synthesis of whey-based hydrogel, pH_{H2O} remained very low (4.17), i.e., with acidification risk. Similar to previous studies, pH swing was observed in our experiment—i.e., pH dropped immediately after the whey-based hydrogel application and steadied near the initial value in the consequent sampling periods. The pH change remained below one pH unit, probably because of a high Ca-content in the solution. Concerning the acidification issue, care should be taken to avoid potential mobilization of soil contamination in acid soils with enhanced contents of pH sensitive trace elements (Cu, Zn, Ni). Trace element concentrations (Al, Cu, Zn, Mn, and Cr) are typically low in whey [31], and hence, the soil application should be proceeded with caution to trace elements in target soils. The temporal pH swing effect together with the ability to stimulate plant growth promotes open possibilities of using whey-based hydrogels as co-substrates for phytoremediation purposes. Also, the positive microbial biomass stimulation of whey makes it a candidate substrate for appropriate in situ bioremediation of different compounds [60]. The pH swing remains also very important because a significant effect of pH on the swelling ratio of natural-based hydrogels was observed under laboratory [36] and field conditions [20].

The nutritional benefits and feasibility of direct land disposal of whey were practically verified [32]. Land application of fresh acid whey does not present an odor issue, but it may turn problematic with time in storage and when co-stored with manure [58]. Considering that utilization of many soil conditioners is limited to seasonal application conditions, preparing stable whey-based hydrogel seems worth the effort. Especially when the nutritional benefits of whey are retained, as was shown in our experimental study with the artificial soil (Figures 3–6). Levels of nutrients in whey-based hydrogel during our experiment showed a shift of chemical composition towards higher contents and availability of three elements (K > P ~ Ca). On the contrary, the results showed hydrogel relative depletion in nitrogen. However, more detailed research on the chemical composition of whey-based hydrogel from multiple sources of acid whey is required because the simultaneous ratios of macro elements play a crucial role when considering the synergic ameliorative effects of the application of whey-based hydrogels. The typical feature of whey's composition is an elevated proportion of P and K. The initial composition of acid whey follows the technological process of milk processing—especially Ca variability in whey from the cheese production (Ca-bond with casein within cheese leading to depletion of Ca in whey) and curd cheese processing (in-soluble Ca-salts enriched the acid whey). Since the nutrient equilibrium is shifted towards K, Ca, and P, the ameliorative effects rather trend to nutrition in K-deficit soils and targeting the situation where the optimal root growth is desirable. In soils with low sorption, K may be extensively pumped off from soils by plants, especially due to high demands of several crops. This process may be enhanced by the seasonal water deficit, and hence, the synergic effects of increasing water availability and K supplementation after hydrogel application may be worth considering. The direct positive nutritional effects were observed for various (semi)natural hydrogels [51] or there were observed the combined effects of hydrogel mixtures with different types of fertilizers based on the traditional NPK [61]. Konzen et al. [61] promoted a positive effect on the growth of *Mimosa scabrella* seedlings after application of combined hydrogel mixture, probably by retaining more water and enabling increased nutrient absorption. In our experiments, we tested both the ability to improve the soil water storing and fertilizing effects under various application rates of hydrogel. Similarly, the chitosan-coated three-layer compound fertilizer prepared from granular NPK coated with chitosan and poly(acrylic acid-co-acrylamide) superabsorbent polymer was used by Wu and Liu [62] and they proved well-controlled release effect and quantitatively described the slow-release mechanism of nutrients. Another multilayer-coated compound fertilizer prepared from NPK granules, polyvinyl alcohol, and chitosan was prepared by Noppakundilograt et al. [63] and they demonstrated increased release of nitrogen, phosphorus, and potassium nutrients. Our results showed similar trends as presented by Rittonga et al. [51] who observed significant changes in cation exchange capacity, levels of nitrogen, phosphorus, and potassium in the soil after introducing hydrogel synthetized by copolymerization of chitosan-TiO_2 composite with polyacrylamide. Various application rates of hydrogel addition led to the differences in soil CEC among variants and in comparison to the control variant with the peak differences up to 38% for the hydrogel-rich mixture. Our results showed that the high application rate (H3 variant) increased immediately CEC by 10% and then remained stable for the duration of the experimental observation of the soil–hydrogel mixture. The differences among CEC values after three hydrogel application rates were not proved statistically significant at $\alpha = 0.05$.

Concerning the potential of whey-based hydrogels to increase the efficiency of water use in soils, synthetic hydrogels proved to be more efficient even under lower doses necessary for water storage increase. Abdallah et al. [64] reported that an amendment of the sandy soil with 0.3% of fine-grained polyacrylamide hydrogel significantly increased the available water capacity. Agaba et al. [49] tested 0.2, and 0.4% w/w polyacrylate hydrogel for its utilization for improving plant available water by an empirical estimation of the water content as the initial and terminal weights of pots in a tree survival study under experimentally induced drought conditions. They found that the addition of either 0.2 or 0.4% hydrogel to the five soil types, prolonged tree survival under water stress compared to the control variants. The higher hydrogel efficiency was observed for the sandy soils compared to loamy and clay soils. Johnson [65] denoted that the addition of super absorbent polymer hydrogel at the dose of 2 $g \cdot kg^{-1}$ improved the water holding capacity of sand from 171% up to 402%. Similarly, Abedi-Koupai et al. [66] proved the effect of hydrogel treatment in the sequence clay < loamy soil < sandy loam soil. On the contrary, Akhter et al. [50] found no significant differences in the amount of plant-available water between loam and sandy loam soil treated with a polyacrylamide–acrylate hydrogel at three application rates (0.1, 0.2, and 0.3%). The experimental results generally showed that well-drained soils (sandy loams, loams) are preferable for hydrogel treatment.

Compared to polyacrylic-based hydrogels, the natural-based hydrogels usually showed a lower yet significant effect on water retention in soil. Song et al. [52] showed that the addition of lignin/sodium alginate hydrogel can increase the maximum water holding capacity of soil with varying intensity from the initial 52.66% up to 55.64%, 58.69% resp. 61.63% following the hydrogel composition (0.375%, 0.650%, resp. 0.975% hydrogel). The whey-based hydrogel used in our study increased maximum water holding capacity from the initial of 44.94% up to 46.75%, 47.67%, and 49.95% following the application rate of hydrogel—1, 2 resp. 3% hydrogel. Despite using different methods for establishing the effects of hydrogel on the soil water holding capacity, Montesano et al. [3] observed the water retention increase depending on the application rate of cellulose-based hydrogel (0.5, 1, and 2%). The same amounts of hydrogel from cellulose derivatives of pharmaceutical grade were used by Demitri (2013) [20]. The use of this hydrogel as water reservoir for the cultivation of vegetables was found to be extremely advantageous for release of water to the soil. The reason for divergences between various types of biodegradable hydrogel could be due to the differences in the macromolecular design of the different gels [49]. Nevertheless, the acrylic hydrogels are macromolecule composites with higher compression strength, which may determine their higher absorption characteristics in comparison to natural-based materials [67]. Since the novel whey-based hydrogel belongs to biodegradable composites with lower compression strength, the water-holding characteristics were measured relatively lower in comparison to the reported efficiency of synthetic hydrogels. According to the chemical compositions, whey-based hydrogels are expected to be higher biodegradable in comparison to synthetic super absorbents, but further investigations should be performed to assess the hydrogel degradation chemistry and kinetics in the soil under various conditions, as well as to evaluate the optimal hydrogel amount to be used for the cultivation of different plant species. Also, there is still a need for some technological solutions for the hydrogel delivery systems for broad field applications.

5. Conclusions

Sustainable agriculture strives to develop cropping systems where the soil sustains the plant productivity for population needs as well as preserves healthy status. Moreover, sustainable development generally turns the economy from linear to circular production, and hence, the efficient solution for eliminating waste and keeping products in circulation has also been developed in crop sciences. The crop vitality highly depends on both water availability and soil nutrition status. Our results showed the perspective of whey-based hydrogel for practical use in agriculture because the natural nutritional benefits of agricultural utilization of whey are enhanced by water retention abilities via the chemical cross-link of whey with citric acid. In the artificial soil experiment, the positive effects of 1%, 2%, and 3% hydrogel addition on both water-holding properties and the temporal availability of nutrients were determined. The results show a clear influence of the whey hydrogel in increasing the water holding capacity with a maximum increase of 15%. At the same time, no adverse side-effects for basic soil properties were observed after different levels of hydrogel had been introduced into the artificial soil. The application of 3% whey-based hydrogel during our experiment increased the available level of phosphorus and potassium up to 50 and 84%, and the nutritional value of whey-based hydrogel with significant effects on elemental compositions may be beneficial during the critical phases in the early crop growth in both agriculture and forestry.

Author Contributions: Conceptualization, V.S., J.Č. and J.D.; methodology, J.Č., S.D. and A.Š.; formal analysis, J.Č.; data curation, J.S.; writing—original draft preparation, J.Č. and J.S.; writing—review and editing, J.S., R.V. and S.D.; visualization, J.Č.; supervision, V.S. and R.V.; project administration, J.Č, V.S. and J.D.; funding acquisition, V.S., J.D. and J.Č. All authors have read and agreed to the published version of the manuscript.

Funding: This research was financially supported by the Ministry of Agriculture of the Czech Republic Project No. QK1910392 and Institutional support MZE-RO0218.

Institutional Review Board Statement: Not applicable.

Informed Consent Statement: Not applicable.

Data Availability Statement: Not applicable.

Conflicts of Interest: The authors declare no conflict of interest.

References

1. Schaller, N. The concept of agricultural sustainability. *Agric. Ecosyst. Environ.* **1993**, *46*, 89–97. [CrossRef]
2. Elshafie, H.S.; Camele, I. Applications of absorbent polymers for sustainable plant protection and crop Yield. *Sustainability* **2021**, *13*, 3253. [CrossRef]
3. Montesano, F.F.; Parentea, A.; Santamaria, A.; Sannino, A.; Serio, F. Biodegradable superabsorbent hydrogel increases water retention properties of growing media and plant growth. *Agric. Agric. Sci. Procedia* **2015**, *4*, 451–458.
4. Zhang, J.; Li, A.; Wang, A. Study on superabsorbent composite XVI. Synthesis, characterization and swelling behaviors of poly(sodium acrylate)/vermiculite superabsorbent composites. *Eur. Polym. J.* **2007**, *43*, 1691–1698. [CrossRef]
5. Chang, C.; Duan, B.; Cai, J.; Zhang, L. Superabsorbent hydrogels based on cellulose for smart swelling and controllable delivery. *Eur. Polym. J.* **2010**, *46*, 92–100. [CrossRef]
6. Klein, M.; Poverenov, E. Natural biopolymer-based hydrogels for use in food and agriculture. *J. Sci. Food. Agric.* **2010**, *100*, 2337–2347. [CrossRef] [PubMed]
7. Lin, C.C.; Metters, A.T. Hydrogels in controlled release formulations: Network design and mathematical modeling. *Adv. Drug. Delivery Rev.* **2006**, *58*, 1379–1408. [CrossRef]
8. Tomášková, I.; Svatoš, M.; Macků, J.; Vanická, H.; Resnerová, K.; Čepl, J.; Holuša, J.; Hosseini, S.M.; Dohrenbusch, A. Effect of different soil treatments with hydrogel on the performance of drought-sensitive and tolerant tree species in a semi-arid region. *Forests* **2020**, *11*, 211. [CrossRef]
9. Miller, V.S.; Naeth, M.A. Hydrogel and organic amendments to increase water retention in anthroposols for land reclamation. *Appl. Environ. Soil Sci.* **2019**, *2019*, 11. [CrossRef]
10. Sim, D.H.H.; Tan, I.A.W.; Lim, L.L.P.; Hameed, B.H. Encapsulated biochar-based sustained release fertilizer for precision agriculture: A review. *J. Clean. Prod.* **2021**, *303*, 127018. [CrossRef]
11. Ibrahim, M.; Abd-Eladl, M.; Abou-Baker, N.H. Lignocellulosic biomass for the preparation of cellulose-based hydrogel and its use for optimizing water resources in agriculture. *J. Appl. Polym. Sci.* **2015**, *132*, 42. [CrossRef]
12. Michalik, R.; Wandzik, I. A mini-review on chitosan-based hydrogels with potential for sustainable agricultural applications. *Polymers* **2020**, *12*, 2425. [CrossRef] [PubMed]
13. Wilske, B.; Bai, M.; Lindenstruth, B.; Bach, M.; Rezaie, Z.; Frede, H.G.; Breuer, L. Biodegradability of a polyacrylate superabsorbent in agricultural soil. *Environ. Sci. Pollut. Res.* **2014**, *21*, 9453–9460. [CrossRef] [PubMed]
14. Cannazza, G.; Cataldo, A.; De Benedetto, E.; Demitri, C.; Madaghiele, M.; Sannino, A. Experimental assessment of the use of a novel superabsorbent polymer (SAP) for the optimization of water consumption in agricultural irrigation process. *Water* **2014**, *6*, 2056–2069. [CrossRef]
15. Chang, L.; Xu, L.; Liu, Y.; Qiu, D. Superabsorbent polymers used for agricultural water retention. *Polym. Test.* **2021**, *94*, 107021. [CrossRef]
16. Shahid, S.A.; Ansar, A.A.; Anwar, F.; Ullah, I.; Rashid, U. Improvement in the water retention characteristics of sandy loam soil using a newly synthesized poly(acrylamide-co-acrylic acid)/AlZnFe$_2$O$_4$ superabsorbent hydrogel nanocomposite material. *Molecules* **2021**, *17*, 9397–9412. [CrossRef]
17. El-Rehim, H.A.; Hegazy, E.S.A.; El-Mohdy, H.A. Radiation synthesis of hydrogels to enhance sandy soils water retention and increaseplant performance. *J. Appl. Polym. Sci.* **2004**, *93*, 1360–1371. [CrossRef]
18. Kapłan, M.; Lenart, A.; Klimek, K.; Borowy, A.; Wrona, D.; Lipa, T. Assessment of the possibilities of using cross-linked polyacrylamide (Agro Hydrogel) and preparations with biostimulation in building the quality potential of newly planted apple trees. *Agronomy* **2021**, *11*, 125. [CrossRef]
19. Rudzinski, W.E.; Dave, A.M.; Vaishnav, U.H.; Kumbar, S.G.; Kulkarni, A.R.; Aminabhavi, T.M. Hydrogels as controlled release devices in agriculture. *Des. Monomers Polym.* **2002**, *5*, 39–65. [CrossRef]
20. Demitri, C.; Scalera, F.; Madaghiele, M.; Sannino, A.; Maffezzoli, A. Potential of cellulose-based superabsorbent hydrogels as water reservoir in agriculture. *Int. J. Polym. Sci.* **2013**, *12*, 1–6. [CrossRef]
21. Billah, S.M.R.; Mondal, M.I.H.; Somoal, S.H.; Pervez, M.N.; Haque, M.O. Enzyme-responsive hydrogels. In *Cellulose-Based Superabsorbent Hydrogels*; Mondal, M., Ed.; Cellulose Polymers and Polymeric Composites: A Reference Series; Springer: Berlin/Heidelberg, Germany, 2019; pp. 309–330.
22. Guilherme, M.R.; Aouada, F.A.; Fajardo, A.R.; Martins, A.F.; Paulino, A.T.; Davi, M.F.T.; Rubira, A.F.; Muniz, E.C. Superabsorbent hydrogels based on polysaccharides for application in agriculture as soil conditioner and nutrient carrier: A review. *Eur. Polym. J.* **2015**, *72*, 365–385. [CrossRef]

23. Warkar, S.G.; Kumar, A. Synthesis and assessment of carboxymethyl tamarind kernel gum based novel superabsorbent hydrogels for agricultural applications. *Polymer* **2019**, *182*, 7.
24. Kruif, C.G.; Anema, S.G.; Zhu, C.; Havea, P. Water holding capacity and swelling of casein hydrogels. *Food Hydrocoll.* **2015**, *44*, 372–379. [CrossRef]
25. Mariotti, M.; Fratini, F.; Cerri, D.; Andreuccetti, V.; Giglio, R.; Angeletti, F.G.S.; Turchi, B. Use of fresh scotta whey as an additive for alfalfa silage. *Agronomy* **2020**, *10*, 365. [CrossRef]
26. Zotta, T.; Solieri, L.; Iacumin, L.; Picozzi, C.; Gullo, M. Valorization of cheese whey using microbial fermentations. *Appl. Microbiol. Biotechnol.* **2020**, *104*, 2749–2764. [CrossRef]
27. Akay, A.; Sert, D. The effects of whey application on the soil biological properties and plant growth. *Eurasian Soil Sci.* **2020**, *9*, 349–355. [CrossRef]
28. Papademas, P.; Kotsaki, P. Technological utilization of whey towards sustainable exploitation. *J. Adv. Dairy Res.* **2021**, *7*, 231.
29. Rocha-Mendoza, D.; Kosmerl, E.; Krentz, A.; Zhang, L.; Badiger, S.; Miyagusuku-Cruzado, G.; Mayta-Apaza, A.; Giusti, M.; Jimenez-Flores, R.; Garcia-Cano, I. Acid whey trends and health benefits. *J. Dairy Sci.* **2021**, *104*, 1262–1275. [CrossRef]
30. Zadona, E.; Blažić, M.; Jambrak, A.R. Whey utilization: Sustainable uses and environmental Approach. *Food Technol. Biotechnol.* **2021**, *59*, 147–161. [CrossRef]
31. Peterson, A.E.; Walker, W.G.; Watson, K.S. Effect of whey applications on chemical properties of soils and crops. *J. Agric. Food. Chem.* **1979**, *27*, 654–658. [CrossRef]
32. Watson, K.S.; Peterson, A.E.; Powell, R.D. Benefits of spreading whey on agricultural land. *J. Water Pollut. Control Fed.* **1977**, *49*, 24–34.
33. Grosu, L.; Fernandez, B.; Grigoras, C.G.; Patriciu, O.I.; Grig-Alexa, I.C.; Nicuta, D.; Ciobanu, D.; Gavrila, L.; Finaru, A.L. Valorization of whey from diary industry for agricultural use as fertilizer: Effects on plant germination and growth. *Environ. Eng. Manag. J.* **2012**, *11*, 2203–2210. [CrossRef]
34. Bielská, L.; Hovorková, I.; Kuta, J.; Machát, J.; Hofman, J. The variability of standard artificial soils: Cadmium and phenanthrene sorption measured by a batch equilibrium method. *Ecotoxicol. Environ. Saf.* **2017**, *135*, 17–23. [CrossRef] [PubMed]
35. Halecki, W.; Stachura, T. Evaluation of soil hydrophysical parameters along a semiurban small river: Soil ecosystem services for enhancing water retention in urban and suburban green areas. *Catena* **2021**, *196*, 104910. [CrossRef]
36. Durpekova, S.; Filatova, K.; Cisar, J.; Ronzova, A.; Kutalkova, E.; Sedlarik, V. A novel hydrogel based on renewable materials for agricultural application. *Int. J. Polym. Sci.* **2020**, *2020*, 13. [CrossRef]
37. International Organization for Standardization. Soil Quality—Effects of Pollutants on Earthworms—Part 1: Determination of Acute Toxicity to Eisenia Fetida/Eisenia Andrei. Available online: https://www.iso.org/standard/53527.html (accessed on 15 September 2021).
38. International Organization for Standardization. Soil Quality—Determination of pH. Available online: https://www.iso.org/standard/40879.html (accessed on 15 September 2021).
39. International Organization for Standardization. Soil Quality—Determination of Organic Carbon by Sulfochromic Oxidation. Available online: https://www.iso.org/standard/23140.html (accessed on 15 September 2021).
40. International Organization for Standardization. Soil Quality—Determination of Total Nitrogen—Modified Kjeldahl Method. Available online: https://www.iso.org/standard/19239.html (accessed on 15 September 2021).
41. Mehlich, A. Mehlich 3 soil test extractant: A modification of Mehlich 2 extractant. *Commun. Soil Sci. Plant Anal.* **1984**, *15*, 1409–1416. [CrossRef]
42. International Organization for Standardization. Soil Quality—Determination of the Potential Cation Exchange Capacity and Exchangeable Cations Using Barium Chloride Solution Buffered at pH = 8.1. Available online: https://www.iso.org/standard/22180.html (accessed on 15 September 2021).
43. European Standard. Soil Improvers and Growing Media—Determination of Physical Properties—Dry Bulk Density, Air Volume, Water Volume, Shrinkage Value and Total Pore Space. Available online: https://standards.iteh.ai/catalog/standards/cen/b487afb1-b8ce-4151-a8d8-fdb7959b2e3d/en-13041-2011 (accessed on 15 September 2021).
44. International Organization for Standardization. Soil Quality—Determination of Particle Density. Available online: https://standards.iteh.ai/catalog/standards/cen/a540c124-0bd3-4c72-8152-b4d18b7c397c/en-iso-11508-2017 (accessed on 15 September 2021).
45. International Organization for Standardization. Soil Quality—Determination of Dry Bulk Density. Available online: https://www.iso.org/standard/68255.html (accessed on 15 September 2021).
46. Skrzypczak, D.; Mikula, K.; Kossińska, N.; Widera, B.; Warchoł, J.; Moustakas, K.; Chojnacka, K.; Witek-Krowiak, A. Biodegradable hydrogel materials for water storage in agriculture—Review of recent research. *Desalin. Water Treat.* **2020**, *194*, 324–332. [CrossRef]
47. Pillai, C.K.S. Challenges for natural monomers and polymers: Novel design strategies and engineering to develop advanced polymers. *Des. Monomers Polym.* **2010**, *13*, 87–121. [CrossRef]
48. Ni, B.; Liu, M.; Lü, S.; Xie, L.; Wang, Y. Environmentally friendly slow-release nitrogen fertilizer. *J. Agric. Food. Chem.* **2011**, *59*, 10169–10175. [CrossRef]
49. Agaba, H.; Orikiriza, L.J.B.; Esegu, J.F.O.; Obua, J.; Kabasa, J.D.; Hüttermann, A. Effects of hydrogel amendment to different soils on plant available water and survival of trees under drought conditions. *Clean Soil Air Water* **2010**, *38*, 328–335. [CrossRef]

50. Akhter, J.; Mahmood, M.; Malik, K.A.; Mardan, A.; Ahmad, M.; Iqbal, M.M. Effects of hydrogel amendment on water storage of sandy loam and loam soils and seedling growth of barley, wheat and chickpea. *Plant Soil Environ.* **2004**, *50*, 463–469. [CrossRef]

51. Ritonga, H.; Basri, M.I.; Rembon, F.S.; Ramadhan, L.O.A.N.; Nurdin, M. High performance of chitosan-co-polyacrylamide-TiO$_2$ crosslinked glutaraldehyde hydrogel as soil conditioner for soybean plant (Glycine max). *Soil Sci. Ann.* **2020**, *71*, 194–204. [CrossRef]

52. Song, B.; Liang, H.; Sun, R.; Peng, P.; Jiang, Y.; She, D. Hydrogel synthesis based on lignin/sodium alginate and application in agriculture. *Int. J. Biol. Macromol.* **2020**, *144*, 219–230. [CrossRef] [PubMed]

53. Graber, E.R.; Fine, P.; Levy, G.J. Soil stabilization in semiarid and arid land agriculture. *J. Mater. Civ. Eng.* **2006**, *18*, 190–205. [CrossRef]

54. Robbins, C.W.; Lehrsch, G.L. Cottage cheese whey effects on sodic soils. *Arid Soil Res. Rehab.* **1992**, *6*, 127–134. [CrossRef]

55. Berry, R.A. The production, composition and utilisation of whey. *J. Agric. Sci.* **1923**, *13*, 192–239. [CrossRef]

56. Radford, J.B.; Galpin, D.B.; Parkin, M.F. Utilization of whey as a fertilizer replacement for dairy pasture. *N. Zealand J. Dairy Sci.* **1986**, *21*, 65–72.

57. Sharratt, W.J.; Peterson, A.E.; Caibek, H.E. Effect of whey on soil and plant growth. *Agron. J.* **1962**, *54*, 359–361. [CrossRef]

58. Ketterings, Q.; Czymmek, K.; Gami, S.; Godwin, G.; Ganoe, K. Guidelines for Land Application of Acid Whey. Available online: http://nmsp.cals.cornell.edu/publications/files/AcidWheyGuidelines2017.pdf (accessed on 3 August 2021).

59. Robbins, C.W.; Hansen, C.L.; Roginske, M.F.; Sorensen, D.L. Bicarbonate extractable K and Soluble Ca, Mg, Na, and K status of two calcareous soils treated with whey. *J. Environ. Qual.* **1996**, *25*, 791–795. [CrossRef]

60. Östberg, T.L.; Jonsson, A.P.; Bylund, D.; Lundström, U.S. The effects of carbon sources and micronutrients in fermented whey on the biodegradation of n-hexadecane in diesel fuel contaminated soil. *Int. Biodeterior. Biodegrad.* **2007**, *60*, 334–341. [CrossRef]

61. Konzen, E.R.; Navroski, M.C.; Friederichs, G.; Ferrari, L.H.; Pereira, M.D.O.; Felippe, D. The use of hydrogel combined with appropriate substrate and fertilizer improve quality and growth performance of *Mimosa scabrella* benth seedlings. *Cerne* **2017**, *23*, 473–482. [CrossRef]

62. Wu, L.; Liu, M. Preparation and properties of chitosan-coated NPK compound fertilizer with controlled-release and water-retention. *Carbohydr. Polym.* **2008**, *72*, 240–247. [CrossRef]

63. Noppakundilograt, S.; Pheatcharat, N.; Kiatkamjornwong, S. Multilayer-coated NPK compound fertilizer hydrogel with controlled nutrient release and water absorbency. *J. Appl. Polym. Sci.* **2015**, *132*, 41249. [CrossRef]

64. Abdallah, A. The effect of hydrogel particle size on water retention properties and availability under water stress. *Int. Soil Water Conserv. Res.* **2019**, *7*, 275–285. [CrossRef]

65. Johnson, M.S. Effect of soluble salts on water absorption by gel-forming soil conditioners. *J. Sci. Food Agric.* **1984**, *35*, 1196–1200. [CrossRef]

66. Abedi-Koupai, J.; Sohrab, F.; Swarbrick, G. Evaluation of hydrogel application on soil water retention characteristics. *J. Plant Nutr.* **2008**, *31*, 317–331. [CrossRef]

67. Lentz, R.D. Polyacrylamide and biopolymer effects on flocculation, aggregate stability and water seepage in a silt loam. *Geoderma* **2015**, *241*, 289–294. [CrossRef]

 *sustainability*

Article

Irrigation with Coalbed Methane Co-Produced Water Reduces Forage Yield and Increases Soil Sodicity However Does Not Impact Forage Quality

Shital Poudyal [1,*] and Valtcho D. Zheljazkov [2,*]

[1] Department of Plant and Soil Science, Texas Tech University, Lubbock, TX 79409, USA
[2] Department of Crop and Soil Science, Oregon State University, Corvallis, OR 97331, USA
* Correspondence: shpoudya@ttu.edu (S.P.); valtcho.jeliazkov@oregonstate.edu (V.D.Z.)

Abstract: The extraction of coalbed methane produces a significant amount of coalbed methane co-produced water (CBMW). Coalbed methane co-produced water is often characterized by high levels of pH, total dissolved solids (TDS), sodium (Na) and bicarbonate (HCO^{-3}) and if used for irrigation without treatment, it may be detrimental to the surrounding soil, plants and environment. CBMW ideally should be disposed of by reinjection into the ground, but because of the significant cost associated, CBMW is commonly discharged onto soil or water surfaces. This study was conducted to elucidate the effect of the CBMW (with TDS value of <1500 ppm) at various blending ratios with fresh water on the yield and quality of representative forage crops [i.e., oat (*Avena sativa*) and alfalfa (*Medicago sativa*)]. Various blends of CBMW with fresh water reduced fresh and dry weight of alfalfa by 21.5–32% and 13–30%, respectively, and fresh and dry weight of oat by 0–17% and 0–14%, respectively. Irrigation with various blends of CBMW and fresh water increased soil pH and soil sodium adsorption ratio. However, forage quality parameters such as crude protein (CP), acid detergent fiber (ADF), neutral detergent fiber (NDF), total digestible nutrients (TDN) and relative feed value (RFV) of both forage crops remained unaffected.

Keywords: sodium adsorption ratio; relative feed value, forage nutritive value; oat; alfalfa; forage crops; alternative water source

Citation: Poudyal, S.; Zheljazkov, V.D. Irrigation with Coalbed Methane Co-Produced Water Reduces Forage Yield and Increases Soil Sodicity However Does Not Impact Forage Quality. *Sustainability* 2021, 13, 3545. https://doi.org/10.3390/su13063545

Academic Editor: Bharat Sharma Acharya

Received: 17 February 2021
Accepted: 19 March 2021
Published: 23 March 2021

Publisher's Note: MDPI stays neutral with regard to jurisdictional claims in published maps and institutional affiliations.

1. Introduction

Coalbed methane is a natural gas produced in underground coal seams through a biogeophysical process and is subsequently extracted by pumping water through the seams. Pumping of water lowers the pressure in the seam to release trapped methane, this gas then moves towards the surface and is collected by a separation technique [1,2]. The resulting pumped water, called coalbed methane co-produced water (CBMW), is a major byproduct of coalbed methane extraction. This discharged water typically has high pH, contains elevated concentrations of sodium (Na), bicarbonates (HCO_3) and total dissolved solids (TDS) and also has a high sodium adsorption ratio (SAR). Trace amounts of barium (Ba), boron (B), sulfate, calcium (Ca), chlorine (Cl), magnesium (Mg) and potassium (K) can also be detected in this water [2–4]. Nonetheless, concentrations of the constituents, as mentioned above, vary by wells and locations [2]. Water with a high level of those substances can be detrimental to the soil, plants and the environment.

Coalbed methane production is rising globally and the U.S has the largest coal reserves in the world [5]. In 2017, the U.S. produced approximately 27.8 billion m^3 of coalbed methane that will likely continue in the future as the U.S. hold 3.4 trillion m^3 of economically recoverable coalbed methane reserves [6,7]. In the same year, the production of CBMW was estimated to be somewhere from 77 to 428 million m^3 [7,8]. The state of Wyoming has the highest water to gas discharge ratio for coalbed methane extraction compared to other wells in the U.S. and thus, produce a significant amount of CBMW [7,8]. The best method to dispose CBMW is through reinjection into the ground. However, the cost associated with this method of disposal is very high; therefore, the Environmental Protection Agency (EPA) has relaxed its guidelines for CBMW disposal [9], leading to CBMW being disposed of on soil or surface waters.

Arid zones are in dire need of continuous fresh water supply for irrigation during the production season. In the U.S., even if the arid states have enough freshwater resources, existing water laws of interstate stream agreements do not allow that state to use its water resources completely. Instead, it has to provide a certain amount of water to other nearby states. Hence, states such as Wyoming experience freshwater shortages [10–12]. To address freshwater shortages and to increase the use of alternative water sources, good quality CBMW could potentially be used for the irrigation of forage crops [13]. However, studies on exploring the use of CBMW for irrigation of forage crops are limited. In some of the few related studies, irrigation with a mixture of CBMW and fresh water in specific ratios was found to be as equally effective as irrigation with tap water alone [14,15]. In addition, irrigation of dill (*Anethum graveolens*) with different ratios of CBMW and fresh water did not affect plant height and plant weight [16]. Thus, there remains the possibility of using CBMW as a source of irrigation for forages.

The majority of farmers in Wyoming grow forages; the state produced approximately 2.5 million tons of hay (forage) in 2017, of which 63.5% was alfalfa [17]. According to a six-year report (2003–2008), a 50% decrease in the yield of alfalfa was observed on non-irrigated land compared to irrigated land [18]. Alfalfa and oat both can tolerate moderate salinity. Alfalfa can also tolerate sodicity, while oat is semi-tolerant to sodicity [19,20] and these are usually considered representative forage crops. Hence, these forage crops make good candidates for production in saline-sodic soil. If irrigation with CBMW alone or at various blend ratios with fresh water can sustain forage yield and quality without impairing soil characteristics, CBMW disposal as irrigation can be well justified. In addition, CBMW produced in the state of Wyoming is of better quality compared to the national average, with TDS value of <1500 ppm and pH value of <8.9 [2]. Therefore, we conducted a greenhouse study to explore the use of CBMW at various blend ratios with fresh water on oat and alfalfa. Fresh water in our study refers to potable water supplied by the Sheridan municipal water plant. The objective of this study was to determine the effects of CBMW irrigation treatments on yield and quality of those forage crop, while also outlining changes in soil characteristics.

2. Materials and Methods

2.1. Plant Materials and Growing Conditions

This experiment was conducted at the University of Wyoming's Sheridan Research and Extension Center in Sheridan, Wyoming, from 3 October 2014 to 10 March 2015. Coalbed methane co-produced water used in this study was hauled each week from BeneTerra LLC in Sheridan, WY, USA. The quality of CBMW was analyzed and the results are presented in Table 1. Alfalfa and oat were grown in a greenhouse, which allowed us to control external conditions and make precise treatment applications. The temperature in the greenhouse was set to 26 °C during the day and 18 °C during the night. The relative humidity ranged from 50 to 60%. These are the optimal growth conditions for both oat and alfalfa [21–23]. Cultivar (cv.) Monida of oat (*Avena sativa*) and cv. WL363HQ of alfalfa (*Medicago sativa*) were used in this study. Certified seeds were obtained from the Sheridan Research and Extension Center and sowed (1.2 cm deep) in plastic containers (11.4 L volume, 25 cm deep and 28 cm diameter) on 3 October 2014 and left to germinate. Germination percentage for oat and alfalfa was 92 and 90, respectively. A Hargreave–Moskee sandy loam was used in this container study, as described in Web Soil Survey [24] Each container was filled with 10 kg of clay loam soil that was high in organic matter (Table 2). Soil was homogenized by digging 15 cm top layer of a field soil and then mixing the soil before filling in the pots. The experiment was carried out in six replications without initial fertilization. On 10 December 2014, 5 g of 20:20:20 general-purpose fertilizer (Jack's

Professional, JR Peters. Inc, Allentown, PA, USA) was applied to both crops mixed with irrigation water. An additional 5 g of fertilizer/pot application applied on 15 January 2015 for alfalfa only. Before complete germination and emergence, irrigation was carried out manually by pouring 600 mL of fresh water into each pot every three days until 13 October 2014. Complete emergence of the oat and alfalfa was observed on 13 October 2014 after which the irrigation treatment was started. The irrigation treatments contained the following mixture of CBMW and fresh water: 0% CBMW (fresh water only); 25% CBMW (25% CBMW and 75% fresh water); 50% CBMW (50% CBMW and 50% fresh water); 75% CBMW (75% CBMW and 25% fresh water); and 100% CBMW. Each pot received a 300 mL aliquot of either 0, 25, 50, 75 or 100% CBMW that was hand-applied every day with few adjustments in water usage based on sunny and cloudy days. Plants in each pot were thinned to 12 plants per container on 31 October 2014 and pots were frequently moved around the greenhouse to maintain uniformity and randomization.

Table 1. Properties of pure coal bed methane co-produced water (CBMW) and fresh water used in greenhouse study at Sheridan, WY, USA, for irrigating oat and alfalfa.

General Parameters	CBMW	Fresh Water	Units	Method Used
pH	8.83	7.8	s.u.	SM 4500 H B
Electrical Conductivity	2.1	0.2	dS/m	SM 2510B
Total Dissolved Solids (180)	1300	77	mg/L	SM 2540
Alkalinity, Total (As CaCO3)	1193	28	mg/L	SM 2320B
Hardness, Calcium/Magnesium (As CaCO3)	16.3	28	mg/L	SM 2340B
Nitrogen, Ammonia (As N)	ND	NA	mg/L	EPA 350.1
Sodium Adsorption Ratio	59.8	1.2		Calculation
Anions				
Alkalinity, Bicarbonate as HCO3	1280	34	mg/L	SM 2320B
Alkalinity, Carbonate as CO3	86	NA	mg/L	SM 2320B
Chloride	7	17	mg/L	EPA 300.0
Nitrate+Nitrite as N	ND	ANC	mg/L	EPA 300.0
Sulfate	ND	9	mg/L	EPA 300.0
Cations				
Calcium	4	8	mg/L	EPA 200.7
Magnesium	1.3	2	mg/L	EPA 200.7
Sodium	555	14	mg/L	EPA 200.7
Cation/Anion-Milliequivalents				
Bicarbonate as HCO3	21	ANC	meq/L	SM 1030E
Carbonate as CO3	2.873	ANC	meq/L	SM 1030E
Hydroxide as OH	ND	ANC	meq/L	SM 1030E
Chloride	0.196	ANC	meq/L	SM 1030E
Fluoride	ND	0.7	meq/L	SM 1030E
Nitrate + Nitrite as N	ND	ANC	meq/L	SM 1030E
Sulfate	ND	ANC	meq/L	SM 1030E
Calcium	0.19	ANC	meq/L	SM 1030E
Magnesium	0.12	ANC	meq/L	SM 1030E
Sodium	24.2	ANC	meq/L	SM 1030E
Cation/Anion Balance				
Cation Sum	24.5	ANC	meq/L	SM 1030E
Anion Sum	24.1	ANC	meq/L	SM 1030E
Cation-Anion Balance	0.87	ANC	%	SM 1030E
Radiochemistry				
Radium 226 (Dissolved)	0.3 ± 0.2	ANC	pCi/L	SM 7500-Ra B
Dissolved Metals				
Aluminum	ND	ANC	mg/L	EPA 200.7
Antimony	ND	ANC	mg/L	EPA 200.8
Arsenic	ND	ANC	mg/L	EPA 200.8
Barium	0.33	ANC	mg/L	EPA 200.8
Beryllium	ND	ANC	mg/L	EPA 200.7
Boron	0.2	ANC	mg/L	EPA 200.7
Cadmium	ND	ANC	mg/L	EPA 200.8
Chromium	ND	ANC	mg/L	EPA 200.7
Copper	ND	0.27	mg/L	EPA 200.8
Iron	ND	ANC	mg/L	EPA 200.7
Lead	ND	ANC	mg/L	EPA 200.8
Manganese	ND	ANC	mg/L	EPA 200.7

Table 1. *Cont.*

General Parameters	CBMW	Fresh Water	Units	Method Used
Mercury	ND	ANC	mg/L	EPA 245.1
Nickel	ND	ANC	mg/L	EPA 200.7
Phosphorus	ND	ANC	mg/L	EPA 200.7
Selenium	ND	ANC	mg/L	EPA 200.8
Zinc	ND	ANC	mg/L	EPA 200.7

ND = Not Detected, ANC = Analysis not conducted.

Table 2. pH, organic matter (OM), electrical conductivity (EC), sodium adsorption ratio (SAR), potassium, calcium, magnesium and sodium concentration of soil used for alfalfa and oat study.

pH	OM	EC	SAR	Potassium	Calcium	Magnesium	Sodium
	%	(dS/m)				ppm	
7.6	4.36	3.74	0.4	799	2667	770	92

2.2. Harvesting and Drying

Crops were harvested at commercial maturity; oat was harvested once on 30 November 2014 and alfalfa thrice on 7 January 2015, 5 February 2015 and on 10 March 2015. Alfalfa was harvested when 10–15% of the plants started flowering and oat was harvested at the early milk stage, as these stages are optimal harvesting stage for highest forage nutritive value [25,26]. Crops were hand harvested with clippers at 5-6 cm above the soil surface and fresh weights were recorded immediately. Both crops were then packed in paper bags and dried in an oven for 24 h at 65 °C to determine dry weight. Samples were then sent to commercial, National Forage Testing Association (NFTA), certified Laboratory (American Agricultural Laboratory, Inc, McCook, NE, USA) for analysis of forage nutritive value. All the nutritive values parameters were determined on dry matter. For alfalfa, only the final harvest samples were sent for analysis of forage nutritive value. After completion of the study, soil samples from the top 15 cm were also sent to the National Forage Testing Association (NFTA) certified Laboratory (American Agricultural Laboratory, Inc., McCook NE, http://www.olsenlab.com/ accessed on 18 March 2021) for analysis.

2.3. Statistical Analysis

The study involved five different CBMW treatments with six replications per treatment and two crop types. All plants were randomly arranged in the greenhouse. A two-factor factorial analysis of variance was performed for CBMW treatments, crop types and CBMW x crop interaction. Coalbed methane co-produced water treatments, crop types and CBMW x crop interaction were significant (p-value < 0.05) for pH, Na, SAR, total fresh weight and total dry weight. Hence, separate statistical analyses were performed by plant types using the one-way analysis of variance. Forage nutrient quality parameters only differed by plant type (p-value < 0.05), hence, all those parameters were pooled for various CBMW treatments. All the statistical analysis were conducted in JMP®, Version 14.0.0 (SAS Institute Inc., Cary, NC, USA). Post-hoc mean comparisons were performed using the Fisher's least significant difference test at p-value < 0.05. Regression analysis was performed on data for fresh and dry weight. Models were verified by checking scatter plots of residuals. Both fresh and dry weight of alfalfa followed quadratic regression model while fresh and dry weight of oat followed simple linear model.

$$Y = a + bX \text{ (Linear model); Oat}$$

where Y is either the fresh weight or dry weight of oat, a is the intercept at Y axis, b is the slope of the line and X is the percentage of CBMW.

$$Y = a + b_1X + b_2X^2 \text{ (Quadratic model); Alfalfa}$$

where Y is either the fresh weight or dry weight of alfalfa, a is the intercept at Y axis, b_1 and b_2 are the slope of the curve and X is the percentage of CBMW.

3. Results

3.1. Alfalfa Yield and Quality

The application of CBMW reduced fresh weight and dry weight of alfalfa, while its forage quality remained unchanged. Alfalfa was harvested three times and the combined fresh weight and dry weight for all three harvests were highest (102.2 g and 22.9 g per pot, respectively) at the 0% CBMW treatment and lowest (67.9 g and 16.5 g per pot, respectively) with the 100% CBMW treatment (Figure 1). Both fresh weight and dry weight gradually decreased with increasing levels of CBMW treatment until 50% CBMW treatment and then plateaued. Coalbed methane co-produced water treatments did not alter forage characteristics such as crude protein, acid detergent fiber, neutral detergent fiber, total digestible nutrients, net energy for maintenance, net energy for lactation, net energy gain and relative feed value (Table 3).

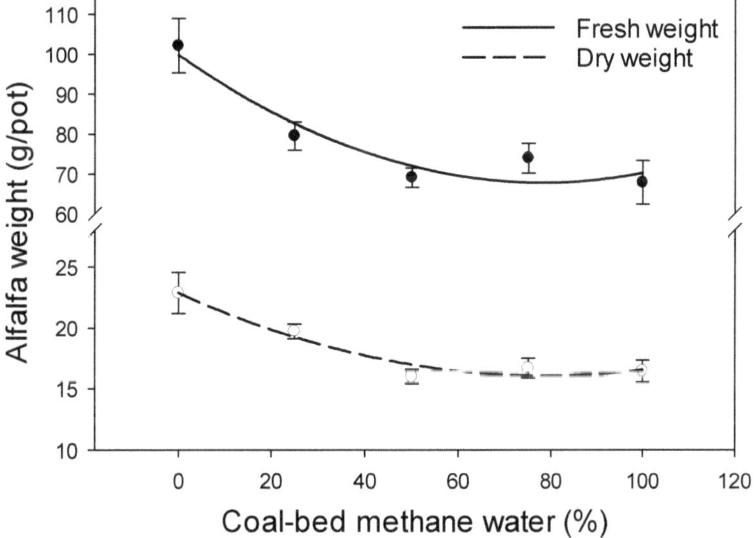

Figure 1. Mean of total fresh weight and dry weight of alfalfa (*Medicago sativa*) for six replicates irrigated with various percentages of coalbed methane co-produced water (CBMW). Standard errors of each mean are drawn as capped lines extended vertically from each mean. Irrigation treatments contained the following mixture of CBMW and fresh water: 0% CBMW (fresh water only); 25% CBMW (25% CBMW and 75% fresh water); 50% CBMW (50% CBMW and 50% fresh water); 75% CBMW (75% CBMW and 25% fresh water); and 100% CBMW. Plants received CBMW treatments for 150 days. Both fresh weight (r-square = 0.54) and dry weight (r-square = 0.55) followed quadratic regression curve with *p*-value of <0.0001. $Y = 100.3 - 0.085x + 0.0055X^2$ and $Y = 23 - 0.181x + 0.0012X^2$ represents equation for fresh weigh and dry weight, respectively; where Y is total weight and x is the % of CBMW.

Table 3. Mean value of crude protein (CP), acid detergent fiber (ADF), neutral detergent fiber (NDF), total digestible nutrients (TDN), net energy lactation (NEL), net energy maintenance (NEM), net energy gain (NEG) and relative feed value (RFV) of alfalfa (*Medicago sativa*) and oat (*Avena sativa*) for six replicates.

Crop type	CP	ADF	NDF	TDN	NEL	NEM	NEG	RFV
			—% of DM—			—(Mj/kg)—		
Alfalafa	28.5a	22.5b	26.8b	74.6a	7.1a	7.4a	4.8a	248.4a
Oat	25.1b	26.2a	42.4a	72.9b	7b	7.2b	4.6b	151b
				ANOVA *p*-value				
Crop	<0.0001	<0.0001	<0.0001	<0.01	<0.01	<0.01	<0.01	<0.0001
CBMW	NS	NS	NS	NS	NS	NS	NS	NS
CBMW x Crop	NS	NS	NS	NS	NS	NS	NS	NS

Mj/kg = Mega joules/kilogram, DM = dry matter, NS = Not significant at *p*-value of <0.05. Mean separations for forage quality parameters were carried out using Fisher's least significant difference test at *p*-values < 0.05. Means within each column followed by the same letters are not significantly different at given *p*-values of 0.05.

Increasing concentration of CBMW treatments increased soil pH. Soil pH was lowest (7.33) at 0% CBMW treatment and increased with increasing levels of CBMW treatments and was highest (pH 9.03) at the 100% CBMW treatments. Coalbed methane co-produced water application did not alter electrical conductivity (EC) or organic matter (OM) of soil with alfalfa. However, SAR and Na concentration in the soil increased drastically from 0.64 and 154.3 ppm, respectively, with 0% CBMW treatment to 12.37 and 2940 ppm, respectively with 100% CBMW treatment. The concentrations of Mg and Ca were highest (867.67 ppm and 2170 ppm, respectively) at 0% CBMW treatment, that slightly decreased with 50% CBMW and then did not change (Table 4).

Table 4. Mean of pH, organic matter (OM), electrical conductivity (EC), sodium adsorption ratio (SAR), potassium, calcium, magnesium and sodium of soils irrigated with various percentages of coalbed methane co-produced water (CBMW) at the end of alfalfa (*Medicago sativa*) study.

CBMW	pH	OM	EC	SAR	Potassium	Calcium	Magnesium	Sodium
		%	(dS/m)			— ppm —		
0	7.33cb	4.10	5.56	0.64d	1042.0	2940.0a	867.7a	154.3e
25	8.16b	3.93	7.27	3.42c	844.7	2890.0ab	781.3ab	806.3d
50	8.20b	3.96	7.08	6.00c	1098.0	2500.0bc	664.0bc	1311.0c
75	8.56ab	3.60	7.20	9.45b	891.7	2306.7c	642.0bc	1985.0b
100	9.03a	4.00	5.75	12.37a	1018.0	2170.0c	591.0c	2514.7a
p-value	<0.001	NS	NS	<0.001	NS	<0.01	<0.05	<0.0001

dS/m = decisiemens per meter; ppm = parts per million; NS = Not significant at *p*-value of <0.05.

Irrigation treatments contained the following mixture of CBMW and fresh water: 0% CBMW (fresh water only); 25% CBMW (25% CBMW and 75% fresh water); 50% CBMW (50% CBMW and 50% fresh water); 75% CBMW (75% CBMW and 25% fresh water); and 100% CBMW. Plants received CBMW treatments for 150 days. Mean separations were carried out using Fisher's least significant difference test at *p*-values < 0.05. Means within each column followed by the same letters are not significantly different at given *p*-values of 0.05.

3.2. Oat Yield and Quality

Similar to alfalfa, CBMW treatments influenced fresh weight and dry weight of oat, but its forage quality remained unchanged. The fresh weight and dry weight were highest (99.3 g and 18.0 g, respectively) for the 25% CBMW treatment and were lowest (78.2 g and 15.4 g, respectively) for the 100% CBMW treatment (Figure 2). Both fresh yield and dry weight showed gradual decreasing trends with increasing levels of CBMW treatment. Similar to the forage quality parameters of alfalfa, there was no significant effect of CBMW treatments on forage quality parameters (Table 3).

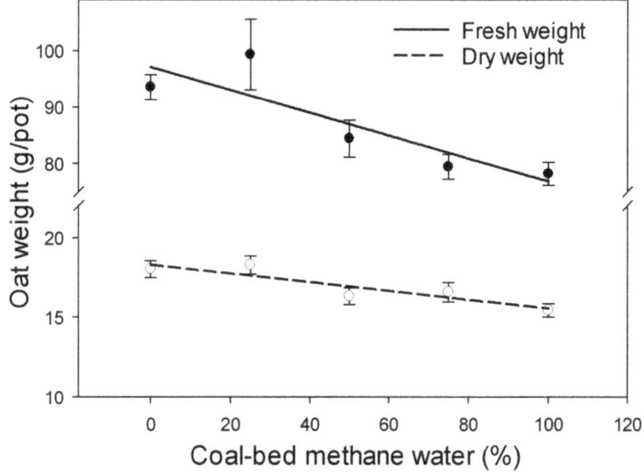

Figure 2. Mean of total fresh weight and dry weight of oat (*Avena sativa*) for six replicates irrigated with various percentages of coal-bed methane co-produced water (CBMW). Standard errors of each mean are drawn as capped lines extended vertically from each mean. Irrigation treatments contained the following mixture of CBMW and fresh water: 0% CBMW (fresh water only); 25% CBMW (25% CBMW and 75% fresh water); 50% CBMW (50% CBMW and 50% fresh water); 75% CBMW (75% CBMW and 25% fresh water); and 100% CBMW. Plants received CBMW treatments for 45 days. Both fresh weight (r-square = 0.4) and dry weight (r-square = 0.37) followed linear regression curve with p-value of <0.0001. $Y = 97.1 - 0.2x$ and $Y = 18.3 - 0.027x$ represents equation for fresh weigh and dry weight, respectively; where Y is total weight and x is the % of CBMW.

The CBMW treatments altered soil characteristics in pots with oats. Soil pH increased from 7.53 at 0% CBMW to 8.10 at 100% CBMW treatments. Electrical conductivity was similar at the 0, 25, 50 and 75% CBMW treatments but was higher at the 100% CBMW treatment. The SAR and Na concentrations were lowest (0.62 and 152 ppm, respectively) for the 0% CBMW treatment and drastically increased (3.38 and 745 ppm, respectively) for the 100% CBMW treatment. Calcium concentration decreased with increasing levels of CBMW treatments, but the concentrations of K and Mg remained unchanged (Table 5).

Table 5. Mean of pH, organic matter (OM), electrical conductivity (EC), sodium adsorption ratio (SAR), potassium, calcium, magnesium and sodium of soil irrigated with various percentage coalbed methane co-produced water (CBMW) at the end of oat (*Avena sativa*) study.

CBMW	pH	OM	EC	SAR	Potassium	Calcium	Magnesium	Sodium
		%	(dS/m)			ppm		
0	7.53c	4.13	3.38b	0.6d	826.3	3020.0a	855.6	152.0d
25	7.73bc	3.83	3.77b	1.5c	843.0	3046.7a	808.8	365.3c
50	7.90ab	3.83	3.43b	2.0bc	993.0	2920.0a	771.7	477.7bc
75	8.03a	3.53	3.12b	2.4b	686.7	2803.0ab	762.9	554.0b
100	8.10a	3.90	4.74a	3.4a	1121.0	2506.7b	693.8	745.0a
p-value	<0.01	NS	<0.01	<0.001	NS	<0.05	NS	<0.0001

dS/m = decisiemens per meter; ppm = parts per million; NS = Not significant at p-value of < 0.05. Irrigation treatments contained the following mixture of CBMW and fresh water: 0% CBMW (fresh water only); 25% CBMW (25% CBMW and 75% fresh water); 50% CBMW (50% CBMW and 50% fresh water); 75% CBMW (75% CBMW and 25% fresh water); and 100% CBMW. Plant received CBMW treatments for 45 days. Mean separations were carried out using Fisher's least significant difference test at p-values < 0.05. Means within each column followed by the same letters are not significantly different at given p-values of 0.05.

4. Discussion

Alfalfa was harvested during the 10–15% blossom stage to optimize forage yield and quality [25]. The maximum fresh and dry weights were obtained with the 0% CBMW treatment. As the level of CBMW application increased, the values of soil pH, Na concentration and SAR increased, resulting in soil that was saline for the 25, 50 and 75% CBMW treatments and saline-sodic for the 100% CBMW treatment [20]. Saline and saline-sodic conditions reduced yield of alfalfa as alfalfa best performs in a soil with pH of 6.5–7.5 [27]. Increasing salinity and sodicity can lower the yield of forage crops including alfalfa [28–30]. In a study performed by Cornacchione and Suarez [30], increasing soil salinity levels proportionally reduced fresh weight of alfalfa. Similarly, the 50, 75 and 100% CBMW treatments in our study also reduced alfalfa fresh yields by 27–30%. Fresh weight is highly correlated with dry weight and this was evident in our study where 50–100% of CBMW treatments reduced dry weight of alfalfa by 28–30%. For alfalfa, dry weight in this study, was 21–25% of the fresh weight, which was slightly lower than the value (26–32%) reported by Khosrowchahli et al. (2013) [31].

Oat was harvested at the start of the early milk stage to obtain desirable forage yield and quality [26]. Forage oat is tolerant to soil salinity with only slight reductions in yield at saline conditions [20,32]. In this study, soil sampling after harvest indicated that the soil was in the category of slightly saline (25, 50 and 75% CBMW) to saline (100% CBMW) conditions [33]. Fresh and dry weights of the forage crop were higher at the 0 and 25% CBMW treatments, which indicated that the oat variety used in this study was tolerant to 25% CBMW treatments; however, CBMW at 50% and above induced salinity stress and reduced fresh and dry yield of oat similar to that found in other studied on oat and alfalfa [34,35]. Chatrath et al. (2000) also found an increase in salinity for various oat varieties reduced photosynthesis rates, which in turn reduced the yield of oat varieties [36]. In another similar study, yield of oat was negatively correlated to soil salinity levels [37].

Crude protein, acid detergent fiber, total digestible nutrients, neutral detergent fiber, net energy for lactation, net energy for maintenance, net energy gain and relative feed value cumulatively represent forage nutritive value and all those forage characteristics remained unchanged across the CBMW treatments, both for oat and alfalfa. Forage nutritive value of alfalfa fell under an excellent category based on forage quality analysis and was not affected by CBMW treatments, which may be due to salt tolerant variety selection and growth in optimum environmental conditions [38–40]. In a similar study, nutrient quality of various forages crops were unaffected by soil salinity, sodicity and minor change in pH, but their yields were reduced [41–43].

For oat, the percentage of total digestible nutrients were more than 71.2% and the relative feed value was more than 140. These values were superior to other reported studies [44,45] and indicate forage of superior quality [26]. Crude protein of oat in our study was very high, which may be explained by early harvesting stages of oat, higher percentage of leaves and calculation of crude protein done for 100% dry matter. Some previous studies have also reported crude protein of oat to be in the range that we observed [46,47].

Soil characteristics under both crops changed with the application of the CBMW treatments, with more significant changes in the alfalfa experiment compared to the oat experiment. Sodium and other salts present in CBMW increased soil pH, SAR and Na concentration. A similar greenhouse study also observed changes in soil characteristics, as was the case in this study [14]. Various other research works conducted to explore the application of CBMW as an alternative source of irrigation have also observed similar trends [15,16,48]. The observed differential response in soil under different crops is most probably due to the fact that alfalfa received CBMW treatments for approximately 150 days, while oat had the treatments for 45 days. Hence, it should be noted that the application of CBMW will significantly change the soil characteristics and therefore, irrigation with CBMW may not be suitable from the perspective of soil quality. We used field soil in our study to accurately determine the consequences of CBMW irrigation in the field; however, greenhouse conditions are different from field conditions. Sodium and other salts accumulated through irrigation with CBMW in the field may leach below the root zone or disperse into the surrounding soil, lowering Na concentration and SAR as seen in a three-year field study [49]. Therefore, long-term field research is needed to identify and explore the full effect of CBMW on soil properties.

5. Conclusions

For alfalfa, both fresh and dry weight decreased with increasing levels of CBMW application and the decrease in fresh weight was between 22 and 33.5%. However, CBMW did not change the forage nutritive value of alfalfa. Thus, the application of CBMW alone or mixed with fresh water may be a suitable alternative for irrigation where fresh water sources are not available. For oat, both fresh and dry weight were very similar at 0 and 25% CBMW, after which the fresh weight started to decrease with increasing levels of CBMW treatments. However, the nutrient value of forage oat was similar in all treatments; therefore, 3:1 blend ratio of CBMW to fresh water could be used for irrigating oat without a significant reduction in yield. There is the possibility of achieving similar results by alternating irrigation with fresh water and CBMW where the dilution is not feasible, however, further research is needed to accurately verify that assumption. In addition, in this study, we grew crops that are slightly to moderately tolerant to saline and sodic conditions and results may differ if sensitive crops are grown using CBMW.

Author Contributions: Conceptualization, investigation, methodology, visualization, writing—original draft preparation and software, S.P.; validation, resources, supervision, funding acquisition, project administration, V.D.Z.; formal analysis, data curation, writing—review and editing, S.P. and V.D.Z. All authors have read and agreed to the published version of the manuscript.

Funding: This research was supported by the University of Wyoming School of Energy funds awarded to Valtcho D. Zheljazkov (Jeliazkov).

Institutional Review Board Statement: Ethical approval was not required by the lead authors institutions.

Informed Consent Statement: Informed consent was obtained from all subjects involved in the study.

Data Availability Statement: Original data is available from the authors.

Acknowledgments: The authors thank Derek Lowe of BeneTerra LLC, in Sheridan, WY, for providing access to coal-bed methane water and Dan Smith the farm manager at the University of Wyoming's Sheridan Research and Extension Center, Sheridan. The article processing charges were paid by the Oregon State University startup funds awarded to Valtcho D. Jeliazkov (Zheljazkov).

Conflicts of Interest: The authors declare no conflict of interest.

References

1. McBeth, I.H.; Reddy, K.J.; Skinner, Q.D. Coalbed methane product water chemistry in three Wyoming watersheds. *J. Am. Water Resour. Assoc.* **2003**, *39*, 575–585. [CrossRef]
2. U.S. EPA. *Coalbed Methane Extraction: Detailed Study Report*; United States Environmental Protection Agency (US EPA): Washington, DC, USA, 2010.
3. Horpestad, A. Water Quality Analysis of the Effects of CBM Produced Water on Soils, Crop Yields and Aquatic Life. Montana Department of Environmental Quality. 2001. Available online: https://deq.mt.gov/Portals/112/Energy/CoalbedMethane/Documents/Criteria-sar-EC-h.pdf (accessed on 22 June 2020).
4. Kinnon, E.C.P.; Golding, S.D.; Boreham, C.J.; Baublys, K.A.; Esterle, J.S. Stable isotope and water quality analysis of coal bed methane production waters and gases from the Bowen Basin, Australia. *Int. J. Coal Geol.* **2010**, *82*, 219–231. [CrossRef]
5. Al-Jubori, A.; Johnston, S.; Boyer, C.; Lambert, S.W.; Bustos, O.A.; Pashin, J.C.; Wray, A. Coalbed methane: Clean energy for the world. *Oilf. Rev.* **2009**, *21*, 4–13.
6. Nuccio, V. *Coal-Bed Methane: Potential and Concerns*; United States Geological Survey: Reston, VA, USA, 2000.
7. U.S. Energy Information Administration. Coalbed Methane Production. Available online: https://www.eia.gov/dnav/ng/ng_prod_coalbed_s1_a.htm (accessed on 22 June 2020).
8. Rice, C.A.; Nuccio, V. *Water Produced with Coal-Bed Methane*; United States Geological Survey: Reston, VA, USA, 2000.
9. U.S. EPA Office of Water. *Final 2012 and Preliminary 2014 Effluent Guidelines Program Plans*; United States Environmental Protection Agency (US EPA): Washington, DC, USA, 2014.

10. Hansen, K.; Nicholson, C.; Paige, G. *Wyoming's Water: Resources and Management*; University of Wyoming Extension: Torrington, WY, USA, 2015.
11. Chaudhry, A.M.; Barbier, E.B. Water and growth in an agricultural economy. *Agric. Econ.* **2013**, *44*, 175–189. [CrossRef]
12. Jacobson, J.J.; Brosz, D.J. Wyoming's Water Resources. Cooperative Extension Service College of Agriculture Wyoming Water Resources. 1993. Available online: http://library.wrds.uwyo.edu/wrp/93-12/93-12.html (accessed on 22 June 2020).
13. ALL Consulting. *Consulting. Handbook on Coal Bed Methane Produced Water: Management and Beneficial Use Alternatives*; ALL Consulting: Tulsa, OK, USA, 2003.
14. Zheljazkov, V.D.; Cantrell, C.L.; Astatkie, T.; Schlegel, V.; Jeliazkova, E.; Lowe, D. The effect of coal-bed methane water on spearmint and peppermint. *J. Environ. Qual.* **2013**, *42*, 1815–1821. [CrossRef] [PubMed]
15. Burkhardt, A. *Effects of Coal Bed Methane Water on Soil Characteristics and Plant Secondary Metabolites*; University of Wyoming: Laramie, WY, USA, 2014.
16. Poudyal, S.; Zheljazkov, V.D.; Cantrell, C.L.; Kelleners, T. Coal-bed methane water effects on dill and its essential oils. *J. Environ. Qual.* **2016**, *45*, 728. [CrossRef] [PubMed]
17. USDA-NASS. Wyoming Agriculture. National Agricultural Statistics Service. 2018. Available online: https://quickstats.nass.usda.gov/ (accessed on 22 June 2020).
18. USDA-NASS. Wyoming Field Office Wyoming Agriculture Statistics: 2013. Wyoming Business Council University Wyoming U.S. Department Agricultural. 2013. Available online: https://www.nass.usda.gov/Statistics_by_State/Wyoming/Publications/Annual_Statistical_Bulletin/bulletin2013.pdf (accessed on 22 June 2020).
19. El-Swaify, S.A. Soil and Water Salinity. In *Plant Nutrient Management in Hawaii's Soils, Approaches for Tropical and Subtropical Agriculture*; Silva, J.A., Uchida, R., Eds.; College of Tropical Agriculture and Human Resources, University of Hawaii at Manoa: Honolulu, HI, USA, 2000; pp. 151–158.
20. McCauley, A.; Jones, C. Salinity & Sodicity Management. Soil Water Management. Module. Montana State University Extension. 2005. Available online: http://landresources.montana.edu/swm/documents/Final_Proof_SW2.pdf (accessed on 22 June 2020).
21. Shanahan, J.F.; Dillon, M.A. *Oat Production*; Colorado State University Extension: Fort Collins, CO, USA, 2003.
22. Mueller, S.C. Considerations for successful alfalfa stand establishment in the central San Joaquin Valley. In Proceedings of the California Alfalfa and Forage Symposium, Reno, Visalia, CA, 12–14 December 2005; pp. 12–14.
23. Rahetlah, V.B.; Randrianaivoarivony, J.M.; Razafimpamoa, L.H.; Ramalanjaona, V.L. Effects of seeding rates on forage yield and quality of oat (*Avena sativa* L.) vetch (*Vicia sativa* L.) mixtures under irrigated conditions of Madagascar. African Journal of Food, Agriculture, Nutrition and Development. 2010, Volume 10. Available online: http://hdl.handle.net/1807/55673 (accessed on 18 March 2021).
24. United States Department of Agriculture. Web Soil Survey. Available online: http://websoilsurvey.nrcs.usda.gov/ (accessed on 30 May 2014).
25. Gardisser, D. Harvesting Alfalfa Hay. University Arkansas Cooperative. Extension. Service. Priniting Service. 2005. Available online: https://www.uaex.edu/publications/PDF/FSA-2005.pdf (accessed on 22 June 2020).
26. Barnhart, S.K. Oats for Forage. IOWA State University. Agronomy. Extension. 2011. Available online: https://crops.extension.iastate.edu/cropnews/2011/06/oats-forage (accessed on 22 June 2020).
27. Redmon, L.A.; Mcfarland, M.L. Soil pH and Forage Production. Texas A&M AgriLife Research and Extension. 2013, pp. 7–9. Available online: https://agrilifecdn.tamu.edu/coastalbend/files/2016/06/Soil-pH-and-Forage-Production.pdf (accessed on 22 June 2020).
28. Rasool, S.; Hameed, A.; Azooz, M.M.; Muneeb-u-Rehman; Siddiqi, T.O.; Ahmad, P. *Ecophysiology and Responses of Plants under Salt Stress*; Ahmad, P., Azooz, M.M., Prasad, M.N.V., Eds.; Springer: New York, NY, USA, 2013; ISBN 978-1-4614-4746-7.
29. Li, R.; Shi, F.; Fukuda, K.; Yang, Y. Effects of salt and alkali stresses on germination, growth, photosynthesis and ion accumulation in alfalfa (*Medicago sativa* L.). *Soil Sci. Plant Nutr.* **2010**, *56*, 725–733. [CrossRef]
30. Cornacchione, M.V.; Suarez, D.L. Emergence, forage production, and ion relations of alfalfa in response to saline waters. *Crop Sci.* **2015**, *55*, 444–457. [CrossRef]
31. Khosrowchahli, M.; Houshang, A.; Moghbeli, H. Selection for moderate salinity stress tolerance in alfalfa (*Medicago sativa* L.) ecotypes. *Int. J. Agric. Crop Sci.* **2013**, *5*, 2868–2870.
32. Hanson, B.R.; Grattan, S.R.; Fulton, A. *Agricultural Salinity and Drainage*; Program, Division of Agriculture and Natural Resources, Irrigation, University of California, Davis: Davis, CA, USA, 2006.
33. Davis, J.G.; Waskom, R.M.; Bauder, T.A. Managing Sodic Soils. Colorado State University Extension Service. 2012, p. 2. Available online: https://extension.colostate.edu/topic-areas/agriculture/managing-sodic-soils-0-504/ (accessed on 22 June 2020).
34. Shah, S.S.; Li, Z.; Yan, H.; Shi, L.; Zhou, B. Comparative study of the effects of salinity on growth, gas exchange, n accumulation and stable isotope signatures of forage oat (Avena sativa l.) genotypes. *Plants* **2020**, *9*, 1025. [CrossRef]
35. Suyama, H.; Benes, S.E.; Robinson, P.H.; Grattan, S.R.; Grieve, C.M.; Getachew, G. Forage yield and quality under irrigation with saline-sodic drainage water: Greenhouse evaluation. *Agric. Water Manag.* **2007**, *88*, 159–172. [CrossRef]
36. Chatrath, A.; Mandal, P.K.; Anuradha, M. Effect of secondary salinization on photosynthesis in fodder Oat (Avena sativa L.) genotypes. *Agron. Crop Sci.* **2000**, *184*, 13–16. [CrossRef]
37. Bole, J.B.; Wells, S.A. Dryland soil salinity: Effect on the yield and yield components of 6-row Barley, 2-row Barley, wheat, and oats. *Can. J. Soil Sci.* **1979**, *59*, 11–17. [CrossRef]

38. Dunham, J.R. Relative Feed Value Measures Forage Quality. Kansas State University Agricultural Experiment Station and Cooperative Extension Service. 2007. Available online: https://www.asi.k-state.edu/doc/forage/fora41.pdf (accessed on 22 June 2020).
39. Lacefield, G.; Ball, D.; Hancock, D.; Andrae, J.; Smith, R. *Growing Alfalfa in the South*; National Alfalfa & Forage Alliance: St. Paul, MN, USA, 2009.
40. W-L Research WL 363HQ: Exceptional yield potential. WL Alfalfa 2013. Available online: https://www.wlalfalfas.com/WLAlfalfas/media/PDF/Seed-Guide_2015.pdf?ext=.pdf (accessed on 22 June 2020).
41. Masters, D.G.; Benes, S.E.; Norman, H.C. Biosaline agriculture for forage and livestock production. *Agric. Ecosyst. Environ.* **2007**, *119*, 234–248. [CrossRef]
42. Grattan, S.R.; Grieve, C.M.; Poss, J.A.; Robinson, P.H.; Suarez, D.L.; Benes, S.E. Evaluation of salt-tolerant forages for sequential water reuse systems: III. Potential implications for ruminant mineral nutrition. *Agric. Water Manag.* **2004**, *70*, 109–120. [CrossRef]
43. Robinson, P.H.; Grattan, S.R.; Getachew, G.; Grieve, C.M.; Poss, J.A.; Suarez, D.L.; Benes, S.E. Biomass accumulation and potential nutritive value of some forages irrigated with saline-sodic drainage water. *Anim. Feed Sci. Technol.* **2004**, *111*, 175–189. [CrossRef]
44. Riveland, N.R.; Erickson, D.O.; French, E.W. An Evaluation of Oat Varieties for Forage. 1974. Available online: https://library.ndsu.edu/ir/bitstream/handle/10365/4431/farm_35_1_4.pdf?isAllowed=y&sequence=1 (accessed on 22 March 2021).
45. Mochon, J.; Conley, S. Wisconsin Oats and Barley Performance Tests—2014. University Wisconsin. 2013. Available online: https://ipcm.wisc.edu/blog/2013/11/wisconsin-oats-and-barley-performance-tests-2014/ (accessed on 22 June 2020).
46. Gardner, F.P.; Allen, R.S. *Dough Stage Best for Oat Silage*; Iowa State University: Ames, IA, USA, 1961.
47. Contreras-Govea, F.E.; Albrecht, K.A. Forage production and nutritive value of oat in autumn and early summer. *Crop. Sci.* **2006**, *46*, 2382–2386. [CrossRef]
48. Stearns, M.; Tindall, J.; Cronin, G.; Friedel, M. Effects of coal-bed methane discharge waters on the vegetation and soil ecosystem in Powder River Basin, Wyoming. *Water Air Soil* **2005**, *168*, 33–57. [CrossRef]
49. Poudyal, S. *Utilization of Coal Bed Methane Water (CBMW) for Irrigation of Forage, Medicinal and Bio-fuel Crops*; University of Wyoming: Laramie, WY, USA, 2015; ISBN 9781339679815.

 *sustainability*

Review

Managing Micronutrients for Improving Soil Fertility, Health, and Soybean Yield

Sushil Thapa [1,*], Ammar Bhandari [2], Rajan Ghimire [3], Qingwu Xue [4], Fanson Kidwaro [5], Shirin Ghatrehsamani [6], Bijesh Maharjan [7] and Mark Goodwin [1]

1   Department of Agriculture, University of Central Missouri, Warrensburg, MO 64093, USA; sgoodwin@ucmo.edu
2   Department of Agriculture, Agribusiness, and Environmental Sciences, Texas A&M University—Kingsville, Kingsville, TX 78363, USA; ammar.bhandari@tamuk.edu
3   Agricultural Science Center, New Mexico State University, Clovis, NM 88101, USA; rghimire@nmsu.edu
4   Texas A&M AgriLife Research and Extension Center, Amarillo, TX 79106, USA; qxue@ag.tamu.edu
5   Department of Agriculture, Illinois State University, Normal, IL 61761, USA; fmkidwa@ilstu.edu
6   Agricultural Systems and Natural Resources, University of Missouri, Columbia, MO 65211, USA; sh.samani@missouri.edu
7   Panhandle Research and Extension Center, University of Nebraska-Lincoln, Lincoln, NE 68583, USA; bmaharjan@unl.edu
*   Correspondence: sthapa@ucmo.edu

Citation: Thapa, S.; Bhandari, A.; Ghimire, R.; Xue, Q.; Kidwaro, F.; Ghatrehsamani, S.; Maharjan, B.; Goodwin, M. Managing Micronutrients for Improving Soil Fertility, Health, and Soybean Yield. *Sustainability* 2021, 13, 11766. https://doi.org/10.3390/su132111766

Academic Editors: Primo Proietti and Emanuele Radicetti

Received: 9 July 2021
Accepted: 12 October 2021
Published: 25 October 2021

Abstract: Plants need only a small quantity of micronutrients, but they are essential for vital cell functions. Critical micronutrients for plant growth and development include iron (Fe), boron (B), manganese (Mn), zinc (Zn), copper (Cu), molybdenum (Mo), chlorine (Cl), and nickel (Ni). The deficiency of one or more micronutrients can greatly affect plant production and quality. To explore the potential for using micronutrients, we reviewed the literature evaluating the effect of micronutrients on soybean production in the U.S. Midwest and beyond. Soil and foliar applications were the major micronutrient application methods. Overall, studies indicated the positive yield response of soybean to micronutrients. However, soybean yield response to micronutrients was not consistent among studies, mainly because of different environmental conditions such as soil type, soil organic matter (SOM), moisture, and temperature. Despite this inconsistency, there has been increased pressure for growers to apply micronutrients to soybeans due to a fact that deficiencies have increased with the increased use of high-yielding cultivars. Further studies on quantification and variable rate application of micronutrients under different soil and environmental conditions are warranted to acquire more knowledge and improve the micronutrient management strategies in soybean. Since the SOM could meet the micronutrient need of many crops, management strategies that increase SOM should be encouraged to ensure nutrient availability and improve soil fertility and health for sustainable soybean production.

Keywords: macronutrient; nutrient deficiency; nutrient uptake; site-specific nutrient management; soil organic matter

---

1. Introduction

Soybean (*Glycine max* L.) is one of the most cultivated legume crops in the world. In 2019, its global production was about 334 million metric tons from a harvested area of 121 million hectares [1]. In the U.S., soybean is the second largest crop after corn (*Zea mays* L.) and is primarily grown in the Midwest region, where about 75% of the total agricultural area (38.5 million hectares) is used for corn and soybean productions [2]. The U.S. Midwest is one of the most productive agricultural regions in the world, producing over 33% of the world's corn and 34% of the world's soybeans [1]. Soybeans belong to the Fabaceae family, and they provide approximately 50% of edible oil around the world [3]. The usage of soybeans ranges from human consumption to animal feed to non-food products. In the U.S., soybeans are planted in May and early June and harvested in late September and early October [4]. Farmers commonly grow soybeans in crop rotation with corn. Although commercial fertilizer is applied to less than 40% of soybean acreage [4], over the years, extensive use of primary macronutrients, especially in corn, has resulted in micronutrient deficiency, poor soil fertility, and low soybean productivity.

Plants need 17 essential nutrients for their growth and production. Three basic elements, hydrogen (H), oxygen (O), and carbon (C) are available from air and water. Nitrogen (N), phosphorus (P), potassium (K), sulfur (S), calcium (Ca), and magnesium (Mg) are considered macronutrients, while micronutrients include iron (Fe), boron (B), manganese (Mn), zinc (Zn), copper (Cu), molybdenum (Mo), chlorine (Cl), and nickel (Ni). Plant micronutrient requirement is lower than the macronutrient requirement. Hence, micronutrient deficiencies are less common than macronutrient deficiencies in soybean, but they are essential for critical cell functions [5,6]. Micronutrient deficiency can reduce plant growth, yield, and quality, thereby affecting the health and productivity of animals and human beings [7–9]. Currently, the micronutrient deficiency in arable soil is a global problem [10,11].

The micronutrients typically studied for soybean are Mn, B, Zn, and Mo [12,13]. The extensive use of N fertilizers, especially after the WW-II on corn, wheat (*Triticum aestivum* L.), and other small grains, has resulted in a high yield, which encouraged researchers and scholars to explore the possibility of yield increase using different nutrients [14]. Interest in micronutrients has increased in recent years because of increased nutrient removal rates by newly developed high-yielding cultivars [15]. The plant uptake of micronutrients largely depends on their availability in the soil [16]. Positive yield responses on various crops, including soybean, were observed when micronutrients were applied with macronutrients [17].

Crop production is affected by multiple environmental stresses, including disease and pest infestations, low soil fertility, and inadequate water supply [18,19]. If we feed the soil, it will feed us; therefore, only productive soil that provides all essential nutrients required by plants can support successful crop production. Maintaining soil fertility to an optimum level is necessary to produce healthy plants, maximize crop yield, and sustain soil health. In recent years, soil quality and soil health have been receiving more attention among the scientific community because healthy soils provide the foundation for food production, water conservation, nutrient cycling, climate change mitigation, and biodiversity conservation [20]. The concept of soil health and soil quality was started in the 1980s as a comprehensive approach beyond fertility management to address soil degradation problems. The Food and Agriculture Organization of the United Nations (FAO) [21] describes soil health as the "capacity of soil to function as a living system, with the ecosystem and land-use boundaries, to sustain plant and animal productivity, maintain or enhance water and air quality, and promote plant and animal health". Successful soil health management involves understanding the need for all the essential nutrients and related soil physical, chemical, and biological processes to produce crops and support the farm economy (Figure 1).

Legume-based cropping systems improve soil fertility and health in different ways, such as the availability of soil organic matter (SOM) rich in N and P concentrations [24]. Legume crops such as soybeans can increase SOM by supplying biomass, organic carbon, and N [25] and increase the population of nodule-forming bacteria, Rhizobia [26]. Benefits of legumes include increasing SOM, improving soil porosity and structure, recycling nutrients, buffering soil pH, diversifying the microscopic soil flora and fauna, and breaking the pest and disease cycle [27]. The SOM plays a crucial role in micronutrient availability and their uptake by plants [28]. The presence of chelating organic compounds in soils could increase the availability and solubility of micronutrients. For example, the chelation of metal elements such as Zn and Fe with SOM is essential for transporting these elements to the root system [29].

There are several reports on micronutrient management in soybean on individual field levels and a few on regional scales [6]. In this paper, results are examined and summarized from existing studies on the use of micronutrients in soybean across the globe. We reviewed the literature on various micronutrients and their availability, discussed their deficiency symptoms, and yield responses to soybeans. Micronutrient application methods and variable application rates are also considered. We focused on studies from the U.S. Midwest, the largest soybean producing region in the world, but also recapped relevant studies from other parts of the world. The objective was to synthesize evidence relating to the application of micronutrients in soybean and better understand its importance for improving fertility and the overall health of soils and soybean grain yield in the U.S. Midwest and beyond.

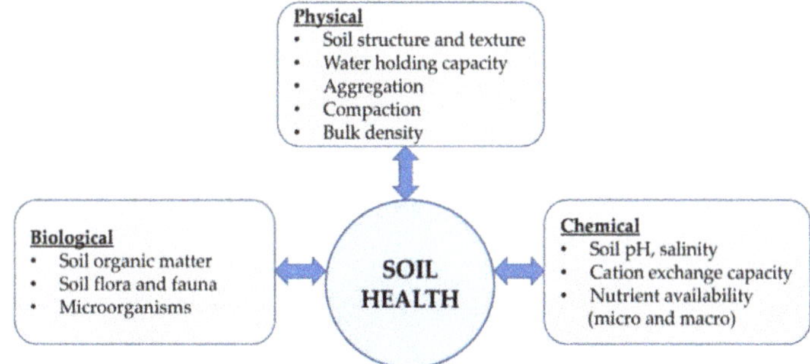

Figure 1. Biological, physical, and chemical components of soil health. (Source: modified from Moebius-Clune [22] and Hills et al. [23]).

2. Micronutrient and Soil Fertility

Two primary objectives of soil nutrient management are to improve soil fertility and meet the nutrient requirements of growing crops. Improving soil fertility is an important agronomic practice for profitable crop production and ensuring soil health. Healthy soils encompass a diverse community of soil flora and fauna that help minimize disease and pests, promote beneficial symbiotic relationships, decompose and recycle essential plant nutrients, and improve soil structure and nutrient holding capacity [21].

A soil nutrient management plan describes the selection of the right source of nutrients for application at the right rate, at the right time, and in the right place [30]. The major role of soil nutrients in soil fertility and ecosystem functions is related to their effect on crop yield. Furthermore, soil fertility and health are influenced by the increased rate of decomposition of high C:N ratio organic matter applied to the soil [31]. Fertilizer application increases microbial activity and enhances the organic matter decomposition, although a few studies reported that added fertilizer did not affect the decomposition of high C:N materials [32]. However, rational use of fertilizers for several years could lead to increased SOM in the soil profile, thereby improving soil fertility, health, and crop yield [33,34]. Studies showed that the increased availability of certain micronutrients, such as Mn and Zn, was highly related to the higher SOM application rates [35–37].

The availability of micronutrients in the soil is also influenced by fertilizers and cropping practices incorporating crop residue in soil [38]. Continuous use of synthetic fertilizers without organic supplements damages the soil's physical, chemical, and biological properties and causes environmental pollution [39]. In contrast, regular use of organic residues significantly improves soil physical, chemical, and biological properties, and hence, soil health [40,41]. Therefore, the balanced application of organic and inorganic fertilizer is recommended in crop management programs for improving soil health and increasing yield [42,43]. The use of organic amendments in the form of compost, farmyard manure (FYM), green manure, and even incorporation of plant residues in soil was noted to be beneficial as they provided some amounts of micronutrients essential for plant growth and development [28,44].

The edaphic and biological factors in the soil, such as redox potential (a measure of electrochemical potential), soil pH, microbial activity, and organic matter, also influence the micronutrient availability to plants [44]. Similarly, micronutrient concentration in crops increased with green manure application along with organic and inorganic fertilizers [45,46]. The SOM acts as a source of nutrients and increases the population of the microbial community, sequestration of soil organic carbon, and nutrient availability to plants [28,47–49]. As such, soil organisms are actively involved in processes such as nutrient cycling, N_2 fixation, decomposition, and mineralization of the SOM [50]. Application of organic manures and crop residues with synthetic fertilizers results in higher fertilizer use efficiency [51].

Micronutrients are generally available in acidic soils and are often unavailable at high pH. Soil pH is a key characteristic that affects the solubility and availability of plant nutrients. As shown in Figure 2, Fe, Mn, B, Cu, and Zn are mostly available between pH 5 and 7, and Mo is mostly available at pH higher than 7. At a low pH (<5), the solubility of Al, and Fe is high but low for molybdenum. Soil pH also affects the microbial community size and activity in the soil, which are responsible for breaking down organic matter and ensuring most chemical transformations in the soil to make nutrients available for plants [27].

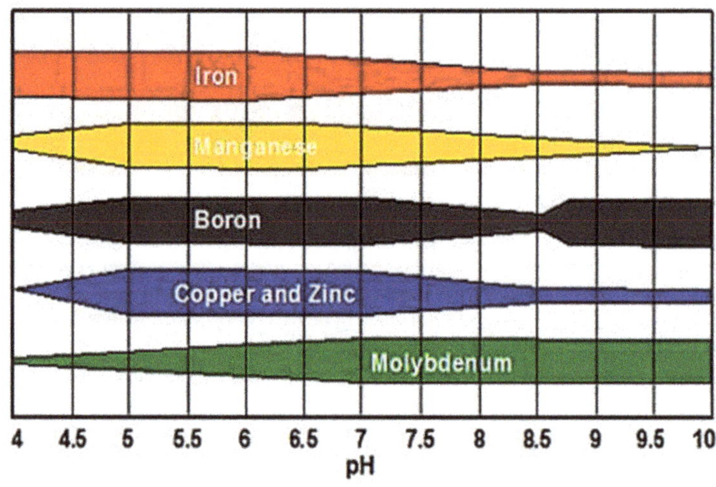

Figure 2. Relative availability of micronutrients by soil pH. (Source: adapted from Truog [52]).

3. Micronutrient Deficiency in Soybean

Micronutrient deficiencies can be detected by visual symptoms on crops and by testing soil samples and plant tissues. Plant symptoms, including stunted growth and leaf chlorosis, may have a variety of causes, including disease/pest infestations, herbicide damage, nutrient deficiencies, or adverse environmental conditions [53]. The deficiency of any one of the 17 essential plant nutrients can limit soybean yield. A nutrient concentration below the sufficiency range or critical value implies deficiency symptoms, and deficient plants respond to the nutrient application [54]. Micronutrients are needed in small amounts, and their adequate concentrations in plants are generally below the 100 ppm level (Table 1).

Many cultivated soils are abundant in Fe, on average, having a total concentration of 20–40 g kg^{-1} soil [56]. However, Fe deficiency is a major problem in soybean production, especially in semiarid environments where carbonates (soluble salts) do not leach easily due to low rainfall [53] (Table 2). The Fe^{+3} becomes less soluble in semiarid environments and slows the conversion to Fe^{+2} for uptake by plants [57]. Iron deficiency results in chlorosis between the veins, especially in younger leaves of soybean. In severe conditions, brown necrotic spots are observed near the leaf margin. However, soybean cultivars differ in their tolerance of Fe deficiency. Some cultivars show complete chlorosis at mild deficiency levels, while others remain normal [58]. Iron is the central component of leghemoglobin for soybean and other legume crops, a nodular component that protects nitrogenase from decomposition by oxygen (O_2) inside root nodules. Therefore, an adequate supply of Fe supports root nodulation and atmospheric N_2 fixation [14].

Table 1. Range of concentrations and adequate concentrations of different nutrient elements in plants (dry weight basis). (Source: Lohry [55]).

Element	Range of Concentrations (ppm)	Adequate Concentration (ppm)
Iron (Fe)	20–600	100
Boron (B)	0.20–800	20
Manganese (Mn)	10–600	50
Zinc (Zn)	10–250	20
Copper (Cu)	2–50	6
Molybdenum (Mo)	0.10–10	0.10
Chlorine (Cl)	10–80,000	100
Nickel (Ni)	0.05–5	0.05

Table 2. Soil conditions that favor micronutrient deficiencies in soybean.

Micronutrient	Soil Characteristics Favoring Deficiency in Soybean	References
Iron (Fe)	Soils with high pH (>7.4), soluble salts, and/or calcium carbonate levels, and low SOM.	Butzen [53], Kaiser et al. [59]
Boron (B)	Alkaline or strongly acidic soils in high rainfall areas or under drought conditions (low rainfall).	Lohry [55]
Manganese (Mn)	Medium and fine-textured soils with high pH (>6.5), low SOM, and poor drainage.	Butzen [53], Ritchey [60], Graham et al. [61], Boring and Thelen [62]
Zinc (Zn)	Sandy soils with a near-neutral pH (6.5), high P levels, low SOM, and cool wet soil conditions.	Bruns [14], Mengel [63], Culman et al. [64]
Copper (Cu)	Alkaline peat muck soil with pH between 7 and 8 and highly leached sandy soils.	Sinclair [58]
Molybdenum (Mo)	Highly acidic soils (pH < 5.8) that are strongly weathered and leached.	Butzen [53], Ritchey [60], Culman et al. [64]
Chlorine (Cl)	Occasionally on sandy soils in dry areas.	Sinclair [58]
Nickel (Ni)	Soils poor in extractable Ni.	Freitas [65]

Boron has very important physiological roles such as enzyme activities, cell elongation, protein synthesis, pollen germination, fruit/grain formation, and yield improvement [66–68]. Boron deficiencies have been observed in several agronomic and horticultural crops, where the soil is alkaline or strongly acidic and in soils high or very low in organic matter content [55] (Table 2). Boron is rarely deficient in clay soils, but coarse and well-drained sandy soils are generally low in B content [69]. Studies have observed a close relationship between B availability and primary cell wall formation. For example, Hanson [70] reported that around 90% of cellular B is localized in the cell wall fraction. Therefore, abnormalities in the cell wall and the organization of middle lamella are the early symptoms of B deficiency in plants [71]. Because soybean plants are relatively tolerant of B deficiency, only small responses to B fertilization have been observed for soybean [72,73]. Boron application can be toxic to soybean with broadcast application rates of about 2.2 kg ha^{-1} [59,74].

Manganese (Mn) availability in soils is influenced by various factors, such as SOM, pH, $CaCO_3$, and redox conditions. Soybean is sensitive to Mn deficiency; hence, its deficiency is a problem in different parts of the world, including the U.S. Midwest [75]. Manganese deficiency is more likely to occur on soils with low moisture content, high soil pH, and low

SOM [53,60,62] (Table 2). Since Mn is relatively immobile in soybean, deficiency symptoms appear on younger leaves as interveinal chlorosis (leaf veins are green, but areas between the veins are yellow). Manganese deficiency in Kentucky has occurred on medium and fine-textured soils with a pH of 6.5 or greater [60], while in Minnesota, soybean responded to Mn when grown on soils with a pH greater than 7.4 [59].

Although soybean is less sensitive to Zn, deficiency symptoms are more common in soybean plants growing on sandy soils with low SOM content [14,63,64] (Table 2). Removal of topsoil through the process of erosion can also increase the Zn deficiency of soil [76,77]. Zinc is essentially immobile in plants, and deficiency symptoms often appear in newer or younger leaves [14]. In soybean, Zn deficiency results in stunted plant growth and chlorotic leaves, with premature lower leaf abscission and poor pod set [14]. An oppositive relationship is identified between Zn and Mn. Severe Zn deficiencies in soybean are often associated with high Mn concentrations in plant tissue, especially in young leaves [78,79].

Copper is specifically required for the synthesis of lignin needed for cell wall strengthening [80]. Like most of the other micronutrients, it is immobile in plants, and deficiency symptoms will first appear in new growth or young leaves. Alkaline peat musk soil, highly leached, and sandy soil with pH between 7 and 8 have a better chance of Cu deficiency [58] (Table 2). Overall light chlorosis, necrotic leaf tips, and loss of turgor in young leaves are the symptoms of Cu deficiency. The problem with Cu toxicity is more common than the toxicity of other micronutrients because Cu is a central element in several pesticides, mainly fungicides [14]. Copper toxicity can result in low protein metabolism as well as low N_2 fixation in soybean [81].

Molybdenum is required by the plants at the lowest concentrations of all essential elements and its level ranges between 0.2 and 5.0 mg kg^{-1} soil [82]. Highly weathered and leached soils that have pH below 5.8 are associated with reduced Mo availability [53,60,62] (Table 2). When the soil pH decreases, it results in the strong adsorption of Mo on oxides of Fe, Al, and Mn [83,84]. The major function of Mo in soybean is to facilitate nodule formation and N_2 fixation by rhizobium bacteria (*B. japonicum*). Thus, biological N_2 fixation will be affected when Mo is deficient [85,86]. Like those of ordinary nitrogen deficiency, general chlorosis of young plants and chlorosis of oldest leaves are the most visible symptoms of Mo deficiency in soybean [58].

Chlorine plays an important role in gas exchange, photosynthesis, and disease resistance in plants. Although Cl deficiency is rarely observed, chlorosis and wilting of young leaves are generally associated with Cl deficiency. Instead of deficiency, Cl toxicity is a serious problem in soybean production in most of the U.S. southern states, where the precipitation is limited. In these areas, the poorly drained soils will accumulate more Cl on the upper soil profile causing Cl toxicity [87,88]. After applying Cl-containing fertilizer, Cl toxicity in soybean was also found in the poorly drained soils of Georgia [89].

Nickel deficiency in soybean occurs in soils poor in extractable Ni [65]. Field-grown soybean plants with Ni deficiency may not show visible leaf symptoms (hidden deficiency) [60]. Therefore, most studies on Ni in plants have been conducted in the context of toxicity than deficiency [90–93]. Plants cannot complete their life cycle without an adequate supply of Ni [94] because it is a structural component of urease [95], the enzyme that is responsible for converting urea to ammonia [96,97]. Therefore, legumes that are dependent on N_2 fixation may be particularly susceptible to an inadequate Ni supply.

4. Micronutrients Management

An appropriate method of micronutrient application depends on the element and its formulation (liquid or dry), the severity of the deficiency, and the plant growth stage at which the deficiency symptoms are being addressed. Soil and foliar application are the major routes of micronutrient application. Soil application is generally preferred for most nutrients if deficiencies are known prior to or at the beginning of the growing season. Micronutrients banded with other fertilizers at planting are usually more effective over a longer period than foliar-applied micronutrients. Soil application also makes the nutrients available to the plant at the earliest [53].

Micronutrients, Cu, Mn, Mo, Fe, Cl, B, and Zn with limited mobility may benefit from band application near soybean roots [98]. However, a small fraction of nutrients reaches the plant system through soil application and the remaining amount goes to waste through leaching in soil, causing land and water pollution [99]. Overall, the utilization of the soil-applied fertilizers is low in calcareous and alkaline soil due to high fixation and less mobility of the nutrients [100,101]. In contrast, foliar spray is the fastest way to cope with the deficiency and translocate micronutrients in plant organs [102]. The foliar spray also improves plant resistance against disease, pests, and drought [103]. Plant leaves not only capture light and CO_2 but also can absorb nutrients through the cuticle, cuticular cracks and imperfections, stomata, trichomes, and lenticels, which have long been recognized and used in nutrient management programs [104,105]. The foliar application also minimizes the leaching loss of nutrients which is more common in soil application [99].

Like in many other grain crops, soybean nutrient demand increases at the time of flowering and grain filling. Foliar application of micronutrients during this period helps to complement soil nutrient supply. Furthermore, most foliar micronutrient supplements can be mixed and sprayed with herbicides and other pesticides. A meta-analysis by Joy et al. [106] suggested that foliar application with Zn was more efficient/cost-effective than soil application for enhancing Zn concentrations in various crops. Foliar applications of Mn were most effective in soybean when applied at the early blossom or early pod set stage or in multiple applications at these stages [107]. Foliar application of B increased its concentration on soybean grains but did not increase the yield [108]. Soybean seed yield showed a small response to soil-applied Mn and Zn, but when micronutrients were foliar-applied, seed yield decreased, likely due to some leaf damage caused by foliar feeding [109]. Therefore, the effectiveness of foliar micronutrient applications varies significantly concerning their solubility and ingredients such as salts, surfactants, complexes, or chelates [110]. There are ongoing discussions about the different effects of micronutrient delivery to plants as soil vs. foliar [106,111].

In recent years, site-specific (or variable rate) nutrient management that uses modern technology and tools, such as remote sensing, Geographic Information System (GIS), and Geographical Positioning System (GPS) is becoming popular among growers [112]. Researchers have mostly focused on the site-specific management of macronutrients, and few investigations have been conducted with respect to site-specific micronutrient management, which is even scarce in the case of soybean. Variations in the micronutrient contents of soils depict an intrinsic nature and properties of soils [113–115]. Foroughifar et al. [114] utilized the Geostatistics and GIS techniques to characterize the spatial variability of soil properties, including micronutrients. They found the spatial distribution of micronutrients varied with soil sedimentation sequence, underground water levels, and underlying pedological and hydrological processes. Geostatistics and GIS techniques were also applied to understand the spatial dependency of bioavailable micronutrients in the rice (*Oryza sativa* L.) field, where the spatial distribution of the micronutrients was significantly correlated to the soil formation factors [116].

Eze et al. [117] compared uncertainties and correlations in spatial process models for the distribution of Zn in the topsoil of a semiarid environment and found geostatistical modeling as a decision-making tool in soil micronutrient management because it could map spatial heterogeneity and uncertainty. Vasu et al. [115] illustrated that the large-scale spatial variability mapping of soil micronutrients is a prerequisite for implementing site-specific nutrient management in the semiarid tropical regions. Wang et al. [118] found some similarity of spatial structure between soil pH and the grain Cu, Fe, Mn, and Zn, and by analogy, similar spatial variation was observed between SOM and DTPA-extractable micronutrients in the soil. Udeigwe et al. [119] examined the fixation pattern and kinetics of plant-available DTPA-extractable Cu, as well as basic soil properties that influence Cu availability in semiarid soils, while Zhao et al. [120] used geostatistical methods for identifying the possible spatial distribution of Cu. Eze et al. [121] used the sequential gaussian simulation (SGS) to map the spatial distribution of Cu concentration and modeled the spatial uncertainties for arable dryland in central Botswana. Furthermore, Kriging-based techniques are used to better understand the spatial variation of micronutrients in different parts of the world [118,122].

5. Soybean Micronutrients Uptake and Yield Response

Micronutrients are essential elements that plants use in small quantities. For most micronutrients, crop uptake is less than 2.0 kg ha^{-1} (Table 3). Despite this low requirement, deficiency of micronutrients limits the critical plant functions, resulting in abnormal plant growth and reduced yield and quality. In this condition, even the use of other inputs such as macronutrients and water can be wasted. This situation is clearly illustrated by Liebig's Law of minimum; every field contains a maximum of one or more and a minimum of one or more nutrients [65]. Crop yields are regulated by the factor in greatest limitation, and yields can be increased only by correction of that limiting factor. When that limitation is overcome, yields are then regulated by the next important limiting factor. This process is repeated, with stepwise yield increases, until there are no remaining limiting factors. All successful agricultural producers use this important principle knowingly or unknowingly [65].

Table 3. Approximate per-hectare micronutrient uptake by soybeans. (Source: adapted from Mengel [63]).

Micronutrient	Nutrient Uptake (kg/ha) by 4000 kg/ha Soybean
Iron (Fe)	1.91
Manganese (Mn)	0.67
Zinc (Zn)	0.22
Boron (B)	0.11
Copper (Cu)	0.11
Molybdenum (Mo)	0.01

In different studies, soybean yield response to micronutrient applications was inconsistent because of different genetic resources, management factors, and environmental conditions such as soil mineralogy, pH, organic matter, moisture, temperature, and aeration. Therefore, identifying genetic and environmental factors affecting soybean micronutrient uptake and crop removal could help growers to implement better nutrient management strategies. Studies, especially on the application of micronutrients such as Zn and Fe, showed an improvement in yield and yield components in various crops [123–125]. The use of micronutrients also helped plants to minimize the impact of drought. For example, foliar spray of Fe and Mo on soybean reduced the damages caused by water deficit conditions and increased the yield compared to the control treatment that did not receive both Fe and Mo [126].

According to Ross et al. [127], when B was applied to the plants grown with low B levels and visual deficiency symptoms, soybean yield increased from 4% to 130% in Arkansas. They reported that an application of 0.28 to 1.12 kg B ha^{-1} was sufficient to produce maximum soybean yield in the area. In Georgia, increasing rates of soil-applied B significantly increased the soil, leaf, and grain B concentration [128]. In Missouri, foliar application of 0.56 kg B ha^{-1} was found to be the appropriate rate for increasing the number of pods per branch, but the application of 1.12 kg B ha^{-1} increased the seed size and resulted in the highest yield per plant [129]. In contrast, Oplinger et al. [130] summarized 29 trials across Missouri, Illinois, Ohio, and Wisconsin and found yield increases only in four sites on B-sufficient soils. The excess use of B (2.24 kg ha^{-1}) reduced soybean yields in the Coastal Plains of the Southeastern U.S. [131]. Bellaloui [132] in Mississippi found increased seed protein and oleic acid and decreased linolenic acids in foliar B-applied soybean.

A study in 40 sites of Iowa showed that foliar application of Zn, Mn, Cu, and B increased their concentration in the trifoliate leaf and seed but did not increase soybean grain yield [108]. In contrast, a study from Australia showed that foliar application of Zn before flowering increased soybean grain yield by 13% to 208% at different locations [133].

In the Mn-deficient soils of Wisconsin, Mn applied both to the soil and the foliage increased yield compared to the soil or foliage independent application [107]. Soybean grain yield increased up to 2518 kg ha^{-1} in the coastal region of Virginia when MnSO$_4$ was foliar applied at the rate of 1.12 kg Mn ha^{-1} at vegetative and reproductive growth stages [134].

A micronutrient mixture that included B, Fe, and Zn sprayed at the five-leaf growth stage of soybean did not show any yield response in 18 sites of Iowa [135], and the application of different levels of B, Cu, and Zn did not significantly affect soybean yield in Virginia [136]. A foliar fertilizer containing Zn, Mn, Fe, and B increased soybean yield by 2.4% within the northern Corn Belt, but resulted in a 0.7% yield decrease within the central Corn Belt [137]. The addition of B, Cl, Mn, and Zn did not increase soybean grain yield and had a marginal impact on soybean grain quality in Minnesota [70]. They also reported that soybean grain protein and oil concentration were only marginally impacted by B or Cl and were not impacted by Mn or Zn.

In the case of Cl, toxicity is more studied than deficiency. Toxicity of Cl is caused by an accumulation of Cl in the upper soil profile, especially on poorly drained soils and with limited precipitation [87,88]. For example, soybean was grown in the poorly drained soils of Georgia and the application of Cl-containing fertilizer exhibited leaf scorching due to Cl toxicity, resulting in reduced grain yield [89]. A study in Missouri showed that the application of Cl fertilizer increased the mean trifoliate Cl concentration in 60 cultivars tested [87]. Knowledge on the effect of Ni on soybean production is still limited, but Ni fertilization increased soybean yield under both the greenhouse [138,139] and field conditions [60].

Soybean yield response to micronutrients has been reported from the other parts of the world, as well. For example, Barbosa et al. [140] conducted a soybean field study in Brazil using different doses of fertilizers containing micronutrients (6.8% Mn, 3.9% Zn, 2.1% Fe, 1.2% Cu, and 1.1% B) and found the increased soybean yield from 33.6% to 79.7% compared to the control. In India, Vyas et al. [141] applied Zn, Mo, and B with FYM in soybean and reported an 18.2% higher grain yield with the combined use of Zn and FYM compared to other treatments. The Zn alone increased the soybean yield by 11.4% compared to the control treatment. Dwivedi et al. [142] applied Cu, Zn, B, and Mo using both soil and foliar methods in soybean–wheat cropping systems and reported a significantly greater yield of soybeans with micronutrients either alone or in the mixture. In another study, Shivakumar and Ahlawat [143] found the application of 5 kg Zn ha^{-1} with crop residue and FYM increased soybean yield. They also found a residual effect of micronutrient application with crop residue and FYM in the following wheat crop with significantly greater yield.

In Iran, Ghasemain et al. [103] applied different rates of Zn, Fe, and Mn in soybean and reported the highest grain number, seed weight per plant, pod number, biomass, and grain yield of soybean in a field study with Zn and Mn applied at 40 kg ha^{-1}. Kobraee et al. [144] used three different rates of Zn (0, 20, and 40 kg ha^{-1}), Fe (0, 25, and 50 kg ha^{-1}), and Mn (0, 20, and 40 kg ha^{-1}) and found significantly higher grain yields at 40 kg ha^{-1} of Zn, 50 kg ha^{-1} of Fe, and 40 kg ha^{-1} of Mn. A study conducted by Gheshlaghi et al. [145] used foliar application of Zn and Mn micronutrients in irrigated soybean and found significantly greater yield with the application of Zn. Furthermore, they found that the micronutrient application, especially Zn, significantly increased seed quality by increasing the oleic, linolenic, and linoleic saturated fatty acids in soybean. A similar result (higher pod number per plant and grain yield using Zn and Fe) was also reported by Heidarian et al. [123]. Kobraei et al. [146] found the accelerated formation of proteins, RNA, and DNA due to the application of Zn.

Zahoor et al. [147] evaluated the response of micronutrients, Fe, Mo, and Co in a soybean field study in Pakistan. They reported that Fe and Mo significantly increased the shoot length, shoot dry weight, nodules per plant, nodules fresh weight, thousand seed weight, and soybean yield (42.3% greater yield compared to the control). The activation of different enzymes in N$_2$-fixing bacteria due to the application of Fe, Mo, and Co might have increased the number of nodules and yield [148]. Eisa et al. [149] conducted a field study in Egypt using three cultivars of soybean and three rates of Fe, Zn, and Mn with phosphorus fertilizer as a foliar application at 30 and 45 days after planting. They reported a significant increase in soybean seed quality (proteins, P, K, Fe, Zn, and Mn content) and yield in all cultivars. The increased yield and quality of soybeans were likely due to the positive and regulatory effect of micronutrients on different enzymes and overall plant metabolism [150]. In Russia, foliar application of liquid fertilizer containing Cu at the early boom stage of soybean increased the 1000 seed weight [151]. In Ukraine, foliar application of fertilizers containing a high concentration of Mo, Mn, and B helped soybean plants to form more pods and seeds, resulting in increased seed weight and yield [152].

6. Conclusions

A good soil fertility management plan for a farm is a long-term strategy. Yield maximization is possible only when the plant nutrients are available to meet the crop demand. Thus, maintaining soil fertility and health is essential to plant health and, consequently, to animal and human health. Studies show that several factors likely influence the soybean response to micronutrients, including the location and soil condition, soil pH, cultivar, seasonal rainfall/irrigation, and the use of SOM. Therefore, basic knowledge of what, how much, when, and how to apply fertilizer is an essential part of a soybean nutrient management plan to ensure a high yield. In many cases, soil or foliar-applied micronutrients at rates recommended according to soil test results was effective for increasing soybean yield. Some studies have indicated the importance of plant tissue analysis, as well, but it is only useful in diagnosing nutritional deficiency problems during the crop growing season. Hence, plant analysis is recommended in conjunction with regular soil testing. Most studies evaluating soybean yield response to foliar application of micronutrients have shown mixed results. Therefore, any additional foliar feeding should be considered under conditions that do not jeopardize the production cost. Studies are also needed to test the feasibility of variable rate technologies in micronutrients on a site-specific basis to increase profits and decrease nutrient loss.

Author Contributions: Conceptualization and methodology: S.T., A.B. and R.G.; investigation: S.T.; resources: S.T. and R.G; writing—original draft preparation: S.T., A.B., R.G. and S.G.; writing— review and editing: S.T., A.B., R.G., Q.X., F.K., S.G., B.M. and M.G.; funding acquisition: S.T. and R.G. All authors have read and agreed to the published version of the manuscript.

Funding: This research was funded by the School of Natural Sciences and the Department of Agriculture at the University of Central Missouri through a startup grant provided to Sushil Thapa. The APC was funded by New Mexico State University through Research Enhancement Fund of Agricultural Science Center at Clovis.

References

1. FAOSTAT. Crops. 2021. Available online: http://www.fao.org/faostat/en/#data/QC (accessed on 3 May 2021).
2. USDA. Agriculture in the Midwest. 2017. Available online: https://www.climatehubs.usda.gov/hubs/midwest/topic/agriculture-midwest (accessed on 5 May 2021).
3. Akparobi, S.O. Evaluation of six cultivars of soybean under the soil of rainforest agro-ecological zones of Nigeria. *Middle East J. Sci. Res.* **2009**, *4*, 6–9.
4. USDA. Oil Crops Sector at a Glance. 2021. Available online: https://www.ers.usda.gov/topics/crops/soybeans-oil-crops/oil-crops-sector-at-a-glance/ (accessed on 5 May 2021).
5. Malakouti, M.J. The effect of micronutrients in ensuring efficient use of macronutrients. *Turk. J. Agric. For.* **2008**, *32*, 215–220.
6. Mallarino, A.P.; Kaiser, D.E.; Ruiz, D.A.; Laboski, C.A.M.; Camberato, J.J.; Vyn., T.J. *Micronutrients for Soybean Production in the North Central Region*; CROP-3135; Iowa State University: Ames, IA, USA, 2017.
7. Cakmak, I. Plant nutrition research: Priorities to meet human needs for food in sustainable ways. *Plant Soil* **2002**, *247*, 3–24. [CrossRef]
8. Marschner, H. *Mineral Nutrition of Higher Plants*, 2nd ed.; Academic Press: London, UK, 1995; ISBN 978-0-12-473542-2.

9. Malakouti, M.J. Zinc is a neglected element in the life cycle of plants: A review. *Middle East. Rus. J. Plant Sci. Biotechnol.* **2007**, *1*, 1–12.

10. Monreal, C.M.; DeRosa, M.; Mallubhotla, S.C.; Bindraban, P.S.; Dimkpa, C. Nanotechnologies for increasing the crop use efficiency of fertilizer-micronutrients. *Biol. Fertil. Soils* **2016**, *52*, 423–437. [CrossRef]

11. Oliver, M.A.; Gregory, P. Soil, food security and human health: A review. *Eur. J. Soil Sci.* **2015**, *66*, 257–276. [CrossRef]

12. Mascarenhas, H.A.A.; Esteves, J.A.D.F.; Wutke, E.B.; Gallo, P.B. Micronutrients in soybeans in the state of São Paulo. *Nucleus* **2014**, *11*, 131–149. [CrossRef]

13. Bender, R.R.; Haegele, J.W.; Below, F. Nutrient Uptake, Partitioning, and Remobilization in Modern Soybean Varieties. *Agron. J.* **2015**, *107*, 563–573. [CrossRef]

14. Bruns, H.A. Soybean Micronutrient Content in Irrigated Plants Grown in the Midsouth. *Commun. Soil Sci. Plant Anal.* **2017**, *28*, 103. [CrossRef]

15. Maharjan, B.; Shaver, T.M.; Wortmann, C.S.; Shapiro, C.A.; Ferguson, R.B.; Krienke, B.T.; Swewart, Z.P. *Micronutrient Management in Nebraska, Nebraska Extension*; University of Nebraska—Lincoln: Lincoln, NE, USA, 2018.

16. Fageria, N.K.; Filho, M.B.; Moreira, A.; Guimarães, C.M. Foliar Fertilization of Crop Plants. *J. Plant Nutr.* **2010**, *32*, 1044–1064. [CrossRef]

17. Rietra, R.P.J.J.; Heinen, M.; Dimkpa, C.O.; Bindraban, P.S. *Effects of Nutrient Antagonism and Synergism on Fertilizer Use*; VFRC Report 2015/5; Virtual Fertilizer Research Center: Washington, DC, USA, 2015.

18. Dimkpa, C.O.; Bindraban, P.S.; Fugice, J.; Agyin-Birikorang, S.; Singh, U.; Hellums, D. Composite micronutrient nanoparticles and salts decrease drought stress in soybean. *Agron. Sustain. Dev.* **2017**, *37*, 5. [CrossRef]

19. Adisa, I.O.; Pullagurala, V.L.R.; Peralta-Videa, J.R.; Dimkpa, C.O.; Elmer, W.H.; Gardea-Torresdey, J.L.; White, J.C. Recent advances in nano-enabled fertilizers and pesticides: A critical review of mechanisms of action. *Environ. Sci. Nano* **2019**, *6*, 2002–2030. [CrossRef]

20. McBratney, A.; Field, D.; Koch, A. The dimensions of soil security. *Geoderma* **2014**, *213*, 203–213. [CrossRef]

21. FAO. *An International Technical Workshop. Investing in Sustainable Crop Intensification: The Case for Improving Soil Health*; Food and Agriculture Organization of the United Nations: Rome, Italy, 2008.

22. Moebius-Clune, B.N.; Moebius-Clune, D.J.; Gugino, B.K.; Idowu, O.J.; Schindelbeck, R.R.; Ristow, A.; van Es, H.M.; Thies, J.E.; Shayler, H.A.; McBride, M.B.; et al. *Comprehensive Assessment of Soil Health—The Cornell Framework, Edition 3.2*; Cornell University: Ithaca, NY, USA, 2016.

23. Hills, K.; Collins, H.; Yorgey, G.; McGuire, A.; Kruger, C. *Safeguarding Potato Cropping Systems in the Pacific Northwest through Improved Soil Health*; Center for Sustaining Agriculture and Natural Resources, Washington State University: Pullman, WA, USA, 2018.

24. Jensen, E.S.; Ambus, P.; Bellostas, N.; Boisen, S.; Brisson, N.; Corre-Hellou, G.; Crozat, Y.; Dahlmann, C.; Dibet, A.; von Fragstein, P.; et al. Intercropping of cereals and grain leg-umes for increased production, weed control, im-proved product quality and prevention of N-losses in European organic farming systems. In *Proceedings of the International Conferences: Joint Organic Congress—Theme 4: Crop Systems and Soils*, Odense, Denmark, 30–31 May 2006.

25. Lemke, R.L.; Zhong, Z.; Campbell, C.A.; Zentner, R.P. Can pulse crops play a role in mitigating greenhouse gases from North American agriculture? *Agron. J.* **2007**, *99*, 1719–1725. [CrossRef]

26. La Favre, J.S.; Focht, D.D. Conservation in soil of H_2 liberated from N_2 fixation by H up-nodules. *Appl. Environ. Microb.* **1983**, *46*, 304–311. [CrossRef]

27. USDA. *Legumes and Soil Quality*; Technical Note No. 6; United States Department of Agriculture, Natural Resources Conservation Service: Washington, DC, USA, 1998; p. 998.

28. Rengel, Z.; Batten, G.; Crowley, D. Agronomic approaches for improving the micronutrient density in edible portions of field crops. *Field Crop. Res.* **1999**, *60*, 27–40. [CrossRef]

29. Schulin, R.; Khoshgoftarmanesh, A.; Afyuni, M.; Nowack, B.; Frossard, E. Effects of soil management on zinc uptake and its bioavailability in plants. In *Development and Uses of Biofortified Agricultural Products*; Banuelos, G., Lin, Z., Eds.; CRC Press: Boca Raton, FL, USA, 2009.

30. Rogers, E. The 4R's of Nutrient Management. Michigan State University Extension. 2019. Available online: https://www.canr.msu.edu/news/the-4r-s-of-nutrient-management (accessed on 21 May 2021).

31. Recous, S.; Robin, D.; Darwis, D.; Mary, B. Soil inorganic nitrogen availability: Effect on maize residue decomposition. *Soil Biol. Biochem.* **1995**, *27*, 1529–1538. [CrossRef]

32. Hobbie, S.E. Contrasting Effects of Substrate and Fertilizer Nitrogen on the Early Stages of Litter Decomposition. *Ecosystem* **2005**, *8*, 644–656. [CrossRef]

33. Ladha, J.K.; Kesava Reddy, C.; Padre, A.D.; van Kessel, C. Role of nitrogen fertilization in sustaining organic matter in culti-vated soils. *J. Environ. Qual.* **2011**, *40*, 1756–1766. [CrossRef] [PubMed]

34. Geiseller, D.; Scow, K.M. Long-term effects of mineral fertilizers on soil microorganisms—A review. *Soil Biol. Biochem.* **2014**, *75*, 54–63. [CrossRef]

35. de Santiago, A.; Quintero, J.M.; Delgado, A. Long-term effects of tillage on the availability of iron, copper, manganese, and zinc in a Spanish Vertisol. *Soil Tillage Res.* **2008**, *98*, 200–207. [CrossRef]

36. Motschenbacher, J.M.; Brye, K.R.; Anders, M.M.; Gbur, E.E. Long-term rice rotation, tillage, and fertility effects on near-surface chemical properties in a silt-loam soil. *Nutr. Cycl. Agroecosyst.* **2014**, *100*, 77–94. [CrossRef]
37. Moreira, S.G.; Prochnow, L.I.; Kiehl, J.D.C.; Pauletti, V.; Martin-Neto, L. Chemical forms in soil and availability of manganese and zinc to soybean in soil under different tillage systems. *Soil Tillage Res.* **2016**, *163*, 41–53. [CrossRef]
38. Wei, X.; Hao, M.; Shao, M.; Gale, W.J. Changes in soil properties and the availability of soil micronutrients after 18 years of cropping and fertilization. *Soil Tillage Res.* **2006**, *91*, 120–130. [CrossRef]
39. Zhong, W.; Cai, Z. Long-term effects of inorganic fertilizers on microbial biomass and community functional diversity in a paddy soil derived from quaternary red clay. *Appl. Soil Ecol.* **2007**, *36*, 84–91. [CrossRef]
40. Chang, E.-H.; Chung, R.-S.; Wang, F.-N. Effect of different types of organic fertilizers on the chemical properties and enzymatic activities of an Oxisol under intensive cultivation of vegetables for 4 years. *Soil Sci. Plant Nutr.* **2008**, *54*, 587–599. [CrossRef]
41. Surekha, K.; Latha, P.C.; Rao, K.V.; Kumar, R.M. Grain Yield, Yield Components, Soil Fertility, and Biological Activity under Organic and Conventional Rice Production Systems. *Commun. Soil Sci. Plant Anal.* **2010**, *41*, 2279–2292. [CrossRef]
42. Weber, J.; Karczewska, A.; Drozd, J.; Licznar, M.; Jamroz, E.; Kocowicz, A. Agricultural and ecological aspects of a sandy soil as affected by the application of municipal solid waste composts. *Soil Biol. Biochem.* **2007**, *39*, 1294–1302. [CrossRef]
43. Kumar, A.; Tripathi, H.P.; Yadav, D.S. Correcting nutrients for sustainable crop production. *Indian J. Fert.* **2007**, *2*, 37–44.
44. Dhaliwal, S.S.; Naresh, R.K.; Mandal, A.; Singh, R.; Dhaliwal, M.K. Dynamics and transformations of micronutrients in agricultural soils as influenced by organic matter build-up: A review. *Environ. Sust. Indic.* **2019**, *1*, 100007.
45. Soni, M.L.; Swarup, A.; Singh, M. Effect of manganese and phosphorus application on yield and nutrition of wheat in reclaimed sodic soil. *Curr. Agric.* **2000**, *24*, 105–109.
46. Singh, V.; Ram, N. Effect of 25 years of continuous fertilizer use on response to applied nutrients and uptake of micronutrients by rice-wheat-cowpea system. *Cereal Res. Commun.* **2005**, *33*, 589–594. [CrossRef]
47. Kowaljow, E.; Mazzarino, M.J. Soil restoration in semiarid Patagonia: Chemical and biological response to different compost quality. *Soil Biol. Biochem.* **2007**, *39*, 1580–1588. [CrossRef]
48. Pedra, F.; Polo, A.; Ribeiro, A.; Domingues, H. Effects of municipal solid waste compost and sewage sludge on mineralization of soil organic matter. *Soil Biol. Biochem.* **2007**, *39*, 1375–1382. [CrossRef]
49. Sebastia, J.; Labanowski, J.; Lamy, I. Changes in soil organic matter chemical properties after organic amendments. *Chemosphere* **2007**, *68*, 1245–1253. [CrossRef] [PubMed]
50. Welbaum, G.E.; Sturz, A.V.; Dong, Z.; Nowak, J. Managing Soil Microorganisms to Improve Productivity of Agro-Ecosystems. *Crit. Rev. Plant Sci.* **2004**, *23*, 175–193. [CrossRef]
51. Lamps, S. Principles of integrated plant nutrition management system. In Proceedings of the Symposium Integrated Plant Nutrition Management, Islamabad, Pakistan, 8–10 November 1999.
52. Truog, E. Soil Reaction Influence on Availability of Plant Nutrients. *Soil Sci. Soc. Am. J.* **1946**, *11*, 305–308. [CrossRef]
53. Butzen, S. Micronutrients for Crop Production. *Pioneer Crop Insights* **2020**, *20*, 1–4.
54. Mundorf, T.; Wortmann, C.; Shapiro, C.; Paparozzi, E. Time of Day Effect on Foliar Nutrient Concentrations in Corn and Soybean. *J. Plant Nutr.* **2015**, *38*, 2312–2325. [CrossRef]
55. Lohry, R. Micronutrients: Functions, sources and application methods. In Proceedings of the Indiana CCA Conference, Indianapolis, IN, USA, 18–19 December 2007.
56. Cornell, R.M.; Schwertmann, U. *The Iron Oxides*, 2nd ed.; Wiley-CH: Weinheim, Germany, 2003.
57. Hansen, N.C.; Schmitt, M.A.; Anderson, J.E.; Strock, J. Iron Deficiency of Soybean in the Upper Midwest and Associated Soil Properties. *Agron. J.* **2003**, *95*, 1595–1601. [CrossRef]
58. Sinclair, J.B. Soybeans. In *Nutrient Deficiencies in Toxicities in Crop Plants*, 3rd ed.; Bennett, W.F., Ed.; American Phytopathological Society: Saint Paul, MN, USA, 1996.
59. Kaiser, D.E.; Fernandez, F.; Wilson, M. *Fertilizing Soybean in Minnesota*; University of Minnesota Extension: Saint Paul, MN, USA, 2020.
60. Ritchey, E.; Lee, C.; Knott, C.; Grove, J. *Plant and Soil Sciences. Soybean Nutrient Management in Kentucky*; University of Kentucky College of Agriculture, Food and Environment Cooperative Extension Service: Lexington, KY, USA, 2014.
61. Graham, M.J.; Nickell, C.D.; Hoeft, R.G. Effect of manganese deficiency on seed yield of soybean cultivars. *J. Plant Nutr.* **1994**, *17*, 1333–1340. [CrossRef]
62. Boring, T.J.; Thelen, K.D. Soybean foliar manganese recommendations on chronically Mn deficient soils. In Proceedings of the 39th North Central Extension-Industry Soil Fertility Conference, Des Moines, IA, USA, 18–19 November 2009.
63. Mengel, D. *Role of Micronutrients in Efficient Crop Production*; Purdue University: West Lafayette, IN, USA, 1990.
64. Culman, S.; Fulford, A.; Camberato, J.; Steinke, K. *Tri-State Fertilizer Recommendations for Corn, Soybeans, Wheat and Alfalfa*; Ohio State University: Columbus, OH, USA, 2020.
65. Freitas, D.S.; Rodak, B.W.; dos Reis, A.R.; Reis, F.D.B.; De Carvalho, T.S.; Schulze, J.; Carneiro, M.A.C.; Guilherme, L.R.G. Hidden Nickel Deficiency? Nickel Fertilization via Soil Improves Nitrogen Metabolism and Grain Yield in Soybean Genotypes. *Front. Plant Sci.* **2018**, *9*, 614. [CrossRef]
66. Dell, B.; Huang, L. Physiological response of plants to low boron. *Plant Soil* **1997**, *193*, 103–120. [CrossRef]
67. Brown, P.H.; Bellaloui, N.; Wimmer, M.A.; Bassil, E.S.; Ruiz, J.; Hu, H.; Pfeffer, H.; Dannel, F.; Römheld, V. Boron in Plant Biology. *Plant Biol.* **2002**, *4*, 205–223. [CrossRef]

68. Fleischer, A.; O'Neill, M.A.; Ehwald, R. The pore size of non-graminaceaous plant cell walls is rapidly decreased by borate ester cross-linking of the pectic polysaccharide rhamnogalacturonan II. *Plant Physiol.* **1999**, *121*, 829–838. [CrossRef] [PubMed]
69. Tisdale, S.L.; Nelson, W.L.; Beaton, J.D. *Soil Fertility and Fertilizers*, 4th ed.; Macmillan Publishing Co.: New York, NY, USA, 1985.
70. Hanson, E.J. Movement of Boron Out of Tree Fruit Leaves. *HortScience* **1991**, *26*, 271–273. [CrossRef]
71. Brown, P.H.; Hu, H. Does boron play only a structural role in the growing tissues of higher plants? *Plant Soil* **1997**, *196*, 211–215. [CrossRef]
72. Sutradhar, A.K.; Kaiser, D.E.; Rosen, C.J. Boron for Minnesota soils. University of Minnesota Extension Publications. 2016. Available online: https://extension.umn.edu/micro-and-secondary-macronutrients/boron-minnesota-soils (accessed on 12 June 2021).
73. Martens, D.C.; Westermann, D.T. Fertilizer Applications for Correcting Micronutrient Deficiencies. In *SSSA Book Series*; Soil Science Society of America and American Society of Agronomy: Madison, WI, USA, 2018; pp. 549–592.
74. Sutradhar, A.K.; Kaiser, D.E.; Behnken, L.M. Soybean Response to Broadcast Application of Boron, Chlorine, Manganese, and Zinc. *Agron. J.* **2017**, *109*, 1048–1059. [CrossRef]
75. Adams, M.L.; Norvell, W.A.; Philpot, W.D.; Peverly, J.H. Spectral Detection of Micronutrient Deficiency in 'Bragg' Soybean. *Agron. J.* **2000**, *92*, 261–268. [CrossRef]
76. Westfall, D.G.; Bauder, T.A. *Zinc and Iron Deficiencies*; Colorado State University Extension: Fort Collins, CO, USA.
77. Grunes, D.L.; Boawn, L.C.; Carlson, C.W.; Viets, R.G. Land Leveling May Cause Zinc Deficiency. In *Micronutrients in Agriculture*; Mortvedt, J.J., Ed.; The Soil Science Society of America Book Series No. 4; The Soil Science Society of America: Madison, WI, USA, 1961.
78. Carter, O.G.; Rose, I.A.; Reading, P.F. Variation in Susceptibility to Manganese Toxicity in 30 Soybean Genotypes. *Crop. Sci.* **1975**, *15*, 730–732. [CrossRef]
79. Parker, M.B.; Harris, H.B.; Morris, H.D.; Perkins, H.F. Manganese Toxicity of Soybeans as Related to Soil and Fertility Treatments. *Agron. J.* **1969**, *61*, 515–518. [CrossRef]
80. Yruela, I. Copper in plants: Acquisition, transport and interactions. *Funct. Plant Biol.* **2009**, *36*, 409–430. [CrossRef]
81. Mortvedt, J.J. Bioavailability of micronutrients. In *Handbook of Soil Science*; Sumner, M.E., Ed.; CRC Press: Boca Raton, FL, USA, 2000; p. 2148.
82. Ritchie, S.W.; Hanway, J.J.; Thompson, H.E.; Benson, G.O. *How a Soybean Plant Develops*; Special Report 53; Revised Edition Service: Ames, IA, USA, 1994.
83. Karimian, N.; Cox, F.R. Molybdenum Availability as Predicted from Selected Soil Chemical Properties. *Agron. J.* **1979**, *71*, 63–65. [CrossRef]
84. Goldberg, S.; Shouse, P.J.; Lesch, S.M.; Grieve, C.M.; Poss, J.A.; Forster, H.S.; Suarez, D.L. Soil boron extractions as indicators of boron content of field-grown crops. *Soil Sci.* **2002**, *167*, 720–728. [CrossRef]
85. Wurzburger, N.; Bellenger, J.P.; Kraepiel, A.M.L.; Hedin, L.O. Molybdenum and Phosphorus Interact to Constrain Asymbiotic Nitrogen Fixation in Tropical Forests. *PLoS ONE* **2012**, *7*, e33710. [CrossRef]
86. Jean, M. E.; Phalyvong, K.; Forest-Drolet, J.; Bellenger, J.-P. Molybdenum and phosphorus limitation of asymbiotic nitrogen fixation in forests of Eastern Canada: Influence of vegetative cover and seasonal variability. *Soil Biol. Biochem.* **2013**, *67*, 140–146. [CrossRef]
87. Yang, J.; Blanchar, R.W. Differentiating Chloride Susceptibility in Soybean Cultivars. *Agron. J.* **1993**, *85*, 880–885. [CrossRef]
88. Rupe, J.C.; Widick, J.D.; Sabbe, W.E.; Robbins, R.T.; Becton, C.B. Effect of Chloride and Soybean Cultivar on Yield and the Development of Sudden Death Syndrome, Soybean Cyst Nematode, and Southern Blight. *Plant Dis.* **2000**, *84*, 669–674. [CrossRef]
89. Parker, M.B.; Gascho, G.J.; Gaines, T.P. Chloride Toxicity of Soybeans Grown on Atlantic Coast Flatwoods Soils. *Agron. J.* **1983**, *75*, 439–443. [CrossRef]
90. Chen, C.; Huang, D.; Liu, J. Functions and Toxicity of Nickel in Plants: Recent Advances and Future Prospects. *CLEAN Soil, Air, Water* **2009**, *37*, 304–313. [CrossRef]
91. Muhammad, B.H.; Shafaqat, A.; Aqeel, A.; Saadia, H.; Muhammad, A.F.; Basharat, A.; Saima, A.B.; Hussain, M.B.; Ali, S.; Azam, A.; et al. Morphological, physiological and biochemical responses of plants to nickel stress: A review. *Afr. J. Agric. Res.* **2013**, *8*, 1596–1602. [CrossRef]
92. dos Reis, A.R.; Barcelos, J.P.D.Q.; Osório, C.R.W.D.S.; Santos, E.F.; Lisboa, L.A.M.; Santini, J.M.K.; dos Santos, M.J.D.; Junior, E.F.; Campos, M.; de Figueiredo, P.A.M.; et al. A glimpse into the physiological, biochemical and nutritional status of soybean plants under Ni-stress conditions. *Environ. Exp. Bot.* **2017**, *144*, 76–87. [CrossRef]
93. Yusuf, M.; Fariduddin, Q.; Hayat, S.; Ahmad, A. Nickel: An Overview of Uptake, Essentiality and Toxicity in Plants. *Bull. Environ. Contam. Toxicol.* **2011**, *86*, 1–17. [CrossRef] [PubMed]
94. Brown, P.H.; Welch, R.M.; Cary, E.E. Nickel: A Micronutrient Essential for Higher Plants. *Plant Physiol.* **1987**, *85*, 801–803. [CrossRef] [PubMed]
95. Dixon, N.E.; Gazzola, C.; Blakeley, R.L.; Zerner, B. Jack bean urease (EC 3.5.1.5). Metalloenzyme. Simple biological role for nickel. *J. Am. Chem. Soc.* **1975**, *97*, 4131–4133. [CrossRef]
96. Polacco, J.C.; Mazzafera, P.; Tezotto, T. Opinion—Nickel and urease in plants: Still many knowledge gaps. *Plant Sci.* **2013**, *199-200*, 79–90. [CrossRef] [PubMed]
97. Witte, C.-P. Urea metabolism in plants. *Plant Sci.* **2011**, *180*, 431–438. [CrossRef] [PubMed]

98. Minor, H.C.; Wiebold, W. *Wheat-Soybean Double-Crop Management in Missouri*; University of Missouri Extension: Columbia, MO, USA, 1998.

99. Neumann, P.M. Late-season foliar fertilization with macronutrients—Is there a theoretical basis for increased seed yields? *J. Plant Nutr.* **1982**, *5*, 1209–1215. [CrossRef]

100. Rashid, A.; Rafique, E.; Ryan, J. Establishment and Management of Boron Deficiency in Field Crops in Pakistan. In *Boron in Plant and Animal Nutrition*; Springer: Boston, MA, USA, 2002; pp. 339–348.

101. Zekri, M.; Obreza, T.A. *Micronutrient Deficiencies in Citrus: Iron, Zinc, and Manganese*; University of Florida: Gainesville, FL, USA, 2003.

102. Boaretto, A.; Boaretto, R.; Muraoka, T.; Filho, V.N.; Tiritan, C.S.; Filho, F.M. Foliar micronutrient application effects on citrus fruit yield, soil and leaf zn concentrations and 65zn mobilization within the plant. *Acta Hortic.* **2002**, *594*, 203–209. [CrossRef]

103. Ghasemian, V.; Ghalavand, A.; Soroosh zadeh, A.; Pirzad, A. The effect of iron, zinc and manganese on quality and quantity of soybean seed. *J. Phytol.* **2010**, *2*, 73–79.

104. Fernández, V.; Eichert, T. Uptake of Hydrophilic Solutes through Plant Leaves: Current State of Knowledge and Perspectives of Foliar Fertilization. *Crit. Rev. Plant Sci.* **2009**, *28*, 36–68. [CrossRef]

105. Marschner, P. *Marschner's Mineral Nutrition of Higher Plants*; Academic Press: New York, NY, USA, 2012.

106. Joy, E.; Stein, A.; Young, S.D.; Ander, E.L.; Watts, M.; Broadley, M.R. Zinc-enriched fertilisers as a potential public health intervention in Africa. *Plant Soil* **2015**, *389*, 1–24. [CrossRef]

107. Randall, G.W.; Schulte, E.E.; Corey, R.B. Effect of Soil and Foliar-applied Manganese on the Micronutrient Content and Yield of Soybeans. *Agron. J.* **1975**, *67*, 502–507. [CrossRef]

108. Enderson, J.T.; Mallarino, A.P.; Haq, M.U. Soybean Yield Response to Foliar-Applied Micronutrients and Relationships among Soil and Tissue Tests. *Agron. J.* **2015**, *107*, 2143–2161. [CrossRef]

109. Widmar, A.; Ruiz Diaz, D.A. *Evaluation of Macro- and Micronutrients for Double-Crop Soybean after Wheat. Kansas Fertilizer Research*; Kansas State University Agricultural Experiment Station and Cooperative Extension Service: Manhattan, KS, USA, 2012.

110. Stewart, Z.P.; Paparozzi, E.T.; Wortmann, C.S.; Jha, P.K.; Shapiro, C.A. Foliar Micronutrient Application for High-Yield Maize. *Agronomy* **2020**, *10*, 1946. [CrossRef]

111. Dimkpa, C.O.; Bindraban, P.S. Fortification of micronutrients for efficient agronomic production: A review. *Agron. Sustain. Dev.* **2016**, *36*, 1–26. [CrossRef]

112. Verma, P.; Chauhan, A.; Ladon, T. Site specific nutrient management: A review. *J. Pharmacogn. Phytochem.* **2020**, *9*, 233–236.

113. Eze, P.N.; Udeigwe, T.K.; Stietiya, M.H. Distribution and potential source evaluation of heavy metals in prominent soils of Accra Plains, Ghana. *Geoderma* **2010**, *156*, 357–362. [CrossRef]

114. Foroughifar, H.; Jafarzadeh, A.A.; Torabi, H.; Pakpour, A.; Miransari, M. Using Geostatistics and Geographic Information System Techniques to Characterize Spatial Variability of Soil Properties, Including Micronutrients. *Commun. Soil Sci. Plant Anal.* **2013**, *44*, 1273–1281. [CrossRef]

115. Vasu, D.; Sahu, N.; Tiwary, P.; Chandran, P. Modelling the spatial variability of soil micronutrients for site specific nutrient management in a semi-arid tropical environment. *Model. Earth Syst. Environ.* **2021**, *7*, 1797–1812. [CrossRef]

116. Liu, X.M.; Xu, J.M.; Zhang, M.K.; Huang, J.H.; Shi, J.C.; Yu, X.F. Application of geostatistics and GIS technique to characterize spatial variabilities of bioavailable micronutrients in paddy soils. *Environ. Geol.* **2004**, *46*, 189–194. [CrossRef]

117. Eze, P.N.; Kumahor, S.K. Gaussian process simulation of soil Zn micronutrient spatial heterogeneity and uncertainty – A performance appraisal of three semivariogram models. *Sci. Afr.* **2019**, *5*, e00110. [CrossRef]

118. Wang, L.; Wu, J.-P.; Liu, Y.-X.; Huang, H.-Q.; Fang, Q.-F. Spatial Variability of Micronutrients in Rice Grain and Paddy Soil. *Pedosphere* **2009**, *19*, 748–755. [CrossRef]

119. Udeigwe, T.K.; Eichmann, M.; Eze, P.N.; Ogendi, G.M.; Morris, M.N.; Riley, M.R. Copper micronutrient fixation kinetics and interactions with soil constituents in semi-arid alkaline soils. *Soil Sci. Plant Nutr.* **2016**, *62*, 289–296. [CrossRef]

120. Zhao, Y.; Xu, X.; Huang, B.; Sun, W.; Shao, X.; Shi, X.; Ruan, X. Using robust kriging and sequential Gaussian simulation to delineate the copper- and lead-contaminated areas of a rapidly industrialized city in Yangtze River Delta, China. *Environ. Geol.* **2007**, *52*, 1423–1433. [CrossRef]

121. Eze, P.N.; Kumahor, S.K.; Kebonye, N.M. Predictive mapping of soil copper for site-specific micronutrient management using GIS-based sequential Gaussian simulation. *Model. Earth Syst. Environ.* **2021**, 1–11. [CrossRef]

122. Eljebri, S.; Mounir, M.; Faroukh, A.T.; Zouahri, A.; Tellal, R. Application of geostatistical methods for the spatial distribution of soils in the irrigated plain of Doukkala, Morocco. *Model. Earth Syst. Environ.* **2019**, *5*, 669–687. [CrossRef]

123. Heidarian, A.R.; Kord, H.; Mostafavi, K.; Lak, A.P.; Mashhadi, F.A. Investigating Fe and Zn foliar application on yield and its components of soybean (*Glycine max* L) at different growth stages. *J. Agric. Biotech. Sustain. Dev.* **2011**, *3*, 189–197.

124. Fox, T.C.; Guerinot, M.L. Molecular biology of cation transport in plants. *Annu. Rev. Plant Biol.* **1998**, *49*, 669–696. [CrossRef] [PubMed]

125. Ekhtiari, S.; Kobraee, S.; Shamsi, K. Soybean yield under water deficit conditions. *J. Biodivers. Environ. Sci.* **2013**, *3*, 46–52.

126. Heidarzade, A.; Esmaeili, M.; Bahmanyar, M.; Abbasi, R. Response of soybean (Glycine max) to molybdenum and iron spray under well-watered and water deficit conditions. *J. Exp. Biol. Agric. Sci.* **2016**, *4*, 37–46.

127. Ross, J.R.; Slaton, N.A.; Brye, K.R.; DeLong, R.E. Boron Fertilization Influences on Soybean Yield and Leaf and Seed Boron Concentrations. *Agron. J.* **2006**, *98*, 198–205. [CrossRef]

128. Touchton, J.T.; Boswell, F.C.; Marchant, W.H. Boron for soybeans grown in Georgia. *Commun. Soil Sci. Plant Anal.* **1980**, *11*, 369–378. [CrossRef]
129. Schon, M.K.; Blevins, D.G. Foliar Boron Applications Increase the Final Number of Branches and Pods on Branches of Field-Grown Soybeans. *Plant Physiol.* **1990**, *92*, 602–607. [CrossRef] [PubMed]
130. Oplinger, E.S.; Hoeft, R.G.; Johnson, J.W.; Tracy, P.W. Boron fertilization of soybeans: A regional summary. In *Foliar Fertilization of Soybeans and Cotton*; Murphy, L.S., Ed.; PPI/FAR Technical Bullet 1993-1; Potash Phosphate Institute: Norcross, GA, USA, 1993.
131. Touchton, J.T.; Boswell, F.C. Effects of B Application on Soybean Yield, Chemical Composition, and Related Characteristics. *Agron. J.* **1975**, *67*, 417–420. [CrossRef]
132. Bellaloui, N. Effect of Water Stress and Foliar Boron Application on Seed Protein, Oil, Fatty Acids, and Nitrogen Metabolism in Soybean. *Am. J. Plant Sci.* **2011**, *2*, 692–701. [CrossRef]
133. Rose, I.; Felton, W.; Banks, L. Responses of four soybean varieties to foliar zinc fertilizer. *Aust. J. Exp. Agric.* **1981**, *21*, 236–240. [CrossRef]
134. Gettier, S.W.; Martens, D.C.; Brumback, T.B. Timing of Foliar Manganese Application for Correction of Manganese Deficiency in Soybean. *Agron. J.* **1985**, *77*, 627–630. [CrossRef]
135. Mallarino, A.P.; Haq, M.U.; Wittry, D.; Bermudez, M. Variation in Soybean Response to Early Season Foliar Fertilization among and within Fields. *Agron. J.* **2001**, *93*, 1220–1226. [CrossRef]
136. Martens, D.C.; Carter, M.T.; Jones, G.D. Response of Soybeans Following Six Annual Applications of Various Levels of Boron, Copper, and Zinc. *Agron. J.* **1974**, *66*, 82–84. [CrossRef]
137. Orlowski, J.M.; Haverkamp, B.J.; Laurenz, R.G.; Marburger, D.A.; Wilson, E.W.; Casteel, S.N.; Conley, S.; Naeve, S.L.; Nafziger, E.D.; Roozeboom, K.L.; et al. High-Input Management Systems Effect on Soybean Seed Yield, Yield Components, and Economic Break-Even Probabilities. *Crop. Sci.* **2016**, *56*, 1988–2004. [CrossRef]
138. Kutman, B.Y.; Kutman, U.B.; Cakmak, I. Nickel-enriched seed and externally supplied nickel improve growth and alleviate foliar urea damage in soybean. *Plant Soil* **2013**, *363*, 61–75. [CrossRef]
139. Lavres, J.; Franco, G.C.; Câmara, G.M.D.S. Soybean Seed Treatment with Nickel Improves Biological Nitrogen Fixation and Urease Activity. *Front. Environ. Sci.* **2016**, *4*, 37. [CrossRef]
140. Barbosa, J.M.; Rezende, C.F.A.; Leandro, W.; Ratke, R.F.; Flores, R.; Da Silva, Á.R.; Goiânia, G.-B. Effects of micronutrients application on soybean yield. *Aust. J. Crop. Sci.* **2016**, *10*, 1092–1097. [CrossRef]
141. Vyas, M.D.; Jain, A.K.; Tiwari, R.J. Long-term effect of micronutrients and FYM on yield of and nutrient uptake by soybean on a typic chromustert. *J. Indian Soc. Soil Sci.* **2003**, *51*, 45–47.
142. Dwivedi, G.K.; Dwivedi, M.; Pal, S.S. Modes of application of micronutrients in acid soil in soybean-wheat crop sequence. *J. Indian Soc. Soil Sci.* **1990**, *38*, 458–463.
143. Shivakumar, B.G.; Ahlawat, I.P.S. Integrated nutrient management in soybean (*Glycine max*)—Wheat (*Triticum aestivum*) cropping system. *Indian J. Agron.* **2008**, *53*, 273–278.
144. Kobraee, S.; Shamsi, K.; Rasekhi, B. Effect of micronutrients application on yield and yield components of soybean. *Ann. Biol. Res.* **2011**, *2*, 476–482.
145. Gheshlaghi, M.Z.; Pasari, B.; Shams, K.; Rokhzadi, A.; Mohammadi, K. The effect of micronutrient foliar application on yield, seed quality and some biochemical traits of soybean cultivars under drought stress. *J. Plant Nutr.* **2019**, *42*, 2715–2730. [CrossRef]
146. Kobraei, S.; Etminan, A.; Mohammadi, R.; Kobraee, S. Effect of drought stress on yield and yield components of soybean. *Ann. Biol. Res.* **2011**, *2*, 504–509.
147. Zahoor, F.; Ahmed, M.; Malik, M.A.; Mubeen, K.; Siddiqui, M.H.; Rasheed, M.; Ansar, R.; Mehmood, K. Soybean (*Glycine max* L.) response to micro-nutrients. *Turkish J. Filed Crops* **2013**, *18*, 134–138.
148. Hegazi, A.; Mohamed, M.A.; Sayed, A.G.S.; Elsherif, M.H.; Gad, N. Reducing N doses by enhancing nodule formation in groundnut plants via Co and Mo. *Australian J. Basic Appl. Sci.* **2011**, *5*, 2568–2577.
149. Eisa, S.A.I.; Mohamed, T.B.; Mohamedin, A.M.A. Amendment of soil fertility and augmentation of the quantity and quality of soybean crop by using phosphorus and micronutrients. *Int. J. Acad. Res.* **2011**, *3*, 1–9.
150. Abd E-Hady, B.A. Effect of zinc application on growth and nutrient uptake of barley plant irrigated with saline water. *J. Appl. Sci. Res.* **2007**, *3*, 431–436.
151. Kolesar, V.; Sharipova, G.; Safina, D.; Safin, R. Use of foliar fertilizers on soybeans in the Republic of Tatarstan. *BIO Web Conf.* **2020**, *17*, 00069. [CrossRef]
152. Novytska, N.; Gadzovskiy, G.; Mazurenko, B.; Kalenska, S.; Svistunova, I.; Martynov, O. Effect of seed inoculation and foliar fertilizing on structure of soybean yield and yield structure in Western Polissya of Ukraine. *Agron. Res.* **2020**, *18*, 2512–2519. [CrossRef]

Article

Modeling Climate Change Effects on Rice Yield and Soil Carbon under Variable Water and Nutrient Management

Zewei Jiang [1], Shihong Yang [1,2,3,*], Jie Ding [1], Xiao Sun [1], Xi Chen [1], Xiaoyin Liu [1] and Junzeng Xu [1,2]

[1]   College of Agricultural Science and Engineering, Hohai University, Nanjing 210098, China; zwaq@hhu.edu.cn (Z.J.); hhudingjie@hhu.edu.cn (J.D.); sx1027@hhu.edu.cn (X.S.); sunrise@hhu.edu.cn (X.C.); lxyin1819@hhu.edu.cn (X.L.); xjz481@hhu.edu.cn (J.X.)

[2]   State Key Laboratory of Hydrology-Water Resources and Hydraulic Engineering, Hohai University, Nanjing 210098, China

[3]   Cooperative Innovation Center for Water Safety & Hydro Science, Hohai University, Nanjing 210098, China

*   Correspondence: ysh7731@hhu.edu.cn; Tel.: +86-25-8378-6015

Abstract: Soil organic carbon (SOC) conservation in agricultural soils is vital for sustainable agricultural production and climate change mitigation. To project changes of SOC and rice yield under different water and carbon management in future climates, based on a two-year (2015 and 2016) field test in Kunshan, China, the Denitrification Decomposition (DNDC) model was modified and validated and the soil moisture module of DNDC was improved to realize the simulation under conditions of water-saving irrigation. Four climate models under four representative concentration pathways (RCP 2.6, RCP 4.5, RCP 6.0, and RCP 8.5), which were integrated from the fifth phase of the Coupled Model Intercomparison Project (CMIP5), were ensembled by the Bayesian Model Averaging (BMA) method. The results showed that the modified DNDC model can effectively simulate changes in SOC, dissolved organic carbon (DOC), and rice yield under different irrigation and fertilizer management systems. The normalized root mean squared errors of the SOC and DOC were 3.45–17.59% and 8.79–13.93%, respectively. The model efficiency coefficients of SOC and DOC were close to 1. The climate scenarios had a great impact on rice yield, whereas the impact on SOC was less than that of agricultural management measures on SOC. The average rice yields of all the RCP 2.6, RCP 4.5, RCP 6.0, and RCP 8.5 scenarios in the 2090s decreased by 18.41%, 38.59%, 65.11%, and 65.62%, respectively, compared with those in the 2020s. The long-term effect of irrigation on the SOC content of paddy fields was minimal. The SOC of the paddy fields treated with conventional fertilizer decreased initially and then remained unchanged, while the other treatments increased obviously with time. The rice yields of all the treatments decreased with time. Compared with traditional management, controlled irrigation with straw returning clearly increased the SOC and rice yields of paddy fields. Thus, this water and carbon management system is recommended for paddy fields.

Keywords: paddy field; soil organic carbon; denitrification decomposition (DNDC); climate change

Citation: Jiang, Z.; Yang, S.; Ding, J.; Sun, X.; Chen, X.; Liu, X.; Xu, J. Modeling Climate Change Effects on Rice Yield and Soil Carbon under Variable Water and Nutrient Management. *Sustainability* **2021**, *13*, 568. https://doi.org/10.3390/su13020568

Received: 20 November 2020
Accepted: 29 December 2020
Published: 8 January 2021

1. Introduction

The carbon cycle is a popular topic in ecological research [1]. Soil organic carbon (SOC) is the largest carbon pool on the planet excluding the ocean's and rock's sediments; thus, small changes in SOC have a great impact on the atmosphere [2]. The carbon pool of the agro-ecosystem is one of the most active parts of the global carbon cycle, in which soil organic carbon storage in farmland accounts for 8–10% of that in all types of land [3]. The SOC in farmland is vulnerable to disturbances from human activities [4], but this SOC can be artificially regulated on a short-time scale [5]. In addition, China has a total paddy soil area of 45.7 Mha, accounting for approximately one-fifth of the total cultivated land area in the world [6]. Thus, paddy fields have a considerable carbon sequestration potential. At the same time, the dynamics of SOC in paddy fields are affected by many factors, such as temperature, precipitation, irrigation, and fertilization [7]. However, few studies on SOC changes in paddy fields have focused on the impact of coupling water-saving irrigation and fertilizer management. In recent years, water-saving irrigation technology has been widely used in China and has changed the soil moisture status and organic carbon content. Thus, evaluating the impact of water and carbon management measures on the dynamic changes in SOC is important to maintaining agricultural productivity.

Moreover, our understanding of climate change as an important factor affecting SOC and rice yield remains limited. Thus, improving our understanding of the impact of environmental change and field management on nutrient cycling and crop growth is necessary. Despite the growing importance of industry, agricultural production, as one of the most sensitive sectors to climate change [8], plays an important role in ensuring food security throughout the world, especially in China [9]. Rice paddies are an important source of both global food production and greenhouse gas (GHG) [10,11], and rice yield is extremely sensitive to agricultural measures, such as irrigation and fertilization [12]. At present, China's sustainable agricultural development is facing challenges in maintaining optimal yields while mitigating environmental impacts [13,14]. Therefore, addressing climate change and optimizing management measures for paddy fields are problems that should be urgently resolved.

The combination of process-based modeling and various experimental data provides opportunities for quantifying the impacts of different management practices and future climate change on soil C dynamics [15]. In fact, comprehensively and accurately evaluating SOC change is difficult due to the low speed of SOC dynamics and time-consuming and laborious on-site sampling; thus, a calibration model is necessary. Agronomists and scientists have worked diligently in the past to devise and promote the use of agricultural practices that can maintain or increase SOC levels. With the continuous development of agricultural research methods in addition to physical sampling and analysis of soil profiles for SOC, dynamic modelling of SOC can be used to effectively monitor soil organic carbon storage under different agricultural management. Among the relatively mature related models, including CENTURY, denitrification and decomposition (DNDC), NCSOIL, and RothC, the DNDC model is widely used due to its simple parameter inputs and accurate result simulation. The DNDC model can satisfactorily simulate SOC conversion in paddy fields and crop growth under climate change [4].

Kunshan is located in the Tai-Lake region in the middle and low reaches of the Yangtze River paddy soil region of China, which is a typical rice production area in the country [16]. Many recent studies have revealed that the paddy soils in this area have high SOC sequestration potential [17,18]. Hence, combining the experimental data from Kunshan with that from the DNDC model is feasible. Although we are encouraged by the tests of and improvements in DNDC for crop yields and environmental impact estimation in the past two decades, the widespread application of this tool in China has several limitations [19]. For instance, the constant 50-cm soil depth leads to overestimation of soil water content [20]. In addition, some soil properties, such as bulk density, porosity, and hydraulic parameters are assumed to be constant across all layers (down to a depth of 50 cm). However, most soil properties vary inherently between layers. Additionally, the traditional flood irrigation mode is the only irrigation mode for paddy fields, which makes it difficult to simulate the increasingly popularized water-saving irrigation mode. These factors may decrease the accuracy of irrigation simulation.

Although the DNDC model has been improved and applied in China through a two-decade effort, only four models exist for paddy fields under flood irrigation, namely, continuous flooding (the field water level is maintained at 10 cm), alternative wet and dry flooding (water level fluctuates between −5 to 5 cm), and rain-fed and empirical parameters. These four existing modes are inconsistent with the water-saving irrigation model in China. In a previous study [21], DNDC was used to simulate methane emissions from paddy fields under medium-term drainage, intermittent irrigation, and continuous flooding. In contrast to the above irrigation methods, under the condition of controlled irrigation, which is widely applied in China, a shallow water layer is reserved on the surface of the field after transplanting seedlings until the regreening stage and the soil remains not flooded on the surface of the irrigation field in each subsequent growth stage, usually 60–80% of the time [22]. The irrigation time and irrigation amount were determined with the root-layer soil moisture as the control index. The existing model cannot simulate the controlled irrigation conditions. Thus, urgently modifying the DNDC model for paddy fields under water-saving irrigation is necessary to decrease site-specific suitability [23]. Given these problems, this research modified the 50-cm soil layer in the model to the depth of the root layer and controlled the upper and lower limits of paddy irrigation with soil moisture content. Additionally, the limits were modified in accordance with the needs of different growth stages of rice to adapt to the local water-saving irrigation mode. We hypothesized that crop growth and SOC dynamics could be simulated by improving the soil moisture module of this model. On this basis, the effects of different water and carbon management on SOC and rice yield in future climate conditions were studied.

Interest is growing in terms of finding ways to simulate climate change by using General Circulation Models (GCMs), which is the main current approach to predict future climate change and its responses. Substantial progress in global and regional modeling at medium to high resolutions and in downscaling methods has provided the basis for an increasing number of studies that attempt to simulate the effect of future climate change [24]. Predicting the dynamic changes in SOC and rice yield in paddy fields in the future is important for formulating agricultural management measures to save water, to increase yield, and to promote sustainable development. We carried out this study on the basis of the modified DNDC model and four climate scenarios under four GCMs weighted by Bayesian model averaging (BMA). The objectives of the study are (1) to validate the relevant parameters and to simulate changes in SOC and rice yield in Kunshan for the next 80 years, and (2) to extend the paddy field irrigation module in the DNDC model to provide a theoretical basis for optimizing field management measures to cope with climate change.

2. Materials and Methods

2.1. Experimental Site

The experiment was conducted in 2015 and 2016 at the State Key Laboratory of Hydrology-Water Resources and Hydraulic Engineering of Hohai University, Kunshan Irrigation and Drainage Experiment Station (31°15′15″ N, 120°57′43″ E), Jiangsu Province, China (Figure 1). The study area has a subtropical monsoon climate with a mean annual precipitation of 1097 mm, an average annual air temperature of 15.5 °C, a sunshine duration of 2086 h, and a frost-free period of 234 days·y^{-1}. The locals are accustomed to a rotation of rice and wheat planting. The paddy soil is classified as a hydragric anthrosol, which has a heavy loam texture, with a bulk density of 1.32 g cm^{-3} at 0–30 cm and an initial pH of 7.4 at 0–18 cm. The organic matter is 21.71 g kg^{-1} for the top 0–18 cm layer, and total K, total P, and total N are 20.86, 1.40, and 1.79 g kg^{-1} for the 0–30 cm layer, respectively.

2.2. Field Management

The experiment was laid out (plot size 150 m^2) in a randomized block design with six treatments in triplicate. The six treatments were a combination of irrigation and fertilizer management systems: the two irrigation managements regimes were flood irrigation (FI) and controlled irrigation (CI), and the three fertilizer managements were wheat straw returning (S), organic fertilizer management (O), and farmer fertilizer practices (FFP). The six treatments were FS (FI and S), FO (FI and O), FF (FI and FFP), CS (CI and S), CO (CI and O), and CF (CI and FFP), with a total of 18 cells. Rain-fed wheat was grown in the plots during the non-rice planting season.

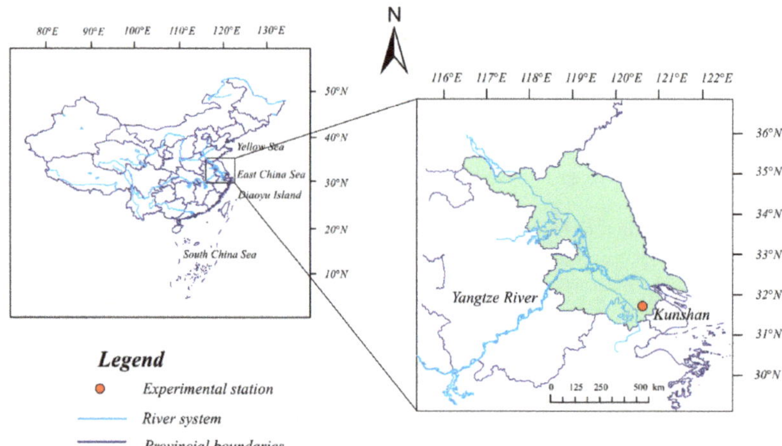

Figure 1. Location of the experimental station.

The rice variety in the experiment was Japonica Rice Nanjing 46. Three or four seedlings per hill were transplanted in late June, with a plant spacing of 13.0 cm × 25.0 cm, and were harvested in late October. Local nitrogen (N) fertilizer was adopted in FFP (Table 1). The chemical fertilizer management of the S treatment was similar to that of the FFP treatment, and 3000 kg ha^{-1} of straw from the previous wheat crop (the organic carbon content of wheat straw was 441 g kg^{-1}, while the C/N ratio was 50:1 and the organic carbon input through wheat straw was 1322 g kg^{-1}) was returned to the S paddy fields both years. Additionally, 7500 kg ha^{-1} of well-decomposed chicken manure (23% moisture content, 16.3 g kg^{-1} N, 261 g kg^{-1} organic carbon, 15.4 g kg^{-1} P_2O_5, and 20.7 g kg^{-1} K_2O (Shijiazhuang Jitian Biotechnology Co., Ltd., China) was applied to the O paddy fields in 2015 and 2016. The base fertilizer and wheat straw were mixed into the muddy soil during tillage, and surface application was adopted for all other fertilizers.

Table 1. Date and rate of nitrogen fertilization during the rice-growing season in farmer fertilizer practice (FFP) (kg N ha^{-1}).

Activity	2015	2016
Base fertilizer (29 and 28 June)	155.2 (72.0CF + 83.2AB)	72.0 (72.0CF)
Tillering fertilizer (16 Jul)	69.3 (U)	97.0 (U)
Panicle fertilizer (9 and 11 Aug)	58.9 (U)	104.0 (U)
Total nitrogen	283.4	273.0

Dates in brackets are when the fertilizer was applied in 2015 and 2016, respectively. CF: compound fertilizer (N, P_2O_5, and K_2O contents were 16.0%, 12.0%, and 17.0% in 2015 and 2016), AB: ammonium bicarbonate (N content was 17.1%), U: urea (N content was 46.2%).

The irrigation water layer of the CI paddy fields was maintained at 5–25 mm in the regreening stage. Irrigation was applied only to keep the soil moist, and standing water was avoided in the other stages except during periods of pesticide and fertilizer application. In accordance with local rice planting habits, a 30–50 mm shallow water layer was retained in the FI paddy fields after transplantation except during the midseason drainage period and the yellow maturity stage of rice. Rainfall was deflected with a canopy to accurately control soil moisture. The root zone soil water content criteria in different rice growth stages for CI are shown in Table 2.

Table 2. Limits for irrigation in different stages of rice under controlled irrigation.

Stages	Re-Greening	Tillering			Jointing and Booting		Heading and Flowering	Milk Maturity	Yellow Maturity
		Former	Middle	Later	Former	Later			
Upper limit [a]	25 mm [b]	θ_{s1}	θ_{s1}	θ_{s1}	θ_{s2}	θ_{s2}	θ_{s3}	θ_{s3}	Drying
Lower limit	5 mm [b]	70% θ_{s1}	65% θ_{s1}	60% θ_{s1}	70% θ_{s2}	75% θ_{s2}	80% θ_{s3}	70% θ_{s3}	Drying
Monitored soil depth/cm	—	0–20	0–20	0–20	0–30	0–30	0–40	0–40	—

θ_{s1}, θ_{s2}, and θ_{s3} represent saturated volumetric soil moisture for the 0–20, 0–30, and 0–40 cm layers, respectively. [a] In the case of pesticide, fertilizer application, and rainfall, standing irrigation water at a depth of up to 5 cm was maintained for less than 5 days. [b] The data show the water depth during the re-greening stage.

2.3. Yield Measurement, Soil Sampling, and Analysis

Rice yield was estimated by artificially harvesting crops per unit area of each plot. Three hills of rice were randomly chosen to evaluate the filled grain number, setting percentage, thousand kernel weight, and panicle number of each treatment.

A total of 108 soil samples were collected from each plot following an S-shaped pattern at 0–10, 10–20, and 20–40 cm depths during the whole growth stage of rice in 2015 (23 June, 12 July, 20 August, 23 August, 21 September, and 25 October) and 2016 (29 June, 27 July, 21 August, 4 September, 21 September, and 25 October). After harvesting with a spiral drill (diameter, 38 mm; length, 1 m), three samples of 0–40 cm soil were randomly collected in each plot and fully stirred. Then, samples from the same depth were homogenized by mixing, separated from visible debris and crop residues, and divided into two parts. One part of the fresh samples was stored at 4 °C, and the other was air-dried, ground, and screened with a sieve of 0.149 mm; 12.5 g of fresh soil samples for dissolved organic carbon (DOC) was placed in a conical flask, immersed into 50 mL of 0.5 mol L^{-1} K$_2$SO$_4$ solution, and shaken for 30 min before the extracts were separated with a 0.45-μm filter. SOC was measured by the potassium dichromate external heating method, and the oxidation correction coefficient was considered. Besides, soil water content was recorded by a time domain reflectometer (Soil Moisture Equipment, Ltd., Corp. USA), and vertical rulers were used to monitor water layer at 8 a.m. everyday. The amount of irrigation water for each plot was calculated by using the water meter

2.4. DNDC Model

2.4.1. Overview of the DNDC Model

The DNDC model is a process-cased biogeochemical model written in Visual C++ 6.0 language for C and N dynamics in agro-ecosystems. This model has evolved over decades of development since it was developed by Li et al. [25]. Various soil hydrological processes were included in the present model. The DNDC model has been used worldwide because of its simple parameter input and accurate simulation results. It was designated as the preferred biogeochemical model in Asia by the International Symposium on Global Change in the Asia-Pacific region in 2000 [26]. The DNDC model has good adaptability in China [27,28], but studies on predicting SOC dynamics in paddy soil under water-saving irrigation and water-carbon coupling based on future climatic conditions are few. Therefore, the present study improved the irrigation module of DNDC95, which is the latest version of the DNDC model, to realize simulation of paddy fields under controlled irrigation and to optimize the irrigation module of paddy fields in the model on the basis of experimental data. More detail can be found in the Supplementary Materials.

2.4.2. Input Data

Daily meteorological data, soil properties, and agricultural management measures were collected to support DNDC simulation. Soil physical and chemical properties, including initial soil C and N content, texture, and field capacity, were obtained through field sampling and laboratory analysis. The value of SOC at surface soil (0–10 cm) used as an input to the model was based on a measured total SOC value (11.1 g C kg^{-1}). The contents of TN, NH_4^+-N, and NO_3^--N served as a pre-fertilization input value of DNDC. Agricultural management measures were obtained on the basis of field records and local farmers' habits. The meteorological data used in this paper were as follows: historical meteorological observation data and GCMs from the Meteorological Information Center of China Meteorological Administration (http://data.cma.gov.cn/). Data included the daily maximum temperature, minimum temperature, radiation, wind speed, and precipitation. The future climate projections were acquired from four GCMs participating in the Coupled Model Intercomparison Project (CMIP5) experiment, including BCC-CSM1.1 (m), MIROC-ESM-CHEM, GFDL-ESM2M, and HadGEM2-ES (Table 3) [24,29]. In accordance with the new emissions scenarios proposed by CMIP5, representative concentration pathways (RCPs) and four climate scenarios, RCP 2.6, RCP 4.5, RCP 6.0, and RCP 8.5, were selected. RCP 2.6 is a low peak-and-decay scenario (the radiation force reaches its maximum near the middle of the 21st century before falling to 2.6 W m^{-2}), RCP 8.5 is a high-emissions scenario (the radiation force rises to 8.5 W m^{-2} by 2100), and RCP 6.0 and RCP 4.5 are two intermediate scenarios (with a radiation force stability in 6.0 W m^{-2} and 2.6 W m^{-2}, respectively, by 2100).

Table 3. Four general climate models used in this study.

Institutions	Models	Approximate Atmospheric Resolution
Beijing Climate Center, China Meteorological Administration	BCC-CSM1.1 (m)	1.125° × 1.125°
Japan Agency for Marine-Earth Science and Technology [1]	MIROC-ESM-CHEM	2.8125° × 2.8125°
Geophysical Fluid Dynamics Laboratory	GFDL-ESM2M	2.5° × 2°
Met Office Hadley Center	HadGEM2-ES	1.875° × 1.24°

[1]: Atmosphere and Ocean Research Institute (The University of Tokyo) and National Institute of Environmental Studies.

2.4.3. BMA Method

As an advanced statistical method based on Bayesian theory and in consideration of model uncertainty, BMA has been proposed to combine multiple climate models to provide good performance models with high weights. BMA has been widely used in multimodel ensemble predictions of future climate. Therefore, four future climate models were predicted using the BMA weighted set in the present study and estimated on the basis of two statistical downscaling methods: back-propagation neural network and Statistical Downscaling Model (SDSM) developed by Wilby et al. [30,31]. Their mathematical expressions are as follows [24]:

Assume that y is the prediction variable, and its posterior distribution is as follows:

$$p(y|f_1, f_2, \ldots, f_k, D) = \sum_{k=1}^{K} p(y|f_k, D)p(f_k|D) \tag{1}$$

On the premise of satisfying the minimum mean squared error, the combined prediction formula on the basis of the basic principle of Bayesian theorem is as follows:

$$E_{BMA}(y|D) = \sum_{k=1}^{K} p(f_k|D)\, E[p_k(y|f_k, D)] = \sum_{k=1}^{K} \omega_k, f_k \tag{2}$$

where $p(f_k | D)$ denotes the posterior probability that model f_k is correct given the training data and is calculated with Bayes' theory; $p(y|f_k, D)$, estimated from the training data, is the predictive probability density function based on model $y|f_k$ alone; and k is the number of models being combined, which is equal to four in this study. This formula uses the posterior probability $p(f_k | D)$ of the model as the weight for all possible model predictions $E(D|f_k, D)$ and obtains the combined predicted value.

Based on the field experimental data, we modified and verified the DNDC model to simulate soil organic carbon in paddy fields under different water and carbon management systems. The controlled irrigation module was added to the irrigation module of DNDC to realize the simulation of paddy fields under controlled irrigation. Then, combined with the climate model and climate scenarios after the BMA-weighted average, the simulation of SOC and rice yield under the corresponding water and carbon management systems in the next 80 years was conducted.

2.5. Data Analysis

Validation of the model results in the current study mainly included the average deviation method, correlation coefficient method, relative error method, and root mean squared method [32]. The absolute root mean squared error ($RMSE_a$), normalized root mean squared error ($RMSE_n$), coefficient of model efficiency (EF), and coefficient of determination (R^2) were used to quantitatively assess the goodness-of-fit between the simulated results and measured (observed) results. Their mathematical expressions are as follows:

$$EF = 1 - \frac{\sum_{i=1}^{n} (SM_i - OBS_i)^2}{\sum_{i=1}^{n} (OBS_i - \overline{OBS})^2} \tag{3}$$

$$EF = 1 - \frac{\sum_{i=1}^{n} (SM_i - OBS_i)^2}{\sum_{i=1}^{n} (OBS_i - \overline{OBS})^2} \tag{4}$$

$$RMSE_n = \frac{100 \times RMSE_a}{OBS_{avg}} \tag{5}$$

$$R^2 = \left(\frac{\sum_{i=1}^{n} (OBS_i - OBS_{avg})(SM_i - SM_{avg})}{\sqrt{\sum_{i=1}^{n} (OBS_i - OBS_{avg})^2 \sum_{i=1}^{n} (SM_i - SM_{avg})^2}} \right)^2 \tag{6}$$

where OBS_i is the observed value, OBS_{avg} is the average observed value, SM_i is the simulated value, SM_{avg} is the average simulated value, and n is the sample size. Higher R^2 and lower $RMSE_n$ indicated a good fit between the simulated and observed data. The smaller the $RMSE_n$ value is, the higher the fitting degree between the simulated value and the observed value. A value less than 10% indicates good consistency between the simulated value and the observed value. The results between 10% and 20% indicate an ordinary simulation effect, and a value higher than 30% indicates an unsatisfactory simulation effect [33,34]. The Taylor diagram is a polar-style graph, which summarizes the three statistical indices, i.e., the correlation coefficient between simulations and observations (R), the root mean squared error ($RMSE$), and the standard deviation (STD) using a single point. Given its comprehensiveness and visibility, Taylor diagrams are particularly beneficial in evaluating the relative accuracy of the different models. The radial distance from the origin reflects STD, the cosine of the azimuth angle denotes R, and the radial distance from the observed points is proportional to the $RMSE$ difference. A main criterion can usually be summarized: the closer a point is to the observed data, the better the fit between the observed and simulated data [35].

Origin 9.1 software (OriginLab Corporation, Northampton, MA, USA) and MATLAB 2017 (MathWorks Corporation, USA) were used to calculate data and construct the relevant charts. Statistical analysis was carried out using standard procedures on a randomized plot design (SPSS 22.0). Significance was calculated on the basis of a Least significant difference (LSD) test at the 0.05 probability level.

The Mann–Kendall trend test, which we used in this study based on MATLAB 2017, is one of the widely used distribution-free tests of trend in time series. A standard normal variate Z is calculated as follows:

$$Z = \begin{cases} \frac{S-1}{\sqrt{Var(S)}}, S > 0 \\ 0, S = 0 \\ \frac{S+1}{\sqrt{Var(S)}}, S < 0 \end{cases} \tag{7}$$

$$UF_k = \frac{S_k - E(S_k)}{\sqrt{Var(S_k)}}, k = 1, 2, \ldots, n \tag{8}$$

$$UB_k = \begin{cases} -UF_k, k = n, n-1, \ldots, 1 \\ 0, k = 1 \end{cases} \tag{9}$$

In a two-sided test for the trend, the null hypothesis of no trend is rejected if $|Z| > Z_{\alpha/2}$ where α is the significance. The calculation method of $Var(S)$ and S can be found in the literature [36], where $Z > 0$ indicates an upward trend and $Z < 0$ indicates a downward trend. In addition, UF is the standardized result of S, which is a statistical sequence calculated in time sequence and obeys normal distribution, while UB is repeatedly calculated in reverse chronological order.

3. Results

3.1. Model Modification and Validation

3.1.1. Model Modification

On the basis of the source code of DNDC95, this study improved the module on paddy field flooding in the farmland management menu. The two methods for the original water management module are the following: continuous flooding (water level is maintained at 10 cm) and alternative irrigation (water level fluctuates between −5 to 5 cm). The problems in the model were solved by improving the following three aspects: (1) the 50-cm constant soil layer assumed in the original DNDC model was adjusted to a value that varied with the depth of the rice root layer; (2) the fluctuation range of the water level was adjusted in accordance with the upper and lower limits of irrigation water controlled by soil moisture content; and (3) the upper and lower limits of irrigation with controlled irrigation were changed with the rice growth period, controlled irrigation with rice growth period was implemented, and the corresponding parameters were adjusted. Controlled irrigation was monitored in accordance with the soil moisture and water layer indicators in Table 2. The amount of irrigation water simulated by DNDC under controlled irrigation and traditional flooding irrigation after the modification was consistent with the observed irrigation water amount (Table 4). Additionally, crop parameters were calibrated in this study. The maximum crop yield, biomass allocation, and C/N ratio of the crops were modified on the basis of the observed results, and some internal parameters were modified to simulate actual conditions in the field. For example, the chromic acid wet oxidation method [37] and the Kjeldahl method [38] were used to estimate the total carbon nitrogen ratio of stems, leaves, and grains at the heading and maturing stages of Nanging 46. The total C/N ratios used for model correction were 55 for the root, 75 for the stem and leaf, and 75 for the grain. The maximum biomass production of grain was modified to 4700 kg C ha^{-1} to stay consistent with our observed data.

Table 4. Comparison of observed and simulated irrigation values of the Denitrification Decomposition (DNDC) model simulation.

Year	Treatments	Observed/mm	Simulated/mm	$RMSE_n$
2015	Controlled irrigation	356.93	346.03	3.08
	Flood irrigation	812.11	789.10	
2016	Controlled irrigation	456.43	468.14	3.77
	Flood irrigation	954.78	919.01	

Notes: Observed and simulated denote the observed irrigation amount and the simulated irrigation amount, respectively.

3.1.2. Model Calibration and Validation

Model Calibration

The comparisons of DOC and SOC measured values and simulated values in the test area in 2015 are shown in Figures 2 and 3. The dynamic changes in SOC and DOC in paddy soil under different water and carbon management systems in one year were well fitted through the modified DNDC model. The simulated values were consistent with the observed values. Tables 5 and 6 reflect the evaluation results of the SOC and DOC simulation values, respectively. The $RMSE_a$ values of the SOC and DOC simulations were 0.35–1.62 g kg^{-1} and 23.63–38.49 mg kg^{-1}, respectively. The $RMSE_n$ values of the SOC and DOC simulations were 3.54–17.59% and 8.79–13.93%, respectively. The regression coefficient R^2 of DOC was 0.80–0.99, and the EF values of SOC and DOC were close to 1. The SOC regression coefficients of the partial treatments (FS and FO) were closer to 1, which indicated that the modified DNDC model can accurately simulate the effects of different water and carbon management systems on SOC and DOC dynamics in paddy soil.

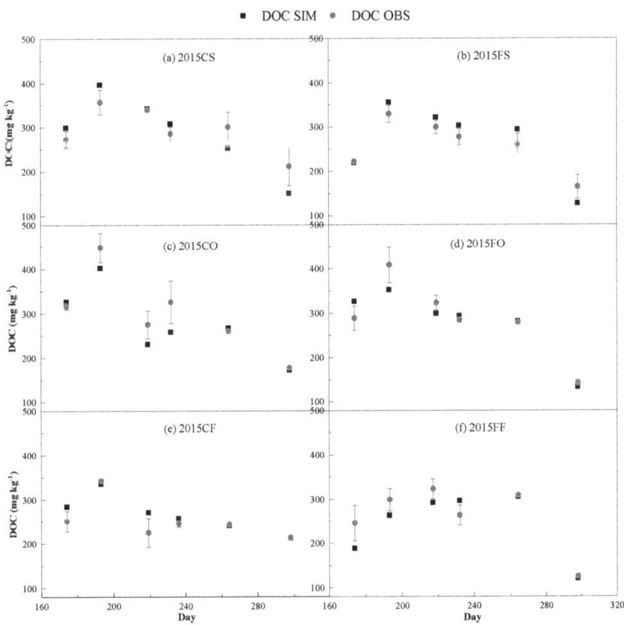

Figure 2. Simulation of dissolved organic carbon (DOC) (0–10 cm soil) change in each treatment during the calibration period (2015), where (**a**–**f**) present the CS, FS, CO, FO, CF, and FF treatments, respectively.

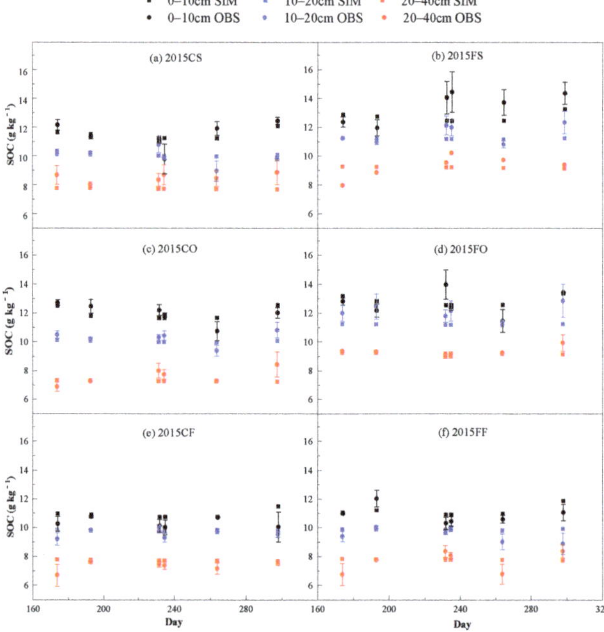

Figure 3. Simulation of soil organic carbon (SOC) change in each treatment during the calibration period (2015), where (**a**–**f**) present the CS, FS, CO, FO, CF, and FF treatments, respectively.

Table 5. Estimation of SOC results for each treatment by using the modified DNDC model during the calibration period (units of SOC: $g\,kg^{-1}$).

Variable	Treatments	N	X_{obs}(SD)	X_{sim}(SD)	$P(t^*)$	α	β	R^2	$RMSE_{\alpha}$	$RMSE_n$	EF
SOC	CF	6	10.31(0.30)	10.94(0.27)	0.03	1.37	2.90	0.86	0.78	7.11	0.91
0–10 cm	CS	6	11.44(0.87)	11.49(0.30)	0.89 *	0.19	10.20	0.72	0.71	6.19	1.00
	CO	6	12.34(1.04)	11.97(0.39)	0.49 *	0.36	7.67	0.74	1.16	9.73	0.99
	FF	6	10.89(0.58)	11.15(0.34)	0.34 *	0.25	9.04	0.81	0.59	5.31	1.00
	FS	6	13.48(0.97)	12.69(0.30)	0.16 *	0.66	4.78	0.94	1.32	10.37	0.99
	FO	6	12.70(0.83)	12.84(0.32)	0.71 *	0.46	7.16	0.89	0.80	6.24	1.00
SOC	CF	6	8.95(0.79)	9.76(0.04)	0.07 *	0.09	9.03	0.82	0.35	3.54	0.98
10–20 cm	CS	6	10.00(0.56)	10.04(0.05)	0.87 *	0.58	4.25	0.93	0.54	5.40	1.00
	CO	6	10.95(1.00)	11.04(0.06)	0.09 *	0.02	9.78	0.87	1.34	13.39	0.99
	FF	6	9.49(0.44)	9.85(0.04)	0.14 *	−0.05	10.38	0.78	0.58	5.86	1.00
	FS	6	11.41(0.86)	11.20(0.04)	0.61 *	0.02	10.94	0.84	0.87	7.74	0.99
	FO	6	12.11(0.47)	11.20(0.04)	0.01	0.06	10.38	0.60	1.01	9.01	0.99
SOC	CF	6	7.13(0.77)	7.13(0.04)	0.14 *	0.05	7.39	0.81	0.96	12.40	0.98
20–40 cm	CS	6	8.51(0.28)	7.72(0.03)	0.01	−0.09	8.51	0.84	0.84	10.86	0.99
	CO	6	7.60(0.52)	7.27(0.04)	0.24 *	−0.06	7.67	0.59	0.65	8.88	0.99
	FF	6	7.68(0.68)	7.78(0.03)	0.76 *	−0.01	7.80	0.81	0.70	8.95	0.99
	FS	6	10.12(1.31)	9.20(0.03)	0.19 *	−0.02	9.41	0.62	1.62	17.59	0.97
	FO	6	9.32(0.31)	9.20(0.03)	0.46 *	−0.05	8.39	0.79	0.35	3.82	1.00

Notes: N is the number of samples; X_{obs} is the average observed value; X_{sim} is the average simulated value; SD is standard deviation; $P(t^*)$ is t-test significance; α and β are the slope and intercept of the linear correlation between simulated values and observed values, respectively; and R^2 is the coefficient of determination between the simulated value and the observed value. In $P(t^*)$, * means that the difference between the simulated value and the observed value is not significant and that the credibility is 95%.

Table 6. Evaluation of DOC simulation results of each treatment by using a modified DNDC model during the calibration period and verification period (units of DOC: mg kg^{-1}).

Period	Treatments	N	X_{obs}(SD)	X_{sim}(SD)	$P(t^*)$	α	β	R^2	$RMSE_\alpha$	$RMSE_n$	EF
Calibration	CF	6	253.43(41.85)	268.65(55.03)	0.83 *	0.80	63.48	0.82	23.63	8.79	0.68
2015	CS	6	294.48(47.33)	291.55(76.59)	0.47 *	1.49	−146.39	0.84	38.12	13.08	0.59
	CO	6	300.76(82.01)	276.36(72.57)	0.13 *	0.83	28.02	0.86	38.49	13.93	0.78
	FF	6	261.05(66.19)	244.55(67.47)	0.26 *	0.92	3.10	0.82	33.19	13.57	0.75
	FS	6	259.38(53.43)	270.43(76.00)	0.36 *	1.41	−94.33	0.98	26.90	9.95	0.75
	FO	6	287.82(78.81)	280.93(70.41)	0.62 *	0.83	41.88	0.86	29.98	10.67	0.86
Validation	CF	6	217.12(43.39)	228.97(49.39)	0.14 *	1.09	−6.69	0.84	19.37	8.46	0.80
2016	CS	6	189.72(50.10)	201.27(75.88)	0.40 *	1.49	−82.10	0.97	30.06	14.94	0.64
	CO	6	222.98(68.81)	232.05(85.68)	0.42 *	1.22	−39.77	0.96	24.77	10.67	0.87
	FF	6	181.52(43.42)	168.36(55.57)	0.30 *	1.15	−39.69	0.80	28.73	17.06	0.56
	FS	6	174.99(45.67)	172.98(46.78)	0.38 *	1.02	−5.48	0.99	4.92	2.84	0.99
	FO	6	176.29(52.26)	176.92(52.97)	0.91 *	0.99	3.01	0.95	12.22	6.91	0.95

Notes: N is the number of samples; X_{obs} is the average observed value; X_{sim} is the average simulated value; SD is standard deviation; $P(t^*)$ is *t*-test significance; α and β are the slope and intercept of the linear correlation between simulated values and observed values, respectively; and R^2 is the coefficient of determination between the simulated value and the observed value. In $P(t^*)$, * means that the difference between the simulated value and the observed value is not significant and that the credibility is 95%.

Validation of Model Parameters

This study validated the modified DNDC model with 2016 data. The comparison between the simulated and observed values of DOC and SOC with different treatments during the verification period is shown in Figures 4 and 5. In most cases, the modified DNDC model with calibration parameters can simulate the dynamics of DOC and SOC in paddy fields under different water and carbon management systems. On the time scale of one year, DOC in paddy fields clearly changed with time, showing an increasing first and then decreasing trend, whereas the SOC content had a negligible change. In addition, the vertical distribution of SOC in paddy fields under different water and carbon management systems was relatively consistent. The SOC in the paddy field decreased as the soil depth increased, and the SOC fluctuation of 0–10 cm was larger than the SOC fluctuations of 10–20 cm and 20–40 cm. These results were essentially consistent with those of previous studies [39]. The results (Figure 6) showed that the simulated values of rice yield under different water and carbon treatments in the calibration and verification periods were close to the observed data, that is, to the line 1:1.

Comparison of Observed and Simulated Values

The parameter evaluation results for DOC (Table 6) and SOC (Table 7) in paddy fields simulated by the modified DNDC model showed the relationship between the simulated and observed values. $RMSE_a$ and $RMSE_n$ were small, indicating that the simulation was good. The model verification results indicated that irrigation and fertilization management had a great impact on SOC and DOC in paddy fields. Irrigation affected the dynamics of SOC and DOC. SOC under controlled irrigation was lower than that under flooding irrigation, but DOC was higher. Controlled irrigation is beneficial to the oxidative decomposition of paddy soil, which may be the cause of this phenomenon. In addition, the SOC contents of the organic fertilizer and straw returning treatments were significantly higher than the SOC content of the conventional fertilizer treatment, indicating that the appropriate fertilization method was beneficial to SOC accumulation in paddy fields.

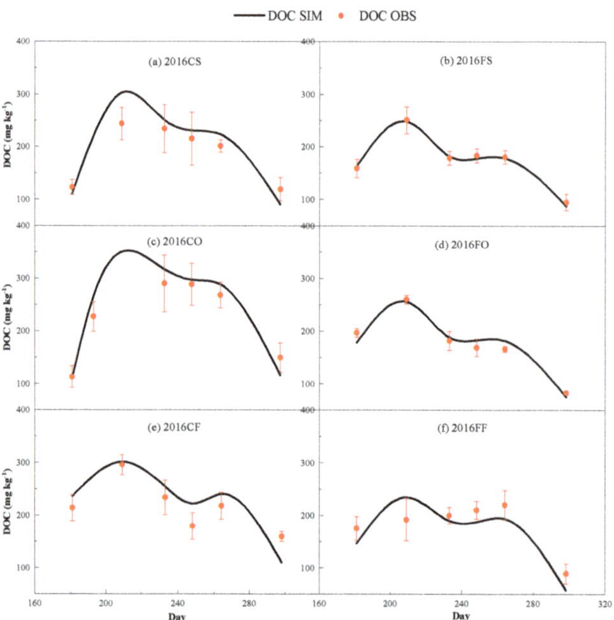

Figure 4. Simulation of DOC (0–10 cm soil) dynamics in each treatment during the verification period (2016), where (**a–f**) present the CS, FS, CO, FO, CF, and FF treatments, respectively.

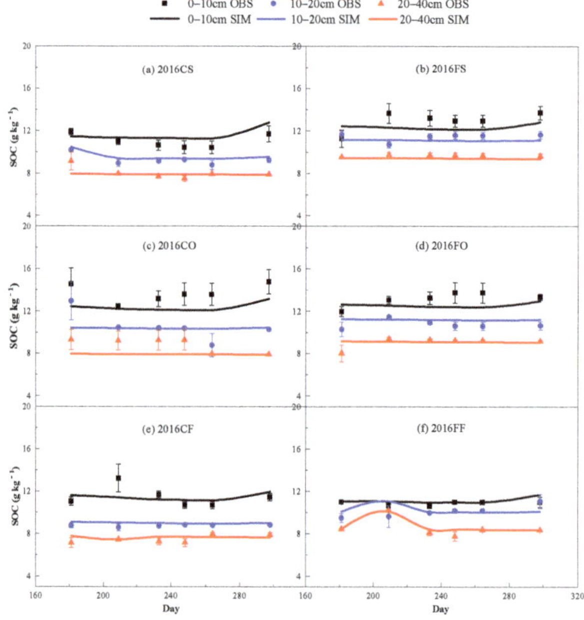

Figure 5. Simulation of SOC changes in each treatment during the verification period (2016), where (**a–f**) present the CS, FS, CO, FO, CF, and FF treatments, respectively.

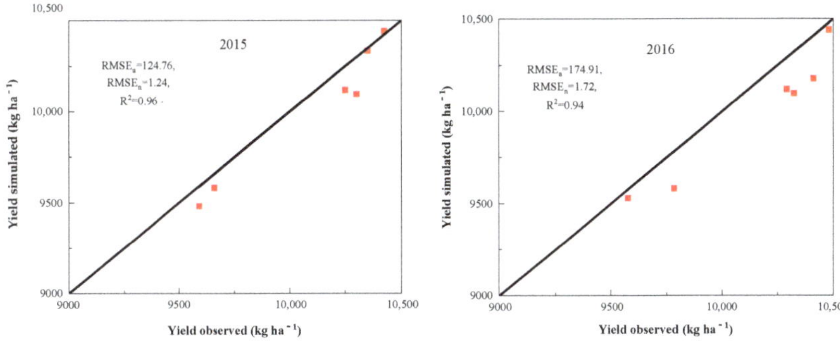

Figure 6. Simulation of yield changes in each treatment during the validation period (2015) and calibration period (2016): the solid line is a 1:1 relationship.

Table 7. Evaluation of SOC simulation results of each treatment by using modified DNDC model during the verification period (units of SOC: g kg^{-1}).

Variable	Treatments	N	X_{obs}(SD)	X_{sim}(SD)	$P(t^*)$	α	β	R^2	$RMSE_\alpha$	$RMSE_n$	EF
SOC	CF	6	11.46(0.87)	11.40(0.29)	0.88 *	0.05	10.69	0.97	0.88	7.72	0.90
0–10 cm	CS	6	11.02(0.59)	11.58(0.54)	0.05 *	0.57	10.00	0.76	0.75	6.46	0.84
	CO	6	13.69(0.78)	12.35(0.38)	0.01	0.35	10.83	0.62	1.45	11.77	0.83
	FF	6	10.88(0.15)	11.11(0.25)	0.09 *	0.52	5.58	0.85	0.34	3.09	0.92
	FS	6	12.96(0.81)	12.35(0.24)	0.16 *	0.82	13.37	0.84	1.02	8.26	0.56
	FO	6	13.18(0.60)	12.56(0.19)	0.10 *	0.09	11.38	0.87	0.93	7.37	0.87
SOC	CF	6	8.76(0.08)	9.01(0.05)	0.01	−0.25	11.75	0.80	0.28	3.10	0.89
10–20 cm	CS	6	9.28(0.45)	9.61(0.39)	0.01	0.82	2.00	0.87	0.37	3.83	0.93
	CO	6	10.56(1.23)	10.41(0.05)	0.79 *	0.02	10.18	0.82	1.21	11.67	0.83
	FF	6	10.10(0.50)	10.24(0.37)	0.69 *	−0.29	17.17	0.73	0.75	7.30	0.83
	FS	6	11.45(0.33)	11.13(0.03)	0.09 *	−0.02	11.41	0.87	0.46	4.16	0.93
	FO	6	10.78(0.36)	11.22(0.03)	0.04	0.01	11.21	0.81	0.57	5.09	0.87
SOC	CF	6	7.48(0.34)	7.66(0.10)	0.33 *	−0.08	8.40	0.73	0.42	5.43	1.00
20–40 cm	CS	6	8.04(0.52)	7.91(0.02)	0.61 *	0.02	7.78	0.84	0.53	6.66	1.00
	CO	6	8.83(0.63)	7.92(0.02)	0.02	0.02	7.76	0.67	1.10	13.83	0.98
	FF	6	8.55(0.77)	8.70(0.62)	0.25 *	0.77	2.25	0.91	0.30	3.42	−1.16
	FS	6	9.70(0.07)	9.42(0.03)	0.00	−0.12	10.83	0.99	0.29	3.03	1.00
	FO	6	9.08(4.07)	9.14(0.03)	0.80 *	−0.03	9.40	0.84	0.49	5.37	1.00

Notes: N is the number of samples; X_{obs} is the average observed value; X_{sim} is the average simulated value; SD is standard deviation; $P(t^*)$ is t-test significance; α and β are the slope and intercept of the linear correlation between simulated values and observed values, respectively; and R^2 is the coefficient of determination between the simulated value and the observed value. In $P(t^*)$, * means that the difference between the simulated value and the observed value is not significant and that the credibility is 95%.

3.2. Projection of SOC and Rice Yield in Paddy Fields Based on BMA and Modified DNDC

3.2.1. BMA Method Evaluation of Predicted Values of Meteorological Parameters Required by DNDC

Different GCMs should be combined to provide detailed and accurate climate data in the context of climate change. In the present study, four GCMs processed by BMA were used to obtain four climate variables as required by the modified DNDC model: maximum temperature, minimum temperature, wind speed, and radiation (Figure 7). The performance of the BMA ensemble multi-model to predict future climate variations was evaluated with a Taylor chart (Figure 8). Numerous studies have shown that the prediction effect of BMA parameters is improved by extending the model training time [40,41]. This study used 40 years (1961–2000) to train BMA weights, and current and future climate parameters were generated in the remaining stages (2001–2099). The comparison between simulated and observed precipitation values in 2015 and 2016 treated by BMA is shown in Figure 9. In the calibration and verification period of the model, the simulated and the observed rainfall values treated by BMA had a good fitting effect. The simulated precipitation value and the observed value were relatively close except for the slightest occurrence of a peak value. In Figure 7, the meteorological parameters generated by BMA were more consistent on the daily scale than at other scales measured by any single model. Figure 8 shows the relative accuracy of the model with a Taylor diagram. The results of the BMA method (point E) were closer to the points marked "observed" than to the data measured by any single model (points A, B, C, and D). Thus, BMA exhibited a good correlation and small RMSE. Except for the analog value matching the effect of wind speed, which was slightly poor (even if R of the BMA method was also approximately 0.7), the prediction of the other meteorological factors was good.

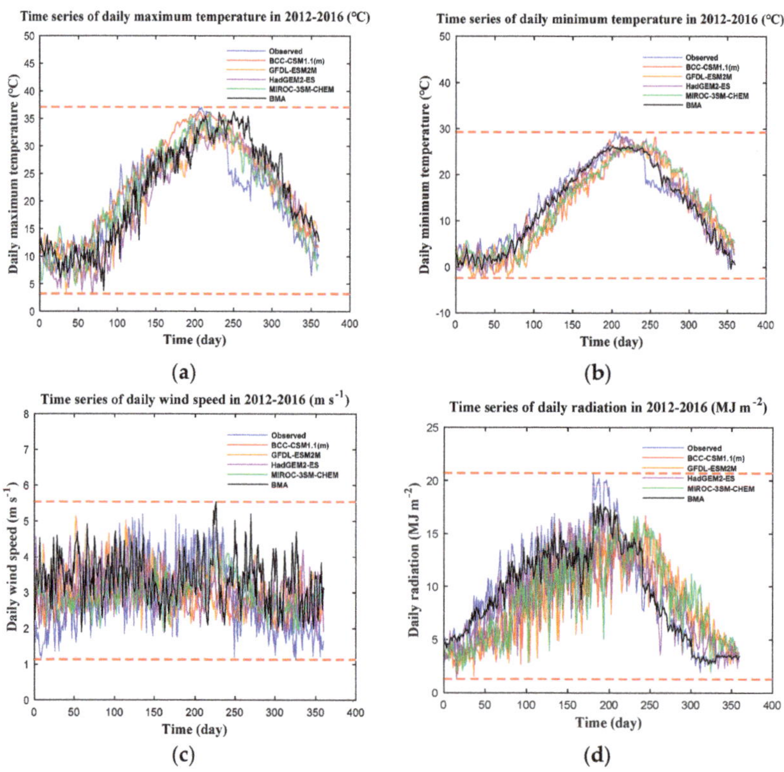

Figure 7. Time series of daily mean maximum temperature (**a**), minimum temperature (**b**), wind speed (**c**), and radiation (**d**) from 2012 to 2016: observed is the measured value, and BCC-CSM1.1 (m), GFDL-ESM2M, HadGEM2-ES, and MIROC2SM-CHEM represent the four climate models in Table 2, respectively. BMA (Bayesian Model Averaging) represents the value after BMA-weighted average.

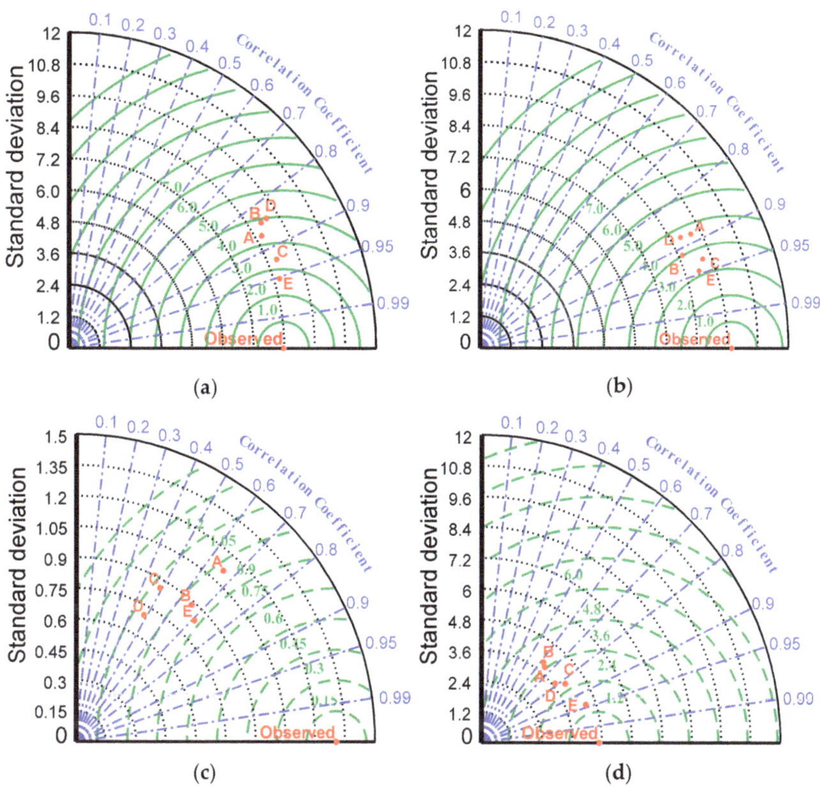

Figure 8. Taylor diagrams for meteorological factors in Kunshan, 2012–2016: this diagram is a comparison between the projected and measured values of four meteorological parameters required by a modified DNDC model. The four figures are as follows: (**a**) maximum temperature, (**b**) minimum temperature, (**c**) wind speed, and (**d**) radiation. Observed is the observed value, A is BCC-CSM1.1 (m), B is GFDL-ESM2M, C is HadGEM2-ES, D is MIROC-3SM-CHEM, and E is the BMA-weighted value.

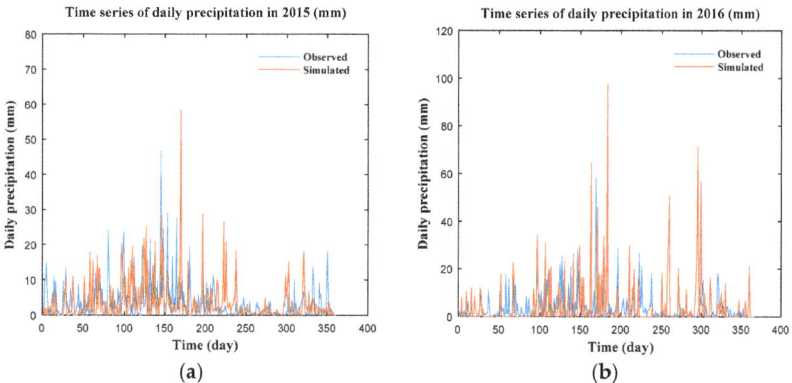

Figure 9. Comparison of simulated and actual precipitation values in 2015 (**a**) and 2016 (**b**) treated by BMA.

3.2.2. SOC Dynamics Prediction in Paddy Fields under Water and Carbon Regulation in Future Climate Conditions

On the basis of the modified DNDC model and the BMA method, this study predicted the SOC changes (0–10 cm) in paddy fields under four climate scenarios (i.e., RCP 2.6, RCP 4.5, RCP 6.0, and RCP 8.5) over the next 80 years (2020–2099), as shown in Figure 10. The average predicted SOC under different climate scenarios consistently occurred in the following order FO > CO > FS > CS > FF > CF. The trend lines of the SOC change in paddy fields under the four climate scenarios were estimated via linear square fitting (Figure 10). This trend indicated that the effect of fertilizer management on SOC changes in paddy fields over the long term was very large in the four scenarios. To some extent, this phenomenon explained the similar results found for the different climate scenarios, i.e., the SOC of the CF and FF treatments decreased with prolonged time, while the CS, CO, FS, and FO treatments showed an increasing trend with an extended time. Fertilizer management obviously affected the long-term trend of SOC in paddy fields under the same irrigation treatment. Irrigation had a certain impact on SOC in paddy fields over a short time, but only a negligible difference was observed over the long term. The overall trend in the SOC changes in paddy fields under flooding irrigation and controlled irrigation treatments was consistent and showed that SOC decreased in the conventional fertilizer treatment and increased in the treatment with organic fertilizer and straw application. In comparison with that in the 2020s, in the 2090s, the average values of the CF and FF treatments decreased by 4.98%, 5.86%, 6.07%, and 7.49% in the RCP 2.6, RCP 4.5, RCP 6.0, and RCP 8.5 scenarios, respectively, while the average values of the other treatments in the 2090s increased by 102.97%, 99.68%, 99.57%, and 97.54%, respectively. In addition, in the first 5 years, the CS and CO treatments showed an unexpected downward trend and then increased rapidly, which was different from the results of the model verification period. This may have been due to the frequent alternation of drying and wetting under controlled irrigation conditions, which promoted soil respiration. Therefore, the SOC of paddy fields decreased in the short term, while the long-term application of organic fertilizer and straw application can offset this carbon loss effect. However, the SOC of the organic fertilizer treatment under the RCP 4.5 and RCP 6.0 scenarios increased in 2100, which were because both the low peak attenuation and high emissions scenarios were not conducive to the accumulation of SOC in paddy fields.

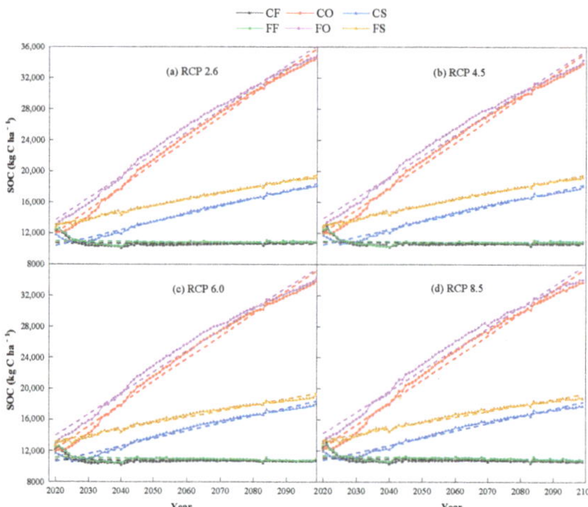

Figure 10. Prediction of SOC change in paddy fields with different treatments in the next 80 years under different climate scenarios (0–10 cm): the dashed lines in different colors in the figure correspond to the corresponding trend lines, and each trend line was derived from a series of annual values. The annual SOC is the final content at the end of the growth period of each treatment in the next 80 years.

In Table 8, the dynamics of SOC every 10 years under different treatments in the next 80 years is reflected by the RCP 2.6 scenario as an example. The results showed that the SOC of the conventional fertilizer treatment decreased rapidly in the first 10 years but gradually flattened. The soil organic carbon levels in the CF and FF treatments decreased by 14.18% and 13.50%, respectively. The SOC of the CS treatment abnormally decreased by 8.13% and increased rapidly. The effect of climate scenario on the SOC in paddy fields was not obvious (Figure 11). The SOC of the organic fertilizer treatment under the various climate scenarios increased with time. Compared with that under baseline conditions (2020), the SOC in the CO treatment under RCP 2.6 increased from 45.89% in 2040 to 149.39% in 2080 and the SOC in the CS treatment under RCP 4.5 increased from 3.07% in 2040 to 41.05% in 2080. In addition, the decline in the SOC in the CF and FF treatments was the largest in the first 20 years and remained unchanged.

Table 8. Changes in the SOC of paddy fields with different treatment in the next 80 years under the RCP 2.6 scenario.

Period	CF	CO	CS	FF	FO	FS
2020–2029	−14.18%	13.97%	−8.13%	−13.50%	18.86%	6.26%
2030–2039	−0.74%	24.33%	10.15%	−3.18%	19.04%	8.74%
2040–2049	4.22%	19.66%	12.47%	5.17%	17.96%	8.36%
2050–2059	0.79%	13.70%	7.67%	0.41%	12.21%	5.36%
2060–2069	0.57%	10.85%	6.61%	0.21%	9.68%	4.66%
2070–2079	0.46%	8.88%	5.07%	−0.04%	7.81%	3.67%
2080–2089	0.66%	6.99%	5.63%	2.04%	6.96%	4.40%
2090–2099	0.71%	6.07%	4.26%	0.11%	5.44%	3.02%

Notes: The values above denote simulated SOC change every 10 years (compared with the baseline 10 years ago) of the CF, CO, CS, FF, FO, and FS treatments in the 2020s, 2030s, 2040s, 2050s, 2060s, 2070s, 2080s, and 2090s.

3.2.3. Projection of Rice Yield Changes

On the basis of the modified DNDC model and BMA method, we predicted rice yield changes under different water and carbon management systems over the next 80 years under four climate scenarios (Figure 12). The relationship between the different treatments was essentially the same under various climate scenarios, which showed that the rice yield of the CS treatment was the highest and that of the CF treatment was the lowest. Thus, the long-term return of straw can significantly promote an increase in rice yield. Similar to the regulation of water and carbon regulation of SOC dynamics in paddy fields, irrigation and carbon management both affected the yield under the same climate conditions while the combination of appropriate fertilization and controlled irrigation evidently increased rice yield. The rice yield in the CS and CO treatments in most cases was higher than that in the FS and FO treatments. This study provides a trend line of each rice yield with time (Figure 12). Overall, the rice yields of the different treatments have good synchronization and almost simultaneously changed at different stages of the 21st century. In comparison with that in the 2020s, the average rice yield of each treatment in the 2090s decreased by 18.41%, 38.59%, 65.11%, and 65.62% in RCP 2.6, RCP 4.5, RCP 6.0, and RCP 8.5, respectively. In addition, the climate scenarios resulted in clear effects on rice yields under the same water and carbon management mode. The rice yield tended to increase in the first 20 years as the radiative force increased. However, under the high emissions scenario of RCP 8.5, the rice yield of the CS treatment initially remained unchanged but declined rapidly with increased time. Taking RCP 2.6 as an example, the results of the Mann–Kendall trend test [42] are shown in Figure 13. The yields of the CF and FF treatments increased in 2020–2023 and 2087 (UF > 0), while the UF values of the CO, CS, FO, and FS treatments were less than zero within the 95% confidence interval, except for the increase in 2020–2023, which indicated that maintaining rice yield via excessive carbon input might be difficult to sustain.

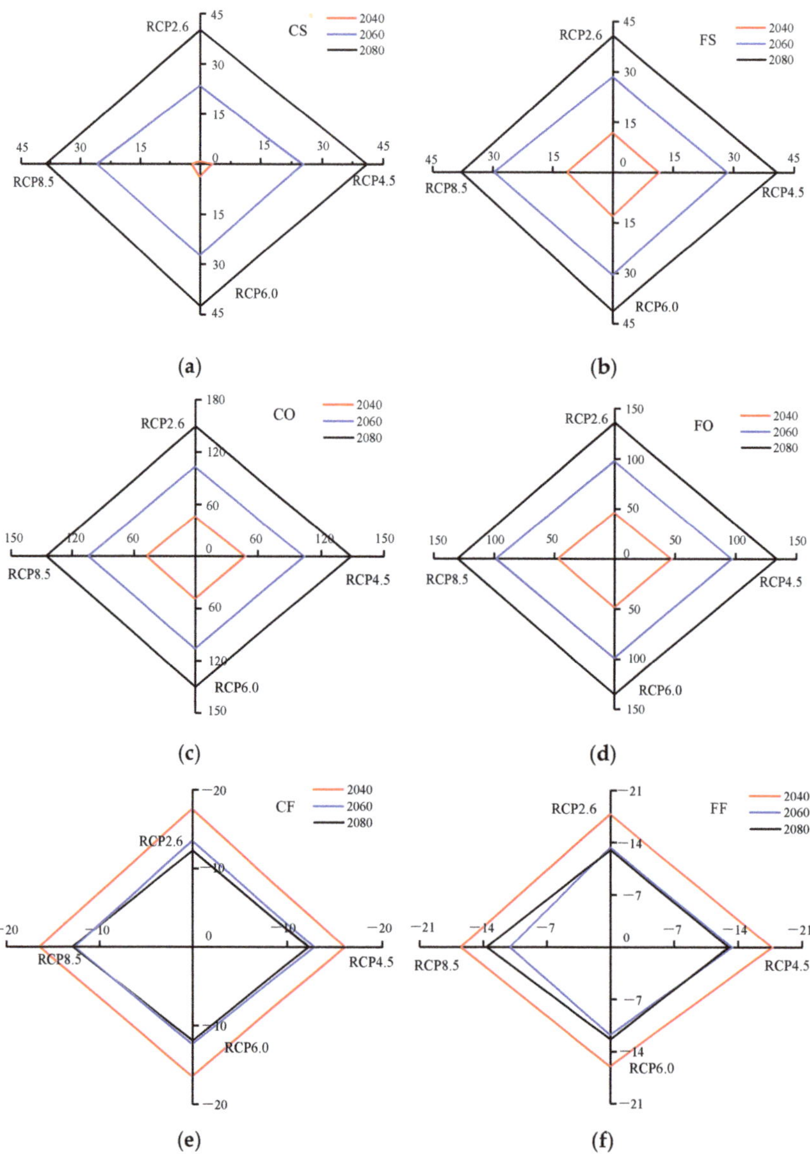

Figure 11. SOC in different treatments in four climate scenarios, where (**a–f**) present the CS, FS, CO, FO, CF, and FF treatments, respectively: the red, blue, and black lines represent the changes of SOC in paddy soil in 2040, 2060, and 2080, respectively, compared with the baseline (2020). The horizontal and vertical coordinates are the percentage values of the changes.

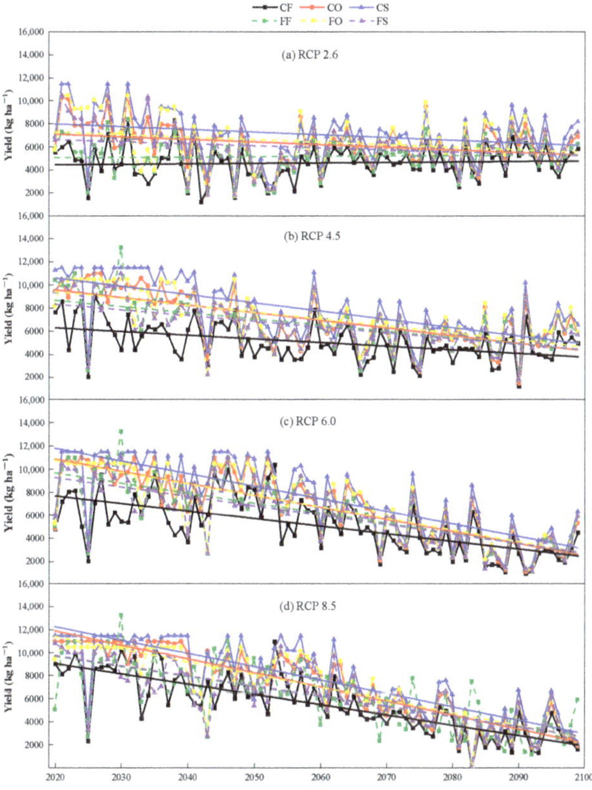

Figure 12. Prediction of rice yield change under different climate scenarios and treatments in the next 80 years: the trend lines in black, red, and blue in the figure represent conventional fertilizer treatment, organic fertilizer treatment, and straw returning treatment, respectively, while the solid and dotted lines represent conventional irrigation and controlled irrigation.

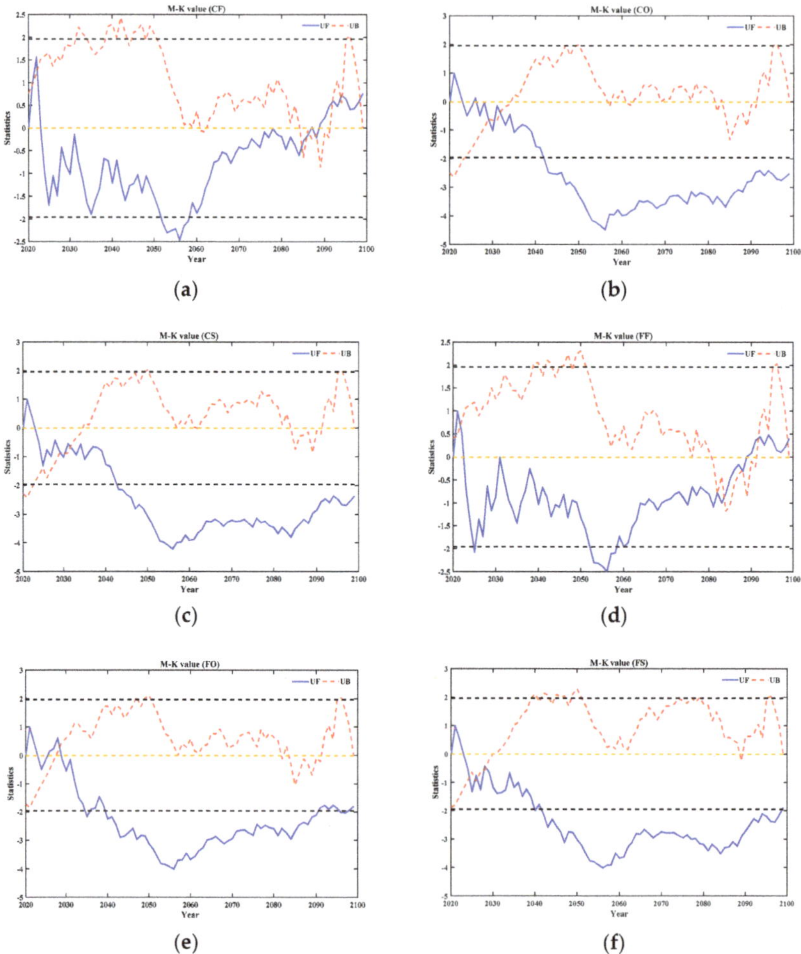

Figure 13. Mann–Kendall test charts of rice yield changes under different water and carbon treatments (with RCP 2.6 as an example), where (**a**–**f**) represent the CF, CO, CS, FF, FO, and FS treatments, respectively. The ordinate axis represents the values of UF and UB. UF > 0 indicates an upward trend, and UF < 0 indicates a downward trend. UB exceeded the upper and lower straight lines, indicating a significant upward or downward trend ($p < 0.05$). A sudden change point is indicated when UF intersects UB and is between the upper and lower lines.

4. Discussion

4.1. Performance of the Modified DNDC Model and Limitations

The default parameters of the DNDC model did not meet the needs of simulating SOC dynamic changes [23], and the model should be calibrated to reduce uncertainties in new systems or environments [20]. The results of this study showed that the modified DNDC model had good adaptability to SOC and yield simulation of paddy fields in the Kunshan area. The modified DNDC model successfully predicted the irrigation situation under water-saving irrigation and flood irrigation, and the effects of different irrigation and fertilization conditions on the SOC, DOC, and rice yield in paddy fields can be simulated. In addition, current research has mainly focused on water consumption and water use efficiency [43] and less on the effect of climate change on SOC in rice fields, and climate factors, such as temperature and precipitation, are important driving forces in SOC change [44], which have a far-reaching impact on agricultural production [9]. In this study, SOC prediction and rice yield were based on the modified DNDC model, local irrigation, fertilization management, and four GCMs integrated with BMA. The results weighted by BMA were closer to the observed points than to any single model in the Taylor diagram; thus, integrating multiple climate models with BMA is reliable, which is consistent with the results of Wang et al. [24]. Interpretation based on the single model was one of the limitations of this study. The uncertainty could be reduced by the method of multi-model ensemble [45]. In addition, it is desirable to calibrate the model results with data from more sites and long-term series of observed data under different water and carbon management.

4.2. Effects of Water and Carbon Management Systems on SOC in Paddy Fields and Rice Yield

The present study found that the combination of irrigation and fertilization patterns can markedly increase SOC and rice yield, which was consistent with the findings of Kamoni et al. [46]. This result may be due to irrigation improving the availability of soil N, thereby increasing productivity. The mechanism of the effects of irrigation on organic carbon remains unclear. Some studies have found that irrigation affects SOC mineralization and transfer [47], while others found that waterlogging affects rice residue input and the decomposition rate of SOC under anaerobic conditions, thus affecting SOC accumulation [48]. For example, Kelliher et al. [49] found that irrigation reduced SOC by 61%, while Houlbrooke et al. [50] found that irrigation had little effect on SOC, which may be related to environmental conditions, soil development stages and types, irrigation water quality, and years. This study found that the SOC of controlled irrigation paddy fields was lower than that of fields with conventional irrigation, which may be due to the frequent dry–wet alternation of controlled irrigation promoting microbial activities, increased soil fertility, and soil respiration, thus increasing soil carbon loss [51]; this finding is different from the results of Zhao et al. [52]. Zhao et al. found that optimized irrigation and fertilization treatments increased SOC in the North China Plain, which may be related to the retention of residue in the experiment every year. In addition, the present study found that controlled irrigation reduced the SOC of paddy fields while reducing irrigation water; the SOC content evidently increased after the combination of irrigation with straw returning or application of organic fertilizer. Thus, applying organic fertilizer or straw returning under controlled irrigation conditions can reduce the water footprint while addressing SOC. Combining controlled irrigation with organic fertilizer and returning straw to the fields, which is a feasible alternative water and carbon management mode, saved a large amount of water resources and increased rice yield and SOC content.

The dynamics of SOC in paddy fields are the net result of organic matter input and output. Irrigation schedules and fertilization affect soil organic carbon in paddy fields by changing the input of energy or material [53]. SOC dynamics are difficult to measure in the short term. This process-based model is a good tool for predicting future trends. The results of long-term simulation of the SOC changes in paddy fields under different water and carbon management systems (Figure 10) showed that the combination of controlled irrigation and suitable organic fertilizer application is a satisfactory water and carbon regulation mode. SOC growth was rapid, and yield was maintained at a high level with prolonged time. In addition, fertilizer management has a considerable impact on the long-term evolution of SOC on farmlands, which was consistent with the results of previous studies. For example, Wan et al. and Wang et al. [40,54] found through model research that an SOC of 0–30 cm on farmland in China would decrease to 7.8–8.2 t ha^{-1} in 2080 without fertilizer management but would increase markedly if organic fertilizer or straw was applied to the field. This study found a synergistic relationship between SOC content and rice yield, and rice yield was high in the treatments with high SOC content, such as the CS and CO treatments, which was similar to the conclusion of Qiu et al. [55].

4.3. Effects of Climate Scenarios on SOC and Rice Yield in Paddy Fields and Possible Countermeasures

Impacts in climate scenarios have a considerable impact on rice yield, but their effect on SOC is less than that of agricultural management measures, which may be because climate change affects the decomposition of SOC, while agricultural management measures affect the soil carbon input quantity [56]; excessive carbon input may mask the impact of SOC decomposition. Additionally, the change in SOC was negatively correlated with initial SOC concentrations [57], and a high carbon input and low initial SOC would increase the pool of soil carbon. Conversely, the conversion of excessive carbon input into soil may offset the carbon loss caused by soil respiration, which explains to some extent why climate scenario impacts have a negligible effect on SOC changes in paddy fields. Unlike the current conclusion that fertilization can maintain high rice yields over the long term, although excessive fertilization can maintain high rice yields in the short term under future climate conditions, rice yields may still decrease in the long term (Figure 12). This phenomenon is attributed to the decline in rice yield caused by high temperatures and water stress that may have exceeded the impacts of promotion by fertilizer. The SOC of the controlled irrigation treatment increased rapidly in the late period but decreased in 2025, 2040, and 2083 in all treatments. This result may have been caused by the impact of climate conditions in such years. The average rice yields in all the treatments after 80 years decreased under the RCP 2.6, RCP 4.5, RCP 6.0, and RCP 8.5 scenarios by 18.41%, 38.59%, 65.11%, and 65.62%, respectively, compared with that in the baseline treatment (2020). In addition, previous studies [58] found that a variety of improvements can offset the decline in rice yield caused by climate warming, which might be a possible strategy to address climate change in the future.

Overall, paddy fields play a significant role in mitigating climate change through carbon sequestration, but the impact of different climate scenarios on SOC changes in paddy fields is less obvious than that of water and carbon management measures. Yu et al. [56] found that maintaining existing farmland management measures can maintain China's paddy soil carbon sequestration potential over the next 20–40 years; however, this result depends on long-term continuation of the current excessive carbon input management, which is closely related to the current policy of vigorously promoting and subsidizing straw returning and organic fertilizer application in China [59]. In accordance with the report released by the agricultural sector, most crop residues were removed from farmlands before the 1980s and used as fuel and animal feed in rural areas. This trend was reversed by the government through a policy in the 1990s to encourage farmers to recycle crop straw as much as possible, and the policy achieved considerable results [60]. At the same time, farmers stopped using crop straw as fuel due to improvements in living standards, which have caused serious environmental pollution in the past [61]. In addition, unreasonable fertilization leads to soil degradation, water pollution, soil acidification, and serious agricultural nonpoint source pollution [62]. Thus, how to promote straw returning in many developing countries across the world and to reduce its pollution is the direction of further study.

In addition, notably, in the future climate model, although water and carbon management will increase production and carbon sequestration, whether it will increase GHGs still needs further study. For example, excessive carbon input may increase greenhouse gases, such as CO_2 and CH_4, while SOC changes are sensitive to CO_2 concentrations. Thus, the benefits of carbon sequestration may be offset. The predicted results showed that the rice yields of all the treatments will decrease in the future, especially after the middle of the 21st century. Although the rice yield decreased under the coupling of controlled irrigation with straw returning and organic fertilizer, the rice yield was always higher than that in conventional fertilizer treatments. Thus, finding an appropriate amount of organic fertilizer or straw application to balance carbon sequestration is necessary to increase production and to reduce greenhouse gas emissions.

5. Conclusions

This study modified the DNDC model to adapt to the common water-saving irrigation mode in China, especially in the middle and lower reaches of the Yangtze River. The parameters related to SOC and rice yield were calibrated. In addition, the dynamics of SOC and rice yield in Kunshan over the next 80 years under different water and carbon management were predicted on the basis of the four climate scenarios synthesized via the BMA method. The results showed that the modified DNDC model had good adaptability to the simulation of SOC and rice yield under different water and carbon management. The $RMSE_n$ values of the SOC and DOC simulations were 3.45% to 17.59% and 8.79% to 13.93%, respectively. The R^2 of DOC was between 0.80 and 0.99, and the model efficiency coefficient EF values of SOC and DOC were all close to 1. In comparison to the single model, the BMA method can better simulate the changes in climate factors. Climate scenarios significantly affect rice yield, but their impact on SOC is less than agricultural management measures. Unfavorable climate will reduce yields in the future climate in spite of long-term over fertilization. Compared with traditional water and carbon management systems, the combination of controlled irrigation and organic fertilizer application or straw returning can obviously increase the SOC content and rice yield in the long-term simulation under the four climate scenarios, and the yield of the straw-returning treatment was higher. The SOC of controlled irrigation paddy fields was lower than that of conventional irrigation, but only a negligible difference was observed over the long term. Therefore, combining controlled irrigation and appropriate organic fertilizer can balance water conservation, can maintain SOC and a stable rice yield in paddy fields, and is the recommended water and carbon management system for paddy fields.

Supplementary Materials: The following are available online at https://www.mdpi.com/2071-105 0/13/2/568/s1, Figure S1: Structure of the DNDC model, Table S1: Input parameters required for regional simulation with DNDC.

Author Contributions: Conceptualization, Z.J. and S.Y.; methodology, Z.J.; software, Z.J.; validation, X.S., J.D. and X.C.; formal analysis, X.L.; investigation, X.S.; resources, J.D.; data curation, X.C.; writing—original draft preparation, Z.J.; writing—review and editing, Z.J. and S.Y.; visualization, Z.J., supervision, J.X.; project administration, X.L.; funding acquisition, S.Y. All authors have read and agreed to the published version of the manuscript.

Funding: This research was funded by the National Natural Science Foundation of China (no. 51879076 and 51579070) and by the Fundamental Research Funds for the Central Universities (no. 2019B67814).

Institutional Review Board Statement: Not applicable.

Informed Consent Statement: Not applicable.

Data Availability Statement: Data is contained within the article or supplementary material.

Acknowledgments: Thanks to Li Changsheng, one of the main founders of the DNDC95 model. Thanks to the National Meteorological Information Center, China Meteorological Administration for offering the meteorological data and the Working Group of the World Climate Research Program on Coupled Modeling, which is responsible for CMIP5.

Conflicts of Interest: The authors declare no conflict of interest.

Abbreviations

BMA	Bayesian Model Averaging
CMIP5	The fifth phase of the Coupled Model Intercomparison Project
CI	Controlled irrigation
CF	Controlled irrigation and farmer fertilizer practices
CO	Controlled irrigation and organic fertilizer management
CS	Controlled irrigation and straw returning
DNDC	Denitrification-Decomposition model

DOC	Dissolved organic carbon, g kg^{-1}
EF	Coefficient of model efficiency
FI	Flood irrigation
FF	Flood irrigation and farmer fertilizer practices
FO	Flood irrigation and organic fertilizer management
FS	Flood irrigation and straw returning
FFP	Farmer fertilizer practices
GCMs	General Circulation Models
LSDs	Least significant differences
R^2	Coefficient of determination
RCPs	Representative concentration pathways
$RMSE_a$	The absolute root mean squared error
$RMSE_n$	The relative root mean squared error
SDSM	Statistical Downscaling Model
SOC	Soil organic carbon, g kg^{-1}
STD	Standard deviation

References

1. Kirkby, C.A.; Richardson, A.E.; Wade, L.J.; Passioura, J.B.; Batten, G.D.; Blanchard, C.; Kirkegaard, J.A. Nutrient availability limits carbon sequestration in arable soils. *Soil Biol. Biochem.* **2014**, *68*, 402–409. [CrossRef]
2. Hutchinson, J.J.; Campbell, C.A.; Desjardins, R.L. Some perspectives on carbon sequestration in agriculture. *Agric. For. Meteorol.* **2007**, *142*, 288–302. [CrossRef]
3. Pan, G.X. Dynamics and Climate Change Mitigation of China. *Adv. Clim. Chang. Res.* **2008**, *4*, 282–289.
4. Zhou, H.Y. A Study on the Dynamic Changes of Cropland Soil Organic Carbon in Hilly Region of Southwest China Based on DNDC Model—A Case Study of Liangping County. Master's Thesis, Southwest University, Chongqing, China, 2016.
5. Pan, G.X.; Zhao, Q.G. Study on evolution of organic carbon stock in agricultural soils of China: Facing the challenge of global change and food security. *Adv. Earth Sci.* **2005**, *4*, 384–393.
6. Liu, Q.-H.; Shi, X.-Z.; Weindorf, D.C.; Yu, D.-S.; Zhao, Y.-C.; Sun, W.-X.; Wang, H.-J. Soil organic carbon storage of paddy soils in China using the 1:1,000,000 soil database and their implications for C sequestration. *Glob. Biogeochem. Cycles* **2006**, *20*, 1–5. [CrossRef]
7. Zhang, L.; Zhuang, Q.; He, Y.; Liu, Y.; Yu, D.; Zhao, Q.; Shi, X.; Xing, S.; Wang, G. Toward optimal soil organic carbon sequestration with effects of agricultural management practices and climate change in Tai-Lake paddy soils of China. *Geoderma* **2016**, *275*, 28–39. [CrossRef]
8. Yang, X.; Asseng, S.; Wong, M.T.F.; Yu, Q.; Li, J.; Liu, E. Quantifying the interactive impacts of global dimming and warming on wheat yield and water use in China. *Agric. For. Meteorol.* **2013**, *182–183*, 342–351. [CrossRef]
9. Piao, S.; Ciais, P.; Huang, Y.; Shen, Z.; Peng, S.; Li, J.; Zhou, L.; Liu, H.; Ma, Y.; Ding, Y.; et al. The impacts of climate change on water resources and agriculture in China. *Nature* **2010**, *467*, 43–51. [CrossRef]
10. Xu, S.; Shi, X.; Zhao, Y.; Yu, D.; Wang, S.; Tan, M.; Sun, W.; Li, C. Spatially explicit simulation of soil organic carbon dynamics in China's paddy soils. *Catena* **2012**, *92*, 113–121. [CrossRef]
11. Yan, X.; Akiyama, H.; Yagi, K.; Akimoto, H. Global estimations of the inventory and mitigation potential of methane emissions from rice cultivation conducted using the 2006 Intergovernmental Panel on Climate Change Guidelines. *Glob. Biogeochem. Cycles* **2009**, *23*, 1–15. [CrossRef]
12. Belder, P.; Spiertz, J.H.J.; Bouman, B.A.M.; Lu, G.; Tuong, T.P. Nitrogen economy and water productivity of lowland rice under water-saving irrigation. *Field Crop. Res.* **2005**, *93*, 169–185. [CrossRef]
13. Zhang, W.L.; Xu, A.G.; Ji, H.J.; Kolbe, H. Estimation of agricultural non-point source pollution in chinaand the alleviating strategies III. A review of policies and practices for agricultural non-point source pollution control in China. *Sci. Agric. Sin.* **2004**, *37*, 1026–1033. (In Chinese with English Abstract)
14. Li, H.; Wang, L.; Li, J.; Gao, M.; Zhang, J.; Zhang, J.; Qiu, J.; Deng, J.; Li, C.; Frolking, S. The development of China-DNDC and review of its applications for sustaining Chinese agriculture. *Ecol. Model.* **2017**, *348*, 1–13. [CrossRef]
15. Muñoz-Rojas, M.; Doro, L.; Ledda, L.; Francaviglia, R. Application of carbon soil model to predict the effects of climate change on soil organic carbon stocks in agro-silvo-pastoral Mediterranean management systems. *Agric. Ecosyst. Environ.* **2015**, *202*, 8–16. [CrossRef]
16. Chen, X.; Wu, H.; Wo, F. Nitrate vertical transport in the main paddy soils of Tai Lake region, China. *Geoderma* **2007**, *142*, 136–141. [CrossRef]
17. Zou, P.; Fu, J.; Cao, Z.; Ye, J.; Yu, Q. Aggregate dynamics and associated soil organic matter in topsoils of two 2000-year paddy soil chronosequences. *J. Soils Sediments* **2014**, *15*, 510–522. [CrossRef]
18. Liao, Q.; Zhang, X.; Li, Z.; Pan, G.; Smith, P.; Jin, Y.; Wu, X. Increase in soil organic carbon stock over the last two decades in China's Jiangsu Province. *Glob. Chang. Biol.* **2009**, *15*, 861–875. [CrossRef]

19. Yu, C.; Huang, X.; Chen, H.; Godfray, H.C.J.; Wright, J.S.; Hall, J.W.; Gong, P.; Ni, S.; Qiao, S.; Huang, G.; et al. Managing nitrogen to restore water quality in China. *Nature* **2019**, *567*, 516–520. [CrossRef]
20. Li, H.; Wang, L.; Qiu, J.; Li, C.; Gao, M.; Gao, C. Calibration of DNDC model for nitrate leaching from an intensively cultivated region of Northern China. *Geoderma* **2014**, *223–225*, 108–118. [CrossRef]
21. Minamikawa, K.; Fumoto, T.; Itoh, M.; Hayano, M.; Sudo, S.; Yagi, K. Potential of prolonged midseason drainage for reducing methane emission from rice paddies in Japan: A long-term simulation using the DNDC-Rice model. *Biol. Fertil. Soils* **2014**, *50*, 879–889. [CrossRef]
22. Yang, S.; Peng, S.; Xu, J.; He, Y.; Wang, Y. Effects of water saving irrigation and controlled release nitrogen fertilizer managements on nitrogen losses from paddy fields. *Paddy Water Environ.* **2015**, *13*, 71–80. [CrossRef]
23. Kröbel, R.; Sun, Q.; Ingwersen, J.; Chen, X.; Zhang, F.; Müller, T.; Römheld, V. Modelling water dynamics with DNDC and DAISY in a soil of the North China Plain: A comparative study. *Environ. Model. Softw.* **2010**, *25*, 583–601. [CrossRef]
24. Wang, W.; Ding, Y.; Shao, Q.; Xu, J.; Jiao, X.; Luo, Y.; Yu, Z. Bayesian multi-model projection of irrigation requirement and water use efficiency in three typical rice plantation region of China based on CMIP5. *Agric. For. Meteorol.* **2017**, *232*, 89–105. [CrossRef]
25. Li, C.; Frolking, S.; Frolking, T.A. A model of nitrous oxide evolution from soil driven by rainfall events: 1. Model structure and sensitivity. *J. Geophys. Res. Atmos.* **1992**, *97*, 9759–9776. [CrossRef]
26. Wang, D. Simulation and Prediction of Doil Organic Carbon Change in Arable Lands of Jilin Based on DNDC Model. Master's Thesis, Chinese Academy of Agricultural Sciences, Beijing, China, 2014.
27. Qiu, J.; Wang, L.; Tang, H.; Li, H.; Li, C. Studies on the Situation of Soil Organic Carbon Storage in Croplands in Northeast of China. *Agric. Sci. China* **2005**, *4*, 594–600, (In Chinese with English Abstract).
28. Smith, P.; Smith, J.U.; Powlson, D.S.; McGill, W.B.; Arah, J.R.M.; Chertov, O.G.; Coleman, K.; Franko, U.; Frolking, S.; Jenkinson, D.S.; et al. A comparison of the performance of nine soil organic matter models using datasets from seven long-term experiments. *Geoderma* **1997**, *81*, 153–225. [CrossRef]
29. Taylor, K.E.; Stouffer, R.J.; Meehl, G.A. An Overview of CMIP5 and the Experiment Design. *Bull. Am. Meteorol. Soc.* **2012**, *93*, 485–498. [CrossRef]
30. Wang, W.; Ding, Y.; Xu, J. Simulation of future climate change effects on rice water requirement and water use efficiency through multi-model ensemble. *J. Hydraul. Eng.* **2016**, *47*, 715–723.
31. Wilby, R.L.; Dawson, C.W.; Barrow, E.M. SDSM—A decision support tool for the assessment of regional climate change impacts. *Environ. Model. Softw.* **2002**, *17*, 147–159. [CrossRef]
32. Zhang, M.; Wei, Y.; Kong, F.; Chen, F.; Hailin, Z. Effects of tillage practices on soil carbon storage and greenhouse gas emission of farmland in North China. *Trans. Chin. Soc. Agric. Eng.* **2012**, *28*, 203–209.
33. Villalobos, F.J.; Hall, A.J.; Ritchie, J.T.; Orgaz, F. OILCROP-SUN: A Development, Growth, and Yield Model of the Sunflower Crop. *Agron. J.* **1996**, *88*, 403–415. [CrossRef]
34. He, M.; Wang, Y.; Ligang, W.; Ping, Z.; Changsheng, L. Using DNDC model to simulate black soil organic carbon dynamics as well as its coordinate relationship with crop yield. *J. Plant Nutr. Fertil.* **2017**, *1*, 9–19.
35. Taylor, E.K. PCMDI Report No. 55. *J. Geophys. Res.* **2001**, *93*, 7183–7212. [CrossRef]
36. Chattopadhyay, G.; Chakraborthy, P.; Chattopadhyay, S. Mann–Kendall trend analysis of tropospheric ozone and its modeling using ARIMA. *Theor. Appl. Climatol.* **2012**, *110*, 321–328. [CrossRef]
37. Walkley, A.; Black, I.A. An examination of the Degtjareff method for determining soil organic matter, and a proposed modification of the chromic acid titration method. *Soil Sci.* **1934**, *37*, 29–38. [CrossRef]
38. Bremner, J.M. Determination of nitrogen in soil by the Kjeldahl method. *J. Agric. Sci.* **2009**, *55*, 11–33. [CrossRef]
39. Alavaisha, E.; Manzoni, S.; Lindborg, R. Different agricultural practices affect soil carbon, nitrogen and phosphorous in Kilombero -Tanzania. *J. Env. Manag.* **2019**, *234*, 159–166. [CrossRef]
40. Wang, W.; Shao, Q.; Yang, T.; Yu, Z.; Xing, W.; Zhao, C. Multimodel ensemble projections of future climate extreme changes in the Haihe River Basin, China. *Theor. Appl. Climatol.* **2014**, *118*, 405–417. [CrossRef]
41. Yang, T.; Wang, X.; Zhao, C.; Chen, X.; Yu, Z.; Shao, Q.; Xu, C.-Y.; Xia, J.; Wang, W. Changes of climate extremes in a typical arid zone: Observations and multimodel ensemble projections. *J. Geophys. Res.* **2011**, *116*, 1–18. [CrossRef]
42. Torsri, K.; Octaviani, M.; Manomaiphiboon, K.; Towprayoon, S. Regional mean and variability characteristics of temperature and precipitation over Thailand in 1961–2000 by a regional climate model and their evaluation. *Theor. Appl. Climatol.* **2012**, *113*, 289–304. [CrossRef]
43. Nkomozepi, T.; Chung, S.-O. Assessing the trends and uncertainty of maize net irrigation water requirement estimated from climate change projections for Zimbabwe. *Agric. Water Manag.* **2012**, *202*, 60–67. [CrossRef]
44. Grace, P.R.; Colunga-Garcia, M.; Gage, S.H.; Robertson, G.P.; Safir, G.R. The Potential Impact of Agricultural Management and Climate Change on Soil Organic Carbon of the North Central Region of the United States. *Ecosystems* **2006**, *9*, 816–827. [CrossRef]
45. Saddique, Q.; Liu, D.L.; Wang, B.; Feng, P.; He, J.; Ajaz, A.; Ji, J.; Xu, J.; Zhang, C.; Cai, H. Modelling future climate change impacts on winter wheat yield and water use: A case study in Guanzhong Plain, northwestern China. *Eur. J. Agron.* **2020**, *119*, 126113. [CrossRef]
46. Kamoni, P.T.; Mburu, M.W.K.; Gachene, C.K.K. Influence of Irrigation and Nitrogen Fertiliser on Maize Growth, Nitrogen Uptake and Yield in a Semiarid Kenyan Environment. *East Afr. Agric. For. J.* **2003**, *69*, 99–108. [CrossRef]

47. Xu, Y.; Zhan, M.; Cao, C.; Ge, J.; Ye, R.; Tian, S.; Cai, M. Effects of irrigation management during the rice growing season on soil organic carbon pools. *Plant Soil* **2017**, *421*, 337–351. [CrossRef]
48. Tian, J.-; Dippold, M.; Pausch, J.; Blagodatskaya, E.; Fan, M.; Li, X.; Kuzyakov, Y. Microbial response to rhizodeposition depending on water regimes in paddy soils. *Soil Biol. Biochem.* **2013**, *65*, 195–203. [CrossRef]
49. Kelliher, F.M.; Condron, L.M.; Cook, F.J.; Black, A. Sixty years of seasonal irrigation affects carbon storage in soils beneath pasture grazed by sheep. *Agric. Ecosyst. Environ.* **2012**, *148*, 29–36. [CrossRef]
50. Houlbrooke, D.J.; Littlejohn, R.P.; Morton, J.D.; Paton, R.J. Effect of irrigation and grazing animals on soil quality measurements in the North Otago Rolling Downlands of New Zealand. *Soil Use Manag.* **2008**, *24*, 416–423. [CrossRef]
51. Wang, Y.; Liu, F.; Andersen, M.N.; Jensen, C.R. Carbon retention in the soil–plant system under different irrigation regimes. *Agric. Water Manag.* **2010**, *98*, 419–424. [CrossRef]
52. Zhao, X.; Hu, K.; Stahr, K. Simulation of SOC content and storage under different irrigation, fertilization and tillage conditions using EPIC model in the North China Plain. *Soil Tillage Res.* **2013**, *130*, 128–135. [CrossRef]
53. Li, Z.; Xu, X.; Pan, G.; Smith, P.; Cheng, K. Irrigation regime affected SOC content rather than plow layer thickness of rice paddies: A county level survey from a river basin in lower Yangtze valley, China. *Agric. Water Manag.* **2016**, *172*, 31–39. [CrossRef]
54. Wan, Y.; Lin, E.; Xiong, W.; Li, Y.E.; Guo, L. Modeling the impact of climate change on soil organic carbon stock in upland soils in the 21st century in China. *Agric. Ecosyst. Environ.* **2011**, *141*, 23–31. [CrossRef]
55. Qiu, J.; Li, C.; Wang, L.; Tang, H.; Li, H.; Van Ranst, E. Modeling impacts of carbon sequestration on net greenhouse gas emissions from agricultural soils in China. *Glob. Biogeochem. Cycles* **2009**, *23*, 1–16. [CrossRef]
56. Yu, Y.; Huang, Y.; Zhang, W. Projected changes in soil organic carbon stocks of China's croplands under different agricultural managements, 2011–2050. *Agric. Ecosyst. Environ.* **2013**, *178*, 109–120. [CrossRef]
57. Li, C.; Zhuang, Y.; Frolking, S.; Galloway, J.N.; Harriss, R.; Moore III, B.; Schimel, D.; Wang, X. Modeling Soil Organic Carbon Change in Croplands of China. *Ecol. Appl.* **2003**, *13*, 327–336. [CrossRef]
58. Liu, L.; Wang, E.; Zhu, Y.; Tang, L. Contrasting effects of warming and autonomous breeding on single-rice productivity in China. *Agric. Ecosyst. Environ.* **2012**, *149*, 20–29. [CrossRef]
59. Tan, D.-X.; Reiter, R.; Manchester, L.; Yan, M.-T.; El-Sawi, M.; Sainz, R.; Mayo, J.; Kohen, R.; Allegra, M.; Hardelan, R. Chemical and Physical Properties and Potential Mechanisms: Melatonin as a Broad Spectrum Antioxidant and Free Radical Scavenger. *Curr. Top. Med. Chem.* **2002**, *2*, 181–197. [CrossRef]
60. Pan, G.X.; Li, L.Q.; Wu, L.S. Storage and sequestration potential of topsoil organic carbon in China's paddy soils. *Glob. Chang. Biol.* **2003**, *10*, 79–92. [CrossRef]
61. Yan, H.; Cao, M.; Liu, J.; Tao, B. Potential and sustainability for carbon sequestration with improved soil management in agricultural soils of China. *Agric. Ecosyst. Environ.* **2007**, *121*, 325–335. [CrossRef]
62. Liu, X.; Zhang, Y.; Han, W.; Tang, A.; Shen, J.; Cui, Z.; Vitousek, P.; Erisman, J.W.; Goulding, K.; Christie, P.; et al. Enhanced nitrogen deposition over China. *Nature* **2013**, *494*, 459–462. [CrossRef]

 *sustainability*

Article

An Optimization Scheme of Balancing GHG Emission and Income in Circular Agriculture System

Sheng Hang [1,2,3], Jing Li [1,2], Xiangbo Xu [1,4], Yun Lyu [5], Yang Li [1,2], Huarui Gong [1,2,3], Yan Xu [1,2] and Zhu Ouyang [2,3,*]

[1] Key Laboratory of Ecosystem Network Observation and Modeling, Institute of Geographical Sciences and Natural Resources Research, Chinese Academy of Sciences, Beijing 100101, China; hang_some@126.com (S.H.); jingli@igsnrr.ac.cn (J.L.); ydxu.ccap@igsnrr.ac.cn (X.X.); liyangyx1991@outlook.com (Y.L.); gonghr.18b@igsnrr.ac.cn (H.G.); xuy.17s@igsnrr.ac.cn (Y.X.)
[2] University of Chinese Academy of Sciences, Beijing 100049, China
[3] Yellow River Delta Modern Agricultural Engineering Laboratory, Institute of Geographic Sciences and Natural Resources Research, Chinese Academy of Sciences, Beijing 100101, China
[4] UN Environment-International Ecosystem Management Partnership (UNEP-IEMP), Beijing 100101, China
[5] Department of Grassland Science, College of Grassland Science and Technology, China Agricultural University, Beijing 100193, China; lvyun@cau.edu.cn
* Correspondence: ouyz@igsnrr.ac.cn

Abstract: With the rapid development of circular agriculture in China, balancing agricultural income and environmental impact by adjusting the structure and scale of circular agriculture is becoming increasingly important. Agriculture is a major source of greenhouse gas and income earned from agriculture drives sustainable agricultural development. This paper built a multi-objective linear programming model based on greenhouse gas emission and agricultural product income and then optimized the structure and scale of circular agriculture using Beiqiu Farm as a case study. Results showed that greenhouse gas emission was mainly from manure management in livestock industry. While the agriculture income increased by 64% after optimization, GHG emission increased by only 12.3%. The optimization made full use of straw, manure and fodder, but also minimized soil nitrogen loss. The results laid a generalized guide for adjusting the structure and scale of the planting and raising industry. Measures for optimizing the management of manure were critical in achieving low agricultural carbon emissions in future agricultural development efforts.

Keywords: structure optimization; carbon footprint; multi-objective linear programming; circular agriculture

Citation: Hang, S.; Li, J.; Xu, X.; Lyu, Y.; Li, Y.; Gong, H.; Xu, Y.; Ouyang, Z. An Optimization Scheme of Balancing GHG Emission and Income in Circular Agriculture System. *Sustainability* **2021**, *13*, 7154. https://doi.org/10.3390/su13137154

Academic Editors: Rajan Ghimire and Bharat Sharma Acharya

Received: 12 April 2021
Accepted: 19 June 2021
Published: 25 June 2021

Publisher's Note: MDPI stays neutral with regard to jurisdictional claims in published maps and institutional affiliations.

1. Introduction

Circular agriculture is the inevitable drive towards sustainable development of agricultural production [1]. The combination of planting and breeding is one good way to realize resource utilization, prevent pollution and reduce application of fertilizers in farmlands. This effort contributes to reducing agricultural non-point source pollution and increasing income from agriculture. In China, circular agriculture is promoted as a top-down national political drive [2]. China has explored several new modes of agricultural production such as recycling wastes via biogas digesters and compost. This has been possible through the combination of planting and breeding during small-scale peasant economy. Based on local conditions, farmers have promoted ecological circular agricultural models, such as rice-fish symbiosis, pig-biogas fruits, and forest economies [3]. Scholars have focused on making circular agricultural systems workable and able to generate more income, ignoring the appropriate size and variety between farming and breeding. As a result, there is still the issue of excessive agricultural waste in circular agriculture that is leading to environmental pollution. Therefore, the optimization of agricultural development by adjusting the structure and scale of planting and breeding for increased agricultural economic benefits remains challenging.

The mathematical model and programming techniques, such as linear programming, dynamic programming, genetic algorithms, particle swarm optimization and multi-objective planning have been widely used in solving the problem of agricultural structure and scale of adjustment [4]. Most of these studies adjust the structure and scale of agriculture based on resource consumption. An agricultural water-energy-food sustainable management (AWEFSM) model which incorporates multi-objective programming, nonlinear programming and intuitionistic fuzzy numbers into a general framework, was developed for sustainable management of limited water-energy-food resource by identifying tradeoffs of water, energy and land resources across various sub-areas and crops [5]. Some studies consider the relationship between food and energy, others focus on the balance across nutritional needs of animals and feed supplies, and yet others analyze labor and water requirements and income [6,7]. Not many studies link agricultural greenhouse gas emission to agricultural economic benefit and agricultural waste disposal. A multi-objective regional optimization model was therefore built to identify optimum land management adaptations to climate change [8]. It is not difficult to find that Pareto-based multi-objective optimization methods are well-suited for explorations of trade-offs and synergies [9].

As a major agricultural country, agricultural development in China has its own problems. The environmental problems caused by the rapid agricultural development and the low agricultural income continue to attract increasing attention. In particular, the growing demand for food will compete with the effort to mitigate Greenhouse Gas (GHG) emissions and adapt to climate change. The Food and Agriculture Organization (FAO) of the United Nations pointed out in its report that the global agricultural GHG emission in 2014 was 5.442 billion tons of carbon dioxide equivalent [10]. Of this, China emitted 708 million tons, accounting for 13.51% of the global agricultural GHG emission and making it the country with the largest agricultural GHG emission in the world. Studies show that the input of nitrogen and phosphorus fertilizers in plantation industry will increase by 2.7–3.4 times in the future. The input of nitrogen fertilizer alone will result in an annual emission of 3 billion tons of CO_2 equivalent [11]. Compared with crop production, the livestock sector contributes more to GHG emissions. As a country with the largest livestock production in the world, China's GHG emissions from the livestock industry was increased from 137.423 million tons to 150.563 million tons from 2000 to 2014, of which emissions from gastrointestinal fermentation of livestock and manure management systems were the two key sources, accounting for 65.58–73.23% [12]. As the agricultural sector is most affected by human activity, GHG emission from crop production and livestock industry could be negligible or even negative under improved agricultural management practices [13–16]. Hence, agriculture systems in China are becoming increasingly important as a global solution to mitigating anthropogenic GHG emission [17,18].

China's development of low-carbon agriculture is aimed at energy saving, emission reduction and waste disposal. The goal is to build a sustainable agricultural development, achieved by adjusting the structure and scale of circular agriculture. Through this drive, the full use of resources can be achieved at the input, operations and waste treatment stages of agricultural production. This can give the level of carbon emission that is important for the development of low-carbon agriculture across the country and improvement of agricultural resource utilization. To take advantage of these measures and promote sustainable agricultural development, it is critical that the concept of "carbon footprint" is used to study GHG emission in the agriculture and animal husbandry [19]. The purpose of this study was to: (1) calculate GHG emission in planting and breeding system modules of a representative circular farm using carbon footprint and find the difference in GHG emission; (2) study differences in economic benefits of agriculture on the basis of input, output and actual conditions; and (3) determine the optimal structure and scale of different industries in circular agriculture in relation to economic benefit, environmental impact and farm waste utilization using a multi-objective linear programming model.

2. Materials and Methods

2.1. Study Area

In this study, Beiqiu Farm was used as case study of combined crop and animal husbandry. The farm is located in Beiqiu village at 37°00′12.4″ N and 116°34′22.3″ E in Yucheng, Dezhou City, Shandong Province. This is at an alluvial plain in the middle and lower reaches of the Yellow River.

Beiqiu Farm was selected in this study because it is a typical agricultural ecosystem that combines planting and breeding industry in the North China Plain. The plain is one of the main agricultural production areas in China, where 50% of the wheat and 30% of the corn are produced in the country. The farm also has waste disposal and feed processing facilities for making fodder and organic fertilizer. Thus, research on the production model could provide a reference for a sustainable agriculture development model in the country. Data were collected mainly on production in 2018 and the related statistics and literature.

An agricultural production anniversary was used to set the research boundary. Carbon emission from the crop industry subsystem mainly included GHG from agricultural input as chemical fertilizers, pesticides, seeds and energy use, agricultural operations and crop growth processes. There is scientific consensus that global warming is driven by the increasing emission of GHG from human activities [20]. Studies show that the agriculture system including the process from the production of agricultural materials to the agricultural harvest is the main source of CO_2, CH_4 and N_2O emission [21,22]. Carbon emission from the agriculture subsystem comes mainly as GHG produced from agricultural inputs, intestinal fermentation in poultry and manure making. GHG emission was analyzed for each major stage of the combined crop and animal industry (Figure 1).

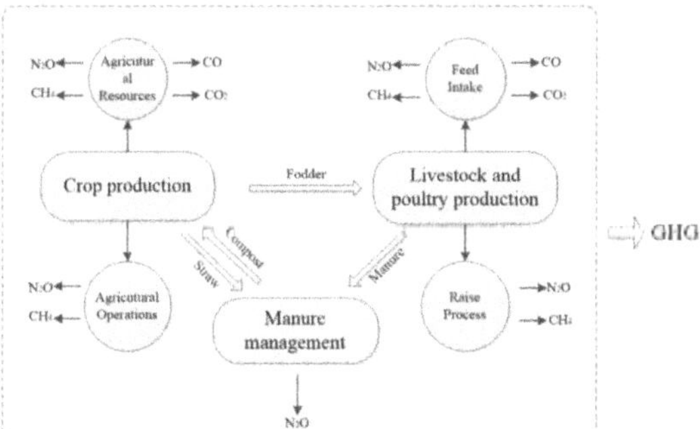

Figure 1. Major greenhouse gas emissions at each stage of the crop and animal production system.

2.2. Multi-Objective Linear Programming Model

The multi-objective linear programming model was the selected optimization method in the study and the Lingo software was used for the calculation runs. Linear programming is a mathematical method used to obtain the optimal solution to objective functions under a set of constraint conditions. It organically combines qualitative and quantitative analyses [23] and is a very effective and simple method for structural adjustment and optimization.

2.3. GHG Emission Calculation

GHG emission is the direct and indirect greenhouse gas emission produced by the products or services in a life cycle (or geographical space). It is an indicator used to measure the carbon emission level and to identify carbon emissions of different functional units. The common methods used in the study of carbon footprint is life cycle assessment. This is a "cradle-to-grave" environmental evaluation approach that accounts for every link of the product or service, including production of raw materials, product manufacturing or processing, product use stages and evaluation processes. In this study, GHG produced via agricultural inputs is calculated using life cycle evaluation of agricultural operations as defined by IPCC Guidelines for National Greenhouse Gas Inventories.

2.3.1. Agriculture GHG Emission Calculation

The input of agricultural resources for the Beiqiu Farm production process in 2018 is listed in Table 1 GHG emission from the agricultural subsystem is driven mainly by agricultural production materials, such as pesticides, fertilizers, electric power and planting processes. According to the data in Table 2, GHG emission factors for agricultural resources are in Table 2.

Table 1. Inputs for crops in farm production.

Item	Unit	Wheat	Maize	Unit	Unheated Greenhouse	Greenhouse
Seeds	kg·ha^{-1}	120	20	Seedings/each	3000	6000
N	kg·ha^{-1}	45	25	kg/each	225	345
P_2O_5	kg·ha^{-1}	37.5	37.5	kg/each	225	345
K_2O	kg·ha^{-1}	37.5	37.5	kg/each	225	345
Manure Compost	N kg·ha^{-1}	15	15	N Kg/each	90	90
Herbicide	kg·ha^{-1}	0.22	0.2	-	-	-
Diesel	L·ha^{-1}	40	40	-	-	-
Electricity	Kwh·ha^{-1}	-	-	K·wh/each	500	900
GHG Emissions	$CO_{2\text{-eq}}$ kg·ha^{-1}	323.583	292.78	CO_2-eq kg/each	1138.75	1806.75

Table 2. Greenhouse gas emission factors for different crops and livestock.

Item	Emission Factor	Unit	Reference
Maize seed	1.93	kg CO_2-eq/kg	Ecoinvent 2.2 [24]
Wheat seed	0.58	kg CO_2-eq/kg	Ecoinvent 2.2
Corn (fodder)	0.79	kg CO_2-eq/kg	CLCD 0.7 [25]
Bean (fodder)	0.84	kg CO_2-eq/kg	CLCD 0.7
Bran (fodder)	0.01	kg CO_2-eq/kg	CLCD 0.7
Pesticide	10.15	kg CO_2-eq/kg	Ecoinvent 2.2
N from fertilizer	1.53	kg CO_2-eq/kg	CLCD 0.7
Manure Compost	0.20	kg CO_2-eq/kg	(Li et al. 2016) [26]
P_2O_5	1.63	kg CO_2-eq/kg	CLCD 0.7
K_2O	0.65	kg CO_2-eq/kg	CLCD 0.7
Electricity	0.527	kg CO_2-eq/K·wh	National Development & Reform Commission [27]
Diesel	4.10	kg CO_2-eq/kg	CLCD 0.7

2.3.2. Agricultural Operation Inventory

The list of the farming subsystem is divided into two parts—one is GHG emission from farm crops and the other from livestock. This also includes GHG emissions during farming and manure management operations. GHG emissions from the soil during farm operations such as nitrogen and organic fertilizer management are both direct and indirect. Direct emission of N_2O was calculated using Equation (1) as follows:

$$DF_{N_2O} = F_{SN} \times EF_1 \times \frac{44}{28} \tag{1}$$

where DF{N_2O} is the direct emission of soil N_2O expressed as equivalent CO_2 emission (kg $CO_{2\text{-eq}}$/hm^2); F_{SN} is the annual application rate of soil nitrogen fertilizer; EF_1 is the emission factor (kg $CO_{2\text{-eq}}$/hm^2) of soil N_2O emission from nitrogen input and 44/28 is the conversion coefficient between nitrogen element and nitrous oxide.

The carbon footprint generated by indirect emission of soil N_2O is calculated using Equation (2) as follows:

$$IDF_{N_2O} = F_{SN} \times Frac_{GASF} \times EF_4 \times \frac{44}{28}$$

(2)

where $IDF\{N_2O\}$ is the soil N_2O indirect emission expressed as CO_2 emission equivalent (kg $CO_{2\text{-eq}}$/hm^2); $Frac\{GASF\}$ is the ratio of nitrogen volatized as NH_3 and NO_X (kg volatile nitrogen/kg nitrogen); EF_4 is the N_2O emission factor (kg N_2O-n/kg (NH_3-n +$NO_{X\text{-N}}$)) of nitrogen in atmospheric deposition on soil and water surface.

Therefore, the carbon footprint for soil N_2O emission is calculated using Equation (3) as follows:

$$CF_{N_2O} = DF_{N_2O} + IDF_{N_2O}$$

(3)

The calculated GHG emission from soil management (Table 3) was added to the GHG emission from farming. Then, the GHG emission from the other sub-system was calculated in the same way as the GHG emission from farming (Table 4).

Table 3. Data for soil N_2O emission from wheat-maize field in the study area.

Items	N from Fertilizer (kg N_2O-N/kg N)	N from Manure Compost (kg N_2O-N/kg N)
Managed soil	60.00	30.00
Direct emission	1.41	-
Indirect emission	-	0.19
Total	-	1.60

Table 4. Greenhouse gas emissions from different farming systems.

Item	Wheat	Maize	Unheated Greenhouse	Greenhouse
Unit	$CO_{2\text{-eq}}$ kg·ha^{-1}	$CO_{2\text{-eq}}$ kg·ha^{-1}	$CO_{2\text{-eq}}$ kg/per	$CO_{2\text{-eq}}$ kg/per
Value	907.91	706.19	2804.86	4090.00

2.3.3. Livestock GHG Emissions Calculation

Based on IPCC research, the calculation of GHG emissions from livestock mainly considers fodder input, methane (CH_4) emission from enteric fermentation and methane and nitrous oxide emissions in manure management (direct and indirect).

CH_4 emission from enteric fermentation of livestock is calculated as follows:

$$TCH_{4-Enteric} = \sum_i EF_i \times N_i$$

(4)

where $TCH_{4Enteric}$ is the total methane emission from enteric fermentation (Gg CH_4 yr^{-1}); EF_i is the emission factor (kg CH_4 animal^{-1} yr^{-1}); Ni is the number of head of livestock category i; i is the species of livestock.

The CH_4 emission from manure management is next calculated as follows:

$$TCH_{4-M} = \sum_i EF_j \times N_i$$

(5)

where TCH_{4-M} is the total CH_4 emission from manure management (kg CH_4 yr^{-1}); EF_j is the emission factor (kg CH_4 animal^{-1} yr^{-1}); Ni is the number of head of livestock category i; i is the species of livestock.

Direct N_2O emission occurs via combined nitrification and denitrification of nitrogen contained in manure. The emission of N_2O from manure during storage and treatment depends on nitrogen and carbon content of manure, and on the duration of storage and type of treatment. Nitrification does not occur under anaerobic conditions. Indirect emission results from volatile nitrogen loss that occurs primarily in the form of ammonia and NO_X. Therefore, the N_2O emission from manure management is calculated as follows:

$$N_2O_{D(mm)} = \left[\sum_S \left[\sum_T (N_i \times Nex_i \times MS_i) \right] \times EF_{3(S)} \right] \times \frac{44}{28} \tag{6}$$

where $N_2O_{D(mm)}$ is the direct N_2O emission from manure management (kg N_2O yr^{-1}); N_i is the number of head of livestock category i; $Nex_{(i)}$ is the annual average N excretion per head of livestock species i (kg N animal^{-1} yr^{-1}); MS_i is the fraction of total annual nitrogen excretion for each livestock category i; EF_3s is the emission factor for direct N_2O emission from manure management (kg N_2O-N/kg N); S is the manure management system; 44/28 is the conversion of N_2O-N(mm) emission to N_2O(mm) emission.

The N loss due to volatilization from manure management is calculated as follows:

$$N_2O_{G(mm)} = (N_{\text{volatilisation}-MMS} \times EF_4) \times \frac{44}{28} \tag{7}$$

where, $N_{\text{volatilisation-MMS}}$ is the amount of manure nitrogen lost due to volatilization of NH_3 and NO_X; EF_4 is the emission factor for N_2O emission from atmospheric deposition of nitrogen on soil and water surfaces (kg N_2O-N(kg NH_3-N+NO_X-N $_{\text{volatilised}}$)$^{-1}$).

The carbon footprint of the entire production farm system is finally calculated as follows:

$$C_i = [CF_{N_2O\text{soil}} + N_2O_{D(mm)} + N_2O_{G(mm)}] \times 298 + [TCH_{4-\text{Enetric}} + TCH_{4(mm)}] \times 21 \tag{8}$$

where C_T is the total greenhouse gas emission from livestock category i; a constant factor of 21 is the coefficient of conversion from CH_4 to CO_2; a constant factor of 298 is the coefficient of conversion from N_2O to CO_2.

Then, GHG emissions from different species on the farm (Table 5) are calculated using the equations above.

Table 5. Greenhouse gas emissions from different livestock industries.

Species	$CH_{4\text{-Enetric}}$	$CH_{4\text{-M}}$	N_2O_D (mm)	N_2O_G (mm)	Bean	Total
Units	kg/head	kg/head	kg/head	kg/head	kg $CO_{2\text{-eq}}$/head	kg $CO_{2\text{-eq}}$/head
Pig	0.33	1.50	3.10	7.75×10^{-2}	6.51×10^{-2}	985.00
Sheep	5.00	0.17	2.19	2.62×10^{-2}	2.20×10^{-2}	768.00
Goose	-	0.02	0.04	1.77×10^{-3}	1.48×10^{-3}	14.10
Layer	-	0.03	0.09	3.46×10^{-3}	2.90×10^{-3}	27.40
Broiler	-	0.02	1.46×10^{-3}	5.85×10^{-5}	4.91×10^{-5}	0.87

3. Model Building

The multi-objective linear programming model is generally composed of more than two objective functions and a number of constraints. The construction of the model in this paper is to achieve the best economic and ecological benefits, keep the agricultural system a virtuous cycle, and promote the sustainable development of agriculture production. Reconfiguration of farming systems to reach various productive and environmental objectives while meeting farm and policy constraints is complicated by the large array of farm components involved and the multitude of interrelations among the components [7]. In this study, two goals were primarily set up in the model—one was the economic benefit target and the other was the ecological benefit target. Nine kinds of agricultural and livestock products were selected in Beiqiu Farm as decision variables of the model and the corresponding data (Table 6) were all from the actual production process on the farm.

Table 6. Parameters for different crops and livestock species.

Items	Variable	Unit	Profit/($¥·Unit^{-1}$)	Land/m^2	Fodder/(kg·Unit^{-1})			
					Corn	Bran	Ensiling	Bean
Wheat	X_1	Ha	5505	10000	-	1050.00	-	-
Maize	X_2	Ha	3780	10000	6990	-	30000	-
Unheated greenhouse	X_3	Each	16,100	667.67	-	-	-	-
Greenhouse	X_4	Each	23,490	1335.34	-	-	-	-
Pig	X_5	Head	300	0.67	−69.70	−19.92	−76.21	−33.32
Sheep	X_6	Head	200	1.50	−63.88	-	−146.00	−41.96
Goose	X_7	Head	40	0.10	−3.60	−1.80	−3.60	−3.00
Layer	X_8	Head	30	0.25	−28.47	−4.38	-	−10.95
Broiler	X_9	Head	15	0.05	−4.84	−0.74	-	−1.86

Objective function: The objective function is set according to the needs of agricultural decision-makers as follows:

$$Y = \sum_{i=1}^{n} a_i \times X_i \qquad i = 1, 2 \quad n \tag{9}$$

where ai is the objective function value of variable Xi which, in this study, is unit profit and greenhouse gas emissions in unit variable; Xi is the scale of production activity, namely the scale of planting and breeding industries.

Specifically, agricultural income is an important factor in promoting sustainable development of agriculture. Thus, the objective function was set as the highest agricultural net profit as follows:

$$\text{Max } f_1 (xi) = \sum_{i=1}^{9} ai \times xi \tag{10}$$

$$\text{Max } f_1 = 5505 \times x_1\text{-}3780 \times x_2 + 16,\!100 \times x_3 + 23,\!100 \times x_4 + 300 \times x_5 + 200 \times x_6$$
$$+ 40 \times x_7 + 30 \times x_8 + 15 \times x_9;$$

Whereas agriculture maintains high profits, the impact of agricultural production on the environment, especially GHG emission, cannot be ignored. Therefore, the objective function was set as the minimum GHG emission as follows:

$$\text{Min } f_2 (xi) = \sum_{i=1}^{9} bi \times xi \tag{11}$$

$$\text{Min } f_2 = 907.91 \times x_1 + 706.19 \times x_2 + 2804.86 \times x_3 + 4090.00 \times x_4 + 985 \times x_5$$
$$+ 768 \times x_6 + 27.4 \times x_7 + 0.873 \times x_8 + 14.1 \times x_9$$

Since economic and ecological benefits have different dimensions, the extreme value of a single objective function is first calculated and used to construct a new dimensionless objective function by linear weighting as a single objective function; thereby eliminating the impact of the dimensions [28].

$$\text{MaxF}(xi) = w_1 \times \frac{f_1}{f_1^*} - w_2 \times \frac{f_2}{f_2^*} \tag{12}$$

where wi is the weight of the economic and ecological benefit; in this research, we give them a weight of 0.5 each. The $f_1{}^\times$ was calculated as 617,391.8 ¥ and $f_2{}^\times$ as 503,542.7 kg $CO_{2\text{-eq}}$.

Currently in the process of production, a planting industry system provides feed and a raising system provides manure as organic fertilizer. However, the problem is the imbalance of the scale of livestock and plant. The soil bearing capacity was too high because of large poultry manure emission and the planting industry provided less feed. The objective of the study was to control the growth of chemical fertilizers, animal manure and GHG emission on the one hand, and to keep planting and breeding industries in a dynamic balance on the other. The adjustment of planting structure proportion in order to make livestock and poultry dung digestible by the internal system needed to be set in the following constraint equations:

Land resource: Beiqiu Farm covers an area of 15 ha; thus, land resource was one of the reasons for limiting the scale of agricultural development. The scale of planting and breeding industry did not exceed 15 ha as:

$$10{,}000x_1 + 10{,}000x_3 + 1335.34x_4 + 0.667x_5 + 1.5x_6 + 0.1x_7 + 0.25x_8 + 0.05x_9 = 133{,}400$$

Otherwise, wheat and corn were rotated in the planting industry so that the area of wheat and that of corn were set equal in the model. At the same time, corn feed came from corn planting and so the area of corn feed was set less than that of corn planting as:

$$x_1 = x_2;\ x_2 = x_{21} + x_{22}$$

where x_{21} is the area of corn silage and x_{22} is the area of corn stalk

Feed resource: The planting industry provides feed for the livestock industry to ensure the quality of meat and reduce the cost. In the case of Beiqiu Farm, it was mainly corn, wheat bran and ensiling. The demand for fodder for the breeding industry in Beiqiu Farm is given in Table 5. Therefore, the constraint was set such that it was less than the supply of farming industry as follows:

Corn feed: $6990 \times x_2 = 69.7 \times x_4 + 63.88 \times x_5 + 3.6 \times x_6 + 4.84 \times x_9 + 28.47 \times x_8$;
Wheat bran feed: $1050 \times x_1 = 19.92 \times x_5 + 1.8 \times x_7 + 0.74 \times x_9 + 4.38 \times x_8$;
Ensiling feed: $30{,}000 \times x_{21} = 76.21 \times x_5 + 146 \times x_6 + 3.6 \times x_7$;

Straw and Manure: Agricultural wastes produced in farm production are mainly straw and dung. Resource utilization was achieved through composting. Therefore, the amount of straw and manure produced matched. Studies show that in the process of composting, the effect is best when the ratio of carbon to nitrogen is 27 [29,30]. The straw demand of different animal manures is given in Table 7.

$$11{,}250 \times x_1 + 30{,}000 \times x_{22} = 398.6 \times x_5 + 213 \times x_6 + 7.03 \times x_7 + 0.98 \times x_9 + 24.98 \times x_8$$

Table 7. The straw demand of different livestock.

Species	Straw Demand	N from Manure Compost
	kg/head	kg/head
Pig	398.59	25.68
Sheep	7.03	16.96
Goose	213.01	0.9
Layer	24.98	0.38
Broiler	0.98	0.02

Amount of organic fertilizer: The amount of organic fertilizer made from straw and manure should meet the daily needs of the farm. These data were derived from the actual production process on Beiqiu Farm.

$$11{,}250 \times x_1 + 30{,}000 \times x_2 + 398.6 \times x_5 + 213 \times x_6 + 7.03 \times x_7 + 0.98 \times x_9 + 24.98 \times x_8$$
$$\geq 3000 \times x_3 + 3000 \times x_4 + 1000 \times x_1;$$

Land bearing capacity: Straw and manure returned to the field via composting to make organic fertilizer, replacing part of potential chemical fertilizer use. In the planting system, the total demand for crop N is fixed, so the total amount of N input from organic fertilizer and chemical fertilizer should be balanced with the demand for crop N as:

$$85.2 \times x_1 + 64.5 \times x_2 = 25.68 \times x_4 + 16.96 \times x_5 + 0.9 \times x_6 + 0.38 \times x_7 + 0.02 \times x_8;$$

Input cost constraints: Currently, the farm has 3 plastic houses and 2 greenhouses. The number of plastic houses and greenhouses are restricted as follows:

$$X_4 \geq 2; X_3, X_4 \leq 10; X_3 \geq 3;$$

4. Results and Discussions

4.1. Optimized Planting/Breeding Structure

The optimal solution for the farm structure optimization was obtained using the Lingo software calculation. Based on the model optimization results (Table 8), the planting area of wheat and maize on the farm are reduced, and the numbers of greenhouse and unheated greenhouse increase because of high economic benefits. For the breeding industry, the scale of geese is reduced, that of sheep and broiler have increased. However, industries of pigs and layers have completely disappeared. In terms of the rate of change, greenhouses and unheated greenhouses had the highest change. There were mainly two reasons for this high change. The first was that the two structures consumed a lot of organic fertilizer which were from manure compost in the system. The next reason was that income from them was much higher than that from food crops. However, GHG emission from the structures was much higher than that from the wheat-maize system. Given the initial input cost, the expansion of greenhouses and unheated greenhouses was restricted. The scale of winter wheat/summer maize cropping pattern was little changed. This was because winter wheat and summer maize rotation can expend compost and supply fodder, straw and silage to livestock industries. In addition, straw can be used with the manure from the breeding system to dispose excrement and urine, both of which promote a circular economy.

Table 8. Model simulated results under circular agriculture optimization.

Item	Variable	Unit	Result	Actual Scale	Rate %
Wheat	X_1	ha	5.67	7.33	−22.73
Maize	X_2	ha	5.67	7.33	−22.73
Unheated greenhouse	X_3	each	10	3	233.33
Greenhouse	X_4	each	9	2	350.00
Pig	X_5	head	0	55	−100.00
Sheep	X_6	head	628	50	1156.00
Goose	X_7	head	183	1000	−81.70
Layer	X_8	head	0	5500	−100.00
Broiler	X_9	head	7635	1000	663.50

However, the results were for specific conditions. Due to large fluctuations of prices of agricultural products, price changes affected the farm optimization effort. There was therefore need to discuss the change of scale under specific circumstances [31]. The optimization models were relevant in assisting cropping and management of agricultural production. It was also applicable in estimating potential gains from the use of integrated systems [32]. The optimization results obtained in this study were according to the conditions of Beiqiu farm and could be used for referrals only.

4.2. Post-Optimization Benefits/Effects

Farmers usually adjust their farming systems evolutionarily for various reasons. It could be for change in market price that farmers adjust the type and size of their farm products. It could also be for policy change that farmers adjust the structure of cropping and animal husbandry. The understanding of farmers about the impact of agricultural production on the environment is relatively weak. Bio-economic farm models have the potential to support information structuring for more insight into the consequences of adjustments of farming systems [33].

Based on the model optimization, GHG emission from the farm was 36,833.39 kg CO_{2-eq} ha^{-1} for an estimated profit of 44,831.3 ¥ ha^{-1}. That is the equivalent of 0.82 kg

$CO_{2\text{-eq}}$ for one ¥ generated. The GHG emission was normalized by the benefit to get ecological efficiency (Figure 2). Sheep industry had the highest ecological efficiency, meaning that sheep industry created one ¥ of profit for a unit release of GHG. Broiler industry was the reverse. In the breeding subsystem, optimization greatly increased the scale of sheep and broilers for a balance to be maintained. This was because sheep which can consume straw and broilers which can consume wheat bran and corn, are indispensable in the circular agriculture system. The main source of GHG emissions was the livestock industry (Figure 3), especially manure management activities (Figure 4). Specifically, CH_4 emission from enteric fermentation of sheep excreta was over 10%, while geese and broilers can be ignored. CH_4 emission from broiler manure management was over 45% (Figure 5), almost the same as direct N_2O emissions from manure management ($N_2O_{D\,(mm)}$). Direct N_2O emissions, which occur via combined nitrification and denitrification of nitrogen contained in the manure, was the main source of GHG emission from sheep and geese industries (Figure 6). The emission of N_2O from manure during storage and treatment depends on the nitrogen and carbon content of manure, and on the duration of the storage and type of treatment. Indirect N_2O emissions due to volatilization of N from manure management ($N_2O_{G\,(mm)}$) can be ignored. Thus, the focus of reducing GHG emissions from circular agriculture was on improving measures for manure management and adjusting the feed structure. There was also need to reform livestock and poultry breeding technologies and management and to change the livestock pattern [34]. Sustainable livestock intensification can be key in reducing GHG emission. It provides synergy across productivity and increases income. The mitigation of climate change was another benefit due to future development of low-carbon agriculture [35].

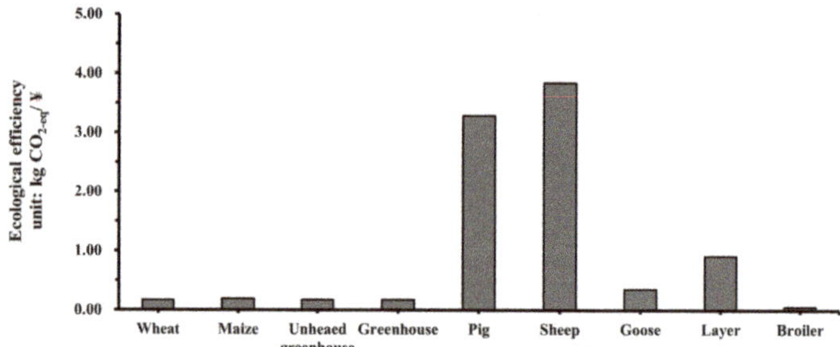

Figure 2. Ecological efficiency for different categories of the farm activities.

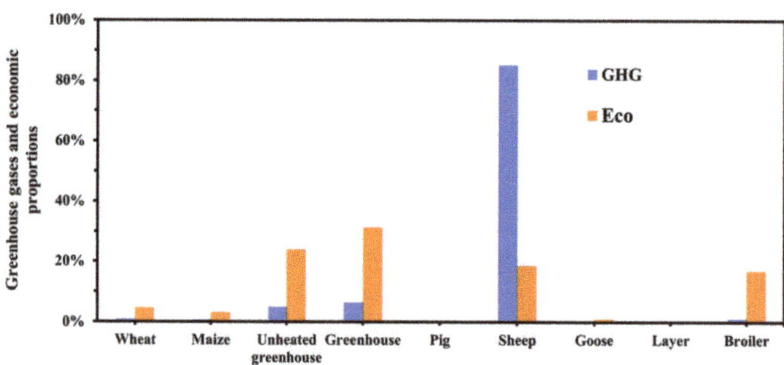

Figure 3. Plot of difference between the economy and environment due to cropping and raising animals.

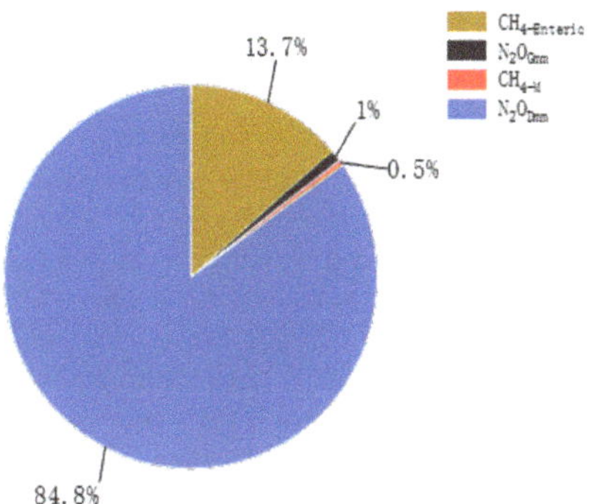

Figure 4. Different sources of GHG emission ($CO_{2\text{-}eq}$) from sheep industry.

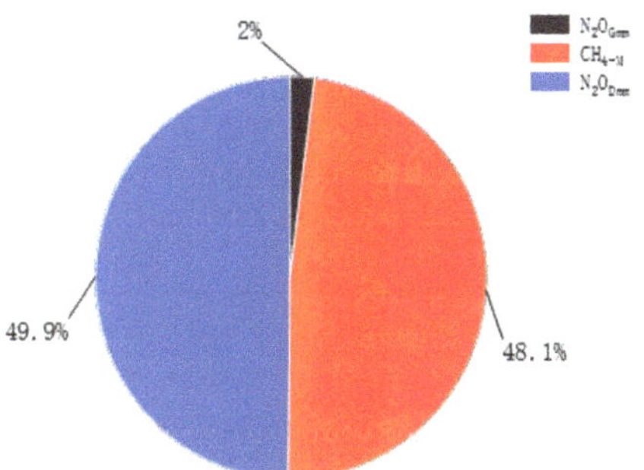

Figure 5. Different sources of GHG emission (CO2-eq) from goose broiler industry.

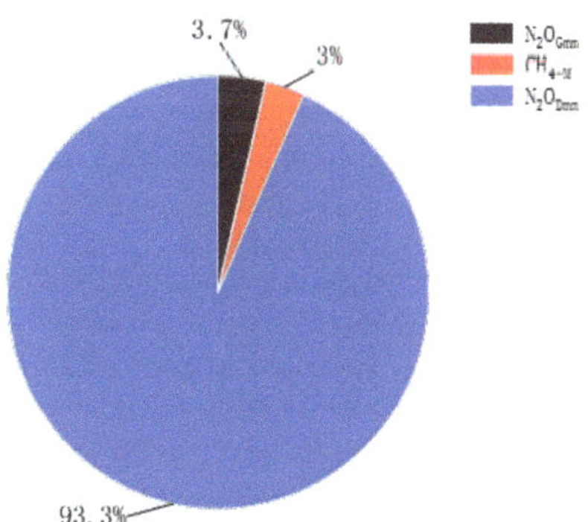

Figure 6. Different sources of GHG emission ($CO_{2\text{-}eq}$) from goose industry.

Crop production accounts for only 13% of GHG emissions in circular agriculture, and it was mainly from the heavy fertilizer use. However, looking at the individual plates, GHG emissions from crop in circular agriculture are far less than that from conventional agriculture. This was because of the use of organic fertilizer from the recycle use of manure and straw. Studies show that using organic fertilizer is a key way of reducing carbon emission from chemical fertilizers [12,36]. China's traditional energy consumption structure has increased the carbon footprint in fertilizer production and agricultural machinery use. Improving the energy efficiency and using cleaner energy can reduce the overall GHG emission [14]. Therefore, reducing carbon emission from agricultural production and maintaining the scales of planting and breeding remains is possible. This can be achieved by reducing inputs of chemical fertilizers and pesticides in cropping systems and utilizing the resource as poultry manure in livestock production.

4.3. Balance in Economy and Environment

In circular agriculture, feed generated in the crop industry is fed to the livestock industry and this helps reduce carbon emission as GHG along the food chain. Part of the maize silage was used in place of concentrate to reduce cost in the livestock industry [37]. Most of the feed was from the agricultural system, which not only ensured quality of agricultural products but also saved cost. This ecological efficiency of the circular agriculture system increased from 0.7 to 0.82 after optimization. Here, part of the environment was sacrificed in terms of GHG emission for economic benefits on the simulated circular agriculture system. On the other hand, however, recycling agriculture waste solved the issue of environmental pollution caused by straw burning and waste emissions.

In addition to economic gains, the use of integrated systems is beneficial to the environment. This is especially so through reuse of resources and the related negative environmental externalities [38]. China is one of the countries with the most abundant straw and dung. Based on statistics, the annual crop straw in China is as high as 900 million tons, and is increasing at a rate of 5-10% every year [39]. In 2016, the amounts of livestock and poultry manure in China hit 3.16×10^9 t. However, the comprehensive utilization rate of the resources was less than 60% [40]. Straw and feces returned to the field is a carbon sequestration measure that can lead to sustainable agricultural development. After optimization, we can make the full use of agricultural waste.

The comprehensive use of solid organic waste on the farmlands is an increasing concern in agricultural production. Straw and livestock manure are rich in organic matter and various nutrient elements, making it suitable for boosting soil fertility and soil organic matter content when applied as organic fertilizer. The main drive for straw and manure resource use is waste utilization, and the basis of it is a dynamic balance between the scale of crop and livestock industries. Composting as a valuable technology, is also widely used in recycling agricultural organic wastes. This converts organic matter into a relatively stable humus-like substance through microbial transformation [41,42]. Soil quality can be improved by using compost in place of chemical fertilizer, which is critical in the development of circular agriculture.

The European Union (2012) encourages the use of bio-waste in agriculture as it improves soil condition and provides valuable nutrients to plants [43]. Composting is one of the most effective processes used in recycling organic waste, applicable to soils as organic amendment [30]. The focus of this study was on the balance between GHG emission and agriculture income, which has a far-reaching influence on agricultural development. Issues such as soil N bearing capacity, agricultural waste disposal, food safety, food health and cost input were considered under controlled conditions. Studies show that, although farmers have more options in making practical decisions, the general focus is often on economic maximization. For strategic decision-making, therefore, farmers account for options that influence long-term performance and indicators associated with sustainability [44]. The model used in this study optimized the structure and scale of circular agriculture on the basis of both economic and environmental outputs, laying the basis for agricultural development [45]. It is difficult to balance economy with environment in terms of agricultural operations. This study provided a feasible pathway for rational decision-making in complex agricultural systems.

5. Conclusions

Building a management model for agricultural planting structure adjustment is a complex engineering task that considers land, water and climate resources in a time-space fabric. The challenge with on-farm research on models is to keep processes and output functions transparent and relevant to farm management [46,47]. There is also concern for market demand and social characteristics in relation to economic growth, environmental protection, etc. The objective of this study was to determine the economic benefit and GHG emission in a typical agricultural farm using a multi-objective linear scale model, ant to optimize the structure and scale of growing crops and raising animals on the farm. The model-driven optimized farm was a strong scientific basis for the adjustment of agricultural structure and the development of circular agriculture in the study area and beyond. The model was strongly operable, flexible and adjustable to set targets or constraint conditions. It therefore provided the needed guidance for the development of circular agriculture with different needs in different social settings.

Author Contributions: Conceptualization, S.H. and Z.O.; methodology, S.H.; data curation, Y.L.; writing—original draft preparation, S.H.; writing—review and editing, J.L., X.X., H.G., Y.X., Y.L. (Yun Lyu), Y.L. (Yang Li); funding acquisition, Z.O. All authors have read and agreed to the published version of the manuscript.

Funding: This research was funded by Chinese Academy of Sciences (Project numbers: XDA23050103 and KFJ-STS-ZDTP-049).

Institutional Review Board Statement: Not applicable.

Informed Consent Statement: Not applicable.

Data Availability Statement: Not applicable.

Acknowledgments: We thank the reviewers and editors very much for the very useful suggestions and comments.

Conflicts of Interest: The authors declare no conflict of interest.

References

1. Wang, H.; Li, H ; Wu, X. Study on Information Needs for Promoting the Development of Circular Agriculture. In *Environmental Technology and Resource Utilization II*; Zhang, L., Ed.; Trans Tech Publications Ltd.: Stafa-Zurich, Switzerland, 2014; pp. 1028–1031.
2. Ghisellini, P.; Cialani, C.; Ulgiati, S. A Review on Circular Economy: The Expected Transition to a Balanced Interplay of Environmental and Economic Systems. *J. Clean. Prod.* **2016**, *114*, 11–32. [CrossRef]
3. Li, B.; Feng, Y.; Xia, X.; Feng, M. Evaluation of China's Circular Agriculture Performance and Analysis of the Driving Factors. *Sustainability* **2021**, *13*, 1643. [CrossRef]
4. Lalehzari, R.; Nasab, S.B.; Moazed, H.; Haghighi, A ; Yaghoobzadeh, M. Simulation–optimization modelling for water resources management using nsgaii-oip and modflow. *Irrig. Drain.* **2020**, *69*, 317–332. [CrossRef]
5. Li, M.; Fu, Q.; Singh, V.P.; Ji, Y.; Liu, D.; Zhang, C.; Li, T. An optimal modelling approach for managing agricultural water-energy-food nexus under uncertainty. *Sci. Total Environ.* **2019**, *651*, 1416–1434. [CrossRef] [PubMed]
6. Allam, M.M.; Eltahir, E.A.B. Water-Energy-Food Nexus Sustainability in the Upper Blue Nile (UBN) Basin. *Front. Environ. Sci.* **2019**, *7*, 7. [CrossRef]
7. Groot, J.; Oomen, G.J.; Rossing, W.A. Multi-objective optimization and design of farming systems. *Agric. Syst.* **2012**, *110*, 63–77. [CrossRef]
8. Klein, T.P.; Holzkämper, A.; Calanca, P.; Seppelt, R.; Fuhrer, J. Adapting agricultural land management to climate change: A regional multi-objective optimization approach. *Landsc. Ecol.* **2013**, *28*, 2029–2047. [CrossRef]
9. Groot, J.C.J.; Rossing, W.A.H. Model-aided learning for adaptive management of natural resources: An evolutionary design perspective. *Methods Ecol. Evol.* **2011**, *2*, 643–650. [CrossRef]
10. FAO. *The State of Food and Agriculture: Climate Change, Agriculture and Food Security*; FAO: Rome, Italy, 2016.
11. Tilman, D.; Balzer, C.; Hill, J.; Befort, B.L. Global food demand and the sustainable intensification of agriculture. *Proc. Natl. Acad. Sci. USA* **2011**, *108*, 20260–20264. [CrossRef]
12. Zhang, X.; Liu, H.; Lai, G. Relationship between fertilizer application and carbon emission reduction of large-scale farmers. *Jiangsu Agric. Sci.* **2018**, *46*, 279–284.
13. Gan, Y.; Liang, C.; Campbell, C.A.; Zentner, R.P.; Lemke, R.L.; Wang, H.; Yang, C. Carbon footprint of spring wheat in response to fallow frequency and soil carbon changes over 25 years on the semiarid Canadian prairie. *Eur. J. Agron.* **2012**, *43*, 175–184. [CrossRef]
14. She, W.; Wu, Y.; Huang, H.; Chen, Z.; Cui, G.; Zheng, H.; Guan, C.; Chen, F. Integrative analysis of carbon structure and carbon sink function for major crop production in China's typical agriculture regions. *J. Clean. Prod.* **2017**, *162*, 702–708. [CrossRef]
15. Liu, W.; Zhang, G.; Wang, X.; Lu, F.; Ouyang, Z. Carbon footprint of main crop production in China: Magnitude, spatial-temporal pattern and attribution. *Sci. Total Environ.* **2018**, *645*, 1296–1308. [CrossRef] [PubMed]
16. Huang, J.; Chen, Y.; Pan, J.; Liu, W.; Yang, G.; Xiao, X.; Zheng, H.; Tang, W.; Tang, H.; Zhou, L. Carbon footprint of different agricultural systems in China estimated by different evaluation metrics. *J. Clean. Prod.* **2019**, *225*, 939–948. [CrossRef]

17. Liu, C.; Lu, M.; Cui, J.; Li, B.; Fang, C. Effects of straw carbon input on carbon dynamics in agricultural soils: A meta-analysis. *Glob. Chang. Biol.* **2014**, *20*, 1366–1381. [CrossRef] [PubMed]
18. Yan, M.; Cheng, K.; Luo, T.; Yan, Y.; Pan, G.; Rees, R.M. Carbon footprint of grain crop production in China—Based on farm survey data. *J. Clean. Prod.* **2015**, *104*, 130–138. [CrossRef]
19. Yang, X.; Gao, W.; Zhang, M.; Chen, Y.; Sui, P. Reducing agricultural carbon footprint through diversified crop rotation systems in the North China Plain. *J. Clean. Prod.* **2014**, *76*, 131–139. [CrossRef]
20. Ledgard, S.F.; Wei, S.; Wang, X.; Falconer, S.; Zhang, N.; Zhang, X.; Ma, L. Nitrogen and carbon footprints of dairy farm systems in China and New Zealand, as influenced by productivity, feed sources and mitigations. *Agric. Water Manag.* **2019**, *213*, 155–163. [CrossRef]
21. West, T.O.; Marland, G. Net carbon flux from agriculture: Carbon emissions, carbon sequestration, crop yield, and land-use change. *Biogeochemistry* **2003**, *63*, 73–83. [CrossRef]
22. Schmidhuber, J.; Tubiello, F.N. Global food security under climate change. *Proc. Natl. Acad. Sci. USA* **2007**, *104*, 19703–19708. [CrossRef] [PubMed]
23. Yu, J.; Wang, R.; Chang, H.; Gao, M.; Wang, Z.; Mo, J.; Gao, C. Optimization of crop and livestock industry in Daxinganling agricultural reclamation based on planting-breeding balance. *J. China Agric. Resour. Reg. Plan.* **2017**, *38*, 228–236.
24. Ecoinvent Database [EB/OL] (2011-05-11). Available online: http://www.ecoinvent.ch (accessed on 20 June 2016).
25. Liu, X.L.; Wang, H.T.; Chen, J.; He, Q.; Zhang, H.; Jiang, R.; Chen, X.X.; Hou, P. Method and basic model for development of Chinese reference life cycle database of fundamental industries. *Acta Sci. Circumstantiate* **2010**, *30*, 2136–2144.
26. Li, J.; Wang, D.; Wang, L.; Wang, Y.; Li, H. Evaluation of nitrogen and water management on greenhouse gas mitigation in winter wheat-summer maize cropland system in North China. *J. Plant. Nutr. Fertitizer* **2016**, *22*, 921–929.
27. National Development and Reform Commission on Climate Change. *Low- Carbon Development and Provincial Greenhous Gas. Inventory Training Materials[R]*; National Development and Reform Commission on Climate Change: Beijing, China, 2013.
28. Niu, K. Studies of multi-objective linear programming model on Chinese agricultural structure adjustment. *Acta Agric. Zhejiangensis* **2011**, *23*, 840–846.
29. Zhou, J.-M. The Effect of Different C/N Ratios on the Composting of Pig Manure and Edible Fungus Residue with Rice Bran. *Compos. Sci. Util.* **2017**, *25*, 120–129. [CrossRef]
30. Pergola, M.; Persiani, A.; Palese, A.M.; Di Meo, V.; Pastore, V.; D'Adamo, C.; Celano, G. Composting: The way for a sustainable agriculture. *Appl. Soil Ecol.* **2018**, *123*, 744–750. [CrossRef]
31. Todman, L.C.; Coleman, K.; Milne, A.E.; Gil, J.D.B.; Reidsma, P.; Schwoob, M.-H.; Treyer, S.; Whitmore, A.P. Multi-objective optimization as a tool to identify possibilities for future agricultural landscapes. *Sci. Total Environ.* **2019**, *687*, 535–545. [CrossRef] [PubMed]
32. Gameiro, A.H.; Rocco, C.; Filho, J.V.C. Linear Programming in the economic estimate of livestock-crop integration: Application to a Brazilian dairy farm. *Rev. Bras. Zootec.* **2016**, *45*, 181–189. [CrossRef]
33. Thornton, P.; Herrero, M. Integrated crop–livestock simulation models for scenario analysis and impact assessment. *Agric. Syst.* **2001**, *70*, 581–602. [CrossRef]
34. Patra, A.K. Accounting methane and nitrous oxide emissions, and carbon footprints of livestock food products in different states of India. *J. Clean. Prod.* **2017**, *162*, 678–686. [CrossRef]
35. Paul, B.K.; Groot, J.C.J.; Birnholz, C.A.; Nzogela, B.; Notenbaert, A.; Woyessa, K.; Sommer, R.; Nijbroek, R.; Tittonell, P. Reducing agro-environmental trade-offs through sustainable livestock intensification across smallholder systems in Northern Tanzania. *Int. J. Agric. Sustain.* **2019**, *18*, 35–54. [CrossRef]
36. Bos, J.F.; Berge, H.F.T.; Verhagen, J.; Van Ittersum, M.K. Trade-offs in soil fertility management on arable farms. *Agric. Syst.* **2017**, *157*, 292–302. [CrossRef]
37. Lyu, Y.; Li, J.; Hou, R.; Zhu, H.; Zhu, W.; Hang, S.; Ouyang, Z. Goats or pigs? Sustainable approach of different raising systems fed by maize silage. *J. Clean. Prod.* **2020**, *254*, 120151. [CrossRef]
38. Accatino, F.; Tonda, A.; Dross, C.; Léger, F.; Tichit, M. Trade-offs and synergies between livestock production and other ecosystem services. *Agric. Syst.* **2019**, *168*, 58–72. [CrossRef]
39. Shi, Z.L.; Jia, T.; Wang, Y.J.; Wang, J.C.; Sun, R.H.; Wang, F.; Li, X.; Bi, Y.Y. Comprehensive utilization status of crop straw and estimation of carbon from burning in China. *J. China Agric. Resour. Reg. Plan.* **2017**, *38*, 32–37.
40. Song, D.; Hou, S.; Wang, X.; Liang, G.; Zhou, W. Nutrient resource quantity of animal manure and its utilization potential in China. *J. Plant Nutr. Fertitizer* **2018**, *24*, 1131–1148.
41. Negi, S.; Mandpe, A.; Hussain, A.; Kumar, S. Collegial effect of maggots larvae and garbage enzyme in rapid composting of food waste with wheat straw or biomass waste. *J. Clean. Prod.* **2020**, *258*, 120854. [CrossRef]
42. Sharma, D.; Yadav, K.D.; Kumar, S. Biotransformation of flower waste composting: Optimization of waste combinations using response surface methodology. *Bioresour. Technol.* **2018**, *270*, 198–207. [CrossRef]
43. Afonso, S.; Arrobas, M.; Pereira, E.L.; Rodrigues, M. Ângelo Recycling nutrient-rich hop leaves by composting with wheat straw and farmyard manure in suitable mixtures. *J. Environ. Manag.* **2021**, *284*, 112105. [CrossRef] [PubMed]
44. Mandryk, M.; Reidsma, P.; Kanellopoulos, A.; Groot, J.C.J.; Van Ittersum, M.K. The role of farmers' objectives in current farm practices and adaptation preferences: A case study in Flevoland, the Netherlands. *Reg. Environ. Chang.* **2014**, *14*, 1463–1478. [CrossRef]

45. Strauch, M.; Cord, A.F.; Pätzold, C.; Lautenbach, S.; Kaim, A.; Schweitzer, C.; Seppelt, R.; Volk, M. Constraints in multi-objective optimization of land use allocation—Repair or penalize? *Environ. Model. Softw.* **2019**, *118*, 241–251. [CrossRef]
46. Sterk, B.; Van Ittersum, M.; Leeuwis, C.; Rossing, W.; Van Keulen, H.; Van De Ven, G.; Van Ittersum, M. Finding niches for whole-farm design models–contradictio in terminis? *Agric. Syst.* **2006**, *87*, 211–228. [CrossRef]
47. Andrieu, N.; Nogueira, D.M. Modeling biomass flows at the farm level: A discussion support tool for farmers. *Agron. Sustain. Dev.* **2010**, *30*, 505–513. [CrossRef]

 *sustainability*

Article

A Portfolio of Effective Water and Soil Conservation Practices for Arable Production Systems in Europe and North Africa

Tshering Choden [1,*] and Bhim Bahadur Ghaley [2]

[1] Dutch International Business Development Cooperative, Bornsesteeg 28B, 6708 PE Wageningen, The Netherlands
[2] Department of Plant and Environmental Sciences, University of Copenhagen, Højbakkegård Alle 30, 2630 Taastrup, Denmark; bbg@plen.ku.dk
* Correspondence: tshering.choden@dibcoop.nl

Abstract: To secure sustainable food production for meeting the growing global demand for food, it is imperative, while at the same time challenging, to make efficient use of natural resources with minimal impact on the environment. The study objective is to provide insights into the multiple benefits and trade-offs of different sustainable agricultural practices that are relevant across pedo-climatic zones in Europe and North Africa, including conservation agriculture, crop diversification, organic agriculture, and agroforestry. Widespread adoption of these practices in specific regions depends on the effectiveness with which their applications and attributes are communicated to farmers, and their suitability to local conditions and opportunities. Scale impacts of the practices range from field to catchment levels, but the best empirical evidence has been generated at field level in on-farm and experimental trials. The outcomes from the application of each of these practices depend on variables specific to each site, including pedo-climatic zone, geography, weather, ecology, culture, and traditions. Each practice has trade-offs and the same practice can have different effects when compared to conventional agriculture. To make site-specific recommendations, a careful assessment of overall benefits must be made. Adoption can be stimulated when farmers have the opportunity to experiment on their own land and discover the advantages and disadvantages of different practices.

Keywords: agroforestry; conservation agriculture; Europe; North Africa; nutrient retention; organic agriculture; soil conservation; water conservation

Citation: Choden, T.; Ghaley, B.B. A Portfolio of Effective Water and Soil Conservation Practices for Arable Production Systems in Europe and North Africa. *Sustainability* **2021**, *13*, 2726. https://doi.org/10.3390/su13052726

Academic Editor: Bharat Sharma Acharya

Received: 7 February 2021
Accepted: 27 February 2021
Published: 3 March 2021

Publisher's Note: MDPI stays neutral with regard to jurisdictional claims in published maps and institutional affiliations.

1. Introduction

One of the major challenges of our time is to secure a steady supply of healthy and nutritious food for a growing world population, while also protecting the environment and mitigating climate change. Agriculture is the main consumer of the world's water and land resources and at the same time an important emitter of greenhouse gases (GHG). Globally, cropland expansion and agricultural intensification continue to be the most widespread form of land use change [1]. The costs of increasing agricultural production has led, in many parts of Europe and elsewhere, to a significant decline in biodiversity across farmlands. Attempts to improve production through large-scale homogenization, including the resulting genetic uniformity, make this type of arable agriculture more vulnerable to pests, diseases, and abiotic stresses [2]. The overuse and misuse of synthetic fertilizers and pesticides has led to very high nutrient emissions to air and water, that damage ecosystems and contribute to GHG emissions. For instance, the discharge of large quantities of nitrates, phosphorus (P), pesticides, soil sediments, and saline drainage from farmlands in water bodies, has caused the eutrophication of habitats, which poses a threat to inland aquatic ecosystems and coastal waters [3]. According to the United Nations World Water Assessment Program report (2015) [4], in the European Union, 38% of water bodies are affected by agriculture pollution and overuse, which is further aggravated by increased sediment runoff and groundwater salinization and the depletion of groundwater supplies. In addition, agricultural intensification is often accompanied by highly intensive tillage practices which has led to reductions in organic soil matter and soil biodiversity, causing very serious soil degradation [5].

At the core of the efforts to address the aforementioned social and environmental shortcomings of conventional agriculture is the promotion of a transition from conventional agriculture—with high levels of use of agro-chemicals, simplification of the agro-ecosystem, and dangerous levels of environmental pollution—to more diversified and sustainable forms of agriculture [6,7]. Such a transition will require more widespread adoption of sound agricultural practices that make more efficient use of water and soil than is currently the case. This shift in practices should minimize water and nutrient inputs (e.g., irrigation and excessive P application), reduce greenhouse gas emissions through more efficient fertilizer and fuel use for operational activities, and prevent environmental pollution while providing stable, and, where feasible, increased crop yields [8].

The major problems resulting from homogenized and intensified agricultural systems are aggravated by increasingly widespread and intense climate change impacts on production systems. The degree of these impacts varies regionally. There is already a perceptible northward shift in European agro-ecological zones, which is predicted to accelerate in the next decades [9]. Crop productivity is projected to increase in northern Europe under multiple climate change scenarios, due to increasing precipitation and warmer temperatures, while it could decrease in southern Europe due to increasing aridity [10,11]. At the same time, climate variability is also increasing in Europe, and extreme weather events are predicted to be more frequent [10]. Productivity gains in the north, therefore, are predicted to be countered by the increasing frequency and duration of extreme weather events that cause lodging and flooding from higher rainfall, adversely affecting field accessibility and diminishing the reliability of achieving target yields [12]. These natural hazards leave farming systems vulnerable to economic stress and threaten food security goals [13,14].

There is consistent agreement among global climate change model projections that the Mediterranean Basin region will experience reductions in precipitation in almost all seasons, but particularly in the summer, except in the most northern parts (where lower precipitation in winter is predicted), causing serious problems of low and fluctuating water supply [15–17]. Increasing temperature is also projected during all seasons throughout the Mediterranean Basin as well as higher daily precipitation extremes [16]. Moreover, overexploitation of ground water for irrigation has caused salinization, leading to soil degradation and loss of soil fertility, particularly in North Africa, while extreme rain events have resulted in severe soil erosion, landslides, and flooding. The implication of these projections of a changing climate is that the effects of certain agricultural practices and the suitability of certain crops will also change over time, necessitating the use of more adaptive and effective management approaches. For example, heavier rainfall in hilly regions will necessitate more terracing and/or better soil coverage to prevent soil erosion. Longer dry periods will require better soil coverage with mulch, crop residues, or vegetation to improve soil moisture retention [16].

Both water deficiency and flooding, the likelihood of which is increasing due to climate change, will variably either destroy crops or limit crop growth and production potential. On the other hand, excessive and inappropriate use of water can increase production costs. The excessive and ill-timed application of fertilizers to cropping systems can cause considerable nutrient flows into the ground water and in the surrounding water bodies, thus, affecting surface and groundwater quality and aquatic ecosystems [17]. To address these challenges, agricultural practices that improve the delivery of an array of ecological, agronomic, and social benefits from farming systems are needed. These practices should not only optimize crop yield per unit of water (water use efficiency) and per unit of nutrient (nutrient use efficiency), but also increase ecosystem services, produce healthy food, and support sustainable farming enterprises (increasing eco-effectiveness) while improving social factors through yield stability and lower input costs (economic parity) [6,18,19].

In this article, the objective is to provide insights into the multiple benefits and trade-offs of different sustainable agricultural practices that are relevant across the pedo-climatic zones of Europe and North Africa. We contextualize these trade-offs according to a set of site relevant ecological, agronomic, and economic indicators. Such an analysis remains urgently needed considering that wider uptake of sustainable practices is still not taking place in many parts of the world, including Europe and North Africa [20], despite research that has demonstrated that multiple benefits (including clear 'win-win' outcomes) can be generated [21]. There are many nuanced reasons for this slow uptake of beneficial techniques, such as policy impediments (e.g., lack of incentivization), economic barriers, and knowledge gaps concerning the complexity of spatial, temporal, and crop component interactions of more dynamics agricultural systems [22]. It is important to note that this article does not present and assess examples and experiences from developing countries nor does it aim to compare the European and North African situations with other countries or continents.

The analysis presented in this paper can assist European and North African farmers to make better informed decisions according to their respective local conditions and opportunities. The practices we analyzed, empirically verified in on-farm and experimental trials, resulted in improvements in multiple key performance indicators, such as primary productivity, water conservation, soil conservation, nutrient regulation, biodiversity, and climate regulation. They have been selected for their implementation *potential*, but with the caveat that local conditions will require agile adaptation at any given site. Our study offers a novel approach by broadening the perspective beyond crop yield only. The analytical results also contribute to policy debate and policy review towards achieving a more sustainable agriculture, which is high on the policy agendas in Europe and North Africa. It has direct policy relevance regarding the implementation of the European Union's revised Common Agriculture Policy (CAP), which now has a very strong environmental and climate focus. At the same time, national and regional governments in Europe and North Africa are working on policy review toward circular economies and a circular agricultural sector, which could benefit from the practical, evidence-based strategies and practices analyzed in this study.

Following this introduction, Section 2 describes the materials and methods; Section 3 presents the results; Section 4 details the discussion; and Section 5 finishes with concluding remarks.

2. Materials and Methods

A thematic literature search on sets of sustainable agriculture practices was conducted for six countries of Europe and North Africa representing diverse climatic conditions—Denmark, the Netherlands, Germany, Italy, Egypt, and Tunisia. The goal was to assess the impact of these agricultural management practices based on six components: soil use, nutrient use, water use, biodiversity, agronomic productivity, and profitability. For each of the six components, a number of indicators was identified (Table 1). The literature search was conducted in the Web of Science database and by using the Google Scholar advanced search, focusing on references published between 2000 and 2020. This search was meant to be neither exhaustive nor systematic, but rather to identify research papers, including meta-analyses, that report on the outcomes of either on-farm or experimental research of sustainable agriculture practices that offer information about management impacts on the six components. The practices selected were not decided a priori, but emerged from a preliminary search using the phrase "soil and water conservation" and "agriculture". The refinement of the initial search terms was made after the preliminary search and limited to relevant terms within the subject areas using the key words "conservation agriculture", "crop diversification", "agroforestry", "organic agriculture", and "intercropping" and either "Europe" or "North Africa."

Table 1. Agriculture system indicators under six components.

Component	Indicator
Soil use	Physical properties
	Chemical properties
Nutrient use	Amount of total N, P, and potassium (K)
	Amount of soil carbon
	Nitrogen use efficiency
	Nitrogen surplus
Water use	Water dynamics and use (quantity and quality)
	Actual evapotranspiration
	Water use efficiency
Biodiversity	Crop and varietal diversity
Productivity	Crop yield, crop yield stability
Profitability	Total agricultural inputs
	Total agricultural outputs

Source: Adapted from a set of SMART agriculture system components and indicators identified by the Water-FARMING project.

3. Results

The results of the literature review on the impact of different sustainable agricultural practices on soil, water, and nutrient conservation are summarized in the following subsections and in Table 2. Table 2 presents the assessment of different sustainable agriculture practices that allow the comparison of the effectiveness of each of the practices on nutrient dynamics, water dynamics, productivity, soil use, biodiversity, and profitability (in terms of positive, negative or neutral outcomes). The table also allows the identification of 'win-win' outcomes and trade-offs between options.

One particular category of practices that includes mechanical practices for soil and water conservation (contour farming, terracing, geotextiles, and the use of earth mats) and for vegetative soil and water conservation practices (buffer strips), which can be used from farm to landscape scales, will not be reviewed in detail. Contour farming and terracing are earth engineering strategies that change the direction of water flow from down slope to along an isoline, following the contour of the terrain. Contour farming is useful for soil and water conservation in hilly and sloping landscapes, especially in areas of high precipitation. It reduces water runoff and, therefore, soil erosion, ultimately improving the downstream water quality. The use of geotextiles in fields can decrease water runoff and soil erosion and may help to reduce soil water evaporation. Buffer strips are patches on the farm set aside to maintain semi-natural vegetation that can serve as wind breaks to reduce soil erosion by wind and to create micro-climates conducive for certain crops/trees. They can serve to protect aquatic ecosystems by buffering the movement of soil, water, and nutrients from the field(s) to waterways.

3.1. Conservation Agriculture

Conservation agriculture (CA) is a farming system designed to save farming resources, sustain farm production, and achieve acceptable profits while maintaining the functional ecology of the farm system and meeting environmental conservation targets [23]. To achieve these objectives, practices are employed to enhance natural biological processes and ecological interactions above and below ground, through optimizing the management of soil, water, and nutrients [18,24,25]. CA is most widely recognized for the positive outcomes of a reduction of both soil erosion and water runoff, and the reduction of pollutants to downstream aquatic systems [18,26].

Table 2. The summary on effects of agricultural practices on soil and water conservation and on other variables. The effects are compared to conventional tillage and use of agrichemicals. (Effect indicators: ↑ positive, / neutral, ↓ negative).

Practice	Soil Conservation	Water Conservation	Other
Cover/catch crops	↑ soil structure ↑ soil erosion ↑ soil organic matter (SOM)	↑ soil moisture ↑ soil water capacity ↑ runoff ↑ water quality downstream	↑/↓ crop yield ↑ nutrient regulation ↓ weeds ↑ biodiversity
Residue retention/mulch	↑ soil erosion ↑ SOM	↑ soil moisture ↑ evapotranspiration ↑ runoff	↑↓ crop yield ↑ weeds ↑ fungus
No-till/direct seeding	↑ soil erosion ↑↓ SOM	↑ soil moisture ↑ soil water capacity ↑ runoff ↑ water quality downstream	↓/↑ crop yield ↑ nutrient regulation ↑ weeds ↑/ pests
Minimum tillage	↑ SOM	↑ soil moisture ↑ soil water capacity	↓ crop yield ↑ weeds
Crop rotation	↑ soil structure ↑/ SOM	↑ soil water capacity	↓↑ crop yield ↑ nutrient regulation ↑ weeds ↑ crop diversity
Intercropping	↑ soil structure ↑/ SOM	↑ soil water capacity	↓↑ crop yield ↑ nutrient regulation ↓ weeds ↑ crop diversity
Organic agriculture	↑ soil structure ↑/ SOM	↑ soil water capacity	↓/ crop yield ↑ nutrient regulation ↓↑ weeds ↑ crop diversity
Agroforestry	↑ soil structure ↓ soil erosion ↑ SOM	↑ soil water capacity ↑ evapotranspiration	↑ crop yield ↑ nutrient regulation ↑ weeds ↑ crop diversity ↑ micro-climate ↑ carbon sequestration

At the same time, in some studies, CA is associated with negative outcomes in the form of reduced crop yield and increased weed population and weed diversity [27]. Pittelkow et al. (2015) [28] found in a global meta-analysis that yields are on average 5% lower than in conventional agriculture and found an overall yield decrease (−8.5%) in a meta-analysis of CA in European countries [24]. CA in dry climates appears to result in the least reduction in crop yields [29]. The lower yield can be caused by a number of factors, such as slower rate of soil fertility buildup, waterlogging during the periods of prolonged rainfall on poorly aerated soils, delayed crop establishment due to occasional wet and cold soils, fertilizer placement, residue management problems, increased weed competition, residue-borne diseases, and soil compaction. In the following sections, outcomes attributed to CA practices are described in more detail under the three principles of CA: permanent soil cover, minimal soil disturbance, and diversification of crop species.

3.1.1. Maintaining Vegetation Cover/Permanent Soil Cover

Maintaining vegetation cover on fields helps to retain nutrients in the root zone, protect the soil surface against erosion and nutrient losses, and minimize the risk of surface runoff by improving infiltration. It also enhances soil water holding capacity, reduces evaporation, and assists in soil structure formation through the maintenance of root systems. Practices that enhance these services include cover cropping or catch cropping, mulching of crop residues, intercropping, and using perennial crops. The use of cover crops is a form of usually asynchronous intercropping, as the cover crop is planted in the field between two crop cycles. The cover crop is not grown for harvest or a commercial yield, but rather as a companion plant that provides a set of ecosystem services [30]. Cover crops, such as alfalfa, clover, lucerne, and vetch, which are grown in Europe and North Africa, can also be produced as a secondary understory crop under an existing crop, where they germinate, establish, and then continue to grow after the main crop is harvested. Among the reasons to use a cover crop are the reduction of nitrogen leaching to groundwater [31] and weed pressure reduction, ultimately diminishing the use of herbicides and/or labor inputs [30,32]. Weed control will also contribute to positive environmental and economic impacts.

Important benefits of cover cropping include reduced soil erosion and soil nutrient loss, improved resilience of the crop system against weather variability and climate hazards, and improved soil chemical and physical properties. Incorporating cover crop biomass in the soil can enhance soil structure and, thus, water infiltration. Legume cover crops (e.g., subterranean clover and common vetch) increase soil organic matter, which can reduce the amount of nitrogen fertilizers used [33]. Increases in yield of the subsequent or companion cash crop are found especially with the use of legume species as a cover crop, and the subsequent incorporation of the cover crop residue in the soil [18,30]. Cover cropping in Mediterranean systems appears to be very effective, notably for sediment retention and runoff reduction, thereby contributing to soil erosion control [18]. However, yields do not always increase [30]. Cover cropping can sometimes result in reduced yield of subsequent cash crops as cover crops compete for soil nutrients and water [16,30]. One study showed that the effects of lacy phacelia, white mustard, and hairy vetch mulching on soil quality, microbial functions, and crop yield, were negatively influenced by variable summer precipitation and temperature [34].

Crop residue retention and mulching help to conserve soil moisture, reduce soil erosion and water runoff and retain nutrients in the field. Based on field research results from Tunisia, mulching (residue retention) is more effective than conventional tillage (under both semi-arid and sub-humid conditions) in enhancing yield of wheat (15% higher), water use efficiency (18% versus 13%), and soil organic carbon accumulation ($0.18 \text{ t ha}^{-1} \text{ y}^{-1}$ versus $0.13 \text{ t ha}^{-1} \text{ y}^{-1}$). This practice is effective in preventing erosion ($1.7 \text{ t ha}^{-1} \text{ y}^{-1}$ versus $4.6 \text{ t ha}^{-1} \text{ y}^{-1}$ of soil loss) [15,18], and it limits evapotranspiration from the soil [35]. The effect of cover crops and residue mulching on weed suppression depends on cover crop species used and the target weed species [36]. Residue mulching is most beneficial in arid and semi-arid environments, where soil moisture is a limiting factor. Its use in cooler and wetter regions can cause too much soil moisture and the growth of fungus and outbreak of diseases [18].

It is important to note that in many farming systems crop residues are taken off the field for use as feed for livestock or fuel for energy [37]. Thus, retention of crop residues on the field carries an opportunity cost that some farmers may not be willing to bear. Operational costs may be another factor that influences adoption, as crop residue incorporation can be more expensive compared to burning and removal of crop residue.

3.1.2. Minimizing Soil Disturbance

CA aims to minimize soil disturbances with the goal of reducing soil erosion and nutrient loss. Practices include no-tillage and minimum tillage. No-tillage or direct seeding is when the soil surface is not broken at any point other than for the drilling of small holes for planting. Sowing occurs directly in the stubble by cutting or "drilling" narrow slots for seed. Whereas minimum tillage in experimental trials often refers to a practice whereby only the near-surface soil (5–10 cm) is physically disturbed with discs, chisels, or a field cultivator, but generally without inverting the soil, resulting in loose topsoil. One objective of no-tillage and minimum tillage is to reduce the disturbance of the microbiological community, such as fungal networks whose exudates help ensure soil aggregate formation and stability [38], while also reducing the exposure of soil and, therefore, lowering erosion potential, moisture loss, and water runoff. The environmental benefits of no and minimum tillage at the farm level include the conservation of soil in the fields, and at the catchment scale, reduced sedimentation and run-off losses to downstream water bodies. The practice reduces labor inputs and the use of heavy machinery (and fuel) and, thus, can lower production costs as well [28].

CA is not only a theoretical concept, but has demonstrated economic and practical benefits in Europe and North Africa. For instance, no-tillage is being adopted at an increasing rate in southwestern Europe and North Africa, as it has produced winter crop yields that are higher or equal to systems using the conventional tillage [27]. Furthermore, no-tillage or minimum tillage, coupled with either 50% or 100% fertilizer application rate and weed control, gave better crop yields than conventional tillage in an experiment on exhausted clay soils in Egypt [39]. However, in Tunisia, different tillage practices (no-tillage combined with rotation and conventional tillage) coupled with different rates of fertilizer did not alter the grain yield, nitrogen content in grain, or nitrogen use efficiency. No-tillage combined with crop rotation enhanced the grain yield, nitrogen content in grain and nitrogen use efficiency [40]. Despite these benefits, other studies have shown that direct drilling with no-tillage resulted in significantly lower crop yields compared to ploughing. The tillage effect on crop yields was consistent across the crop species, though differences exist between crop species [41–43]. In one study of no-tillage use, weed infestation was found to be about 2–20 times higher compared to conventional tillage [41]. In another study it was shown that, with differing degrees of tillage, the effect on weed density varied. Under conventional tillage, the weed density decreased up to 71% while in minimum tillage and no-tillage the weed density reduction was up to 85% and 61%, respectively [44].

Other studies comparing no-tillage to conventional tillage have reported significantly higher levels of soil organic matter and total nitrogen in the top 30 cm soil layer, but the bulk density and total soil porosity of soil was enhanced only in the top 10 cm [45,46]. In contrast, studies carried out in northern Tunisia showed no increase in soil organic matter or soil organic carbon, even after 15 years under both no-tillage and CA [47]. A similar result was obtained after four years of experimental trials [48] under both no-tillage and CA. A study carried out in Egypt showed significant differences in soil physical properties with increased soil infiltration in minimum tillage (with a depth of 15 cm) that was 48% and 65% higher than that obtained when tillage depth increased to 20 cm and when compared with conventional tillage (with a depth of 25 cm), respectively; furthermore, the runoff and soil loss were lower (4.91 mm and 0.65 t ha^{-1}) under minimum tillage compared to conventional tillage (11.36 mm and 1.66 t ha^{-1}) [49].

In another study under semi-arid conditions in North Africa, after four years, no-tillage significantly improved soil nutrient content, especially K, potassium oxide, phosphorus pentoxide, and N [48]. In that study, the improvement of soil properties gained via no-tillage depended on factors, such as tillage management, sites (climate and soil type), and crop succession (crop species and cover crop residue). Conservation tillage, especially coupled with residue retention and cover cropping, has positive effects on soil water storage in semi-arid, rainfed conditions. The soil water storage capacity increases with application of conservation tillage systems and with increasing aridity. For instance, one study showed that grain yield and stability were positively correlated with soil water storage, especially in years with lower mean yield production [50]. Wheat water use efficiency was highest under no-tillage with residue retention, according to a study done in Tunisia [15]. In a study carried out in Denmark, the combination of no-tillage with cover crops demonstrated a better soil friability [51]. The reviewed studies demonstrated the capacity of no-tillage CA to improve soil properties that in turn have positive downstream consequences on water flow, water storage, and resulting plant productivity.

Although there are clearly demonstrated benefits of no-tillage practices, it remains important to analyze if there are tradeoffs when deciding management actions. There are a variety of contextual factors, such as degree of tillage and residue retention, that determine the likely outcomes of management actions. The desired outcomes, from biodiversity to productivity, must, therefore, be considered when deciding on any suite of management actions as summarized here. For instance, the use of occasional shallow inversion tillage in an otherwise no-tillage system can help incorporate organic matter into the soil, improve the topsoil structure for increasing water infiltration and water holding capacity, control weeds, and increase soil carbon stocks [29]. The retention of crop residues on the soil surface with no-tillage has also led to more active soil organic carbon dynamics in the top layer (0–5 cm) under the conservation system, while in conventional farming, ploughing helped to incorporate the residues and soil organic carbon accumulation at the 30–50 cm depth [52]. Minimum tillage in combination with organic farming was found to be an effective agricultural strategy to enhance soil microbial biomass, microbial residues, and bacterial and fungal abundances [53]. In a study conducted using conventional tillage (25 cm depth), reduced tillage (7.5 cm depth), and no-tillage, the highest crop yield observed was up to 2.4 t ha^{-1} under reduced tillage and the lowest yield of 1.1 t ha^{-1} under no-tillage [44]. These experimental findings confirm the need to carefully assess the relative benefits and costs of each tillage option.

3.1.3. Crop Rotation

The third principle of CA involves crop rotation, which is the sequential cultivation of crops in the same field instead of planting the same crop species, either from one year to the next, or within a year. Crop rotation increases field level diversity and enables the genetic, temporal and spatial benefits of crop diversification [54]. As different crops have different nutrient requirements, interactions with soil micro-organisms, rooting strategies, and physiology, crop rotation can help balance site dynamics by regulating soil nutrients, building soil structure, suppressing weeds, and decreasing pests and diseases. By choosing a suitable crop sequence, the pre-crop can help to meet the nutritional requirements for the growth of following crop(s) [54]. Moreover, the success of crop rotation systems to suppress weeds is based on the use of crop sequences that create different patterns of resource competition, allelopathic interference, soil disturbance, and temporal field coverage, all of which can prevent the proliferation of a particular weed species [55]. For instance, planting deep-rooted cover crops in rotations helps to distribute P and K and capture the unused nutrients. Planting a shallow-rooted crop, followed by a deeper-rooted crop can help to recover nutrients that were not used by the shallow feeder and which might have leached by irrigation or rainfall to lower layers of the soil profile [56].

Crop rotation is also beneficial for the control of pests and diseases. In Tunisia and the Mediterranean region at large, the use of break crops, such as faba bean and vetch as pre-crops, was demonstrated to be effective for reducing *Fusarium culmorum* inoculum in the soil and the pathogen in wheat roots and stem bases [57]. When durum wheat was cultivated following legumes or vegetables, it showed greater N uptake, but with only a minor effect on its conversion to grains [58]. Cowpea and gaur pre-crops cultivated before lentil gave the greatest reduction in disease severity. Lentils cultivated after cowpeas produced the highest seed yield followed by gaur and millet. Lowest lentil seed yield production was recorded when plants were cultivated after soybean, followed by sesame and groundnut [59]. Findings from Denmark showed that when a continuous winter wheat cropping was compared with a winter oilseed rape–winter wheat–winter wheat–winter barley crop rotation system, the former had higher grain yield (83.5 grain yield ha^{-1}) than the latter (72.9 grain yield ha^{-1}) [60]. In contrast, when a winter oilseed rape–winter wheat–winter rye–peas–winter wheat–winter barley crop rotation system was compared with a winter oilseed rape–winter barley–alfalfa/clover/grass-mixture–winterrye–silage maize–winter wheat crop rotation system in Germany, the grain yield for the latter crop rotation system was higher yielding (72.9 grain yield ha^{-1}) than the former (70.0 grain yield ha^{-1}) [60].

These findings suggest that the proper sequencing of crops in time and space and the informed use of nitrogen fixing crops and cover crops are important to obtain optimum results. The inclusion of legumes in crop rotations can supply biologically fixed atmospheric nitrogen due to the capability of legumes, making it available to the succeeding crops in rotation, thus, reducing the necessity for external inputs [61,62]. However, the long-term effects of these practices on various indicators are relatively unknown. A rare exception is the work by Götze et al. (2016) [63] that concluded that after 41 years, there were no significant differences in total carbon and microbial biomass carbon content among different crop rotations. More of such longer-term studies could be of great help to improve the design of effective crop rotations.

3.1.4. Intercropping

Arable production systems characterized by fields of annual crops are dominated by the use of monocultures, where a single crop species is produced at any given time (i.e., "sole crop"). A variation on sole cropping is intercropping, where two or more crop species are grown simultaneously in the same field for some duration [64]. Intercropping designs vary by crop species and varieties, combination proportions, spatial layout and densities, fertilizer inputs, and timing of sowing and harvesting. All these variables affect the impact on the key indicators of sustainable agriculture, further depending on climate, soil type, surrounding vegetation, and land use history. In general, intercropping has been found to increase crop productivity and crop yield stability [65], though not always significantly [66], while also reducing fertilizer inputs [67,68] and the use of herbicides [31,66]. One critical measure of the advantages of intercropping is the land equivalent ratio (LER) [69], which is a ratio of the summed productivity of the intercropped components compared to the component sole crops. An LER >1.0 indicates higher overall yield of the intercropped crops per unit of land compared to sole cropping. Intercropping is used most notably in organic farming systems, as it reduces fertilizer needs through exploitation of interspecific complementarity, conserves soil moisture, and provides strong competition with weeds [64]. It is, therefore, suggested that cereal-grain legume intercropping is most advantageous in systems with low nitrogen availability [64].

In intercropping, the yield improvements are mainly due to facilitation (positive interspecific interactions) in resource use, resource sharing, and niche complementarity between different crops, thus, enabling the achievement of higher yields than in monoculture. Another advantage of intercropping is the reduced downstream consequences of the biodynamic system on plant health. Work by Hauggaard-Nielsen et al. 2008 [66], has demonstrated the decrease of disease in intercrops in the range of 20–40% when compared to sole crops. Research findings from Italy have demonstrated that the soil carbon retention was promoted by preceding intercropped legumes making mineral nitrogen available and enabling the succeeding crop to achieve adequate yield [70]. In some cases, intercropping has shown an increase of carbon sequestration in above ground biomass and soil carbon under a 2:1 barley and pea organic system in Europe [71]. Intercropping can also increase water use efficiency in arid and semi-arid regions [72], but results vary. In experimental trials in Egypt, intercropping was found to result in an LER slightly above 1.0, but it did not increase the water use efficiency (WUE) when compared to conventional cropping [73]. In contrast, one study has shown that intercropping winter wheat with clover led to a decrease of wheat grain yield by 10–25% compared to the wheat sole crop [74]. The different advantages and tradeoffs of intercropping, therefore, merit further consideration, which can be assessed with the use of crop-system modelling that can appropriately simulate crop combinations and outputs under given environmental and management contexts.

3.2. Organic Farming

Organic farming is characterized by the prohibition of the use of synthetic pesticides and fertilizers, with the objective of balancing productivity with environmental sustainability [75]. Unlike conventional farming, organic farming practices depend on a long-term, integrated, and cyclical approach of nutrient management. Organic agriculture in its purest form is a closed system based on the circular dependence of livestock and plant crops [76]. The green manure grown on the farm and the crop residues produced are returned to the soil. Upon decay, they enrich the soil with organic matter and to a lesser extent with nutrients, such as N and P. Most commonly, organic farmers practice a six-year crop rotation, which meets the requirement for soil health, weed suppression, and protection against pests and diseases. Organic farming can be combined with CA, but the generally poor performance of CA on weed and pest control necessitates the use of labor-intensive biological and mechanical control measures for weeds and pests [75].

Integrated organic farming systems strive to reduce the need for off-farm inputs, which consequently lowers greenhouse gas emissions [77]. In organic agriculture, the introduction of a legume in the crop rotation as well as in an intercropping system improves soil fertility (soil organic carbon, humus content, and making N and P available). However, the N contribution by legumes varies considerably depending on legumes species and the local soil and climatic conditions [77]. For example, in a conventionally managed system, both N input and output were observed to be higher when compared to the organic farming system, in which grass-clover was rotated with the main crop, and when compared to a grain-legume organic cropping system [78]. However, it has also been shown that total soil organic matter content between the soil depth of 0 and 20 cm was higher for organic farming system (24 g kg^{-1}) compared to conventional farming system (15 g kg^{-1}) [79]. The total organic carbon and soil N pool increased over a longer period of organic farming practices as well [80,81]. The N loss from the organic farming was comparable to conventional farming, and the amount of N loss through leaching was dependent on soil type, climate, the use of catch crops, and the amount of soil organic matter [81]. However, various studies have demonstrated that the crop yields under organic systems are generally 20–50% lower when compared to conventional cropping systems, but results are highly dependent on the particularity of the systems and site characteristics, such as available nutrients, soil management history, climatic conditions, and technology [7,82–84]. Yield quality, however, is usually higher than under conventional cropping systems.

3.3. Agroforestry

Agroforestry includes practices that integrate, preserve, and manage perennial woody species in productive systems of annual or perennial agricultural crops, either in orderly plantations or where a diversity of crop types and species are combined in highly integrated, multistrata systems. Agroforestry in arable farming systems, or silvo-arable systems, involves the establishment of alleys of trees, usually fruit, timber, or bioenergy, with annual crop intercropping that usually uses cereals [85]. Agroforestry systems deal with yield diversification and the production of a short-term return on land while the planted trees are still small. Integrating trees in arable farming systems in Europe has multiple advantages, which have been assessed by the Agforward project, among others [86–89]. These include increasing diversification of crops, which can reduce risks in the event of a single crop failure; obtaining revenue from annual crops that can subsidize the investment in the future tree crops; and a host of ecosystem services, including modified microclimates, reduced soil erosion, improved soil structure, reduced N leaching, improved nutrient cycling, and above ground and below ground carbon sequestration. Additional benefits include improving the landscape aesthetic, increasing biodiversity, and providing bio-mulch to suppress weeds.

Agroforestry in arable systems can also increase or maintain grain crop yields when compared to cereal monocultures per unit area of land. For example, the grain yield of some barley varieties was higher in an agroforestry system of walnut trees and cereal than as a monocrop [88]. Agroforestry systems are particularly beneficial in arable and semi-arable climates. Although trees compete with adjacent crops for water and nutrients, the deeper rooting systems of trees can capture drainage water, including during the fallow time [90]. Furthermore, through the process of hydraulic lift, water and nutrients from deeper soil horizons can be drawn up and released in the upper horizons, benefiting shallower rooted plants. Another advantage of agroforestry is the capacity to modify the microclimate through temperature regulation from tree shading [91]. For instance, a study conducted in Germany [91] found increasing soil temperature in a soil depth of 50 cm as the distance from the trees increased, with the lowest value within the trees and the highest at 7 m distance. The effect of trees was strongest during summer months, with diminishing effect during the winter months.

When a high input conventional farming system was converted to an agroforestry system, dry matter yield and land-use efficiency were negatively affected. However, when a low-input organic farming system was converted to agroforestry, dry matter yield and land-use were higher [92]. However, despite having many potential benefits, due to higher overall management costs and mechanization problems compared to conventional agriculture systems, farmers are reluctant to adopt agroforestry systems [8]. As a result, across Europe and the Mediterranean region, agroforestry systems remain a niche practice.

4. Discussion

4.1. Multiple Benefits and Trade-Off Analysis

Based on the literature review and building upon the study carried out by Wezel et al. (2014) [20], it is evident that all the sustainable agricultural practices analyzed provide multiple benefits, such as water conservation, soil conservation, nutrient regulation, biodiversity, crop yield, pest, disease and weed control, and climate regulation. However, when comparing practices, there are trade-offs to be considered, in terms of benefit type, level, visibility, and time and duration of occurrence, as well as inputs required. Some of the practices described here may be directly economically advantageous to the farm, for instance, by increasing crop yield and/or by reducing labor requirements or chemical inputs. Others are beneficial in ways that are more nuanced and/or have long-term or larger-scale benefits. Most sustainable agricultural practices also have non-monetary advantages, such as reducing greenhouse gas emissions, reducing soil erosion, and the consequent sedimentation of downstream aquatic ecosystems. The reduction in greenhouse gas emissions through no till farming, for example, has far-reaching global and long-term benefits, but it may not directly benefit the farm itself in the short term. Considering the variation in pedo-climate, geography, weather, ecology, cultures, traditions, seed quality, and subtle differences in implementation of the practices at experimental and field-trial locations, generalizing about the effectiveness of these practices for any given region is difficult. This is an important point for further study and analysis, for example, by means of improvement of crop system models that can allow decision-making to be critically evaluated given desired outcomes and pedo-climatic contexts. The incorporation of more data in crop system models, especially considering the importance of developing accurate tools for simulating intercrop dynamics, is an important avenue for the increased uptake and adoption of sustainable agricultural practices.

4.2. Constraints of Implementation of Sustainable Agricultural Practices

Implementing the sustainable agriculture practices analyzed here is not limited by one explicit obstacle or a particular constraint. The constraints are more a consequence of several issues that interact and are case-specific. In the following section, an overview of constraints is given. The constraints are (i) high costs of implementation, (ii) ineffective incentives and lack of transition schemes, (iii) low levels of education and awareness; and (iv) lack of and limited field demonstrations.

4.2.1. High Costs of Implementation

Transiting toward sustainable farming practices requires fundamental changes in soil-crop-landscape system management, including design, implementation, and monitoring. Farmers must adjust the nature and scale of farm operations and the degree of mechanization. The scale and degree of mechanization required for profitable farms usually leads to large up-front adoption costs of these practices, which locks farmers into a status quo preference, as they cannot afford any risk given their input and potential debt. Being (un)able to afford the expense of transitioning management practices to novel methods is one of the common barriers to the adoption of sustainable farming practices [93,94]. Other barriers toward the transition to more sustainable practices are the high labor intensity and high labor costs, particularly in Europe [91]. Additionally, low profitability is still a major concern of farmers and landowners considering adoption of these practices [95]. There is, therefore, a need to convince farmers of the risks involved and provide them tools that help depict the consequence of their actions.

4.2.2. Ineffective Incentives and Lack of Transition Schemes

Although in Europe several incentives (schemes) have been designed, such as incentives to promote agroforestry through Pillars I and II of the CAP, they are often limited in practice. The Pillars I and II incentives are only for establishing new agroforestry systems and not for adjusting/improving those already in existence [95]. Likewise, there are market-based incentives that promote the adoption of intercropping and residue mulching, but not for adoption of no-tillage [96]. The externalities generated by conventional practices, and the difficulty of quantifying them and the context of ecosystem services, have contributed to this lack of emphasis on more nuanced and larger-scale benefits, such as those achieved by no-tillage. Under the current CAP provision, the payment for a single farm is not linked to any cropping system or soil management protocol or to any scheme that offers payment for environmental services [96]. Overall, there is an absence of adequate policy and institutional support for the farmers to transit from current unsustainable farming practices to sustainable farming systems. Better designed, more flexible, and more diverse options are required, given the contribution of current agricultural practices to global GHG emissions [97].

Moreover, there is a need for low threshold incentives that make it easier for farmers who have not yet adopted any sustainable agricultural practices to transition to a new practice(s) and eventually to a new farming system. Another constraint is the complexity of some of the regulatory incentives, such as watershed management programs and forest laws, which demotivate farmers to adopt or even consider adoption. Given the necessary scale for profitable farms and incumbent mechanization, the adoption of sustainable agricultural practices often has very high initial investment costs for which there are no good adequate financing options. This makes it very difficult for less well-resourced farmers to adopt. Although there are top-down incentives that are mandatory, such as cross-compliance and greening requirements of the CAP Pillar 1, the Natura 2000 sites (Habitats and Birds Directives), protection of water against pollution caused by nitrates (Nitrate Directive) and voluntary incentives, such as Agri-environment payment schemes or Payment for Ecosystem Services, which are offered to farmers, these incentives do not take into account a household's short-term economic needs [96].

4.2.3. Low Level of Education and Awareness

There is still limited scientific knowledge and limited practical on-farm experiences concerning sustainable agricultural practices, which contributes to a lack of awareness amongst farmers of practices, such as direct seeding and relay intercropping, and the use of natural pesticides, no-tillage, and application of biopesticides and agroforestry [20,98]. The education of policy makers, producers, and consumers is essential to discover, understand, and promote sustainable farming practices for future adoption. For example, intercropping practices in particular offer many advantages, but improved understanding of the ecological mechanisms associated with different crop components, planned temporal and spatial diversity, and the benefits derived from associated diversity, is needed to better identify and quantify all the possible benefits. Moreover, the application of the different sustainable agricultural practices presented in this article implies modifying the farming system, either at crop management scale or at the cropping or farming system scale. When a much larger part of the system has to be reorganized or redesigned, it requires in-depth knowledge about different crop choices and their complementarity in resource use that can enhance yields without compromising the yield of the main crop [61,99]. Thus, to enhance the rate of adoption, technical assistance, training to improve farmers' capacity to adapt technologies to their own situation, and training of extension agents to share the knowledge and practical aspects, are crucial.

4.2.4. Lack of and Limited Field Demonstrations

This study has provided a comprehensive analysis of multiple benefits and trade-offs of the most popular sustainable production practices in European and North African contexts. Further research into this area should provide field-based evidence of the multiple benefits and trade-offs. This could be done by establishing demonstration sites under different pedo-climatic and socio-economic conditions where farm managers and farmers can see the benefits of upscaling the adoption of sustainable agricultural methods for tangible effects at farm, national and regional and continental scales. There is a need to expand the focus of agricultural research from field and plot research to landscape research and assess the costs of the production transition by internalizing environmental costs [100]. On-field demonstration and farmer-to-farmer exchange programs are useful for motivating farmers to shift their practices in this regard. This kind of citizen (research) involvement is a very practical way to contribute to the implementation of the revised European Union CAP and its core proposed measures (Eco-schemes; Enhanced conditionality; and Agri-environment-climate measures) [101]. The on-farm agronomic and environmental research can be complemented by research of new marketing modalities and opportunities (e.g., niche product development, 'buying local' initiatives, product certification schemes), the roles of agri-food sector companies in sustainable supply chain [102], and the use of digital technologies (e.g., real-time price updates, on-line fresh products ordering, e-payment options, tracing and tracking of products). All these research efforts can contribute to overcome the main barriers to increased adoption of sustainable agricultural practices and to reap the benefits of these practices.

5. Conclusions

Farmers in Europe and North Africa have been slow to adopt CA, organic agriculture, agroforestry, and other forms of sustainable agricultural practices, despite the multiple benefits they generate. The soil, water, and nutrient conservation practices reviewed in this paper are imperative to reduce runoff, soil erosion, improve soil quality, water quality, and moisture conservation, and enhance overall crop productivity. Together, these practices in Europe and North Africa can reduce operational costs, increase soil, water, and nutrient conservation, while also increasing crop yields. Moreover, adopting these practices can increase sustainable food production to meet food security.

There are evident scale impacts of adopting sustainable agricultural practices ranging from field to catchment levels, but the best empirical evidence has been generated at the field level. Despite the demonstrated benefits of these practices in different contexts, there are tradeoffs to consider as well. For example, CA practices (such as no-tillage and the application of residues/mulch) in regions of negative winter water deficits (i.e., water surplus), such as northern and western Europe, can result in problems of water-logged soil, harboring of disease and pests, and cold soil hindering germination. On the other hand, the high positive summer water deficits in southwestern Europe and North Africa do merit the use of CA practices for retention of soil water. Therefore, time-space bound water deficit values at a site can guide decisions on the application of a certain practice or practices. Similarly, the effects of these practices on the indicators will vary, as they are influenced by factors such as crop species, aridity index, irrigation use/rainfed, and soil type. Yield performance and stability, operating costs, climate change, and environmental policies and incentives, education and awareness programs, and field demonstrations, will likely be the major driving forces defining the direction and rate of adoption of sustainable agricultural practices. The adoption of sustainable farming practices requires significant efforts from farmers and support of government at national and local levels. To make site-specific recommendations, a careful assessment of overall benefits must be made.

This can be achieved via crop-system modelling, which is one method of addressing the exigent needs of improving our understanding of the complex spatial, temporal, and crop component interactions. As Malézieux et al. 2009 [22] discuss, there is an evident need to design agricultural systems, especially complex ones like intercropping, using models. These models must be applicable to multispecies systems in order to account for the interactions between components, long-term cumulative effects, and multifaceted objectives of crops. The complexity of the interactions necessitates the development of new knowledge; this cannot simply be achieved via studies on plant components independently [103], nor can it be achieved via traditional factorial experimental approaches [22]. Continued investment in crop-system modelling can help to give farmers tools and knowledge to tailor their management to their given context, at a variety of scales. Adoption of sustainable agricultural practices can further be stimulated when farmers have the opportunity to experiment on their own land and discover the advantages and disadvantages of different practices. Research and the expertise of more "advanced" farmers can help guide the decision process as well, utilizing peer-to-peer networks to spread the much-needed transition.

Author Contributions: T.C. and B.B.G. researched, selected and reviewed the literature. T.C. designed and drafted the article. B.B.G. did the final editing and mobilized the funding for the manuscript. All authors have read and agreed to the published version of the manuscript.

Funding: This research was funded by the WaterFARMING (grant agreement no: 689271) project and SustainFARM project (Grant Agreement No: 652615) which are acknowledged for support for the preparation of the manuscript.

Institutional Review Board Statement: Not applicable.

Informed Consent Statement: Not applicable.

Data Availability Statement: Data are contained within the article.

Acknowledgments: We thank Robin Sears for assistance with the literature search and review and Ronnie Vernooy of the Alliance of Bioversity International and CIAT for feedback on the draft article, and Reed John Cowden for the review of the revised article. We thank the anonymous reviewers for the useful comments and suggestions.

Conflicts of Interest: The authors declare no conflict of interest.

References

1. Kremen, C.; Iles, A.; Bacon, C. Diversified farming systems: An agroecological, systems-based alternative to modern industrial agriculture. *Ecol. Soc.* **2012**, *17*. [CrossRef]
2. Frison, E.A. From Uniformity to Diversity: A Paradigm Shift from Industrial Agriculture to Diversified Agroecological Systems. 2016. Available online: https://www.researchgate.net/publication/303737887 (accessed on 7 January 2021).
3. Parris, K. Impact of agriculture on water pollution in OECD countries: Recent trends and future prospects. *Int. J. Water Res. Dev.* **2011**, *27*, 33–52. [CrossRef]
4. WWAP. *The United Nations World Water Development Report 2015: Water for a Sustainable World*; United Nations World Water Assessment Programme (WWAP), United Nations Educational, Scientific and Cultural Organization: Paris, France, 2015.
5. Kopittke, P.M.; Menzies, N.W.; Wang, P.; McKenna, B.A.; Lombi, E. Soil and the intensification of agriculture for global food security. *Environ. Int.* **2019**, *132*, 105078. [CrossRef]
6. Tamburini, G.; Bommarco, R.; Wanger, T.C.; Kremen, C.; van der Heijden, M.G.A.; Liebman, M.; Hallin, S. Agricultural diversification promotes multiple ecosystem services without compromising yield. *Sci. Adv.* **2020**, *6*, eaba1715. [CrossRef]
7. Timsina, J. Can Organic Sources of Nutrients Increase Crop Yields to Meet Global Food Demand? *Agronomy* **2018**, *8*, 214. [CrossRef]
8. Beillouin, D.; Ben-Ari, T.; Makowski, D. Evidence map of crop diversification strategies at the global scale. *Environ. Res. Lett.* **2019**, *14*, 123001. [CrossRef]
9. Ceglar, A.; Zampieri, M.; Toreti, A.; Dentener, F. Observed Northward Migration of Agro-Climate Zones in Europe Will Further Accelerate under Climate Change. *Earth's Future* **2019**, *7*, 1088–1101. [CrossRef]
10. Iglesias, A.; Garrote, L.; Quiroga, S.; Moneo, M. A regional comparison of the effects of climate change on agricultural crops in Europe. *Clim. Chang.* **2012**, *112*, 29–46. [CrossRef]
11. Stagge, J.H.; Kingston, D.G.; Tallaksen, L.M.; Hannah, D.M. Observed drought indices show increasing divergence across Europe. *Sci. Rep.* **2017**, *7*, 14045. [CrossRef] [PubMed]

12. Trnka, M.; Hlavinka, P.; Semenov, M.A. Adaptation options for wheat in Europe will be limited by increased adverse weather events under climate change. *J. R. Soc. Interface* **2015**, *12*, 20150721. [CrossRef]
13. Porter, J.R.; Xie, L.; Challinor, A.J.; Cochrane, K.; Howden, S.M.; Iqbal, M.M.; Lobell, D.B.; Travasso, M.I. Food security and food production systems. In *Climate Change 2014: Impacts, Adaptation, and Vulnerability. Working Group II contribution to the Fifth Assessment Report of the Intergovernmental Panel on Climate Change*; Field, C.B., Barros, V.R., Dokken, D.J., Mach, K.J., Mastrandrea, M.D., Bilir, T.E., Chatterjee, M., Ebi, K.L., Estrada, Y.O., Genova, R.C., et al., Eds.; Cambridge University Press: Cambridge, UK, 2014; pp. 485–533. Available online: https://www.ipcc.ch/site/assets/uploads/2018/02/WGIIAR5-Chap7_FINAL.pdf (accessed on 7 January 2021).
14. Ray, D.K.; Gerber, J.S.; MacDonald, G.K.; West, P.C. Climate variation explains a third of global crop yield variability. *Nat. Commun.* **2015**, *6*, 5989. [CrossRef]
15. Bahri, H.; Annabi, M.; Cheikh M'Hamed, H.; Frija, A. Assessing the long-term impact of conservation agriculture on wheat-based systems in Tunisia using APSIM simulations under a climate change context. *Sci. Total Environ.* **2019**, *692*, 1223–1233. [CrossRef]
16. Lagacherie, P.; Álvaro-Fuentes, J.; Annabi, M.; Bernoux, M.; Bouarfa, S.; Douaoui, A.; Grünberger, O.; Hammani, A.; Montanarella, L.; Mrabet, R.; et al. Managing Mediterranean soil resources under global change: Expected trends and mitigation strategies. *Reg. Environ. Chang.* **2018**, *18*, 663–675. [CrossRef]
17. Lewandowski, J.; Meinikmann, K.; Nützmann, G.; Rosenberry, D.O. Groundwater—The disregarded component in lake water and nutrient budgets. Part 2: Effects of groundwater on nutrients. *Hydrol. Process.* **2015**, *29*, 2922–2955. [CrossRef]
18. Lee, H.; Lautenbach, S.; Paula, A.; García-Nieto, A.; Bondeau, A.; Cramer, W.; Geijzendorffer, I.R. The impact of conservation farming practices on Mediterranean agro-ecosystem services provisioning—A meta analysis. *Reg. Environ. Chang.* **2019**. [CrossRef]
19. Czyżewski, B.; Matuszczak, A.; Muntean, A. Approaching environmental sustainability of agriculture: Environmental burden, eco-efficiency or eco-effectiveness. *Agric. Econ.* **2019**, *65*, 299–306. [CrossRef]
20. Wezel, A.; Casagrande, M.; Celette, F.; Vian, J.F.; Ferrer, A.; Peigné, J. Agroecological practices for sustainable agriculture. A review. *Agron. Sustain. Dev.* **2014**, *34*, 1–20. [CrossRef]
21. Garbach, K.; Milder, J.C.; DeClerck, F.A.; Montenegro de Wit, M.; Driscoll, L.; Gemmill-Herren, B. Examining multi-functionality for crop yield and ecosystem services in five systems of agroecological intensification. *Int. J. Agric. Sustain.* **2017**, *15*, 11–28. [CrossRef]
22. Malézieux, E.; Crozat, Y.; Dupraz, C.; Laurans, M.; Makowski, D.; Ozier-Lafontaine, H.; Rapidel, B.; De Tourdonnet, S.; Valantin-Morison, M. Mixing plant species in cropping systems: Concepts, tools and models: A review. *Sustain. Agric.* **2009**, 329–353. [CrossRef]
23. Palm, C.; Blanco-Canqui, H.; DeClerck, F.; Gatere, L.; Grace, P. Conservation agriculture and ecosystem services: An overview. *Agric. Ecosyst. Environ.* **2014**, *187*, 87–105. [CrossRef]
24. Van den Putte, A.; Govers, G.; Diels, J.; Gillijns, K.; Demuzere, M. Assessing the effect of soil tillage on crop growth: A meta-regression analysis on European crop yields under conservation agriculture. *Eur. J. Agron.* **2010**, *33*, 231–241. [CrossRef]
25. Scopel, E.; Triomphe, B.; Affholder, F.; Da Silva, F.A.M.; Corbeels, M.; Xavier, J.H.V.; Lahmar, R.; Recous, S.; Bernoux, M.; Blanchart, E.; et al. Conservation agriculture cropping systems in temperate and tropical conditions, performances and impacts. A review. *Agron. Sustain. Dev.* **2013**, *33*, 113–130. [CrossRef]
26. Kassam, A.; Friedrich, T.; Derpsch, R. Global spread of Conservation Agriculture. *Int. J. Environ. Stud.* **2019**, *76*, 29–51. [CrossRef]
27. Soane, B.D.; Ball, B.C.; Arvidsson, J.; Basch, G.; Moreno, F.; Roger-Estrade, J. No-till in northern, western and south-western Europe: A review of problems and opportunities for crop production and the environment. *Soil Tillage Res.* **2012**, *118*, 66–87. [CrossRef]
28. Pittelkow, C.M.; Linquist, B.A.; Lundy, M.E.; Liang, X.; van Groenigen, K.J.; Lee, J.; van Gestel, N.; Six, J.; Venterea, R.T.; van Kessel, C. When does no-till yield more? A global meta-analysis. *Field Crop. Res.* **2015**, *183*, 156–168. [CrossRef]
29. Cooper, J.; Baranski, M.; Stewart, G.; Nobel-de Lange, M.; Bàrberi, P.; Fließbach, A.; Peigné, J.; Berner, A.; Brock, C.; Casagrande, M.; et al. Shallow non-inversion tillage in organic farming maintains crop yields and increases soil C stocks: A meta-analysis. *Agron. Sustain. Dev.* **2016**, *36*, 22. [CrossRef]
30. Verret, V.; Gardarin, A.; Pelzer, E.; Médiène, S.; Makowski, D.; Valantin-Morison, M. Can legume companion plants control weeds without decreasing crop yield? A meta-analysis. *Field Crops Res.* **2017**, *204*, 158–168. [CrossRef]
31. Fan, X.; Vrieling, A.; Muller, B.; Nelson, A. Winter cover crops in Dutch maize fields: Variability in quality and its drivers assessed from multi-temporal Sentinel-2 imagery. *Int. J. Appl. Earth Obs. Geoinf.* **2020**, *91*, 102139. [CrossRef]
32. Gerhards, R.; Schappert, A. Advancing cover cropping in temperate integrated weed management. *Pest Manag. Sci.* **2020**, *76*, 42–46. [CrossRef] [PubMed]
33. Tarricone, L.; Debiase, G.; Masi, G.; Gentilesco, G.; Montemurro, F. Cover Crops Affect Performance of Organic Scarlotta Seedless Table Grapes under Plastic Film Covering in Southern Italy. *Agronomy* **2020**, *10*, 550. [CrossRef]
34. Marinari, S.; Mancinelli, R.; Brunetti, P.; Campiglia, E. Soil quality, microbial functions and tomato yield under cover crop mulching in the Mediterranean environment. *Soil Tillage Res.* **2015**, *145*, 20–28. [CrossRef]
35. Zayton, A.; Guirguis, A.; Allam, K. Effect of mulching type and duration on the productivity and water use efficiency of potato. *J. Soil Sci. Agric. Eng.* **2015**, *6*, 719–733. [CrossRef]

36. Kruidhof, H.M.; Bastiaans, L.; Kropff, M.J. Cover crop residue management for optimizing weed control. *Plant Soil* **2009**, *318*, 169–184. [CrossRef]
37. Pannell, D.J.; Llewellyn, R.S.; Corbeels, M. The farm-level economics of conservation agriculture for resource-poor farmers. *Agric. Ecosyst. Environ.* **2014**, *187*, 52–64. [CrossRef]
38. Rillig, M.C.; Mummey, D.L. Mycorrhizas and soil structure. *New Phytol.* **2006**, *171*, 41–53. [CrossRef] [PubMed]
39. Harb, O.; El-Hay, A.; Hager, M.; El-Enin, A. Studies on conservation agriculture in Egypt. *Ann. Agric. Sci.* **2015**, *60*, 105–112. [CrossRef]
40. Souissi, A.; Bahri, H.; Cheikh M'hamed, H.; Chakroun, M.; Benyoussef, S.; Frija, A.; Annabi, M. Effect of Tillage, Previous Crop, and N Fertilization on Agronomic and Economic Performances of Durum Wheat (*Triticum durum* Desf.) under Rainfed Semi-Arid Environment. *Agronomy* **2020**, *10*, 1161. [CrossRef]
41. Gruber, S.; Pekrun, C.; Möhring, J.; Claupein, W. Long-term yield and weed response to conservation and stubble tillage in SW Germany. *Soil Tillage Res.* **2012**, *121*, 49–56. [CrossRef]
42. Zikeli, S.; Gruber, S. Reduced Tillage and No-Till in Organic Farming Systems, Germany—Status Quo, Potentials and Challenges. *Agriculture* **2017**, *7*, 35. [CrossRef]
43. Koch, H.-J.; Dieckmann, J.; Büchse, A.; Märländer, B. Yield decrease in sugar beet caused by reduced tillage and direct drilling. *Eur. J. Agron.* **2009**, *30*, 101–109. [CrossRef]
44. Weber, J.F.; Kunz, C.; Peteinatos, G.G.; Zikeli, S.; Gerhards, R. Weed Control Using Conventional Tillage, Reduced Tillage, No-Tillage, and Cover Crops in Organic Soybean. *Agriculture* **2017**, *7*, 43. [CrossRef]
45. Chen, H.; Marhan, S.; Billen, N.; Stahr, K. Soil organic-carbon and total nitrogen stocks as affected by different land uses in Baden-Württemberg (southwest Germany). *J. Plant Nutr. Soil Sci.* **2009**, *172*, 32–42. [CrossRef]
46. Jemai, I.; Ben Aissa, N.; Ben Guirat, S.; Ben-Hammouda, M.; Gallali, T. Impact of three and seven years of no-tillage on the soil water storage, in the plant root zone, under a dry subhumid Tunisian climate. *Soil Tillage Res.* **2013**, *126*, 26–33. [CrossRef]
47. Bahri, H.; Annabi, M.; Saoueb, A.; M'Hamed, H.C.; Souissi, A.; Chibani, R.; Bahri, B.A. Can Conservation Agriculture Sequester Soil Carbon in Northern Tunisia in the Long Run? Springer International Publishing: Cham, Switzerland, 2018; Volume 87, pp. 85–87. [CrossRef]
48. Ben Moussa-Machraoui, S.; Errouissi, F.; Ben-Hammouda, M.; Nouira, S. Comparative effects of conventional and no-tillage management on some soil properties under Mediterranean semi-arid conditions in northwestern Tunisia. *Soil Tillage Res.* **2010**, *106*, 247–253. [CrossRef]
49. Salem, H.M.; Valero, C.; Muñoz, M.Á.; Gil-Rodríguez, M. Effect of integrated reservoir tillage for in-situ rainwater harvesting and other tillage practices on soil physical properties. *Soil Tillage Res.* **2015**, *151*, 50–60. [CrossRef]
50. Lampurlanés, J.; Plaza-Bonilla, D.; Álvaro-Fuentes, J.; Cantero-Martínez, C. Long-term analysis of soil water conservation and crop yield under different tillage systems in Mediterranean rainfed conditions. *Field Crop. Res.* **2016**, *189*, 59–67. [CrossRef]
51. Abdollahi, L.; Munkholm, L.J. Tillage System and Cover Crop Effects on Soil Quality: I Chemical, Mechanical, and Biological Properties. *Soil Sci. Soc. Am. J.* **2014**, *78*, 262–270. [CrossRef]
52. Piccoli, I.; Chiarini, F.; Carletti, P.; Furlan, L.; Lazzaro, B.; Nardi, S.; Berti, A.; Sartori, L.; Dalconi, M.C.; Morari, F. Disentangling the effects of conservation agriculture practices on the vertical distribution of soil organic carbon. Evidence of poor carbon sequestration in North- Eastern Italy. *Agric. Ecosyst. Environ.* **2016**, *230*, 68–78. [CrossRef]
53. Ghaley, B.B.; Rusu, T.; Sandén, T.; Spiegel, H.; Menta, C.; Visioli, G.; O'Sullivan, L.; Gattin, I.T.; Delgado, A.; Liebig, M.A.; et al. Assessment of benefits of conservation agriculture on soil functions in arable production systems in Europe. *Sustainability* **2018**, *10*, 794. [CrossRef]
54. Steinmann, H.-H.; Dobers, E.S. Spatio-temporal analysis of crop rotations and crop sequence patterns in Northern Germany: Potential implications on plant health and crop protection. *J. Plant Dis. Prot.* **2013**, *120*, 85–94. [CrossRef]
55. Rasmussen, I.A.; Askegaard, M.; Olesen, J.E.; Kristensen, K. Effects on weeds of management in newly converted organic crop rotations in Denmark. *Agric. Ecosyst. Environ.* **2006**, *113*, 184–195. [CrossRef]
56. Mechri, M.; Patil, S.; Saidi, W.; Hajri, R.; Jarrahi, T.; Gharbi, A.; Jedidi, N. Soil organic carbon and nitrogen status under fallow and cereal-legume species in a Tunesian semi-arid conditions. *Eur. J. Earth Environ.* **2016**, *3*, 1–13.
57. Khemir, E.; Chekali, S.; Moretti, A.; Gharbi, M.S.; Allagui, M.B.; Gargouri, S. Impacts of previous crops on inoculum of Fusarium culmorum in soil, and development of foot and root rot of durum wheat in Tunisia. *Phytopathol. Mediterr.* **2020**, *59*, 187–201. [CrossRef]
58. Ben Zekri, Y.; Barkaoui, K.; Marrou, H.; Mekki, I.; Belhouchette, H.; Wery, J. On farm analysis of the effect of the preceding crop on N uptake and grain yield of durum wheat (*Triticum durum* Desf.) in Mediterranean conditions. *Arch. Agron. Soil Sci.* **2019**, *65*, 596–611. [CrossRef]
59. Abdel-Monaim, M.F.; Abo-Elyousr, K.A.M. Effect of preceding and intercropping crops on suppression of lentil damping-off and root rot disease in New Valley—Egypt. *Crop Prot.* **2012**, *32*, 41–46. [CrossRef]
60. Deike, S.; Pallutt, B.; Melander, B.; Strassemeyer, J.; Christen, O. Long-term productivity and environmental effects of arable farming as affected by crop rotation, soil tillage intensity and strategy of pesticide use: A case-study of two long-term field experiments in Germany and Denmark. *Eur. J. Agron.* **2008**, *29*, 191–199. [CrossRef]
61. Nemecek, T.; von Richthofen, J.-S.; Dubois, G.; Casta, P.; Charles, R.; Pahl, H. Environmental impacts of introducing grain legumes into European crop rotations. *Eur. J. Agron.* **2008**, *28*, 380–393. [CrossRef]

62. Ouma, G.; Jeruto, P. Sustainable horticultural crop production through intercropping: The case of fruits and vegetable crops: A review. *Agric. Biol. J. N. Am.* **2010**, *1*, 1098–1105. [CrossRef]
63. Götze, P.; Rücknagel, J.; Jacobs, A.; Märländer, B.; Koch, H.-J.; Holzweißig, B.; Steinz, M.; Christen, O. Sugar beet rotation effects on soil organic matter and calculated humus balance in Central Germany. *Eur. J. Agron.* **2016**, *76*, 198–207. [CrossRef]
64. Bedoussac, L.; Journet, E.-P.; Hauggaard-Nielsen, H.; Naudin, C.; Corre-Hellou, G.; Jensen, E.S.; Prieur, L.; Justes, E. Ecological principles underlying the increase of productivity achieved by cereal-grain legume intercrops in organic farming. A review. *Agron. Sustain. Dev.* **2015**, *35*, 911–935. [CrossRef]
65. Raseduzzaman, M.; Jensen, E.S. Does intercropping enhance yield stability in arable crop production? A meta-analysis. *Eur. J. Agron.* **2017**, *91*, 25–33. [CrossRef]
66. Hauggaard-Nielsen, H.; Jørnsgaard, B.; Kinane, J.; Jensen, E.S. Grain Legume–Cereal Intercropping: The Practical Application of Diversity, Competition and Facilitation in Arable and Organic Cropping Systems. *Renew. Agric. Food Syst.* **2008**, *23*, 3–12. Available online: https://www.jstor.org/stable/44473559 (accessed on 7 January 2021). [CrossRef]
67. Jensen, E.S.; Carlsson, G.; Hauggaard-Nielsen, H. Intercropping of grain legumes and cereals improves the use of soil N resources and reduces the requirement for synthetic fertilizer N: A global-scale analysis. *Agron. Sustain. Dev.* **2020**, *40*, 5. [CrossRef]
68. Rodriguez, C.; Carlsson, G.; Englund, J.-E.; Flöhr, A.; Pelzer, E.; Jeuffroy, M.-H.; Makowski, D.; Jensen, E.S. Grain legume-cereal intercropping enhances the use of soil-derived and biologically fixed nitrogen in temperate agroecosystems. A meta-analysis. *Eur. J. Agron.* **2020**, *118*, 126077. [CrossRef]
69. Xu, Z.; Li, C.; Zhang, C.; Yu, Y.; van der Werf, W.; Zhang, F. Intercropping maize and soybean increases efficiency of land and fertilizer nitrogen use: A meta-analysis. *Field Crop. Res.* **2020**, *246*, 107661. [CrossRef]
70. Scalise, A.; Tortorella, D.; Pristeri, A.; Petrovičová, B.; Gelsomino, A.; Lindström, K.; Monti, M. Legume–barley intercropping stimulates soil N supply and crop yield in the succeeding durum wheat in a rotation under rainfed conditions. *Soil Biol. Biochem.* **2015**, *89*, 150–161. [CrossRef]
71. Chapagain, T.; Riseman, A. Barley–pea intercropping: Effects on land productivity, carbon and nitrogen transformations. *Field Crop. Res.* **2014**, *166*, 18–25. [CrossRef]
72. Yin, W.; Chai, Q.; Zhao, C.; Yu, A.; Fan, Z.; Hu, F.; Fan, H.; Guo, Y.; Coulter, J.A. Water utilization in intercropping: A review. *Agric. Water Manag.* **2020**, *241*, 106335. [CrossRef]
73. Kubota, A.; Safina, S.A.; Shebl, S.M.; Mohamed, A.E.-D.H.; Ishikawa, N.; Shimizu, K.; Abdel-Gawad, K.; Maruyama, S. Evaluation of intercropping system of maize and leguminous crops in the Nile Delta of Egypt. *Trop. Agric. Dev.* **2015**, *59*, 14–19. [CrossRef]
74. Thorsted, M.D.; Olesen, J.E.; Weiner, J. Width of clover strips and wheat rows influence grain yield in winter wheat/white clover intercropping. *Field Crop. Res.* **2006**, *95*, 280–290. [CrossRef]
75. Peigné, J.; Casagrande, M.; Payet, V.; David, C.; Sans, F.X.; Blanco-Moreno, J.M.; Cooper, J.; Gascoyne, K.; Antichi, D.; Bàrberi, P.; et al. How organic farmers practice conservation agriculture in Europe. *Renew. Agric. Food Syst.* **2016**, *31*, 72–85. [CrossRef]
76. Oelofse, M.; Jensen, L.S.; Magid, J. The implications of phasing out conventional nutrient supply in organic agriculture: Denmark as a case. *Org. Agric.* **2013**, *3*, 41–55. [CrossRef]
77. Pandey, A.; Li, F.; Askegaard, M.; Olesen, J.E. Biological nitrogen fixation in three long-term organic and conventional arable crop rotation experiments in Denmark. *Eur. J. Agron.* **2017**, *90*, 87–95. [CrossRef]
78. Pandey, A.; Li, F.; Askegaard, M.; Rasmussen, I.A.; Olesen, J.E. Nitrogen balances in organic and conventional arable crop rotations and their relations to nitrogen yield and nitrate leaching losses. *Agric. Ecosyst. Environ.* **2018**, *265*, 350–362. [CrossRef]
79. Pulleman, M.; Jongmans, A.; Marinissen, J.; Bouma, J. Effects of organic versus conventional arable farming on soil structure and organic matter dynamics in a marine loam in the Netherlands. *Soil Use Manag.* **2003**, *19*, 157–165. [CrossRef]
80. Shahin, R.R.; Khater, H.A. Quality and quantity of soil organic matter as affected by the period of organic farming in Sekem farm, Egypt. *Eurasian J. Soil Sci.* **2020**, *9*, 275–281. [CrossRef]
81. Knudsen, M.T.; Kristensen, I.S.; Berntsen, J.; Petersen, B.M.; Kristensen, E.S. Estimated N leaching losses for organic and conventional farming in Denmark. *J. Agric. Sci.* **2006**, *144*, 135–149. [CrossRef]
82. Biernat, L.; Taube, F.; Vogeler, I.; Reinsch, T.; Kluß, C.; Loges, R. Is organic agriculture in line with the EU-Nitrate directive? On-farm nitrate leaching from organic and conventional arable crop rotations. *Agric. Ecosyst. Environ.* **2020**, *298*, 106964. [CrossRef]
83. De Olde, E.M.; Oudshoorn, F.W.; Bokkers, E.A.M.; Stubsgaard, A.; Sørensen, C.A.G.; De Boer, I.J.M. Assessing the Sustainability Performance of Organic Farms in Denmark. *Sustainability* **2016**, *8*, 957. [CrossRef]
84. Conti, S.; Villari, G.; Faugno, S.; Melchionna, G.; Somma, S.; Caruso, G. Effects of organic vs. conventional farming system on yield and quality of strawberry grown as an annual or biennial crop in southern Italy. *Sci. Hortic.* **2014**, *180*, 63–71. [CrossRef]
85. Palma, J.; Crous-Duran, J.; Graves, A.; Burgess, P.J. Database of Agroforestry System Descriptions; Agforward Project: EU. 2015. Available online: https://www.repository.utl.pt/handle/10400.5/14796 (accessed on 7 January 2021).
86. Camilli, F.; Pisanelli, A.; Seddaiu, G.; Franca, A.; Bondesan, V.; Rosati, A.; Moreno, G.; Pantera, A.; Hermansen, J.; Burgess, P. Benefits and constraints associated to agroforestry systems: The case studies implemented in Italy within the Agforward project. In Proceedings of the 3rd European Agroforestry Conference Montpellier (2016 EURAF), Montpellier, France, 23–25 May 2016; Available online: https://www.repository.utl.pt/bitstream/10400.5/17410/1/EURAFIIIConf_Camilli_F_et_all_page_20_23.pdf (accessed on 7 January 2021).

87. Brunori, E.; Maesano, M.; Moresi, F.V.; Matteucci, G.; Biasi, R.; Scarascia Mugnozza, G. The hidden land conservation benefits of olive-based (*Olea europaea* L.) landscapes: An agroforestry investigation in the southern Mediterranean (Calabria region, Italy). *Land Degrad. Dev.* **2020**, *31*, 801–815. [CrossRef]
88. Beuschel, R.; Piepho, H.-P.; Joergensen, R.G.; Wachendorf, C. Similar spatial patterns of soil quality indicators in three poplar-based silvo-arable alley cropping systems in Germany. *Biol. Fertil. Soils* **2019**, *55*, 1–14. [CrossRef]
89. Moreno, G.; Arenas, G.; López-Díaz, M.L.; Bertomeu, Y.C.; Juarez, E. *Cereal Production beneath Walnut for Quality Timber Production in Spain*; Agforward Project; EU: Brussels, Belgium, 2015.
90. Eichhorn, M.P.; Paris, P.; Herzog, F.; Incoll, L.D.; Liagre, F.; Mantzanas, K.; Mayus, M.; Moreno, G.; Papanastasis, V.P.; Pilbeam, D.J.; et al. Silvoarable Systems in Europe—Past, Present and Future Prospects. *Agrofor. Syst.* **2006**, *67*, 29–50. [CrossRef]
91. Markwitz, C.; Knohl, A.; Siebicke, L. Evapotranspiration over agroforestry sites in Germany. *Biogeosciences* **2020**, *17*, 5183–5208. [CrossRef]
92. Lin, H.-C.; Hülsbergen, K.-J. A new method for analyzing agricultural land-use efficiency, and its application in organic and conventional farming systems in southern Germany. *Eur. J. Agron.* **2017**, *83*, 15–27. [CrossRef]
93. Casagrande, M.; Peigné, J.; Payet, V.; Mäder, P.; Sans, F.X.; Blanco-Moreno, J.M.; Antichi, D.; Bàrberi, P.; Beeckman, A.; Bigongiali, F.; et al. Organic farmers' motivations and challenges for adopting conservation agriculture in Europe. *Org. Agric.* **2016**, *6*, 281–295. [CrossRef]
94. Long, T.B.; Blok, V.; Coninx, I. Barriers to the adoption and diffusion of technological innovations for climate-smart agriculture in Europe: Evidence from the Netherlands, France, Switzerland and Italy. *J. Clean. Prod.* **2016**, *112*, 9–21. [CrossRef]
95. Hernández-Morcillo, M.; Burgess, P.J.; Mirck, J.; Pantera, A.; Plieninger, T. Scanning agroforestry-based solutions for climate change mitigation and adaptation in Europe. *Environ. Sci. Policy* **2018**, *80*, 44–52. [CrossRef]
96. Piñeiro, V.; Arias, J.; Dürr, J.; Elverdin, P.; Ibáñez, A.M.; Kinengyere, A.; Opazo, C.M.; Owoo, N.; Page, J.R.; Prager, S.D.; et al. A scoping review on incentives for adoption of sustainable agricultural practices and their outcomes. *Nat. Sustain.* **2020**, *3*, 809–820. [CrossRef]
97. Kassam, A.H.; Friedrich, T.; Derpsch, R. Conservation agriculture in the 21st century: A paradigm of sustainable agriculture. In Proceedings of the European Congress on Conservation Agriculture 2010, Madrid, Spain, 4–7 October 2010; Volume 10, pp. 4–6.
98. Gómez, J.A.; Guzmán, G.; Giráldez, J.V.; Fereres, E. The influence of cover crops and tillage on water and sediment yield, and on nutrient, and organic matter losses in an olive orchard on a sandy loam soil. *Soil Tillage Res.* **2009**, *106*, 137–144. [CrossRef]
99. Palma, J.; Graves, A.; Burgess, P.J.; Van Der Werf, W.; Herzog, F. Integrating environmental and economic performance to assess modern silvoarable agroforestry in Europe. *Ecol. Econ.* **2007**, *63*, 759–767. [CrossRef]
100. Vlek, P.L.; Khamzina, A.; Azadi, H.; Bhaduri, A.; Bharati, L.; Braimoh, A.; Martius, C.; Sunderland, T.; Taheri, F. Trade-offs in multi-purpose land use under land degradation. *Sustainability* **2017**, *9*, 2196. [CrossRef]
101. European Commission. Working with Parliament and Council to Make the CAP Reform Fit for the European Green Deal; European Commission: 2020. Available online: https://ec.europa.eu/info/sites/info/files/food-farming-fisheries/key_policies/documents/factsheet-cap-reform-to-fit-european-green-deal_en.pdf (accessed on 7 January 2021).
102. Fortunati, S.; Morea, D.; Mosconi, E.M. Circular economy and corporate social responsibility in the agricultural system: Cases study of the Italian agri-food industry. *Agric. Econ.* **2020**, *66*, 489–498. [CrossRef]
103. House, J.I.; Archer, S.; Breshears, D.D.; Scholes, R.J. NCEAS Tree–Grass Interactions Participants. Conundrums in mixed woody–herbaceous plant systems. *J. Biogeogr.* **2003**, *30*, 1763–1777. [CrossRef]

MDPI
St. Alban-Anlage 66
4052 Basel
Switzerland
Tel. +41 61 683 77 34
Fax +41 61 302 89 18
www.mdpi.com

Sustainability Editorial Office
E-mail: sustainability@mdpi.com
www.mdpi.com/journal/sustainability

www.ingramcontent.com/pod-product-compliance
Lightning Source LLC
LaVergne TN
LVHW070152170726
843515LV00007B/1901